Design for Six Sigma

A Roadmap for Product Development

Kai Yang

Basem El-Haik

McGraw-Hill

New York   Chicago   San Francisco   Lisbon   London   Madrid
Mexico City   Milan   New Delhi   San Juan   Seoul
Singapore   Sydney   Toronto

The McGraw·Hill Companies

Library of Congress Cataloging-in-Publication Data

Yang, Kai.
    Design for Six Sigma : a roadmap for product development / Kai Yang, Basem El-Haik.
        p.   cm.
    ISBN 0-07-141208-5
    1. Quality control—Statistical methods. 2. Experimental design. I. El-Haik, Basem.
II. Title.

TS156.Y33   2003
658.5'62—dc21   2003046326

3 4 5 6 7 8 9 0 DOC/DOC 0 9 8 7 6 5

ISBN 0-07-141208-5

*The sponsoring editor for this book was Kenneth P. McCombs, the editing
supervisor was David E. Fogarty, and the production supervisor was
Pamela A. Pelton. It was set in Century Schoolbook following the MHT
design by Joanne Morbit of McGraw-Hill Professional's Hightstown, N.J.,
composition unit.*

Printed and bound by RR Donnelley.

This book was printed on recycled, acid-free paper containing
a minimum of 50% recycled, de-inked fiber.

*To our parents, families, and friends
for their continuous support.*

Contents

ABOUT THE AUTHORS

KAI YANG, Ph.D., has extensive consulting experience in many aspects of quality and reliability engineering. He is also Associate Professor of Industrial and Manufacturing Engineering at Wayne State University, Detroit.

BASEM S. EL-HAIK, Ph.D. and Doctorate of Engineering, is the Director of Enterprise Excellence, Textron Inc., and is presently spearheading the effort to deploy Design for Six Sigma (DFSS) across the enterprise. In addition to Six Sigma deployment, Dr. El-Haik has worked extensively, consulted, and conducted seminars in the field of Six Sigma, lean manufacturing, axiomatic design, TRIZ, reliability, and quality engineering. Dr. El-Haik has received his Ph.D. in Industrial Engineering from Wayne State University and his Doctorate of Engineering in Manufacturing from the University of Michigan—Ann Arbor.

Preface

The success of the Six Sigma movement has generated enormous interest in business world. By quoting one of our friends, Subir Chowdhury, "people's power" and "process power" are among the keys for the success of Six Sigma. The people's power means systematic organization support led from the top, and rigorous training for Six Sigma team members. The process power means the rigor of Six Sigma deployment and project management processes, and a wide array of statistically based methods. It is our belief that unlike other quality improvement movements, where the focus is primarily on the quality of the product or service to external customers, Six Sigma is focusing on the *whole quality* of a business enterprise. The whole quality includes not only the product or service quality to external customers, but also the *operation quality* of all internal business processes, such as accounting, billing, and so on. The business enterprises that have high levels of whole quality will not only provide high quality product or services, but also they will have much lower cost and high efficiency because all their business processes are optimized.

Compared with the "regular" Six Sigma that is featured by "DMAIC" (define-measure-analysis-improve-control), the new wave of Six Sigma is called *Design for Six Sigma* (DFSS). The regular Six Sigma is also called *Six Sigma improvement*, that is to improve a process without design or completely redesign the current system. Design for Six Sigma puts a lot of focus on design and it tries to "do things right at the first time." In our understanding, the ultimate goal of DFSS is to make a process or a product to: (1) Do the right things; and (2) Do things right all the time.

Do the right things means achieving absolute excellence in design, be it in designing a product, a manufacturing process, a service process or a business process. Superior product design will deliver superior products that deliver right product functions to generate great customer excitement. Superior manufacturing process design will generate a process that delivers the product in a most efficient, economic, and flexible manner. Superior service process design will generate a process that fits customer desires and provides service with quality and low cost. Superior business process design will generate the most efficient, effective, and economical business process.

Do the right thing all the time means that not only should we have superior design, but the actual product or process that we build according to our design, will always deliver what it is supposed to do. For example, if a company can develop some very superior products sometimes, but it also develops some poor products, then this company does not do the right thing all the time. If people buy cars from a world-class brand-name, they really expect all the cars from that brand-name to perform well and that these cars will perform consistently during their useful life; that is what we mean by 'do things right all the time'. *Do things right all the time* means high consistency and extremely low variation in performance. The term *Six Sigma* actually means very high consistency and low variation. Nowadays, high consistency is not only necessary for product performance and reputation; it is also a matter of survival. For example, the dispute between Ford and Firestone tires only involves an extremely small fraction of tires, but the negative publicity and litigation brought a giant company like Ford into an unpleasant experience.

Implementing DFSS, as previously stated, will involve (1) *doing the right things and* (2) *doing things right all the time* by using "people's power" and "process power." The people's power involves organizational leadership and support, as well as a tremendous amount of training. The process power involves a sophisticated implementation process and a big collection of methods. Compared to regular Six Sigma (DMAIC), many new methods are introduced in DFSS. Examples are axiomatic design, design for X, and theory of inventive problem solving (TRIZ). Transfer functions and scorecards are really powerful concepts and methods to create superior designs, that is, to do the right things. DFSS also brings another class of powerful methods, Taguchi's methods, into its tool box. The fundamental objective of the Taguchi methods is to create a superior product or process that can perform highly consistently despite many external disturbances and uncertainties. In other words, Taguchi methods create a *robust product or process*, thus achieving *do things right all the time*. The implementation of DFSS will take more effort and training than that of DMAIC, but it will be more rewarding and provide better results.

This book's main objective is to give a complete picture of DFSS to readers:

1. To provide an in-depth and clear coverage of all the important, philosophical, organizational, implementation, and technical aspects of DFSS to readers.

2. To discuss and illustrate very clearly the whole DFSS deployment and execution process.

3. To discuss and illustrate very clearly all major methods used in DFSS.

4. To discuss the theory and background of each method clearly with examples and illustrations.

5. To give the detailed step-by-step implementation process of each DFSS method.

6. To help develop practical skills in applying DFSS in real world implementation.

The background required to study this book is some familiarity with simple statistical methods, such as normal distribution, mean, variance, and simple data analysis techniques.

Chapter 1 begins with a discussion about "what is quality?" It lists (1) *do the right things* and (2) *do things right all the time* as the key tasks to bring superior quality for product and processes. It discusses the relationship between different quality tasks and tools and different stages of product/process development. Finally, this chapter discusses the Six Sigma quality concept, the *whole quality* and business excellence.

Chapter 2 discusses "What is Six Sigma?" and the differences between regular Six Sigma and DFSS. It also discusses the importance of process management in Six Sigma practice.

Chapter 3 provides a high-level description of DFSS, its stages and major tasks, and where and how to use DFSS in a company.

Chapter 4 discusses the people aspects of DFSS, such as how to organize DFSS teams, the roles of master black belt, black belt, and green belt, and how to deploy DFSS initiatives in a company along with highlights of financial aspects of DFSS projects.

Chapter 5 is a very detailed description of the DFSS project implementation process. We use the term *DFSS algorithm* to describe this process. The term *algorithm* is used to emphasize a repeatable and reproducible DFSS project execution. This chapter is very important because it gives a flowchart about how we can turn factors such as product/process development tasks, DFSS teams, and all DFSS methodologies into an executable process. We recommend that the reader revisit this chapter after all methodology chapters.

Chapters 6 to 18 are the DFSS methodology chapters. Chapter 6 introduces all aspects of the transfer function and DFSS project scorecards. Transfer functions and scorecards are unique Six Sigma tools. A transfer function includes the clear mathematical relationships between "causes" (which are often design parameters or process variables) and "effects" (which are often product/process performance metrics). By knowing a transfer function relationship, we are able to optimize the design to achieve superior performance. Scorecards are unique Six Sigma design evaluation worksheets where historical data are recorded and project progress on metrics is tracked.

Chapter 7 presents the quality function deployment method, a powerful method to guide and plan design activities to achieve customer desires. QFD was originally developed in Japan and is now widely used all over the world.

Chapter 8 introduces the axiomatic design method. The axiomatic design method is a relatively new method developed at MIT. It gives some very powerful guidelines (axioms) for "what is a good system design" and "what is a weak system design." Weak designs are often featured by complicated mutual interactions, coupling, nonindependence, and excessive complexity. Good designs are often featured by clear and simple relationship between design parameters and product functions, and elegant simplicity. Axiomatic design principles can help DFSS project to reduce design vulnerabilities and therefore to achieve optimized designs.

Chapter 9 presents the theory of inventive problem solving (TRIZ), which was developed in the former Soviet Union. TRIZ is a very powerful method that makes innovation a routine activity. It is based on an enormous amount of research worldwide on successful patents and inventions. It has a wide selection of methods and knowledge base to create inventive solutions for difficult design problems. This chapter provides a very detailed description of TRIZ and a large number of examples. TRIZ can help the DFSS team to think "outside of the box" and conceive innovative design solutions.

Chapter 10 discusses "Design for X" which includes "design for manufacturing and assembly," "design for reliability," and many others. Design for X is a collection of very useful methods to make sound design for all purposes.

Chapter 11 discusses failure mode and effect analysis (FMEA). FMEA is a very important design review method to eliminate potential failures in the design stage. We discuss all important aspects of FMEA, and also the difference between design FMEA and process FMEA. The objective of FMEA is to mitigate risks to improve the quality of the DFSS project.

Chapter 12 gives a very detailed discussion of a powerful and popular statistical method , design of experiment method (DOE). DOE can be used for transfer function detailing and optimization in a DFSS project. In this chapter, we focus our discussion on the workhorses of DOE, that is, the most frequently used DOE methods, such as full factorial design and fractional factorial design. In this chapter, detailed step-by-step instructions and many worked out examples are given.

Chapters 13 to 15 discuss the Taguchi method. Chapter 13 discuss Taguchi's orthogonal array experiment and data analysis. Chapter 14 gives very detailed descriptions on all important aspects of the Taguchi method, such as loss function, signal-to-noise ratio, inner-outer array, control factors, and noise factors. It also gives a detailed description on how to use Taguchi parameter design to achieve robustness in design.

Chapter 15 discusses some recent development in Taguchi methods, such as ideal functions, dynamic signal-to-noise ratio, functional quality, and robust technology development.

Chapter 16 is a very comprehensive chapter on tolerance design or specification design. It gives all important working details on all major tolerance design methods, such as worst case tolerance design, statistical tolerance design, cost based optimal tolerance design, and Taguchi tolerance design. Many examples are included.

Chapter 17 discusses the response surface method (RSM), which can be used as a very useful method to develop transfer functions and conduct transfer function optimization. We provide fairly complete and comprehensive coverage on RSM.

Chapter 18 is a chapter discussing design validation. We introduce the process of three important validation activities: design validation, process validation, and production validation. In design validation, we discuss in detail the roles of design analysis, such as computer simulation and design review, validation testing in design validation, the guideline to plan design validation activities, and the roles of prototypes in validation. We also discuss many important aspects of process validation, such as process capability validation.

This book's main distinguishing feature is its completeness and comprehensiveness. All important topics in DFSS are discussed clearly and in depth. The organizational, implementation, theoretical, practical aspects of both DFSS process and DFSS methods are all covered very carefully in complete detail. Many of the books in this area usually only give superficial description of DFSS without any details. This is the only book so far to discuss all important DFSS methods, such as transfer functions, axiomatic design, TRIZ, and Taguchi methods in great detail. This book can be used ideally either as a complete reference book on DFSS or a complete training guide for DFSS teams.

In preparing this book we received advice and encouragement from several people. For this we express our thanks to Dr. G. Taguchi, Dr. Nam P. Suh, Dr. K. Murty, Mr. Shin Taguchi, and Dr. O. Mejabi. We are appreciative of the help of many individuals. We are very thankful for the efforts of Kenneth McCombs, Michelle Brandel, David Fogarty, and Pamela A. Pelton at McGraw-Hill. We want to acknowledge and express our gratitude to Dave Roy, Master Black Belt of Textron, Inc. for his contribution to Chapters 7 and 11. We want to acknowledge to Mr. Hongwei Zhang for his contribution to Chapter 9. We are very thankful to Invention Machine Inc. and Mr. Josh Veshia, for their permission to use many excellent graphs of TRIZ examples in Chapter 9. We want to acknowledge Miss T. M. Kendall for her editorial support of our draft. We want to acknowledge the departmental secretary of the Industrial and Manufacturing Engineering Department of Wayne State University, Margaret Easley, for her help in preparing the manuscript.

Readers' comments and suggestions would be greatly appreciated. We will give serious consideration to your suggestions for future editions. Also, we are conducting public and in-house Six Sigma and DFSS workshops and provide consulting services.

Kai Yang
ac4505@wayne.edu

Basem El-Haik
basemhaik@hotmail.com

Quality Concepts

Profitability is one of the most important factors for any successful business enterprise. High profitability is determined by strong sales and overall low cost in the whole enterprise operation. Healthy sales are to a great extent determined by high quality and reasonable price; as a result, improving quality and reducing cost are among the most important tasks for any business enterprise. Six Sigma is a new wave of enterprise excellence initiative which would effectively improve quality and reduce cost and thus has received much attention in the business world. However, quality is a more intriguing concept than it appears to be. To master quality improvement, it is very important to understand exactly "what is quality."

1.1 What Is Quality?

"Quality: an inherent or distinguishing characteristic, a degree or grade of excellence."
    (American Heritage Dictionary, 1996)

"Quality: The totality of characteristics of an entity that bear on its ability to satisfy stated and implied needs"   (ISO 8402)

"Quality: Do the right thing, and do things right all the time."

When the word *quality* is used, we usually think in terms of an excellent product or service that fulfills or exceeds our expectations. These expectations are based on the intended use and the selling price. For example, the performance that a customer expects from a roadside motel is different from that of a five-star hotel because the prices and expected service levels are different. When a product or service surpasses our expectations, we consider that its quality is good. Thus, quality is related to perception. Mathematically, quality can be quantified as follows:

$$Q = \frac{P}{E} \tag{1.1}$$

where Q = quality
   P = performance
   E = expectations

The perceived "performance" is actually "what this product can do for me" in the eyes of customers. The American Society for Quality (ASQ) defines quality as "A subjective term for which each person has his or her own definition. In technical usage, quality can have two meanings: 1. the characteristics of a product or service that bear on its ability to satisfy stated or implied needs. 2. a product or service free of deficiencies."

By examining the ASQ's quality definition, we can find that "on its ability to satisfy stated or implied needs" means that the product or service should be able to deliver potential customers' needs; we call it "doing the right things," and "free of deficiencies" means that the product or service can deliver customer's needs consistently. We can call this "Doing things right all the time."

However, when we try to further define "what is quality" in detail, we would easily find that quality is also an intangible, complicated concept. For different products or services, or different aspects thereof—for different people, such as producers, designers, management, and customers, even for different quality gurus—the perceived concepts of quality are quite different.

According to David A. Garvin (1988), quality has nine dimensions. Table 1.1 shows these nine dimensions of quality with their meanings and explanations in terms of a slide projector.

There are also many other comments about quality (ASQ Website: www.asq.org):

- Quality is "wow"!
- Quality is not a program; it is an approach to business.
- Quality is a collection of powerful tools and concepts that have proved to work.
- Quality is defined by customers through their satisfaction.
- Quality includes continuous improvement and breakthrough events.
- Quality tools and techniques are applicable in every aspect of the business.

TABLE 1.1 The Dimension of Quality

Dimension	Meaning and example
Performance	Primary product characteristics, such as brightness of the picture
Features	Secondary characteristics, added features, such as remote control
Conformance	Meeting specifications or industry standards, quality of work
Reliability	Consistency of performance of time, average time for the unit to fail
Durability	Useful life, including repair
Service	Resolution of problems and complaints, ease of repair
Response	Human-to-human interface, such as the courtesy of the dealer
Aesthetics	Sensory characteristics, such as exterior finish
Reputation	Past performance and other intangibles, such as ranking first

SOURCE: Adapted from David A Garvin, *Managing Quality: The Strategic and Competitive Edge,* Free Press, New York, 1988.

- Quality is aimed at performance excellence; anything less is an improvement opportunity.

- Quality increases customer satisfaction, reduces cycle time and costs, and eliminates errors and rework.

- Quality isn't just for businesses. It works in nonprofit organizations such as schools, healthcare and social services, and government agencies.

- Results (performance and financial) are the natural benefits of effective quality management.

It is clear that all of these sound very reasonable. We can clearly see that the word *quality* has many meanings and therefore is very intriguing. As the life cycle of the product or service might be a long and complicated process, the meaning of *quality* during different stages of the life cycle could be very different. Therefore, to fully comprehend the meaning of quality, it is very important to understand some basic aspects of product life cycle.

1.2   Quality Assurance and Product/Service Life Cycle

To deliver quality to a product or service, we need a system of methods and activities, called *quality assurance,* which is defined as all the planned and systematic activities implemented within the quality system that can be demonstrated to provide confidence that a product or service will fulfill requirements for quality.

Because quality is a way of doing business, it must be related to a specific product or service. For any product and service, its lifespan includes its creation, development, usage, and disposal. We call this whole lifespan the *product/service life cycle*. A good quality assurance program should act on all stages of the life cycle.

Figure 1.1 illustrates a typical product/service life cycle. The earlier stages of the cycle are often called "upstream"; the latter stages are often called "downstream." We will briefly review each stage of the cycle and the role of quality in each stage.

1.2.1   Stage 0: Impetus/ideation

The product or service life cycle begins with impetus/ideation. The impetus of a new product or service could be the discovery of a new technology, such as the invention of semiconductors, with or without clarity in advance as to how it might be commercialized—a great market opportunity identified through some form of market research, or an obvious need to retire an existing product that has been eclipsed by the competition, such as the annual redesign of automobile models; or a new idea using existing technologies, such as "selling books via the Internet." Once the impetus is identified and it is determined that a viable product or service can be subsequently developed, the ideation phase will follow. The ideation phase focuses on stating the possible product or service and setting a general direction, including identifying plausible options for new products or services.

There are several keys for success in this phase, including the lead time to discover the possible new product or service idea and determine its viability, the lead time to formulate its viable new product or service option, and the quality of formulation.

For new product development based on new technology, there are many cases in which the technology works well in the lab but may encounter great difficulties in commercialization. A very new quality method called "robust technology development" can be applied to reduce those difficulties.

1.2.2   Stage 1: Customer and business requirements study

Customer and business requirements study is the first stage. During both initial concept development and product definition stages, customer research, feasibility studies, and cost/value research should be performed. The purpose of customer research is to develop the key functional elements which will satisfy potential customers and therefore eventually succeed in the market. The purpose of feasibility study and cost/value study is to ensure that the new product or service is competitive in the future market. In this stage, modeling, simulation, and optimization may be employed to evaluate and refine the product concept in order to achieve the best possible functionality and lowest possible cost.

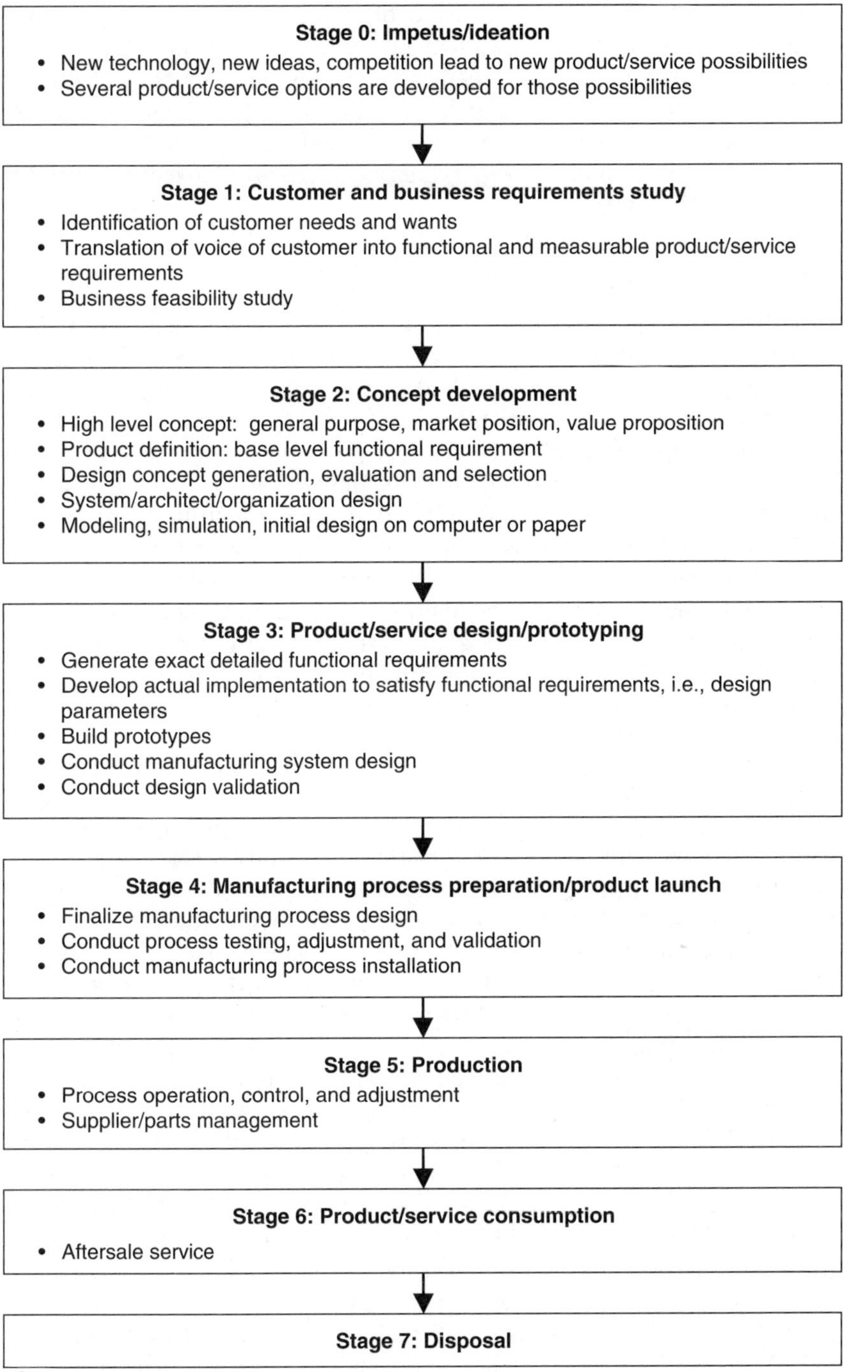

Stage 0: Impetus/ideation
- New technology, new ideas, competition lead to new product/service possibilities
- Several product/service options are developed for those possibilities

↓

Stage 1: Customer and business requirements study
- Identification of customer needs and wants
- Translation of voice of customer into functional and measurable product/service requirements
- Business feasibility study

↓

Stage 2: Concept development
- High level concept: general purpose, market position, value proposition
- Product definition: base level functional requirement
- Design concept generation, evaluation and selection
- System/architect/organization design
- Modeling, simulation, initial design on computer or paper

↓

Stage 3: Product/service design/prototyping
- Generate exact detailed functional requirements
- Develop actual implementation to satisfy functional requirements, i.e., design parameters
- Build prototypes
- Conduct manufacturing system design
- Conduct design validation

↓

Stage 4: Manufacturing process preparation/product launch
- Finalize manufacturing process design
- Conduct process testing, adjustment, and validation
- Conduct manufacturing process installation

↓

Stage 5: Production
- Process operation, control, and adjustment
- Supplier/parts management

↓

Stage 6: Product/service consumption
- Aftersale service

↓

Stage 7: Disposal

Figure 1.1 A typical product/service life cycle. Stage 0–5: Product/service development cycle.

According to the ASQ definition of quality mentioned above, the characteristics of the new product or service should have the "ability to satisfy stated or implied needs"; therefore, one key task of quality assurance activity in this stage is to ensure that the newly formulated product/service functions (features) should be able to satisfy customers. Quality function deployment (QFD) is an excellent quality method for this purpose.

1.2.3   Stage 2: Concept development

Product/service concept development is the second stage. This stage starts with the initial concept development phase. It involves converting one or more options developed in the previous stage into a high-level product concept, describing the product's purpose, general use, and value proposition. Next is the product definition phase. It clarifies product requirements, which are the base-level functional elements necessary for the product to deliver its intended results.

Several quality methods, such as design of experiment (DOE), response surface method (RSM), axiomatic design, and TRIZ (theory of inventive problem solving) are also very helpful in the product concept development stage for enhancing functionality and reducing expected cost. Those methods are also helpful in developing a robust product concept to ensure a final product that is free of deficiencies.

1.2.4   Stage 3: Product/service design/prototyping

The third stage is product design/prototyping. In this stage, product/service scenarios are modeled and design principles are applied to generate exact detailed functional requirements, and their actual implementation and design parameters. For product design, design parameters could be dimension, material properties, and part specifications. For service design, design parameters could be detailed organization layout and specifications. The design parameters should be able to provide all the detail necessary to begin construction or production. For product development, after product design, prototypes are built to test and validate the design. If the test results are not satisfactory, the designs are often revised. Sometimes, this build-test-fix cycle is iterated until satisfactory results are achieved. Besides physical prototyping, computer-based modeling and simulation are also often used and sometimes preferred because they are less costly and more time-efficient. During this stage, manufacturing system design for the product is also conducted to ensure that the product can be manufactured economically.

For quality assurance, it is clear that the key task of this product design/prototyping stage is to formulate the set of design parameters in order to deliver the product's intended functions. By using axiomatic design terminology, product design is a mapping from function space to design parameter space. Therefore, the key task for quality in design is to ensure that the designed product is able to deliver the desired product functions over its useful life. The quality methods used in this stage include robust design (Taguchi method) (Taguchi 1986), design of experiment (DOE), response surface methods (RSMs), Design for X, axiomatic design, TRIZ, and some aspects of reliability engineering.

1.2.5 Stage 4: Manufacturing process preparation/product launch

The fourth stage is manufacturing process preparation/product launch. During this stage, the manufacturing process design will be finalized. The process will undergo testing and adjustment, so there is another set of build-test-fix cycles for the manufacturing process. After iterations of cycles, the manufacturing process will be validated and accepted and installed for production. Using axiomatic design terminology, this stage is a mapping between product design parameters to process variables.

For quality assurance, clearly the key task for this stage is to ensure that the manufactured product should be consistent with product design; that is, the product design on paper or computer can be realized in the manufacturing process. The process is able to produce the real product consistently, economically, and free of defects. The quality methods used in this stage include robust design, DOE, manufacturing troubleshooting and diagnosis, and the Shainin method.

1.2.6 Stage 5: Production

The fifth stage is the full-scale production. In this stage, the product will be produced and shipped to the market. Some parts or subassemblies might be produced by suppliers. During production, it is very important that the manufacturing process be able to function consistently and free of defect, and all parts and subassemblies supplied by suppliers should be consistent with quality requirements.

For quality assurance at this stage, the key task is to ensure that the final product is in conformance with product requirements; that is, all products, their parts, subassemblies should be conformant with their designed requirement; they should be interchangeable and consistent. The quality methods used in this stage include statistical process control (SPC), quality standard and acceptance inspection for suppliers, and production troubleshooting and diagnosis methods.

The combined activities from stage 1 through stage 5 is also called the *product development cycle*.

1.2.7   Stage 6: Product/service consumption

The sixth stage is the product consumption and service. During this stage, the products are consumed by customers. This stage is really the most important to the consumer, whose opinion will eventually determine the success or failure of the product and brand name. When customers encounter problems in using the product during consumption, such as defects, warranty and service are important to keep the product in use and the customer satisfied.

For quality assurance in this stage, it is impossible to improve the quality level for the products already in use, because they are already out of the hands of the producer. However, a good warrantee and service program will certainly help keep the product in use by repairing the defective units and providing other aftersale services. Usually, warranty and service programs are very expensive in comparison with "doing things right the first time." The warranty and service program can also provide valuable information to improve the quality of future production and product design.

1.2.8   Stage 7: Disposal

The seventh stage is product disposal. With increasing concern over the environment, this stage is receiving increasing attention. Once a product has been on the market for a while, a variety of techniques can be used to determine whether it is measuring up to expectations, or if opportunities exist to take the product in new directions. Executives and product managers can then determine whether to stand put, perform minor design refinements, commence a major renovation, or move forward to ideation, beginning the cycle for a new product. The ability to determine the right time to make the leap from an old product to a new one is an important skill.

In terms of quality assurance, and according to the definition of quality, it is clear that the word *quality* has many different meanings, and the quality assurance activities and methods are all different at different stages of the product life cycle. Table 1.2 summarizes the relationship between quality and product life cycle.

1.3   Development of Quality Methods

The history of quality assurance and methods is as old as the industry itself. However, modern quality methods were developed after the industrial revolution. In this section, we review the historical development of quality methods and major quality leaders in chronologic order.

TABLE 1.2 Product Life Cycle and Quality Methods

Product/service life-cycle stages	Quality assurance tasks	Quality methods
0. Impetus/ideation	Ensure new technology and/or ideas to be robust for downstream development	Robust technology development
1 Customer and business requirements study	Ensure new product/service concept to come up with right functional requirements which satisfy customer needs	Quality function deployment (QFD)
2 Concept development	Ensure that the new concept can lead to sound design, free of design vulnerabilities Ensure the new concept to be robust for downstream development	Taguchi method/robust design TRIZ Axiomatic design DOE Simulation/optimization Reliability-based design
3 Product/service design/prototyping	Ensure that designed product (design parameters) deliver desired product functions over its useful life Ensure the product design to be robust for variations from manufacturing, consumption, and disposal stages	Taguchi method/robust design DOE Simulation/optimization Reliability-based design/testing and estimation
4 Manufacturing process; preparation/product launch	Ensure the manufacturing process to be able to deliver designed product consistently	DOE Taguchi method/robust design Troubleshooting and diagnosis
5 Production	Produce designed product with a high degree of consistency, free of defects	SPC Troubleshooting and diagnosis Inspection
6 Product/service consumption	Ensure that the customer has a satisfactory experience in consumption	Quality in aftersale service
7 Disposal	Ensure trouble-free disposal of the product or service for the customer	Service quality

Before the industrial revolution, quality was assured by the work of individual crafters. The production is rather like an art, and crafters were trained and evinced similar behavior to that of artists. A crafter was often the sole person responsible for the entire product. Quality was controlled by the skill of the crafter, who usually had a long training period.

The assembly line and specialization of labor were introduced during the industrial revolution. As a result, the production process became more productive, more routine, and also more complicated. Compared with artistic production, where a single worker makes the whole product and the worker's skill is very important, the new production process employs many workers, each making only a portion of the product with very simple operations, and the worker's skill level became less important. Thus the quality can no longer be assured by an individual worker's skill. In the modern production system, the volume and number of parts in the production increased greatly; therefore, the variation in assembly and variation in part quality became a major impediment in production because it destroyed the consistency of product and part interchangeability. Also, modern production assembles parts from many suppliers; even a small number of defective parts can ruin a big batch of production, and the rework is usually very costly. Therefore, there is an urgent need to control the variation and sort out defective parts from suppliers. This need is the impetus for the creation of modern quality system and quality methods.

The historic development of the modern quality method actually started at the last stage of the product development cycle: production.

1.3.1 Statistical process control (1924)

Statistical process control (SPC) is the application of statistical techniques to control a process. In 1924, Walter. A. Shewhart of Bell Telephone Laboratories developed a statistical control chart to control important production variables in the production process. This chart is considered as the beginning of SPC and one of the first quality assurance methods introduced in modern industry. Shewhart is often considered as the father of statistical quality control because he brought together the disciplines of statistics, engineering, and economics. He described the basic principles of this new discipline in his book *Economic Control of Quality of Manufactured Product.*

1.3.2 Acceptance sampling (1940)

In the production stage, quality assurance of incoming parts from other suppliers is also important, because defective parts could certainly make a defective final product. Obviously, 100% inspection of all incoming parts may identify defective parts, but this is very expensive. *Acceptance sampling,* which was developed to solve this problem, is the inspection of a sample from a lot to decide whether to accept or reject that lot. Acceptance sampling could consist of a simple sampling in which only one sample in the lot is inspected; or multiple sampling, in which a sequence of samples are taken and the accept/reject decision is based on statistical rules.

The acceptance sampling plan was developed by Harold F. Dodge and Harry G. Romig in 1940. Four sets of tables were published in 1940: single-sampling lot tolerance tables, double-sampling lot tolerance tables, single-sampling average outgoing quality limit tables, and double-sampling average outgoing quality limit tables.

1.3.3   Design of experiment (late 1930s)

Design of experiment (DOE) is a very important quality tool in current use. DOE is a generic statistical method which guides design and analysis of experiments in order to find the cause-and-effect relationship between "response" (output) and factors (inputs). This relationship is derived from empirical modeling of experimental data. DOE can also guide the experimenter to design efficient experiment and conduct data analysis to get other valuable information such as identification and ranking of important factors.

DOE was initially developed to study agricultural experiments. In the 1930s, Sir Ronald Fisher, a professor at the University of London, was the innovator in the use of statistical methods in experimental design. He developed and first used analysis of variance (ANOVA) as the primary method in the analysis in experimental design. DOE was first used at the Rothamsted Agricultural Experimental Station in London. The first industrial applications of DOE were in the British textile industry. After World War II, experimental design methods were introduced in the chemical and process industries in the United States and Western Europe.

1.3.4   Tools for manufacturing diagnosis and problem solving (1950s)

Statistical process control (SPC) is a process monitoring tool. It can discern whether the process is in a state of normal variation or in a state of abnormal fluctuation. The latter state often indicates that there is a problem in the process. However, SPC cannot detect what the problem is. Therefore, developing tools for process troubleshooting and problem solving is very important. There are many tools available today for troubleshooting; however, Kaoru Ishikawa's seven basic tools for quality and Dorian Shainin's statistical engineering deserve special attention.

Seven tools of quality. Tools that help organizations understand their processes to improve them are the cause-and-effect diagram, the checksheet, the control chart, the flowchart, the histogram, the Pareto chart, and the scatter diagram (see individual entries).

One of the Japanese quality pioneers, Kaoru Ishikawa, is credited for the development of and dissemination of the seven tools of quality. Ishikawa promoted the "democratizing statistics," which means the universal use of simple, effective statistical tools by all the workforce, not just statisticians, for problem solving and process improvement.

Shanin method. Dorian Shanin developed a discipline called *statistical engineering.* In his statistical engineering, he promoted many effective problem-solving methods such as search by logic, multi-variate chart, and data pattern recognition. He was in charge of quality control at a large division of United Technologies Corporation and later did consulting work for more than 900 organizations. Shanin also was on the faculty of the University of Chicago and wrote more than 100 articles and several books.

1.3.5   Total quality management (TQM) (1960)

After 1960, first in Japan and later in the rest of the world, more and more people realized that quality could not be assured by just a small group of quality professionals, but required the active involvement of the whole organization, from management to ordinary employees. In 1960, the first "quality control circles" were formed in Japan and simple statistical methods were used for quality improvement. Later on, a quality-oriented management approach, *total quality management* (TQM), was developed. TQM is a management approach to long-term success through customer satisfaction and is based on the participation of all members of an organization in improving processes, products, services, and the culture in which they work. The methods for implementing this approach are found in the teachings of such quality leaders as W. Edwards Deming, Kaoru Ishikawa, Joseph M. Juran, and many others.

Dr. W. Edwards Deming. Deming was a protégé of Dr. Walter Shewhart; he also spent one year studying under Sir Ronald Fisher. After Deming shared his expertise in statistical quality control to help the U.S. war effort during World War II, the War Department sent him to Japan in 1946 to help that nation recover from its wartime losses. Deming published more than 200 works, including the well-known books *Quality, Productivity, and Competitive Position* and *Out of the Crisis.*

Dr. Deming is credited with providing the foundation of the Japanese quality miracle. He developed the following 14 points for managing the improvement of quality, productivity, and competitive position:

1. Create constancy of purpose for improving products and services.
2. Adopt the new philosophy.
3. Cease dependence on inspection to achieve quality.
4. End the practice of awarding business on price alone; instead, minimize total cost by working with a single supplier.
5. Improve constantly and forever every process for planning, production, and service.
6. Institute training on the job.
7. Adopt and institute leadership.
8. Drive out fear.
9. Break down barriers between staff areas.
10. Eliminate slogans, exhortations, and targets for the workforce.
11. Eliminate numerical quotas for the workforce and numerical goals for management.
12. Remove barriers that rob people of pride in their work, and eliminate the annual rating or merit system.
13. Institute a vigorous program of education and self-improvement for everyone.
14. Put everybody in the company to work to accomplish the transformation.

Deming's basic quality philosophy is that productivity improves as variability decreases, and that statistical methods are needed to control quality. He advocated the use of statistics to measure performance in all areas, not just conformance to product specifications. Furthermore, he thought that it is not enough to meet specifications; one has to keep working to reduce the variations as well. Deming was extremely critical of the U.S. approach to business management and was an advocate of worker participation in decision making.

Kaoru Ishikawa Ishikawa is a pioneer in quality control activities in Japan. In 1943, he developed the cause-and-effect diagram. Ishikawa published many works, including *What Is Total Quality Control?, The Japanese Way, Quality Control Circles at Work,* and *Guide to Quality Control.* He was a member of the quality control research group of the

Union of Japanese Scientists and Engineers while also working as an assistant professor at the University of Tokyo.

Kaoru Ishikawa's quality philosophy can be summarized by his 11 points:

1. Quality begins and ends with education.
2. The first step in quality is to know the requirements of the customer.
3. The ideal state of quality control is when quality inspection is no longer necessary.
4. Remove the root cause, not symptoms.
5. Quality control is the responsibility of all workers and all divisions.
6. Do not confuse means with objectives.
7. Put quality first and set your sights on long-term objectives.
8. Marketing is the entrance and exit of quality.
9. Top management must not show anger when facts are presented to subordinates.
10. Ninety-five percent of the problem in a company can be solved by seven tools of quality.
11. Data without dispersion information are false data.

Joseph Moses Juran.   Juran was born in 1904 in Romania. Since 1924, Juran has pursued a varied career in management as an engineer, executive, government administrator, university professor, labor arbitrator, corporate director, and consultant. Specializing in managing for quality, he has authored hundreds of papers and 12 books, including *Juran's Quality Control Handbook, Quality Planning and Analysis* (with F. M. Gryna), and *Juran on Leadership for Quality*. His major contributions include the Juran trilogy, which are three managerial processes that he identified for use in managing for quality: quality planning, quality control, and quality improvement. Juran conceptualized the Pareto principle in 1937. In 1954, the Union of Japanese Scientists and Engineers (JUSE) and the Keidanren invited Juran to Japan to deliver a series of lectures on quality that had profound influence on the Japanese quality revolution. Juran is recognized as the person who added the "human dimension" to quality, expanding it into the method now known as *total quality management* (TQM).

1.3.6   Errorproofing (poka-yoke) (1960)

In Japanese, *poke* means inadvertent mistake and *yoke* means prevent. The essential idea of poka-yoke is to design processes in such a way that mistakes are impossible to make or at least are easily detected and corrected.

Poka-yoke devices fall into two major categories: prevention and detection. A prevention device affects the process in such a way that it is impossible to make a mistake. A detection device signals the user when a mistake has been made, so that the user can quickly correct the problem.

Poka-yoke was developed by Shigeo Shingo, a Japanese quality expert. He is credited for his great contribution for tremendous Japanese productivity improvement.

1.3.7   Robust engineering/Taguchi method (1960s in Japan, 1980s in the West)

Dr. Genich Taguchi and his system of quality engineering is one of the most important milestones in the development of quality methods. Taguchi method, together with QFD, extended the quality assurance activities to the earlier stages of the product life cycle. Taguchi's quality engineering is also called the *robust design method,* which has the following distinctive features:

1. Focus on the earlier stages of product life cycle, which include concept design, product design, and manufacturing process design, and preparation. More recently, Taguchi and his colleagues have extended that to the technology development stage, which is even earlier than the impetus/ideation stage. He thinks that the earlier involvements in the product development cycle can produce a bigger impact and better results and avoid costly engineering rework and firefighting measures.

2. Focus on the design of the engineering system which is able to deliver its intended functions with robustness. *Robustness* means insensitivity to variations caused by noise factors, which may include environmental factors, user conditions, and manufacturing disturbances. He pointed out an important fact that variation can be reduced by good design.

3. Promote the use of Taguchi's system of experimental design.

1.3.8   Quality function deployment (QFD) (1960 in Japan, 1980s in the West))

Quality function deployment (QFD) is an effective quality tool in the early design stage. It is a structured approach to defining customer needs or requirements and translating them into specific plans to produce products to meet those needs. The "voice of the customer" is the term used to describe these stated and unstated customer needs or requirements. The voice of the customer is captured in a variety of ways, including direct discussion or interviews, surveys, focus groups, customer specifications, observation, warranty data, and field reports. This understanding of the customer needs is then summarized in a product planning matrix or "house of quality." These matrices are used to translate higher-level "whats" or needs into lower-level "hows"—product requirements or technical characteristics to satisfy these needs.

Quality function deployment matrices are also a good communication tool at each stage in the product development cycle. QFD enables people from various functional departments, such as marketing, design engineering, quality assurance, manufacturing engineering, test engineering, finance, and product support, to communicate and work together effectively.

QFD was developed in the 1960s by Professors Shigeru Mizuno and Yoji Akao. Their purpose was to develop a quality assurance method that would design customer satisfaction into a product before it was manufactured. Prior quality control methods were aimed primarily at fixing a problem during or after manufacturing.

1.3.9   TRIZ (1950s in Soviet Union, 1990s in the West)

TRIZ is another tool for design improvement by systematic methods to foster creative design practices. TRIZ is a Russian acronym for the theory of inventive problem solving (TIPS).

TRIZ is based on inventive principles derived from the study of more than 1.5 million of the world's most innovative patents and inventions. TRIZ provides a revolutionary new way of systematically solving problems on the basis of science and technology. TRIZ helps organizations use the knowledge embodied in the world's inventions to quickly, efficiently, and creatively develop "elegant" solutions to their most difficult product and engineering problems.

TRIZ was developed by Genrich S. Altshuller, born in the former Soviet Union in 1926 and serving in the Soviet Navy as a patent expert in the 1940s. Altshuller screened over 200,000 patents looking for inventive problems and how they were solved. Altshuller distilled the problems, contradictions, and solutions in these patents into a theory of inventive problem solving which he named TRIZ.

1.3.10   Axiomatic design (1990)

Axiomatic design is a principle-based method that provides the designer with a structured approach to design tasks. In the axiomatic design approach, the design is modeled as mapping between different domains. For example, in the concept design stage, it could be a mapping of the customer attribute domain to the product function domain;

in the product design stage, it is a mapping from the function domain to the design parameter domain. There are many possible design solutions for the same design task. However, on the basis of its two fundamental axioms, the axiomatic design method developed many design principles to evaluate and analyze design solutions and gave designers directions to improve designs. The axiomatic design approach can be applied not only in engineering design but also in other design tasks such as the organization system. N. P. Suh is credited for the development of axiomatic design methods (Suh 1990).

In summary, modern quality methods and the quality assurance system have developed gradually since the industrial revolution. There are several trends in the development of quality methods.

1. The first few methods, SPC and acceptance sampling, were applied at production stage, which is the last stage, or downstream in the product development cycle.

2. More methods were developed and applied at earlier stages, or upstream of product development cycle, such as QFD and the Taguchi method.

3. Quality methods and systems are then integrated into companywide activities with participation of top management to ordinary employees, such as TQM.

4. Aftersale service has also gained attention from the business world.

However, the implementation of modern quality methods in the business world has not always been smooth. It is a rather difficult process. One of the main reasons is that many business leaders think that quality is not the only important factor for success. Other factors, such as profit, cost, and time to market, are far more important in their eyes, and they think that in order to improve quality, other important factors have to be compromised.

The newest quality movement is the introduction and widespread implementation of Six Sigma, which is the fastest growing business management system in industry today. Six Sigma is the continuation of the quality assurance movement. It inherited many features of quality methods, but Six Sigma attempts to improve not only product quality itself but also *all* aspects of business operation; it is a method for *business excellence.*

1.4   Business Excellence, Whole Quality, and Other Metrics in Business Operations

Business excellence is featured by good profitability, business viability, growth in sales, and market share, on the basis of quality (Tom Peters, 1982). Achieving business excellence is the common goal for all business leaders and their employees. To achieve business excellence, only the product quality itself is not sufficient; quality has to be replaced by "whole quality," which includes quality in business operations. To understand business excellence, we need to understand business operation per se and other metrics in business operation.

1.4.1 Business operation model

Figure 1.2 shows a typical business operation model for a manufacturing based company. For service-oriented and other types of company, the business model could be somewhat different. However, for every company, there is always a "core operation," and a number of other business elements. The core operation is the collection of all activities to provide products or services to customers. For example, the core operation of an automobile company is to produce cars, and the core operation of Starbucks is to provide coffee service throughout the world. Core operation runs all activities in the product/service life cycle.

For a company to operate, the core operation alone is not enough. Figure 1.2 listed several other typical elements needed in order to make a company fully operational, such as the business process and business management. The success of the company depends on the successes of all aspects of business operation.

Before Six Sigma, quality was narrowly defined as the quality of product or service of the company provided to external customers; therefore, it relates only to the core operation. Clearly, from the point of view of a business leader, this "quality" is only part of the story,

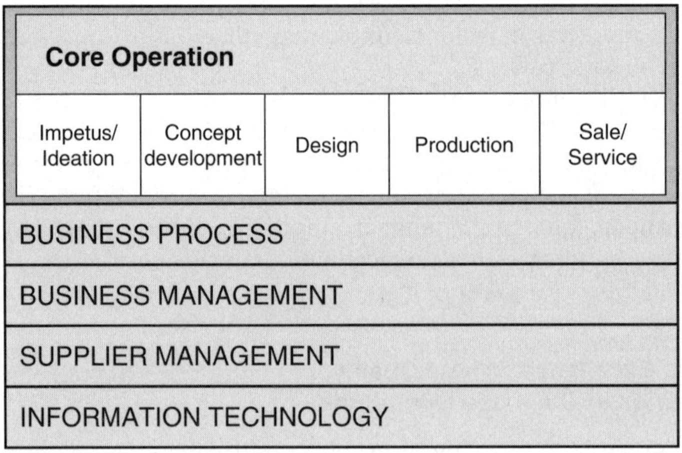

Core Operation				
Impetus/ Ideation	Concept development	Design	Production	Sale/ Service
BUSINESS PROCESS				
BUSINESS MANAGEMENT				
SUPPLIER MANAGEMENT				
INFORMATION TECHNOLOGY				

Figure 1.2 Business operation model.

because other critical factors for business success, such as cost, profit, time to market, and capital acquisition, are also related to other aspects of business operation. The key difference between Six Sigma and all other previously developed quality systems and methods, such as TQM, is that Six Sigma is a strategy for the *whole quality,* which is the drastic improvement for the *whole business operation.*

We will show that improving whole quality will lead to business excellence, because improving whole quality means improving all major performance metrics of business excellence, such as profit, cost, and time to market.

1.4.2 Quality and cost

Low cost is directly related to high profitability. Cost can be roughly divided into two parts: life-cycle costs related to all products and/or services offered by the company and the cost of running the supporting functions within the company, such as various noncore operations-related departments. For a particular product or service, life-cycle cost includes production/service cost, plus the cost for product/service development.

The relationship between quality and cost is rather complex; the term *quality* here refers to the product/service quality, not the *whole quality.* This relationship is very dependent on what kind of quality strategy is adopted by a particular company. If a company adopted a quality strategy heavily focused on the downstream of the product/service life cycle, such as firefighting, rework, and error corrections, then that quality will be very costly. If a company adopted a strategy emphasizing upstream improvement and problem prevention, then improving quality could actually reduce the life-cycle cost because there will be less rework, less recall, less firefighting, and therefore, less product development cost. In the manufacturing-based company, it may also mean less scrap, higher throughput, and higher productivity.

If we define *quality* as the whole quality, then the higher whole quality will definitely mean lower cost. Because *whole quality* means higher performance levels of all aspects of business operation, it means high performance of all supporting functions, high performance of the production system, less waste, and higher efficiency. Therefore, it will definitely reduce business operation cost, production cost, and service cost.

1.4.3 Quality and time to market

Time to market is the speed in introducing new or improved products and services to the marketplace. It is a very important measure for marketplace competitiveness. For two companies that provide similar product/services with comparable functions and price, the one with a shorter time to market will achieve tremendous competitive position, because the psychological effect—the increase of customer expectation to other competitors—will be very difficult to overcome by latecomers.

Many techniques are used to reduce time to market, such as

- Concurrency: encouraging multitasking and parallel working
- Complexity reduction
- Project management: tuned for product development and life-cycle management

In the Six Sigma approach and according to the whole quality concept, improving the quality of managing the product/service development cycle is a part of the strategy. Therefore, improving whole quality will certainly help reduce time to market.

1.5   Summary

1. *Quality* is defined as the ratio of performance to expectation. Performance is determined by how well a product or service can deliver a good set of functions which will achieve maximum customer satisfaction, and how well the product or service can deliver its function consistently. The customer's expectation is influenced by price, time to market, and many other psychological factors.

2. The best quality assurance strategy is "Do the right things, and do things right all the time." "Do the right thing" means that we have to design absolutely the best product or service for customers' needs with low cost, or "quality in design" to ensure that the product or service will deliver the right set functions to the customers. If we do not do the right thing, such as good design, there is no way that we can succeed. "Do things right all the time" means not only that we have good design but also that we will make all our products and services perform consistently so that all customers will be satisfied at all times.

3. Quality assurance strategy is closely related to product/service life cycle and the product development cycle. The historical development of quality methods started downstream in the product development cycle. Modern quality assurance strategy is a systematic approach covering all stages of product life cycle, with more emphasis on upstream improvement.

4. Six Sigma extends the scope of quality from the product quality alone to the quality of all aspects of business operation. Therefore the "Do the right things, and do things right all the time" strategy is applied to all aspects of business operation. The Six Sigma approach is the approach for business excellence.

2

Six Sigma Fundamentals

2.1 What Is Six Sigma?

Six Sigma is a methodology that provides businesses with the tools to improve the capability of their business processes. For Six Sigma, a process is the basic unit for improvement. A process could be a product or a service process that a company provides to outside customers, or it could be an internal process within the company, such as a billing or production process. In Six Sigma, the purpose of process improvement is to increase performance and decrease performance variation. This increase in performance and decrease in process variation will lead to defect reduction and improvement in profits, to employee morale and quality of product, and eventually to business excellence.

Six Sigma is the fastest growing business management system in industry today. It has been credited with saving billions of dollars for companies since the early 1990s. Developed by Motorola in the mid-1980s, the methodology became well known only after Jack Welch from GE made it a central focus of his business strategy in 1995.

The name "Six Sigma" derives from statistical terminology; Sigma (σ) means *standard deviation*. For normal distribution, the probability of falling within a ±6 sigma range around the mean is 0.9999966. In a production process, the "Six Sigma standard" means that the defectivity rate of the process will be 3.4 defects per million units. Clearly Six Sigma indicates a degree of extremely high consistency and extremely low variability. In statistical terms, the purpose of Six Sigma is to reduce variation to achieve very small standard deviations.

Compared with other quality initiatives, the key difference of Six Sigma is that it applies not only to product quality but also to all aspects of business operation by improving key processes. For example, Six Sigma may help create well-designed, highly reliable, and consistent customer billing systems; cost control systems; and project management systems.

2.2 Process: The Basic Unit for the Six Sigma Improvement Project

Six Sigma is a process-focused approach to business improvement. The key feature is "improving one process at a time." The process here could be a production system, a customer billing system, or a product itself.

What is a *process?* Caulkin (1995) defines it as being a "continuous and regular action or succession of actions, taking place or carried on in a definite manner, and leading to the accomplishment of some result; a continuous operation or series of operations." Keller et al. (1999) define the process as "a combination of inputs, actions and outputs." Anjard (1998) further defines it as being "a series of activities that takes an input, adds value to it and produces an output for a customer." This view is summarized in Fig. 2.1.

Many products are also processes; the inputs of a product could be user intent, energy, or other factors. For example, a TV set takes a user control signal, TV signals, and electrical energy, and transforms these into desired TV images. The outputs of a product are functions delivered to the consumer. There are several process models available, but for a product or a manufacturing process, a process model is often represented by a process diagram, often called a *P-diagram* (Fig. 2.2).

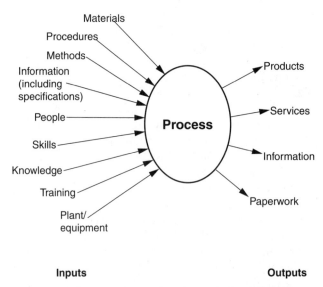

Figure 2.1   A diagram of a process. [*From Oakland (1994).*]

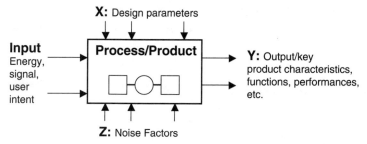

Figure 2.2 P-diagram.

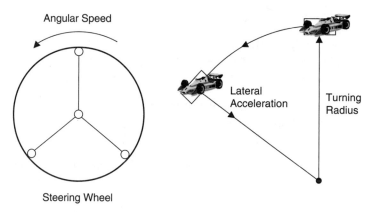

Figure 2.3 Vehicle steering system.

A P-diagram is a very common process model for the Taguchi method. Y is the set of outputs, usually a set of characteristics related to product performance or functions, or customer desired characteristics. X is a set of design parameters, or control factors; X will influence Y, and they both can be adjusted and controlled. Z is a set of "noise factors." Z will also influence Y but cannot be sufficiently controlled. For example, Z could be environmental variations, user conditions, or manufacturing variations. The output Y is usually a function of design parameters X and noise factors Z.

$$Y = f(X, Z) \tag{2.1}$$

A good example of a P-diagram is given by Taguchi and Wu (1986).

Example 2.1. Automobile Steering System (Taguchi and Wu 1986) In an automobile steering system (see Fig. 2.3) when the driver turns the steering wheel, the vehicle's direction should change according to the degree of turn and how quickly the driver is turning the wheel.

The inputs, process, and outputs of the steering system are as follows:

Inputs. User intent: steering wheel turning angular speed; energy: mechanical energy

Process. Vehicle steering system

Design parameters X. Column design parameters, linkage design parameters, material properties, and so on

Noise factors Z. Road conditions, tire air pressure, tire wear, load in vehicle, load position, and so on

Outputs. Turning radius, lateral acceleration of the vehicle

Figure 2.4 shows the P-diagram of the steering system.

Another useful process model is the supplier-input-process-output-customer (SIPOC) diagram (Fig. 2.5).

A SIPOC diagram is one of the most useful models for business and service processes. It can also be used as a model for a manufacturing process. The acronym SIPOC derives from the five elements in the diagram.

1. *Supplier.* The person or group that provides key information, materials, and/or other resources to the process

2. *Input.* The "thing" provided

3. *Process.* The set of steps that transform and, ideally, add value to the input

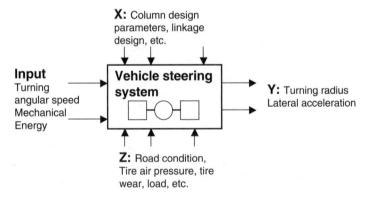

Figure 2.4 P-diagram of vehicle steering system.

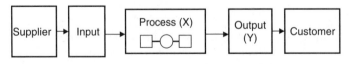

Figure 2.5 SIPOC diagram.

4. *Output.* The final product of the process

5. *Customer.* The person, group, or process that received the output

Customers usually have explicit or implicit requirements for the outputs, which we call *customer requirements.* These requirements are often listed in the SIPOC model as well.

Example 2.2.   An Academic Teaching Program of a University Department

Suppliers. Book publishers and bookstores, university administrators and facility support people, lab equipment suppliers, accreditation board, tuition payers, and so on.

Inputs. Books, classrooms and facilities, labs and facilities, academic program standards, and tuition.

Process. The academic program, which includes the curriculum system, degree program setup, courses, professors, and counselors. The process transform inputs to a system of courses, academic standards (quality control system), and academic records; under this system, incoming students are processed into graduating students in many steps (coursework).

Output. Graduating students with degrees.

Customers. Employers of future students and the students themselves.

Key requirements for output. Excellent combination of knowledge for future career, high and consistent learning qualities, and so on.

Figure 2.6 is the SIPOC diagram of an academic department.

Several other process modeling and process analysis methods are available, such as process mapping and value stream analysis, which are very useful tools for Six Sigma. We will discuss them in Sec. 2.3.

Similar to a product, a process also has its development cycle and life cycle, which is illustrated in Fig. 2.7.

The process life cycle and development cycle illustrated in Fig. 2.7 are more appropriate for service processes or internal business processes of a company.

There are many similarities between products and processes:

1. They all have customers. For product or service processes, the customers are external customers; for internal business processes, the

Figure 2.6   SIPOC diagram of an academic teaching program.

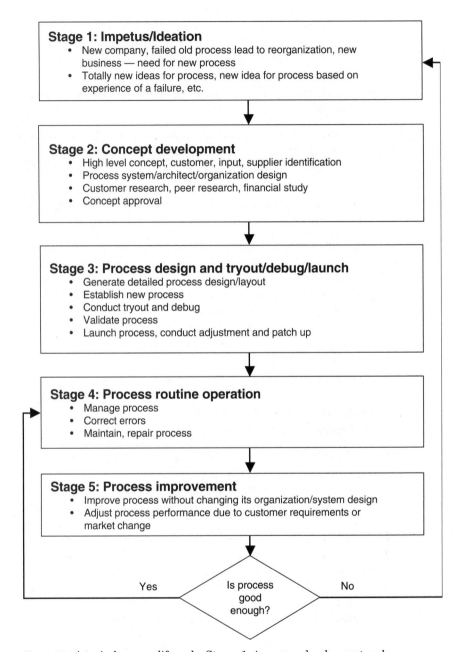

Figure 2.7   A typical process life cycle. Stages 1–4: process development cycle.

customers are internal customers; they are the users of the process, and they work for the same company. For example, the customers for personnel department are people of other functional departments.

2. Both products and processes have to deliver functions; they all do what they should do for their customers. Their mission is to achieve their customers' maximum satisfaction.

3. Both product and process need performance consistency. For example, if a personnel department sometimes hires good employees, and other times hires incompetent employees, it is not a good department.

4. Both products and processes go through similar development cycles and life cycles.

Of course, there are differences between products and processes. For example, after a product is delivered to the user, it is very difficult to make changes, improvements, or remodeling for the product itself.

The similarities between products and processes indicate that many methods and strategies for product quality assurance can be modified and applied to processes. In addition to all quality methods, Six Sigma uses many existing methods for process modeling and process analysis. These methods include process mapping, value stream mapping, and process management.

2.3   Process Mapping, Value Stream Mapping, and Process Management

2.3.1   Process mapping

A process map is a schematic model for a process. "A process map is considered to be a visual aid for picturing work processes which show how inputs, outputs and tasks are linked" (Anjard 1998). Soliman (1998) also describes the process mapping as the "most important and fundamental element of business process re-engineering." There are a number of different methods of process mapping. The two that are most frequently used are the IDEF0 method and the IDEF3 method.

The IDEF0 mapping standard.   The IDEF0 (International DEFinition) is a method designed to model the decisions, actions, and activities of an organization or system. It was developed by the U.S. Department of Defense, mainly for the use of the U.S. Air Force during the 1970s. Although it was developed over thirty years ago, the Computer Systems Laboratory of the National Institute of Standards and Technology (NIST) released IDEF0 as a standard for function modeling in FIPS Publication 183, December 1993. Computer packages (e.g., AI0

WIN) have been developed to aid software development by automatically translating relational diagrams into code.

An IDEF0 diagram consists of boxes and arrows. It shows the function as a box and the interfaces to or from the function as arrows entering or leaving the box. Functions are expressed by boxes operating simultaneously with other boxes, with the interface arrows "constraining" when and how operations are triggered and controlled. The basic syntax for an IDEF0 model is shown in Fig. 2.8.

Mapping using this standard generally involves multiple levels. The first level, the high-level map, identifies the major processes by which the company operates (Peppard and Rowland 1995). The second-level map breaks each of these processes into increasingly fine subprocesses until the appropriate level of detail is reached. For example, Fig. 2.9

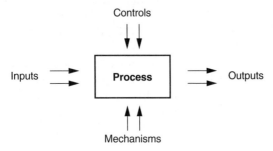

Figure 2.8 A basic IDEF0 process map template.

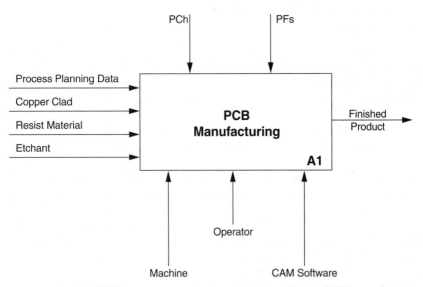

Figure 2.9 Level 1 IDEF0 process mapping for printed-circuit-board (PCB) manufacturing.

shows the first-level and Fig. 2.10 the second-level IDEF0 process map of a printed-circuit-board (PCB) manufacturing process, and you can go further down with every subprocess.

A number of strengths and weaknesses are associated with IDEF0. The main strength is that it is a hierarchical approach enabling users to choose the mapping at their desired level of detail. The main weakness of IDEF0 is that it cannot precisely illustrate "sequence" and transformation. To overcome this weakness, IDEF3 was developed.

IDEF3 process description capture method.   IDEF0 diagrams are used for process mapping and visualization of information in a form of inputs, controls, outputs, and mechanisms (ICOM). Even though the method efficiently supports the hierarchical decomposition of activities, it cannot express process execution sequence, iteration runs, selection of paths, parallel execution, and conditional process flow. The IDEF3 diagrams have been developed to overcome the above-mentioned weaknesses by capturing all temporal information, such as precedence and

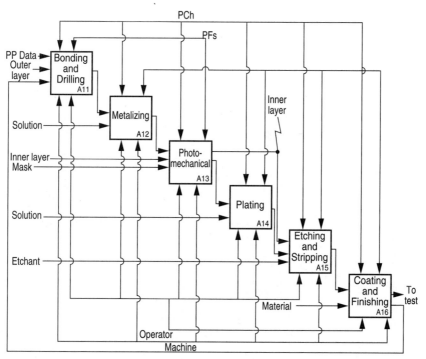

Figure 2.10 Level 2 IDEF0 process mapping for printed-circuit-board (PCB) manufacturing process.

causality relationships associated with the enterprise processes and thus providing a basis for constructing analytical design models.

There are two IDEF3 description modes, process flow and the object state transition network. A *process flow* description captures knowledge of "how things work" in an organization, such as the description of what happens to a part as it flows through a sequence of manufacturing processes. The object state transition network description summarizes the allowable transitions that an object may undergo throughout a particular process. Figure 2.11 shows a sample process flow description diagram.

Object state transition network (OSTN) diagrams capture object-centered views of processes which cut across the process diagrams and summarize the allowable transitions. Figure 2.12 shows a sample OSTN diagram.

Object states and state transition arcs are the key elements of an OSTN diagram. In OSTN diagrams, object states are represented by circles and state transition arcs are represented by the lines connecting the circles. An *object state* is defined in terms of the facts and constraints that need to be valid for the continued existence of the object in that state and is characterized by entry and exit conditions. The entry conditions specify the requirements that need to be met before

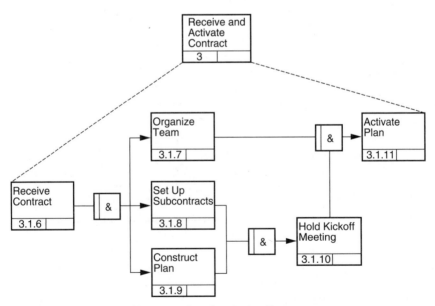

Figure 2.11   Example IDEF3 process flow description diagram.

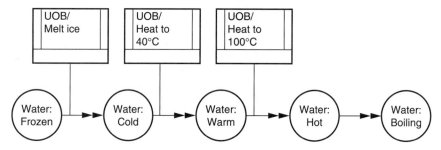

Figure 2.12 Example IDEF3 object state transition network diagram.

an object can transfer into a state. The exit conditions characterize the conditions under which an object can transfer out of a state. The constraints are specified by a simple list of property/value pairs or by a constraint statement. The values of the attributes must match the specified values for the requirements to be met.

State transition arcs represent the allowable transitions between the focus object states. It is often convenient to highlight the participation of a process in a state transition. The importance of such a process constraint between two object states can be represented in IDEF3 by attaching a UOB (unit of behavior) referent to the transition arc between the two object states.

2.3.2 Value stream mapping

Process mapping can be used to develop a value stream map to analyze how well a process works. Once a process map is established at an appropriate level of detail, the flows of products/programs/services, material, information, money, and time can be mapped. For example, Fig. 2.13 shows an example value stream map which maps not only material flows but also the information flows that signal and control the material flows.

After a value stream map is developed, value-adding steps are identified for each type of flow, such as material and information flow. Non-value-adding steps (waste), value inhibitors, costs of flow, and risks to flow are also exposed and their implications to overall process performance are analyzed.

After identifying the problems in the existing process by value stream mapping, process revision or redesign can be initiated to eliminate the deficiencies. In manufacturing processes, the revision can be made by elimination of non-value-adding steps and redesigning the layout and sequence of the subprocesses, thus reducing cost and cycle time. In office processes, the revision can be made by redesigning the organizational structure, reporting mechanisms, building layout, and

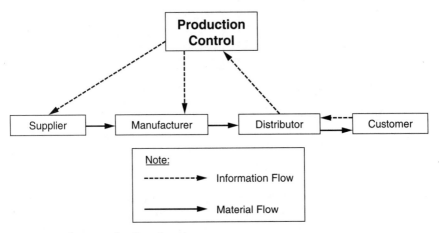

Figure 2.13   An example of a value stream map.

functional responsibilities of various departments in order to reduce non-value-added steps, paperwork travel time, and mistakes, thus reducing waste and improving efficiency.

On the basis of this analysis, an ideal value stream map is created, in which all waste and value inhibitors are removed; the cost and risk for flow are similarly reduced to a minimum level, and we call it the "ideal" state. The full implementation of the ideal state may not be feasible, but it often leads to a much-improved process.

The following is a case example of a value stream mapping project (Bremer 2002) which involves a manufacturing-oriented company.

Example 2.3.   The companywide information flow is illustrated by the following two value stream maps. The first map (Fig. 2.14) shows how management thinks that the information flows in their business; management thinks that the flows are simple and straightforward.

The second map (Fig. 2.15) shows how information really flows. It's a lot more complicated. It shows that many process steps add no value and that they actually impede the production process. Also there are huge amount of hidden information transactions in the process that add no value to a business from a customer's perspective.

Because of these problems, the company usually takes over 34 days to go from raw material to delivery to customers. After identifying these problems, the company redesigns the business process. The improved process is able to reduce most of the hidden transactions, and the company is able to move from raw material to delivery in 5 days after the improvement.

Process mapping and value stream mapping are often used as an efficient tool in the area of process management.

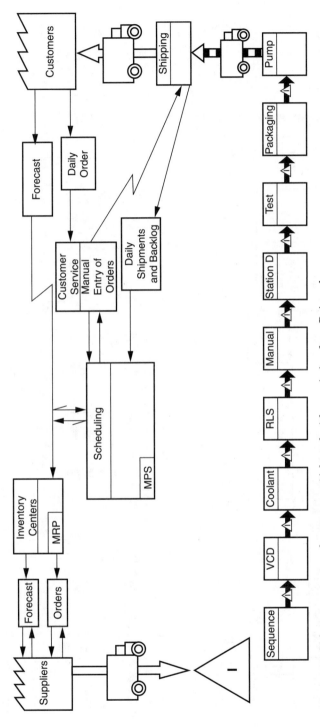

Figure 2.14 Value stream map of a company. *(Adapted with permission from Rainmakers, The Cumberland Group.)*

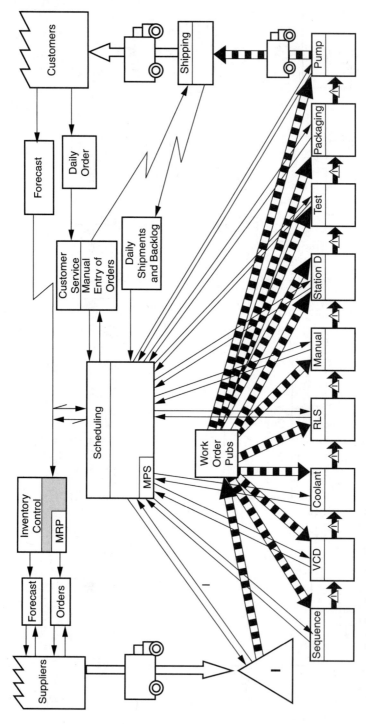

Figure 2.15 Actual value stream map of the company. (*Adapted with permission from Rainmakers, The Cumberland Group.*)

2.3.3 Process management

Process management is a body of knowledge for process improvement. By enhancing efficiency and effectiveness, process management offers the potential to improve customer satisfaction and, ultimately, to offer increased profits, high growth, and long-term business. Most organizations are motivated to manage their processes through one of several dimensions. Fuglseth and Gronhaug (1997) proposed these dimensions as being quality, throughput, efficiency, response time, work-in-progress, and process cost. In order to maximize profits, an organization will need to reduce process cost, increase throughput, and at the same time improve quality.

Process management involves five phases: (1) process mapping, (2) process diagnosis, (3) process design, (4) process implementation, and (5) process maintenance. The process mapping element of this, as mentioned above, involves a definition of the process and captures the issues that will drive the process design and improvement activities. Once the documentation of the objectives and the process have been completed, diagnosis can proceed.

Six Sigma and process management. Process management shares many common goals with Six Sigma. However, process management does not apply a vast array of quality methods for process improvement. Process management focuses mainly on such measures as cost, efficiency, and cycle time, but it does not pay enough attention to process performance consistency or process capability. Process capability is actually the starting point for Six Sigma. Nowadays, Six Sigma will use the methods from both process management and quality assurance to improve process performance and process capability.

2.4 Process Capability and Six Sigma

2.4.1 Process performance and process capability

Process performance is a measure of how well a process performs. It is measured by comparing the actual process performance level versus the ideal process performance level. For a power supply unit, its performance may be measured by its output voltage, and its ideal performance level could be 6 V. For a customer billing process, the performance could be measured by the number of errors per month; in this case, the ideal performance level is "zero error."

For most processes, performance level is not constant. For example, a customer billing process may have very little error in some months, but somewhat more errors in other months. We call this variation the *process variability.* If process performance can be measured by a real number, then the process variability can usually be modeled by normal distribution, and the degree of variation can be measured by standard deviation of that normal distribution.

If process performance level is not a constant but a random variable, we can use process mean and process standard deviation as key performance measures. Mean performance can be calculated by averaging a large number of performance measurements. For example, in the customer billing service, we can collect the number of errors per month for the last 3 years and take the average value, and this is the process mean performance. For the power supply unit, we can measure a large number of units and average their output voltages to get a process mean performance level.

If processes follow the normal probability distribution, a high percentage of the process performance measurements will fall between ±3σ of the process mean, where σ is the standard deviation. In other words, approximately 0.27 percent of the measurements would naturally fall outside the ±3σ limits and the balance of them (approximately 99.73 percent) would be within the ±3σ limits. Since the process limits extend from −3σ to +3σ, the total spread amounts to about 6σ total variation. This total spread is often used to measure the range of process variability, also called the *process spread.*

For any process performance measure, there are usually some performance specification limits. For example, if the output voltage of power supply unit is too high or too low, then it will not function well. Suppose that its deviation from the target value, 6 V, cannot be more than 0.5 V; then its specification limits would be 6 ± 0.5 V, or we would say that its specification spread is (5.5, 6.5), where 5.5 V is the lower specification limit (LSL) and 6.5 V is the upper specification limit (USL).

If we compare process spread with specification spread, we typically have one of these three situations:

Case I: A Highly Capable Process.   The process spread is well within the specification spread (Fig. 2.16).

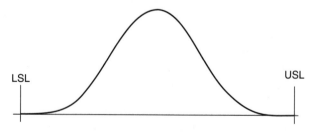

LSL   USL

Figure 2.16 Normal curve of a capable process.

$$6\sigma < (USL - LSL)$$

The process is capable because there is little probability that it will yield unacceptable performance.

Case II: A Marginally Capable Process: The process spread is approximately equal to specification spread (Fig. 2.17).

$$6\sigma = (USL - LSL)$$

When a process spread is nearly equal to the specification spread, the process is capable of meeting specifications, but barely so. This suggests that if the process mean moves to the right or to the left a bit, a significant amount of the output will exceed one of the specification limits.

Case III: An Incapable Process The process spread is more than specification limit (Fig. 2.18).

$$6\sigma > (USL - LSL)$$

When the process spread is greater than the specification spread, a process is not capable of meeting specifications, so it will frequently produce unacceptable performance.

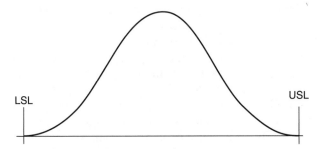

Figure 2.17 Normal curve of a marginal process.

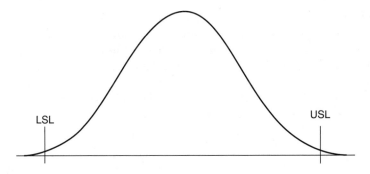

Figure 2.18 Normal curve of an incapable process.

2.4.2 Process capability indices

Capability indices are simplified measures that briefly describe the relationship between the variability of a process and the spread of the specification limits.

The capability index C_p. The equation for the simplest capability index C_p is the ratio of the specification spread to the process spread; the latter is represented by six standard deviations or 6σ.

$$C_p = \frac{\text{USL} - \text{LCL}}{6\sigma}$$

C_p assumes that the normal distribution is the correct model for the process. C_p can be translated directly to the percentage or proportion of nonconforming product outside specifications, if the mean of the process performance is at the center of the specification limit.

When $C_p = 1.00$ (3σ level), approximately 0.27 percent of the parts are outside the specification limits (assuming that the process is centered on the midpoint between the specification limits) because the specification limits closely match the process UCL and LCL. We say that this is about 2700 parts per million (ppm) nonconforming.

When $C_p = 1.33$ (4σ level), approximately 0.0064 percent of the parts are outside the specification limits (assuming the process is centered on the midpoint between the specification limits). We say that this is about 64 ppm nonconforming. In this case, we would be looking at normal curve areas beyond $1.33 \times 3\sigma = \pm4\sigma$ from the center.

When $C_p = 1.67$ (5σ level), approximately 0.000057 percent of the parts are outside the specification limits (assuming that the process is centered on the midpoint between the specification limits). We say that this is about 0.6 ppm nonconforming. In this case, we would be looking at normal curve areas beyond $1.67 \times 3\sigma = \pm5\sigma$ from the center of the normal distribution.

The capability index C_{pk}. The major weakness in C_p is that, for many processes, the processes mean performance is not equal to the center of the specification limit; also many process means will drift from time to time. When that happens, the probability calculation about nonconformance will be totally wrong when we still use C_p. Therefore, one must consider where the process mean is located relative to the specification limits. The index C_{pk} is created to do exactly this.

$$C_{pk} = \min \left\{ \frac{\text{USL} - \mu}{3\sigma} \text{ and } \frac{\mu - \text{LSL}}{3\sigma} \right\}$$

We have the following situation. The process standard deviation is $\sigma = 0.8$ with USL = 24, LSL = 18, and the process mean $\mu = 22$ (Fig. 2.19).

$$C_{pk} = \min\left\{ \frac{24 - 22}{3 \times 0.8} \text{ and } \frac{22 - 18}{3 \times 0.8} \right\} = \min\{0.83 \text{ and } 1.67\} = 0.83$$

If the process mean was exactly centered between the specification limits, then $C_p = C_{pk} = 1.25$.

The capability index C_{pm}. C_{pm} is called the *Taguchi capability index* after the Japanese quality guru, Genichi Taguchi. This index was developed in the late 1980s and takes into account the proximity of the process mean to the ideal performance target T.

$$C_{pm} = \frac{\text{USL} - \text{LSL}}{6\sqrt{\sigma^2 + (\mu - T)^2}}$$

When the process mean is centered between the specification limits and the process mean is on the target T, then $C_p = C_{pk} = C_{pm}$. When a process mean departs from the target value T, there is a substantive effect on the capability index. In the C_{pk} example above, if the target value were $T = 21$, C_{pm} would be calculated as

$$C_{pm} = \frac{24 - 18}{6\sqrt{0.8^2 + (22 - 21)^2}} = 1.281$$

Motorola's Six Sigma quality. In 1988, the Motorola Corporation was the winner of the Malcolm Baldrige National Quality Award. Motorola bases much of its quality effort on its Six Sigma program. The goal of this program was to reduce the variation in every process to such an extent that a spread of 12σ (6σ on each side of the mean) fits within the

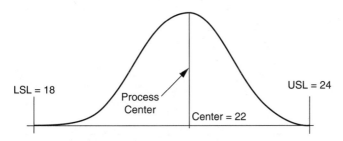

LSL = 18

Process Center

Center = 22

USL = 24

Figure 2.19 Capability index C_{pk}.

process specification limits (Fig. 2.20). Motorola allocates 1.5σ on either side of the process mean for shifting of the mean, leaving 4.5σ between this safety zone and the respective process specification limit.

Thus, even if the process mean strays as much as 1.5σ from the process center, a full 4.5σ remains. This ensures a worst-case scenario of 3.4 ppm nonconforming on each side of the distribution (6.8 ppm total). If the process mean were centered, this would translate into a $C_p = 2.00$. Motorola has made significant progress toward this goal across most processes, including many office and business processes as well.

Six Sigma and process capability. The concept of process capability indicates that in order to achieve high process capability, the following two tasks must be accomplished:

1. The actual process mean performance should be as close to ideal performance level, or target value, as possible.

2. Process performance spread should be small relative to functional limits.

Therefore, it is again a "Do the right thing, and do things right all the time" rule. Accomplishing Six Sigma process capability is a very difficult but necessary task. If a process can produce good performance "on average," for example, a company can make a good profit some years but could lose a lot in other years, then this inconsistency will severely damage the image and morale of the company.

Six Sigma is a strategy that applies to all the quality methods and process management available to a full process life-cycle implementation. The goal for any Six Sigma project is to make the process able to accomplish all key requirements with a high degree of consistency.

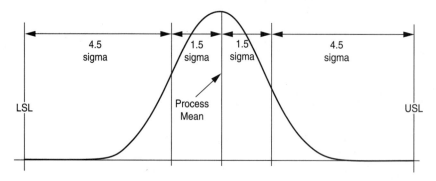

Figure 2.20 Interpretation of Motorola's Six Sigma program.

There are two ways to do this: Six Sigma process improvement and design for Six Sigma (DFSS).

Six Sigma process improvement. Six Sigma process improvement is actually the Six Sigma method that most people refer to. This strategy does not involve any changing or redesigning of the fundamental structure of the underlying process. It involves finding solutions to eliminate the root causes of performance problems in the process and of performance variation, while leaving the basic process intact. If we refer to the process life cycle illustrated by Fig. 2.4, Six Sigma process improvement applies only to stage 4 and after.

Design for Six Sigma (DFSS). Design for Six Sigma is a Six Sigma approach which will involve changing or redesigning of the fundamental structure of the underline process. If we refer to the process life cycle illustrated by Fig. 2.7, DFSS applies to stages 1 through 3. Therefore, it is an upstream activity. The goal of DFSS is to design or restructure the process in order for the process to intrinsically achieve maximum customer satisfaction and consistently deliver its functions.

Design for Six Sigma is needed when

- A business chooses to replace, rather than repair, one or more core processes
- A leadership or Six Sigma team discovers that simply improving an existing process will never deliver the level of quality customers are demanding
- The business identifies an opportunity to offer an entirely new product or services

DFSS is not a quick fix; it will take more effort in the beginning, but it will pay off better than the regular Six Sigma process improvement in the end.

2.5   Overview of Six Sigma Process Improvement

In a Six Sigma project, if the Six Sigma team selects the regular Six Sigma process improvement strategy, then a five-stage process will be used to improve an existing process. These five stages are

- Define the problem and customer requirements
- Measure the defects and process operation
- Analyze the data and discover causes of the problem

- Improve the process to remove causes of defects
- Control the process to make sure defects don't recur

(See also Table 2.1.) This five-step strategy is also called *DMAIC* (define-measure-analyze-improve-control). We will briefly describe the five steps. Here we assume that the process follows a SIPOC model.

2.5.1   Stage 1: Define the project and customer requirements (D or define step)

When we need to launch a Six Sigma process improvement project, the process under improvement seldom performs satisfactorily; at least, we believe that this process has a lot of room for improvement. Usually the "define" (D) stage can be done in the following three steps:

Step 1: Draft project charter, which includes

1. Business case
2. Goals and objectives of the project
3. Milestones
4. Project scope, constraints, and assumptions
5. Team memberships
6. Roles and responsibilities
7. Preliminary project plan

Step 2: Identify and document the process

1. *Identify the process.*   In a Six Sigma process improvement project, usually a team works on one process at a time. The process being identified is usually
   - A core process in the company, such as product development, marketing, or customer service, so it is a very important process for the company
   - A support process, such as a human resource or information system, but this process becomes a bottleneck or a waste center of the company
2. *Document the process.*   After a process is identified, an appropriate process model will be used to model and analyze the process, such as a P-diagram model or a SIPOC model.

After the process model is determined, the major elements of process model, suppliers, inputs, process map, process output, and customer

TABLE 2.1 Process Life Cycle and Six Sigma Approach

Product/service life-cycle stages	Six Sigma tasks	Six Sigma strategy	Six Sigma tools
1. Impetus/Ideation	Identify project scope, customers, suppliers, customer needs	DFSS	Customer research, process analysis, Kano analysis, QFD
2. Concept development	Ensure new process concept to come up with right functional requirements which satisfy customer needs Ensure that the new concept can lead to sound system design, free of design vulnerabilities Ensure the new concept to be robust for downstream development	DFSS	QFD Taguchi method/robust design TRIZ Axiomatic design DOE Simulation/optimization Reliability-based design
3. Process design/tryout/debug/launch	Ensure process to deliver desired functions Ensure process to perform consistently and robust Validate process for performance and consistency	DFSS	Taguchi method/robust design DOE Simulation/optimization Reliability-based design/testing and estimation Statistical validation
4. Process routine operation	Ensure process to perform consistently	Six Sigma process improvement	SPC Troubleshooting and diagnosis Errorproofing
5. Process improvement	Improve to satisfy new requirements	Six Sigma process improvement	DMAIC strategy Customer analysis, Kano analysis, QFD Statistical measurement system DOE, Shanin method, multivariate analysis, regression analysis Process analysis, value stream mapping SPC

base should be defined. In this step, we will only stay at the top-level process model, or at most one level lower, because we do not want to be buried in details at the beginning.

Step 3: Identify, analyze, and prioritize customer requirements

1. *Identify customer requirements.*   There are two kinds of customer requirements:

   - *Output requirements.* These are the features of final product and service delivered to the customer at the end of the process. For example, if the output voltage of a power supply unit is 6 V, the output requirement in the customers' eyes could be "Voltage should not be neither too high nor too low." The numerical requirements could be expressed as "between 5.5 and 6.5 V." For a complicated product or process, such as an automobile or a power plant, the list of outputs and its related requirements could be very long.
   - *Service requirement.* These are the more subjective ways in which the customer expects to be treated and served during the process itself. Service requirements are usually difficult to define precisely.

2. *Analyze and prioritize customer requirements.*   The list of customer requirements could be very long for a complicated product or process and is often hierarchical. For example, there could be many customer requirements for an automobile, such as drivability, appearance, and comfort while driving. For drivability, it could include many items, such as acceleration, braking performance, and steering performance. For each of these, you can further break down to the next level of details.

The list of requirements can be long, but not all requirements are equal in customers' eyes. We need to analyze and prioritize those requirements. This step can be done by Kano analysis or QFD, which is covered in detail in Chap. 4. The list of high-priority customer requirements is often called characteristics critical-to-quality (CTQ).

2.5.2   Stage 2: Measuring process performance

Measure is a very important step. This step involves trying to collect data to evaluate the current performance level of the process, and provide information for analysis and improvement stages.

This stage usually includes the following steps:

1. *Select what needs to be measured.* Usually, we measure the following:

- *Input measures*
- *Output measure.* CTQs, surrogates of CTQs, or defect counts.
- *Data stratification.* This means that together with the collection of output measures *Y*, we need to collect corresponding information about the variables which may have cause-and-effect relationship with *Y*, that is, *X*. If we do not know what *X* is, we may collect other information that may relate to *X*, such as stratification, region, time, and unit factors, and by analyzing the variation in performance level at different stratification factors, we might be able to locate the critical *X* which may influence *Y*.

2. *Develop a data collection plan.* We will determine such issues as sampling frequency, who will perform the measurement, the format of data collection form, and measurement instruments. In this step, we need to pay attention to the

- *Type of data (discrete or continuous).* There are two types of data: discrete and continuous. *Discrete measures* are those that enable one to sort items into distinct, separate, nonoverlapping categories. Examples include car model types and types of credit cards. *Continuous measures* are applied for quantities that can be measured on an infinitely divisible continuum or scale, such as price, cost, and speed. Discrete data are usually easy to collect, easy to interpret, but statistically, are not efficient, and more data need to be collected in data analysis.
- Sampling method.

3. *Calculate the process sigma level.* For continuous data, we could use the methods in process capability calculation described in the last section. For discrete data, we could directly calculate defective rate, and then translate it to sigma level.

2.5.3. Stage 3: Analyze data and discover causes of the problem

After data collection, we need to analyze the data and process in order to find how to improve the process. There are two main tasks in this stage:

Data analysis. Using collected data to find patterns, trends, and other differences that could suggest, support, or reject theories about the cause and effect, the methods frequently used include

- Root cause analysis
- Cause–effect diagram
- Failure modes–effects analysis (FMEA)
- Pareto chart
- Validate root cause

- Design of experiment
- Shanin method

Process analysis. This involves a detailed look at existing key processes that supply customer requirements in order to identify cycle time, rework, downtime, and other steps that don't add value for the customer. We can use process mapping, value stream mapping, and process management methods here.

2.5.4   Stage 4: Improve the process

We should have identified the root causes for the process performance problem after completing stage 3.

If the root causes of process performance problems are identified by process analysis, the solutions are often featured by techniques such as process simplification, parallel processing, and bottleneck elimination. If the root causes are identified by applying data analysis, then sometimes finding the solution to performance problem is easy. There are some circumstances that finding the solution is very difficult, because many "obvious" solutions may potentially solve the problem, but will have harmful effects on other aspects of the process. In this case, creative solutions need to be found. Brainstorming and TRIZ may be used here.

2.5.5   Stage 5: Control the process

The purpose of this stage is to hold on to the improvement achieved from the last stage. We need to document the change made in the improvement stage. If the improvement is made by process management methods, such as process simplification, we need to establish a new process standard. If the improvement is made by eliminating the root causes of low performance, we need to keep track of process performance after improvement and control the critical variables relating to performance, by using control charts.

2.6   Six Sigma Goes Upstream: Design for Six Sigma (DFSS)

Design for Six Sigma (DFSS) is the Six Sigma strategy working on early stages of the process life cycle. It is not a strategy to improve a current process with no fundamental change in process structure. It will start at the very beginning of the process life cycle and utilize the most powerful tools and methods known today for developing optimized designs. These tools and methods are ready to plug directly into your current product development process, or design/redesign of a service process or internal business process.

The rest of this book is devoted exclusively to design for Six Sigma (DFSS). Chapter 3 is the introductory chapter for DFSS, giving the overviews for DFSS theory, DFSS process, and DFSS application. Chapter 4 gives detailed descriptions about how to deploy DFSS in a company, such as the training of DFSS teams, organization support, financial management, and deployment strategy. Chapter 5 gives a very detailed "flowchart" of the whole DFSS process, which includes very detailed description of DFSS stages, task management, scorecards, and how to integrate all methods into DFSS stages. Chapters 6 through 17 give detailed descriptions on all the major methods used in DFSS.

2.7 Summary

1. Six Sigma is a methodology that provides businesses with the tools to improve the capabilities of their business processes. Compared with other quality initiatives, the key difference of Six Sigma is that it applies not only to product quality but also to all aspects of business operation. Six Sigma is a method for business excellence.

2. Process is the basic unit for a Six Sigma improvement project. Process could be a product itself, a service/manufacturing process, or an internal business process. Process mapping, value stream mapping, and process management are effective tools for improving overall performance. Six Sigma process improvement strives to improve both process performance and process capability.

3. Process capability is a measure of process consistency in delivering process performance. Six Sigma capability is a world-class capability.

4. Six Sigma process improvement is a method for improving process performance and capability without process redesign. Design for Six Sigma (DFSS) is a Six Sigma method that works on the early stage of product/process life cycle.

Design for Six Sigma

3.1 Introduction

This chapter is designed for use as an introduction to Design for Six Sigma (DFSS) theory, process, and application. The material presented here is intended to give the reader an understanding of DFSS per se and its uses and benefits. Following this chapter, readers should have a sufficient knowledge of DFSS to assess how it could be used in relation to their jobs and identify their needs for further learning.

Customer-oriented design is a development process of transforming customers' wants into design solutions that are useful to the customer. This process is carried over several phases starting from a conceptual phase. In this phase, conceiving, evaluating, and selecting good design solutions are difficult tasks with enormous consequences. Design and manufacturing companies usually operate in two modes: *fire prevention,* conceiving feasible and healthy conceptual entities, and *firefighting,* problem solving such that the design entity can live up to its committed potentials. Unfortunately, the latter mode consumes the largest portion of the organization's human and nonhuman resources. The design for Six Sigma theory highlighted in this book is designed to target both modes of operation.

The most recent trends in design research are featured by two streams of development for enhancing the design process in order to create better design solutions. The first stream is concerned with improving design performance in the usage environment; the second is related to conceptual methods. The method of robust design as suggested by G. Taguchi belongs in the first stream (Taguchi 1986, 1993, 1994). In this method, a good design is the one that provides a robust solution for stated functional objectives and can be accomplished through a design process constituted from three phases: the concept (system) design, parameter design, and tolerance design. For the second stream, a huge body of research* in the design methodology arena has been published in German on the design practice. Most of these efforts are listed in the *German Guidelines VDI (Varian Destscher Ingenieure)*, 2221 (Hubka 1980, Phal and Beitz 1988). The latest development in the second stream is the scientifically based design as suggested by Suh (1990). A major concern of the design principles is the design vulnerabilities that are introduced in the design solution when certain principles are violated. These vulnerabilities can be resolved or at least reduced by the efficient deployment of basic design principles called the *axioms*. For example, the functional coupling (lack of controllability) will be created in the design solution when the independence axiom is not satisfied.

3.2   What Is Design for Six Sigma Theory?

The theory of Design for Six Sigma (DFSS) is defined in this book as a scientific theory comprising fundamental knowledge areas in the form of perceptions and understandings of different fields, and the relationships between these fundamental areas. These perceptions and relations are combined to produce consequences in the design entity, which can be, but are not necessarily, predictions of observations. DFSS fundamental knowledge areas include a mix of propositions and hypotheses, categorizations of phenomena or objects, ideation, and conception methods such as axiomatic design (Chap. 8) and TRIZ (Chap. 9) and a spectrum of empirical statistical and mathematical models.† Such knowledge and relations constitute our DFSS theory. In the conception arena, this theory builds on the theoretical system of other methods and can be one of two types: axioms or hypotheses, depending on the way in which the fundamental knowledge areas are treated. Fundamental knowledge that can't be tested, yet is generally accepted as truth, will be treated as *axioms*. If the fundamental knowledge areas are being tested, they are treated as hypotheses. Design axioms (Suh 1990) and TRIZ (Altshuller 1988) hypotheses are examples of fundamental knowledge in DFSS theory.

The major objective of DFSS is to "design it right the first time" to avoid painful downstream experiences. The term "Six Sigma" in the context of DFSS can be defined as the level at which design vulnerabilities are *not effective* or minimal. Generally, two major design vulnerabilities may affect the quality of a design entity:

---

*The reader is encouraged to visit the Appendix of this chapter for more literature on design theory.

†Abstract of observations of real-world data.

- Conceptual vulnerabilities that are established because of the violation of design axioms and principles.

- Operational vulnerabilities due to the lack of robustness in the use environment. Elimination or reduction of operational vulnerabilities is the objective of quality initiative including Six Sigma.

The objective of the DFSS when adopted upfront is to "design it right the first time" by anticipating the effect of both sources of design vulnerabilities. This requires that companies be provided by the analytical means to achieve this noble objective and sustain it. Many deploying companies of the Six Sigma philosophy are devising their in-house views of DFSS. It is the authors' perception that most of the thinking about DFSS in many companies who are leading the DFSS deployment is geared toward different packaging of the DMAIC methodology plus "voice of the customer" tools. Their proposed deployment of DFSS is concentrated around phasing DMAIC methods in the development process boosted, however, with dosages of tool complexity (e.g., multiple regression instead of simple linear regression). This track does not guarantee the achievement of Six Sigma capability in the design entity. Additionally, because of unavailability of data in the early design phase, most of the current Six Sigma tools may be useless. In this context, the proposed DFSS strategy in this book has the view depicted in Fig. 3.1. Accordingly, the DFSS should be based on new

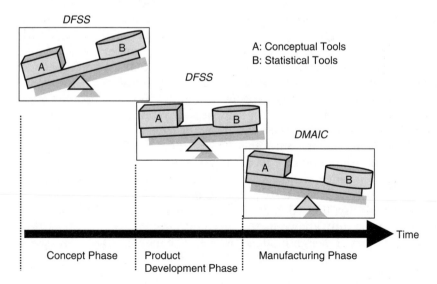

Figure 3.1 The DFSS strategy at design entity phases.

tools that should take into consideration the unique nature of the design process itself.

On the hard side, in order to "design it right the first time," the DFSS theory presented in this book is designed to attack both types of design vulnerabilities to have Six Sigma yield of the designed entity. This objective can be achieved not only by targeting the entity itself but also by extending DFSS deployment to the developmental processes that produce it. On the soft side, DFSS drives for cultural change in the deploying company by shaking current and old paradigms, building success one project at a time, changing and motivating people, and building new paradigms for a decision-making culture, a rich Six Sigma culture.

3.3   Why "Design for Six Sigma"?

The objective of this book is to present the DFSS theory, consisting of concepts and tools that eliminate or reduce both the conceptual and operational types of vulnerabilities of designed entities* and releases such entities at Six Sigma quality levels in all of their requirements, that is, to have all functional requirements at 6 times the standard deviation on each side of the specification limits. This target is called Six Sigma, or 6σ for short, where the Greek letter stands for the standard deviation.

Operational vulnerabilities takes variability reduction and mean adjustment of the critical-to-quality requirements, the CTQs, as an objective and have been the subject of many fields of knowledge such as the method of robust design advanced by Taguchi (Taguchi 1986, Taguchi and Wu 1986, Taguchi et al. 1989), DMAIC Six Sigma (Harry 1994, 1998), and tolerance design/tolerancing techniques. Tolerance research is at the heart of operational vulnerabilities as it deals with the assignment of tolerances in the design parameters and process variables, the assessment and control of manufacturing processes, the metrological issues, as well as the geometric and cost models.

On the contrary, the conceptual vulnerabilities are usually overlooked because of the lack of a compatible systemic approach to find ideal solutions, ignorance of the designer, the pressure of schedule deadlines, and budget limitations. This can be attributed partly to the fact that traditional quality methods can be characterized as after-the-fact practices since they use lagging information to developmental activities such as bench tests and field data. Unfortunately, this practice drives design toward endless cycles of design-test-fix-retest, creating what is broadly known as the "firefighting" mode of operation, that is, the creation of design hidden factories. Companies who follow these

---

*A product, service, or process.

practices usually suffer from high development costs, longer time to market, lower quality levels, and marginal competitive edge. In addition, corrective actions to improve the conceptual vulnerabilities via operational vulnerability improvement means are only marginally effective if at all useful. In addition, these corrective actions are costly and hard to implement as design entity progresses in the development process. Therefore, implementing DFSS in the conceptual phase is a goal and can be achieved when systematic design methods are integrated with quality concepts and methods upfront. Specifically, in this book, we developed a DFSS theory by borrowing from the following fundamental knowledge arenas: quality engineering (Taguchi 1986), TRIZ (Altshuller 1988), axiomatic design principles (Suh 1990), and theory of probability and statistical modeling. The DFSS objective is to attack the design vulnerabilities, both conceptual and operational, by deriving and integrating tools and methods for their elimination and reduction.

In general, most of the current design methods are empirical in nature. They represent the best thinking of the design community that, unfortunately, lacks the design scientific base while relying on subjective judgment. When a company suffers as a result of detrimental behavior in customer satisfaction, judgment and experience may not be sufficient to obtain an optimal Six Sigma solution. This is another motivation to devise a DFSS method to address such needs.

Attention begins to shift from improving the performance during the later phases of the design life cycle to the front-end phases where product development take place at a higher level of abstraction, namely, prevention versus solving. This shift is also motivated by the fact that the design decisions made during the early stages of the design life cycle have the greatest impact on total cost and quality of the system. It is often claimed that up to 80 percent of the total cost is committed in the concept development phase (Fredriksson 1994). The research area of manufacturing including product development is currently receiving increasing focus to address industry efforts to shorten lead times, cut development and manufacturing costs, lower total life-cycle cost (LCC), and improve the quality of the design entities in the form of products, services, and/or processes. It is the experience of the authors that at least 80 percent of the design quality is also committed in the early phases as depicted in Fig. 3.2. The potential defined as the difference between the impact, the influence, of the design activity at certain design phases and the total development cost up to that phase. The potential is positive, but decreases as design progresses, implying reduced design freedom overtime. As financial resources are committed (e.g., buying production machines and facilities, hiring staff), the potential starts changing signs going from positive to negative. In the consumer's

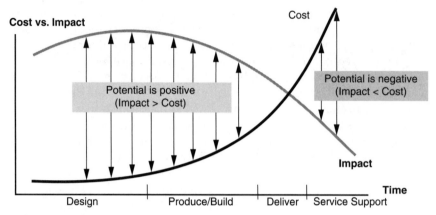

Figure 3.2 Effect of design phases on life cycle.

hand, the potential is negative and the cost overcomes the impact tremendously. At this phase, design changes for corrective actions can be achieved only at high cost, including customer dissatisfaction, warranty, and marketing promotions, and in many cases under the scrutiny of the government (e.g., recall costs).

3.4   Design for Six Sigma (DFSS) Phases

Design for Six Sigma has the following four phases:

- **I**dentify requirements
- **C**haracterize the design
- **O**ptimize the design
- **V**erify the design

We will use the notation ICOV for short to indicate the four DFSS phases, presented below.

DFSS as defined in this book has two tracks: deployment and application. By *deployment,* we mean the strategy adopted by the adopting deploying entity to select, scope, and prioritize projects for application. The aspects of DFSS deployment are presented in Chap. 4. In what follows, we assume that the deployment strategy is in place as a prerequisite for application and project execution.

3.4.1   Phase 1: Identify requirements (I)

DFSS projects can be categorized as design or redesign of an entity. "Creative design" is the term that we will be using to indicate new design,

design from scratch, and incremental design for redesign or design from a datum design. In the latter case, some data can be used to refine design requirements. The degree of deviation of the redesign from datum is the key factor on deciding on the usefulness of relative data.

Step 1: Draft project charter. This is almost the same as that of the DMAIC improvement project. However, project duration is usually longer and initial cost is usually higher. Longer project duration is due to the fact that the company is designing or redesigning a different entity, not merely patching up the holes of an existing one. Higher initial cost is due to the fact that there are many more customer requirements to be identified and studied, since one needs to identify all important critical-to-satisfaction (CTS) metrics to conceive and optimize better designs. For the DMAIC case, one may only work on improving a very limited subset of the CTSs.

Step 2: Identify customer and business requirements. In this step, customers are fully identified and their needs collected and analyzed, with the help of quality function deployment (QFD) and Kano analysis. Then the most appropriate set of CTSs metrics are determined in order to measure and evaluate the design. Again, with the help of QFD and Kano analysis, the numerical limits and targets for each CTS are established.

In summary, following is the list of tasks in this step. Detailed explanations will be provided in later chapters.

- Identify methods of obtaining customer needs and wants.

- Obtain customer needs and wants and transform them into the voice-of-customer (VOC) list.

- Translate the VOC list into functional and measurable requirements.

- Finalize requirements:
  Establish minimum requirement definitions.
  Identify and fill gaps in customer-provided requirements.
  Validate application and usage environments.

- Identify CTSs as critical-to-quality (CTQ), critical-to-delivery (CTD), critical-to-cost (CTC), and so on.

- Quantify CTSs.
  Establish metrics for CTSs.
  Establish acceptable performance levels and operating windows.
  Perform flowdown of CTSs.

DFSS tools used in this phase include:

- Market/customer research

- Quality function deployment
- Kano analysis
- Risk analysis

3.4.2 Phase 2: Characterize design (C)

Step 1: Translate customer requirements (CTSs) to product/process functional requirements. Customer requirements, CTSs, give us ideas about what will satisfy the customer, but they can't be used directly as the requirements for product or process design. We need to translate customer requirements to product/process functional requirements. QFD can be used to add this transformation. Axiomatic design principle will also be very helpful for this step.

Step 2: Generate design alternatives. After the determination of the functional requirements for the new design entity (product, service, or process), we need to characterize (develop) design entities that will be able to deliver those functional requirements. In general, there are two possibilities:

1. The existing technology or known design concept is able to deliver all the requirements satisfactorily; this step then becomes almost a trivial exercise.

2. The existing technology or known design is not able to deliver all requirements satisfactorily; then a *new* design concept needs to be developed. This new design could be "creative" or "incremental," reflecting the degree of deviation from the baseline design. The TRIZ method (Chap. 8) and axiomatic design (Chap. 7) will be helpful in generating many innovative design concepts in this step.

Step 3: Evaluate design alternatives. Several design alternatives might be generated in the last step. We need to evaluate them and make a final determination on which concept will be used. Many methods can be used in design evaluation, including the Pugh concept selection technique, design reviews, design vulnerability analysis (El-Haik 1996, Yang and Trewn 1999), and FMEA. After design evaluation, a winning concept will be selected. During the evaluation, many weaknesses of the initial set of design concepts will be exposed and the concepts will be revised and improved. If we are designing a process, process management techniques will also be used as an evaluation tool.

The following DFSS tools are used in this phase:

- TRIZ
- QFD

- Axiomatic design
- Robust design
- Design for X
- DFMEA and PFMEA (design and performance failure mode–effect analysis)
- Design review
- CAD/CAE (computer-aided design/engineering)
- Simulation
- Process management

3.4.3   Phase 3: Optimize the design (O)

The result of this phase is an optimized design entity with all functional requirements released at the Six Sigma performance level. As the concept design is finalized, there are still a lot of design parameters that can be adjusted and changed. With the help of computer simulation and/or hardware testing, DOE modeling, Taguchi's robust design methods, and response surface methodology, the optimal parameter settings will be determined. Usually this parameter optimization phase, in product DFSS projects, will be followed by a tolerance optimization step. The objective is to provide a logical and objective basis for setting manufacturing tolerances. If the design parameters are not controllable, which is usually the case on the DFSS product projects, we may need to repeat phases 1 to 3 of DFSS for manufacturing process design.

The following DFSS tools are used in this phase:

- Design/simulation tools
- Design of experiment
- Taguchi method, parameter design, tolerance design
- Reliability-based design
- Robustness assessment

3.4.4   Phase 4: Validate the design (V)

After the parameter and tolerance design is completed, we will move to the final verification and validation activities.

Step 1: Pilot test and refining.   No product or service should go directly to market without first piloting and refining. Here we can use design failure mode–effect analysis (DFMEA) as well as pilot and small-scale implementations to test and evaluate real-life performance.

Step 2: Validation and process control.   In this step we will validate the new entity to make sure that the end result (product or service) as designed meets the design requirements and that the process controls in manufacturing and production are established in order to ensure that critical characteristics are always produced to specification of the optimization (O) phase.

Step 3: Full commercial rollout and handover to new process owner.   As the design entity is validated and process control is established, we will launch full-scale commercial rollout and the new entity, together with the supporting processes, can be handed over to design and process owners, complete with requirements settings and control and monitoring systems.

The following DFSS tools are used in this phase:

- Process capability modeling
- DOE
- Reliability testing
- Poka-yoke, errorproofing
- Confidence analysis
- Process control plan
- Training

3.5   More on Design Process and Design Vulnerabilities

An efficient DFSS application can be achieved when analytical tools are combined with science-based design tools such as axiomatic design (Suh 1990), where modeling of the "design structure" carries great importance. Design structure is the set of interrelationships that characterize the design requirements, design parameters, and process variables. Depending on the context, different formats to convey the structure such as block diagrams, process mappings, and functional trees are used, some more popular than others. While some of the modeling is endorsed in Six Sigma, with the DMAIC approach, like cause–effect matrices and process mapping, the need is more pronounced in DFSS, in particular the characterization (C) phase. Such modeling will reveal how the design is coupled in the functional requirements (FRs). Coupling indicates the lack of independence between the FRs. Coupling of the FRs is a design vulnerability that negatively affects controllability and adjustability of the design entity. In addition, coupling will result in reduced reliability and robustness of the design entity and will complicate finding satisfactory solutions that meet customer attributes at release and over time. It will surely impair the Six Sigma design endeavor to achieve unprecedented customer satisfaction capability. Many negative scenarios are produced by coupling. In a traditional design dilemma, the designer tries to resolve a detrimental problem on a certain CTS by adjusting some of the process variables (PVs) without paying attention to the effect of the adjustment on other FRs delivered by the design entity. This ignorance or negligence complicates the situation and results in trial-and-error inertia toward compromised or wrong solutions of the initial problem. In this situation, the creation of new symptoms in the design entity is not a remote possibility.

The integration of Six Sigma philosophy with scientific design methods yields a robust DFSS strategy in both theory and application with many advantages. For example, the employment of abstraction at high levels of the design structure facilitates decision making toward healthy concepts, while the use of mathematical formulation and/or empirical testing at low levels of the structure facilitates the variability reduction and design controllability as the "zigzagging" method of axiomatic design is used. Axiomatic design provides rules to structure and select design entities that are robust from a conceptual perspective when the axioms are obeyed. The optimization (O) and validation (V) phases of DFSS will be easier to execute when a coupling-free design is conceived. In coupled concepts, this flexibility is slim.

Unfortunately, axiom obedience is not always feasible, usually for technological, organizational culture, cost, or other constraints. A design organization may find itself forced to live with some degree of coupling in some or all of its designed family, at least in the short term, even when the technology is capable of resolving coupling due, mainly, to cost constraints. Therefore, the need to improve the capabilities of a coupled design is badly needed, especially when the effect of sources of variation are anticipated to have a detrimental effect on the FRs.

The design process involves three mappings between four domains (Fig. 3.3). The first mapping involves the mapping between customer attributes (CAs) and functional requirements (FRs). This mapping is very critical as it yields the high-level minimum set of functional requirements needed to accomplish the design objective from the customer perspective. It can be performed by the means of quality function deployment (QFD). Once the minimum set of FRs are defined, the *physical mapping* (matrix A) starts. This mapping involves the FR domain and the design parameter (DP) domain. It represents the development activities and can be represented by design matrices as the high-level set of FRs cascade down to the lowest level of decomposition. The collection of design matrices forms the conceptual functional structure

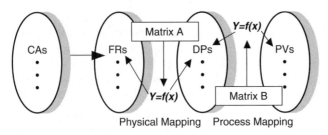

Figure 3.3 The design mappings.

that reveals coupling and provides a means to track the propagation of design changes for if-then scenario analysis.

The *process mapping* (matrix B) is the last mapping and involves the DP domain and the process variables (PV) domains. This mapping can be represented by matrices as is the case with the physical mapping and provides the process structure needed to translate the DPs to process variables.

3.6   Differences between Six Sigma and DFSS

In a design assignment or a problem-solving assignment, whether the black belt is aware or not, design mappings, in terms of matrices **A** and **B**, do exist (Fig. 3.3). In a DFSS project, the three mappings need to be performed sequentially, as the output of one is the input to the next mapping. When the last two mappings follow the design axioms, the possibility of establishing the Six Sigma capability in the design entity is created using conceptual methods. However, the type of project (i.e., creative or incremental) is the deciding factor of whether to modify existing mappings of the datum design or develop new ones.

Many design practices, including DMAIC, drive for finding solutions in the manufacturing environment, the last mapping, for a problematic CTS. However, these practices don't employ the sequential mappings, design decomposition, and design principles in pursuing a solution that is usually obtained with no regard to the coupling vulnerability, that is, solving a design problem with process means by simply employing the process variables, as the x variable[*] The conceptual framework of current Six Sigma can be depicted as shown in Fig. 3.4, thus ignoring the DPs. Additionally, we have the following remarks in the context of Fig. 3.4:

---

[*]Recall the $y = f(x)$ notation in DMAIC.

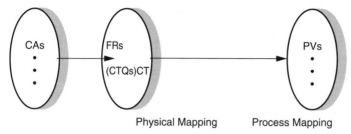

Figure 3.4   The design process according to Six Sigma.

- The black belt may blindly overlook the need for design changes (altering the DP array) when adjusting the PVs is not sufficient to provide a satisfactory solution, that is, when the current process reaches its entitlements. The risk in this scenario occurs when the black belt may introduce a major manufacturing change, namely, altering the PV array, when it is unnecessary.

- Another disadvantage is concerned with ignorance of coupling which may introduce new symptoms in CTSs other than the ones intended when the solution to a problem is institutionalized.

- On the other hand, taking the PVs as the x variable is usually cheaper than taking the DPs as the x variable since the latter involves design change and a process change while the former calls only for process changes. The adjustment of process variables may or may not solve the problem depending on the sensitivities in the physical and process mapping.

Solutions to a design or process problem can be implemented using alterations, *changes,* in independent variables, the x varible. The independent variables may be the DPs or the PVs according to the mapping of interest and where the solution is sought. A "change" can be either soft or hard. *Soft* changes imply adjusting the nominal values within the specified tolerances, changing the tolerance ranges, or both. *Hard* changes imply eliminating or adding DPs or PVs in the concerned mapping and accordingly their subsequent soft changes. For example, in manufacturing, soft process changes can be carried out by parametric adjustment within the permitted tolerances while hard changes may require PV alteration. On the redesign side, design changes to reduce or eliminate a detrimental behavior of an FR may call for dramatic changes in both the design entity and manufacturing processes when soft changes cannot produce the desired result.

Mathematically, let the concerned FR (CTS) be expressed using $y = f(x)$ as FR = $f(\mathbf{DP})$, where $\mathbf{DP}$ is an array of mapped-to DPs of size m. Let each DP in the array be written as $DP_i = g(\mathbf{PV}_i)$, where $\mathbf{PV}_i$, $i = 1,...,m$

is an array of process variables that are mapped to DP$_i$. Soft changes may be implemented using sensitivities in physical and process mappings. Using the chain rule, we have

$$\frac{\partial FR}{\partial PV_{ij}} = \left(\frac{\partial FR}{\partial DP_i} \right) \times \left(\frac{\partial DP_i}{\partial PV_j} \right) = f'\,[g(PV_i)]\,g'(PV_{ij}) \qquad (3.1)$$

where PV$_{ij}$ is a process variable in the array **PV**$_i$ that can be adjusted (changed) to improve the problematic FR. The first term represents a design change; the second, a process change. An efficient DFSS methodology should utilize both terms if all FRs are to be released at Six Sigma performance levels.

3.7   What Kinds of Problems Can be Solved by DFSS?

A design entity of a process or a product can be depicted in a P-diagram as in Fig. 3.5. The useful output is designated as the array of FRs **y,** which in turn is affected by three kinds of variables: the signals represented by the array **m,** the design parameters represented by the array **x,** and the noise factors represented by array **z.** Variation in **y** and its drift from its targeted performance are usually caused by the noise factors. The norms of **m** and **y** arrays are almost equal when they are expressed in terms of energy in dynamic systems. In this context, the objective of DFSS is to reduce the difference array norm $|\Delta| = |\mathbf{y}| - |\mathbf{m}|$ between both array norms to minimum, when the target is zero, and reduce the variability around that minimum. Variability reduction can be achieved by utilizing the interaction $\mathbf{x} \times \mathbf{z}$. In a DFSS project, we are concerned with an FR, say, y_j, which suffers from symp-

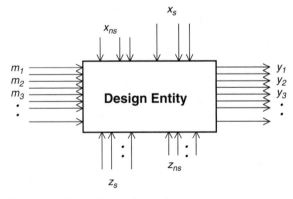

Figure 3.5   The P-diagram.

tomatic behavior as perceived by the customer. Let the array **x** be split into significant and nonsignificant factors denoted as {**x**$_s$,**0**} and non-significant factors {**0**, **x**$_{ns}$}, respectively; that is, **x** = {**x**$_s$,**0**} + {**0**, **x**$_{ns}$}. Also, let the array **z** be spilt into significant and nonsignificant factors denoted as {**z**$_s$,**0**} and nonsignificant factors {**0**, **z**$_{ns}$}, respectively; that is, **z** = {**z**$_s$,**0**} + {**0**, **z**$_{ns}$}. Of course, significance and nonsignificance are subject to physics by derivation or empirically from experimentation.

Usually the nonsignificant factors are numerous while the significant factors are few, assuming their existence. There are four possibilities of a DFSS project from the standpoint of design-versus-noise classifications in the context of this section. They are listed in Table 3.1. The effects of nonsignificant factors, whether design parameters or noise factors, are usually weak and sparse in a manner that bears creditability to the Pareto principle. As such, their existence does not add to the complexity of the problem and/or its solution.

Only when the significant **x**$_s$ array does exist is there a potential for the DFSS method to produce a Six Sigma capability in the concerned FR. The conceptual change in the third classification is to be conducted following the DFSS strategy proposed in this book.

3.8   Design for a Six Sigma (DFSS) Company

A company is Six Sigma–capable when each of its products and services achieves 3.4 defects per million long-term capabilities, a Six Sigma level assuming normality. Over the course of a life cycle, including development, thousands of decisions are taken to design, produce, release, and service the entity. A critical few are taken at the milestones, but the trivial many are taken on a daily basis. A design decision may be conforming or nonconforming. *Nonconformance* in the development process, as used here, has a broader meaning than the intuitive definition of *error.* Nonconformity occurs in design development when a decision produces less-than-ideal results. It could be a wrong action or a missed inaction. In either case, it can be committed as one of the following scenarios:

TABLE 3.1   Possibilities of a Six Sigma Problem

x ╱ z	z$_s$ exists	z$_s$ does not exist
x$_s$ exists	Six Sigma and DFSS have potentials	Trivial problem—may be solved by a DOE only
x$_s$ does not exist	Need conceptual change—DFSS has potential while Six Sigma has *no* potential	No problem—such design entity may not exist

when a necessary assurance decision is not executed, when a necessary measure is pursued inadequately, and when an important matter is not addressed or is addressed inappropriately. These decisions and an assessment of their goodness in light of the nonconformance definition cited above must be recorded. This will allow the company to decide on the right direction and to avoid future pitfalls. A DFSS company has 3.4 nonconfirming decisions per million. A vehicle, for example, has thousands of parts, which translates into millions of decisions concerning concept, performance, quality, appearance, cost, and other variables of the design entity over the course of different development phases. A company can chart its performance in a p-chart over the development milestones. The centerline, $\bar{p}$, can be established from company history. Let the total number of decisions taken at milestone k equal n_k and the number of nonconfirming decisions equal to D_k; then the milestone proportion nonconfirming is given as $\hat{p} = (D_k/n_k)$. The Z-score sigma control limits are given by

$$\bar{p} \pm Z \sqrt{\frac{\bar{p}\,(1-\bar{p})}{n_k}}$$

For example, a 3σ (6σ) limit can be obtained when Z equals 3 (6), respectively. Charting starts by drawing the control limits and plotting the current milestone $\hat{p}$.

3.9   Features of a Sound DFSS Strategy

The DFSS team will have a first-time real-world useful feedback about their design efforts in the prototype phase based on testing and performance in the working environment. This usually happens almost after the middle of the development cycle. As such, there is not much room to make hard changes in the design entity if unpleasant issues do arise. Our objective is still to design it right the first time. It should be understood that using the right tools is not a guarantee to establish a Six Sigma capability in the design entity, especially when weak concepts were conceived and pushed into the development pipeline. The challenge in a DFSS project is the unavailability of useful information to lead design activity upfront, where most influential decisions are to be made. Therefore, a sound DFSS strategy should provide design principles that directionally lead to good concepts. In a totality, the DFSS strategy should have the following to "design right the first time":

- Customer driven and focused
- Measures of the effectiveness of the design process
- Measures to compare performance versus requirements

- Achieve key business objectives
- Effective use of human resources and knowledge
- Adherence to teamwork
- Upfront development of robustness and testing for verification
- Foster learning organization
- Forces paradigm shift from find-fix-test to prevention
- Handles changes without affecting customer
- Insensitive to development processes noises
- Concurrent methods, consistent training—everyone knows how
- Uses integrated approach to design from concept to validation
- Allows useful decision to be taken with absence of data (e.g., in the conceptual phase)
- Is capable of checking the feasibility of having the Six Sigma capability *analytically* in the design entity
- Pinpoints where it is easier to implement changes when needed
- Increases the potential for high reliability and robustness
- Provides ample room to establish Six Sigma capability by conceptual means upfront
- Uses optimization to set tolerances when design concept is finalized

The conception of a coupling-free design does not automatically guarantee that Six Sigma capability can be obtained. However, a coupling-free design has a better chance to establish such capability. In later chapters, this possibility can be assured when an optimal solution is obtained in the parameter and tolerance optimization stages. In addition, the task of implementing changes to obtain the Six Sigma capability is easier, to a large degree, in an uncoupled as opposed to a coupled design.

The DFSS deployment is best suited when it is synchronized with the design life cycle. The next section highlights a generic design process that will provide some foundation. The authors realize that many variants to what is proposed do exist; some more than others, stress certain aspects that best suit their industry. This variance is more prominent in long-term, low-volume industries versus short-term development cycles with high production volume.

Appendix: Historical Development In Design

The research in the design arena started in Europe. Above all, the Germans developed over the years some design guidelines that continued to improve at a consistent pace. A huge body of research has been published in German on the design practice. Unfortunately, only a limited portion has been translated into English. Most of these efforts are listed in Hubka (1980) and Phal and Beitz (1988). The Germans' design schools share common observations. For example, a good design entity can be judged by its adherence to some design principles; the design practice should be decomposed to consecutive phases, the need for methods for concept selection, and so on. Besides the Germans, the Russians developed an empirical inventive theory with promises to solve difficult and seemingly impossible engineering problems, the so-called TRIZ or theory of inventive problem solving (TIPS). TRIZ is an example of the basic principles for synthesis of solution entities (Altshuler 1988, 1990), Rantanen (1988), Arciszewsky (1988), Dovoino (1993), Tsourikov (1993), Sushkov (1994), and Royzen (2002). TRIZ, the Russian acronym to TIPS, is based on inventive principles devised from the study of more than 1.5 million of the world's most innovative patents and inventions. TRIZ was conceptualized by Dr. Gerikh S. Altshuler, a brilliant Russian inventor, in 1946. TRIZ is an empirical theory that was devised along the lines of inventive problem solving, functional analysis, technology prediction, and contradiction elimination. *Contradiction* is synonymous with the coupling vulnerability in axiomatic design (Suh 1990).

The concern of reducing vulnerability to foster customer satisfaction in the design entities continued with the work of the English researcher Pugh (1991, 1996). Pugh proposed a matrix evaluation technique that subjectively weighs each concept against the important technical criteria and customer concerns from a *total* perspective. Pugh (1991) discussed the role of systematic design and concept selection for both conventional and nonconventional (creative) product situations.

Morphological approaches to synthesis developed by Zwicky (1948) and Hubka and Eder (1984) are very similar to the different effects and analogies presented in TIPS. In these approaches, a complex design problem can be divided into a finite number of subproblems. Each solution of a subproblem can be considered separately. Solutions are then arranged in charts and tables. The morphological charts and matrices have been developed to suggest possible solutions or available effects that can be used in a certain situation. Most of the charts and matrices are developed for mechanical product design, and may be difficult to use outside their intended fields. Hubka and Eder (1984) and Ramachandran et al. (1992) researched the synthesis problem and focused on automating the synthesis process. To automate synthesis, most researchers have limited their applications to a certain field. In doing so, only a few principles are covered. Many automated approaches have been implemented as tools using artificial intelligence (AI). They are, however, specific to one or a few engineering principles. It appears to be difficult to find solutions based on other principles, using these tools, such as the building block approach of Kota (1994). The approach to analyze a solution in most product development research is based on comparison. Matrices are commonly used to represent the engineering situation. The matrices can be arranged in different ways: the comparative criteria on one axis and the solution on the other, functional requirements on one axis and the proposed solution on the other axis, or solution decomposition on both axes. The comparative approaches of Clausing (1994), Pugh (1991), Ullman (1992), and Phal and Beitz (1988) are most commonly used. These matrices can be used in situations where solutions to be evaluated originate from the same principles and the same objectives.

In the axiomatic design approach suggested by Suh (1990), evaluation can be made by analyzing how well-proposed solutions are fulfilling the functional requirements. This approach enables evaluation of solutions based on different principles. The main advantage of evaluating matrices with selected solutions on both axes is the possibility of sequencing or scheduling design activities. In this area much research has been conducted by McCord and Eppinger (1993), and Pimmler and Eppinger (1994). Algorithms for optimizing and resequencing project structure are some of the results of this research category. The strength of this evaluation technique is in the sequencing and optimization of engineering projects. In these situations only limited support is offered by sequencing methods to the synthesis of new solutions.

In the United States, and since the early 1970s, there have been progressive research efforts in the design arena, particularly in the field of mechanical design. Engineering design research was motivated by the shrinking market share of the United States. The engineering design started to take its esteemed position as a central theme in society. The late realization of the importance of engineering design led Dixon (1966) and Penny (1970) to place engineering design at the center of the cultural and the technical streams of the society. Ullman (1992) stated that the activities of design research were accelerating along the following trends: artificial intelligence computer-based models, the design synthesis (configuration), cognitive modeling, and design methodologies. In addition, there were considerable efforts in developing rule-based design processes. The ample yield since the 1970s is mature enough to allow classification and comparison. However, these tasks are difficult because there is a minimal set of agreed-on guidelines, as is the case with the European design schools. However, the functionality of the design is a unifying concept. The topic of function and the concept of value are extensively discussed in the context of value engineering (Park 1992).

4

Design for Six Sigma Deployment

4.1 Introduction

The extent to which DFSS produces the desired results is a function of the adopted deployment strategy. This chapter introduces the elements of such strategy by highlighting the key elements for successful deployment.

History tells us that sound initiative, concepts, or ideas become successful and promoted to norms in many companies when commitment is secured from involved people at all levels. DFSS is no exception. A successful DFSS deployment relies on active participation of people on almost every level, function, and division, including the customer.

The traditional Six Sigma initiative, the DMAIC method, is usually deployed as a top-down approach reflecting the critical importance of securing the buy-in from the top leadership level. This has been successful so far and should be benchmarked for DFSS deployment. The black belts and green belts make up the brute force of deployment under the guidance of the champions. Success is measured by increase in revenue and customer satisfaction and the extent to which cash flow is generated in both long and short terms (soft and hard) with each project. These benefits can't be harvested without a sound strategy with the long-term vision of establishing the Six Sigma culture. In the short term, deployment success is dependent on motivation, management commitment, project selection, and scoping, an institutionalized reward and recognition system, and optimized resources allocation.

4.2 Black Belt–DFSS Team: Cultural Change

The first step in a DFSS project endeavor is to establish and maintain a DFSS project team (for both product/service and process) with a shared vision.

The purpose is to establish and maintain a motivated team. The success of development activities depends on the performance of this team, which is selected according to the project charter. The team should be fully integrated, including internal and external members (suppliers and customers).

Special efforts may be necessary to create a multinational, multicultural team that collaborates to achieve a Six Sigma–level design. Roles, responsibilities, and resources are best defined upfront, collaboratively, by all team members. The black belt is the team leader. A key purpose is to establish the core project team and get a good start, with very clear direction derived from the program from which the project was conceived. Projects targeting subsystems or subprocesses are the lower level of deployment. In the initial deployment stage, DFSS projects, some of which are called "pilot" projects, are scaled to subsystem level or equivalent. As the DFSS gains momentum, it can be enlarged to program-level DFSS scale. It is very important to "get it right the first time" to avoid costly downstream errors, problems, and delays.

Once the team has been established, however, it is just as important to the black belt to *maintain* the team so as to continuously improve its performance. This first step, therefore, is an ongoing effort throughout the project's full cycle of planning, development, manufacturing/production, and field operations.

The DFSS teams emerge and grow through systematic efforts to foster continuous learning, shared direction, interrelationships, and a balance between intrinsic motivators (a desire which comes from within) and extrinsic motivators (a desire stimulated by external actions). Constant vigilance at improving and measuring team performance throughout a project life cycle will be rewarded with ever-increasing commitment and capability to deliver winning design entities.

Winning is usually contagious. Successful DFSS teams foster other teams. The growing synergy arising from ever-increasing numbers of motivated teams accelerates improvement throughout the deploying company or enterprise. The payback for small, upfront investments in team performance can be enormous. Team capability to deliver benchmark Six Sigma quality-level design that customers will prefer to a company's toughest competitors will increase as members learn and implement the contemporary processes and practices suggested in this DFSS book.

In his/her DFSS endeavor, the black belt will interface with many individuals with a wide spectrum of personality. In addition to the technical training, the black belt should have enough ammunition of soft skills to handle such interfaces. Many companies now have Six Sigma training programs, allocating time in the training curricula to educate black belts about the cultural change induced by Six Sigma.

DFSS deployment will shake many guarded and old paradigms. People's reaction to change varies from denial to pioneering, passing through many stages. In this venue, the objective of the black belt is to develop alliances for his/her efforts as (s)he progresses. We depict the different stages of change in Fig. 4.1 The stages are linked by what is called the "frustration curves." We suggest that the black belt draw such a curve periodically for each team member and use some or all of the strategies below to move his/her team members to the positive side, the "recommitting" phase.

There are several strategies for dealing with change. To help decelerate (reconcile), the black belt needs to listen with empathy, acknowledge difficulties, and define what is over and what isn't. To help phase out the old paradigm and reorient the team to the DFSS paradigm, the black belt should encourage redefinition, utilize management to provide structure and strength, rebuild sense of identity, gain sense of control and influence, and encourage opportunities for creativity. To help recommit (accelerate) the team in the new paradigm, the black belt should reinforce the new beginning, provide a clear purpose, develop a detailed plan, be consistent in the spirit of Six Sigma, and celebrate success.

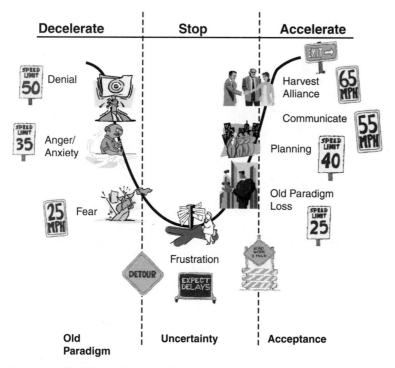

Figure 4.1 The "frustration curve."

4.3   DFSS Deployment Prerequisites

Deployment is, of course, key. The impact of DFSS initiative depends on the effectiveness of deployment, specifically, how well the Six Sigma design principles and tools are practiced by the DFSS team (including the black belts and green belts together with the subject matter experts). Intensity and constancy of purpose beyond the norm are required to constantly improve deployment. In long races, whether leading or lagging, those who go the fastest win. Rapid deployment of DFSS plus commitment, training, and practice characterize winning companies.

Successful DFSS deployment requires the following prerequisites:

1. *Top- and medium-level management commitment.*   If top and medium management teams are not on board with deployment, DFSS initiative will eventually fade away.

2. *Existence of a program management system.*   Our observation is that a project roadmap, or a design algorithm, is required for successful DFSS deployment. The algorithm works as a campus leading black belts to closure by laying out the full picture of the DFSS project. We would like to think of this algorithm as a recipe that can be further tailored to the customized application within the company's program management system that spans the design life cycle.* We usually encounter two venues at this point:

- Develop a new program management system (PMS)† to include the proposed DFSS algorithm. The algorithm is best fit after the R&D stage and prior to customer use. It is the experience of the authors that many companies lack such universal discipline from a practical sense. This venue is suitable for such companies and those practicing a variety of PMSs hoping that alignment will evolve. The PMS should span the design life cycle presented in Chap. 1.

- Integrate with the current PMS by laying this algorithm over and synchronize when and where needed.

In either case, the DFSS project will be paced at the speed of the leading program from which the project was derived in the PMS. Initially, a high-leverage project should target subsystems to which the business and the customer are sensitive. A sort of requirement flowdown, a cascading method, should be adopted to identify these

---

*Design life cycle spans R&D, development, manufacturing, customer, and postcustomer stages (e.g., service and aftermarket).

†Examples are advanced product quality planning (APQP) in the automotive industry, integrated product and process design (IPPD) in the aerospace industry, the Project Management Institute (PMI) process, and the James Martin process in software industry.

subsystems. Later, when DFSS becomes the way of doing business, program-level DFSS deployment becomes the norm and the issue of synchronization with PMS eventually diminishes. Actually, the PMS is crafted to reflect the DFSS learning experience that the company gained over the years of experience.

3. *DFSS project sources.* The successful deployment of the DFSS initiative within a company is tied to projects derived from the company's scorecards. In Six Sigma terminology a *scorecard* is a unified approach to visualize how companies gauge their performance internally and externally. In other words, scorecards are tools used to measure the health of the company. The scorecard usually takes the form of a series of linked worksheets bridging customer requirements with product and process performance at all stages of the product, process, or service development. The reader may conclude that to satisfy such a prerequisite indicates the existence of an active measurement system for internal and external metrics in the scorecard. The measurement system should pass a gauge R&R (repeatability and reproducibility) study in all used metrics.

4. *Establishment of deployment structure.* A premier deployment objective can be that black belts are used as a taskforce to improve customer satisfaction, company image, and other strategic long-term objectives of the deploying company. To achieve such objectives, the deploying division should establish a deployment structure formed from deployment directors, and master black belts (MBBs) with defined roles and responsibilities, long-term and short-term planning. The structure can take the form of a council with definite recurring schedule. We suggest using DFSS to design the DFSS deployment process and strategy. The deployment team should

- Develop a green belt structure of support to the black belts in every department.

- Ensure that the scope of each project is under control and that the project selection criteria are focused on the company's objectives such as quality, cost, customer satisfiers, and delivery drivers.

- Hand off (match) the appropriately scoped projects to black belts.

- Support projects with key upfront documentation such as charters or contracts with financial analysis highlighting savings and other benefits, efficiency improvements, customer impact, project rationale, and other factors. Such documentation will be reviewed and agreed on by primary stakeholders (deployment champions, design owners, black belts, and finance).

- Allocate black belt resources optimally across many divisions of the company targeting high-impact projects first, and create a long-term allocation mechanism to target a mix of DMAIC/DFSS projects to be revisited periodically. In a healthy deployment, the number of DFSS projects should increase as the number of DMAIC projects decreases over time. However, this growth in the number of DFSS projects should be engineered. A growth model, an *S curve,* can be modeled over time to depict this deployment performance. The initiating condition of where and how many DFSS projects will be targeted is a significant growth control factor. This is a very critical aspect of deployment, particularly when the deploying company chooses not to separate the training track of the black belts to DMAIC and DFSS and train the black belt in both methodologies:

- Available external resources will be used, as leverage when advantageous, to obtain and provide the required technical support.
- Promote and foster work synergy through the different departments involved in the DFSS projects.

4.4   DFSS Deployment Strategy

A DFSS deployment strategy should be developed to articulate the basic DFSS deployment mission, guiding principles, goals, key result areas, and strategies for its management and operations to guide and direct its activities. It should be part of the total Six Sigma initiative and deployment. Usually, companies embark on the DMAIC method prior to deploying DFSS. Other companies chose to deploy both simultaneously, adding more deployment mass and taking advantage of the successful deployment in Six Sigma pioneering companies. Companies in this category use benchmarking to avoid deployment failure modes.

The deployment vision is to create a long-term Six Sigma design culture. This long-term vision can be achieved by taking short-term and calculated deployment steps forward, usually annually. The combined effectiveness of these steps is the right-hand side of the deployment equation. By *deployment equation,* we mean the momentum (MO). The momentum can be expressed as the deployment velocity (DV) times deployment mass (DM) or

$$MO = DV \times DM \qquad (4.1)$$

This equation has some scalability depending on the deployment entity. A deployment entity may be scaled up to the enterprise level or down to a DFSS team and pass through a business unit or a company, a division, or a department. For example, at the division level it means the *total* number of projects closed and the *average* velocity with which these projects are successfully closed (ended). Initially, in the DFSS pilot phase, both DM and DV are factors with variation. The pilot phase is company-dependent and is measured in years. Once deployment reaches steady state, the variation can be driven to minimum. When Six Sigma becomes the way of doing business, day-to-day and project-to-project DV and DM can be approximated by constants. Note that deployment mass and velocity can be increased by attacking project failure modes. A deployment FMEA is very useful in order to document and track corrective actions. A prominent failure mode is incorrectly scoping a project, that is, a "hidden factory." The champions' role is significant to foster growth in deploying momentum.

4.4.1 DFSS deployment momentum

The health of DFSS deployment can be measured by momentum. In addition to its meaning in physics, *momentum* is a commonly used term in performance. A DFSS team that possesses the momentum is on the move and would require some effort to stop. We can apply the same analogy to deploying companies. In physics, *momentum* refers to an object's mass in motion; here, this term applies to a deploying company, division, or black belt team. A DFSS team which is "on the move" has the momentum.

The amount of momentum that a black belt has is dependent on two variables: how many projects are moving and how rapidly they are successfully closing (ending). Momentum depends on the variables of mass and velocity. While mass is a scalar quantity, velocity is not. It is a vector with magnitude and direction, the successful project's closure. In terms of an equation, the momentum is equal to the mass of the object times the velocity of the object as expressed in Eq. (4.1).

To calculate a deployment entity momentum, all black belts who have finished their training are considered. Usually in massive deployments, black belts update their projects in the company tracking systems and on a timely basis, offering a measurement system for momentum calculation. The mass and velocity can be pulled out from such a system and applied to the momentum equation. The lowest deployment entity's (a black belt's) momentum is calculated first. The results of these calculators are then aggregated and rolled up to the next-higher deployment entity (a division). The process is repeated up to the enterprise level. This method provides an estimate on deployment health and should be revisited periodically.

4.4.2 Black belt momentum

The *black belt momentum* (BBMO) variable is used and defined as the product of velocity V times the mass M, or BBMO $= V \times M$. The mass M is the *weighted* sum of two types of mass (see Fig. 4.3).

- The *business mass* weighted by, say, 30 percent, and measured by metrics such as

  Improvement (Δ) business metric 1 (ΔBM1), for example, repairs at a weighted target of 50 percent

  Improvement (Δ) business metric 2 (ΔBM2), such as savings (Δ) gained versus the target per a project (e.g., $250,000) weighted at 50 percent

  These two masses are used to illustrate the assessment calculation. Deploying companies have the option to expand.

- The *customer mass* weighted at 70 percent and measured by

  Improvement (Δ) in customer satisfaction metric 1 (50 percent)
  Improvement (Δ) in customer satisfaction metric 2 (30 percent)
  Improvement (Δ) in customer satisfaction metric 3 (20 percent)

Note that we chose only three masses (metrics) for illustration purposes. Also note that some variance in desirability is reflected by the weights and that we give higher weight to customer satisfaction metrics.

The overall BBMO is the sum of the product of mass and velocity over all the black belt's projects. Mathematically, let i be the black belt index and j be the project index handled by the black belt; then

$$\text{BBMO}_{ij} = M_{ij} \times V_{ij} \qquad (4.2)$$

The mass M_{ij} is given by

$$M_{ij} = 0.3 \sum \text{business mass} + 0.7 \sum \text{customer mass} \qquad (4.3)$$

$$= 0.3(0.5\ \Delta\text{BM1} + 0.5\ \Delta\text{BM2})$$

$$= +\ 0.7(0.5\ \Delta\text{CSM1} + 0.3\ \Delta\text{CSM2} + 0.2\ \Delta\text{CSM3})$$

where ΔBM and ΔCSM, respectively, indicate improvement in business and customer satisfaction metrics gained from the project. The velocity V is the velocity of closing the project minus the targeted closure date, say, 4 months.

Deployment momentum is a good measure of strategy that targets DFSS deployment. The deployment strategy should include the mission statements of the deploying company and how DFSS will help achieve this mission, and address the specific issues and needs both internally and externally. As a whole, the strategy provides a framework for deployment that includes assessment of the current environment, resources, and timing as well as a statement of commitment, short- and long-term planning, and directions for the future.

A sound DFSS deployment strategy should include the principles, goals, key results, and short- and long-term planning.

4.4.3 Principles of DFSS deployment strategy

On the principles side of DFSS deployment, we suggest that the DFSS community (black belts, green belts, champions, and deployment directors) will commit to the following:

- Support their company image and mission as a highly motivated producer of choice of world-class, innovative complete product, process, or service solutions that lead in quality and technology and exceed customer expectations in satisfaction and value.

- Take pride in their work and in their contributions, both internally within the company and externally to the industry.

- Constantly pursue "Do it right the first time" as a means of reducing the cost to their customers.

- Strive to be treated as a resource, vital to both current and future development programs and management of operations.

- Establish and foster a partnership with subject matter experts, the technical community within their company.

- Treat lessons learned as a corporate source of returns and savings through replicating solutions and processes to other relevant entities.

- Promote the use of DFSS principles, tools, and concepts where possible at both project and day-to-day operations and promote the data-driven decision culture, the crust of Six Sigma culture.

4.5 DFSS Deployment Strategy Goals

A variation of the following goals can be adopted:

- Maximize the utilization of a continually growing DFSS community by successfully closing most of the matured projects approaching the targeted completion dates.

- Leverage projects that address the company's objectives, in particular the customer satisfaction targets.

- Cluster the green belts (GBs) as a network around the black belts for synergy and to increase the velocity of deployment [see Eq. (4.1)].

- Ensure that DFSS lessons learned are replicated where possible, that is, that common issues are addressed with minimal resources, thereby maximizing momentum.

- Train some targeted levels of green belts and black belts.

- Maximize black belt certification turnover (set target based on maturity).

4.5.1 Key result areas

- *Product/service/service.* Pursue excellence in quality and customer satisfaction of the designed entities.

- *Relationships.* Achieve and maintain working relationships with all parties involved in DFSS projects to promote an atmosphere of cooperation, trust, and confidence between them.

- *Architecture.* Develop and maintain Six Sigma BB and GB architecture which is efficient, responsive, and supportive of the deployment strategy.

- *Human resources.* Maintain a highly qualified, motivated, and productive DFSS community capable of, and committed to, achieving the goals of strategy.

- *Deployment velocity (DV).* Close the DFSS projects in a timely and cost-effective manner.

- *Deployment mass (DM).* Maximize the number of projects closed per each black belt in his/her DFSS life.

- *Technology.* Track and employ DFSS concepts, tools, and technologies that provide opportunities to enhance design and data-driven decision-making practice in the company.

4.5.2 Project identification, selection, scoping, and prioritization

Project champions, together with seasoned black belts and master black belts (MBBs), should hold periodic project selection and scoping meetings. These meetings should be chaired by the designated deployment director or vice president leading the deployment. Of course, champions should already have received the appropriate training. A selection and identification approach for DFSS projects should be developed and enhanced on the basis of experience. This approach should be followed by the appropriate scoping method to scale the project to the right size for black belts on the basis of workload, certification target, and project criteria. The project champions will propose, screen, and concur on the BB projects. MBBs should concur. Black belts can also propose project ideas to their champions, in particular, DMAIC projects that reached their entitlements. The list of feasible and possible ideas will fill the project pipeline and should be documented while awaiting data, funding, and/or approval protocols. A prioritization scheme should also be devised to schedule project launch.

In summary, and on the DFSS project side, the following should be part of the deployment strategy:

- *Number and criteria of projects on the redesign side.*   Short-term DFSS project source.

- *Number and criteria of projects on the design side.*   Long-term DFSS project source.

- *Project complexity.*   Depending on the interrelationships and coupling (*coupling* is an axiomatic design term; see Chap. 8) within and among design entities, projects size will increase and having more than one black belt, say, a black belt team, would be more than justified. This will avoid suboptimizing components and subprocesses reaching out to higher hierarchical levels and stressing system engineering thinking. From this perspective, project matching and scalability appropriate to the black belts will become more significant. In the team approach, the following must be decided:

  Black belt team size.

  Black belt workload.

  Team dynamics to reflect the complex interrelationships so that design decisions are negotiated among the team to achieve the best resolution and/or compromise based on some sequence judged by coupling and complexity.

  Black belt certification.

  Interests of the stockholders involved should be identified and dealt with accordingly by the black belt team. We found that Venn diagrams are useful identification tools to facilitate the analysis (see Fig. 4.2). The team should capitalize on the common interests represented by the common intersection set of interests.

- *Project identification and selection approach.*

- *Project scoping approach.*   The usual linear CTQ flowdown used in DMAIC is successful only when coupling absence is assured, that is, one-to-one mapping between requirements and design parameters is accomplished per axiom 1 of the axiomatic design method (see Chap. 8). Otherwise, more involved scoping schemes must be employed.

- *Project prioritization scheme.*   The selection and scoping approaches usually yield project ideas that are further refined in a project charter. Depending on the size of the project list, a prioritization scheme may be needed. These schemes range in complexity and involvement. Simple schemes like the one depicted in Fig. 4.3 are usually effective. High-leverage projects are in the upper right quadrant denoted as

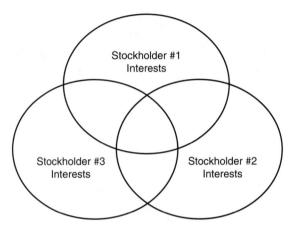

Figure 4.2 Stockholder Venn diagram analysis.

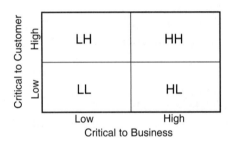

Figure 4.3 Project prioritization scheme.

"HH." Other schemes may take the form of a matrix where business-specific criteria of DFSS projects are displayed in the rows and projects in the columns. The scoping team then rates each project against each criterion using Pugh methods (see Sec. 5.6) and prioritizes accordingly.

4.5.3 DFSS project target setting

The black belts will be responsible for delivering closed projects that will target both business and customer metrics, mainly customer satisfaction in addition to financial benefits to the deploying entity (selection criteria are listed in Fig. 4.4). For example, a deployment entity, say, a division, may have the following annual objectives:

- A 5 percent customer satisfaction improvement (metric 1)
- A 46 percent customer satisfaction improvement (metric 2)
- A 12.2 percent business objective improvement (metric 1)
- A $34 million savings (from DFSS projects)

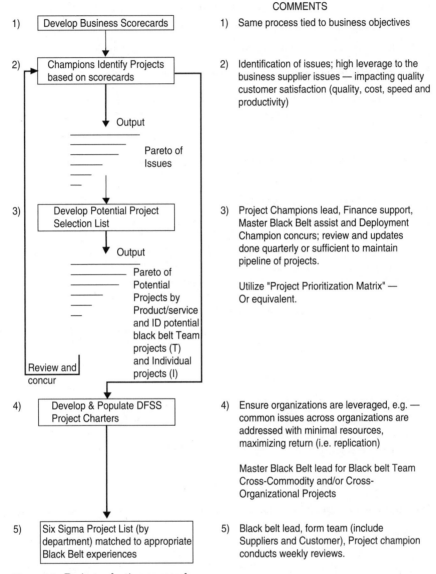

Figure 4.4 Project selection approach.

The DFSS community within the deploying entity can be chartered a fair proportion of this set of objectives. The ratio of people, specifically, the black belt population relative to the division population boosted by 30 percent, is considered appropriate. Assume that the population of black belts is 66 and the division or company employee population is 1000. The proportion will be $(66/1000) \times 1.3 = 0.0858$. In this case the black belt population targets are

- A 0.429 percent customer satisfaction improvement (metric 1)
- A 3.95 percent customer satisfaction improvement (metric 2)
- A 1.047 percent improvement in business objective (metric 1)
- A $34 million savings (from DFSS projects)

Assuming that only 44 black belts finished their training and are eligible to be considered in the target setting calculations, the black belt share will be as follows:

- A 0.001 percent customer satisfaction improvement (metric 1)
- A 0.0898 percent customer satisfaction improvement (metric 2)
- A 0.0238 percent business objective improvement (metric 1)
- A $0.773 million DFSS project savings

The BB can achieve these targets utilizing a targeted annual count of successfully closed projects; let us assume 3. Then, the improvement target Δ per BB project can be calculated as

- A 0.00033 percent customer satisfaction improvement (metric 1)
- A 0.02993 percent customer satisfaction improvement (metric 2)
- A 0.00793 percent business objective improvement (metric 1)
- A $0.258 million DFSS project savings[*]

DFSS projects that don't satisfy these targets can be rejected. The pool of projects with projections above these limits should be refined and approved for start (execution). Of course, priority should be given to those projects that achieve optimum levels in such criteria.

4.5.4   DFSS project types

We suggest two project types for deployment from project perspective:

1. *Type 1 DFSS project.*   Scaled design project that spans stages 1 to 6 (see Fig. 4.5) of the design cycle for both product and service. This type of project is used for initial deployment, usually on the redesign side.

2. *Type 2 DFSS project.*   A project that spans the whole life cycle and is adopted at a mature deployment stage. It includes the creative innovative design projects.

---

[*]Usually companies require $0.250 million both soft and hard savings.

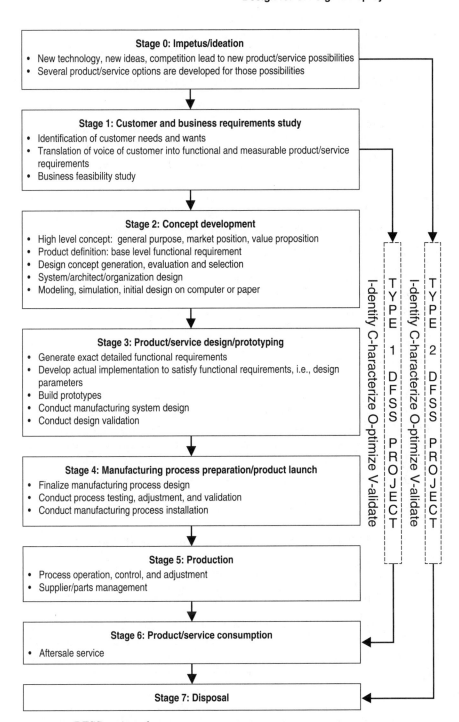

Figure 4.5 DFSS project phases.

This distinction is made to gain momentum before embarking on more challenging DFSS projects and to provide projects to the black belts that is paced with the training. Training projects offer data to exercise the plethora of DFSS tools available in the DFSS algorithm (Chap. 5).

4.6 Six Sigma Project Financial Management

In general, DFSS project financial savings can be categorized as *hard* or *soft* savings and are mutually calculated or assessed by the black belt and the assigned financial analyst (FA) to the project. The FA assigned to a DFSS team should act as the lead in quantifying the savings related to the project "actions" at the initiation and closure phases, assist in identification of "hidden factory" savings, support the black belt (BB) on an ongoing basis, and if financial information is required from areas outside the BB's area of expertise, the FA should direct the BB to the appropriate contacts, follow up, and ensure that the BB receives the appropriate data. The analyst, at project closure, should ensure that the appropriate offices concur with the savings. This primarily affects manufacturing costs, engineering expense, and nonrevenue items for rejects not directly led by black belts from those organizations. In essence, the analyst needs to provide more than an audit function.

"Hard savings" are defined as measurable savings associated with improvements in repairs, rework, scrap, inspection, material cost, warranty, labor savings (achievable or collectable through work rebalances), revenue associated with reductions in customer dissatisfaction, cash flow savings (i.e., inventory), and other values of lost customer satisfaction. Hard savings are calculated against present operating levels, not against a budget or a plan. They represent the bottom-line saving that directly affects the company's income statement and cash flow and are the result of measurable product, service, and process improvements. The effect on company financial statements will be determined off line by the appropriate company office.

"Soft" savings are less direct in nature and include projects that open plant floor space (as a side benefit), which may allow for the location of future operations; projects that reduce vehicle weight, which may enable other design actions to delete expensive lightweight materials; and cost avoidance. *Cost avoidance* is usually confused with *cost savings*; for example, employing robot welding instead of manual welding is an avoidance of costs, whereas reducing scrap is avoidance rather than saving.

The finance analyst should work with the black belt to assess the projected annual financial savings on the basis of the information available at that time (scope, expected outcome, etc.). This is not a detailed review, but a start approval. These estimates are usually revised as the project progresses and more accurate data become available. The project should have the potential to achieve the annual target, usually $250,000. The analyst confirms the business rationale for the project where necessary.

4.7 DFSS Training

Specific training sessions for leadership, champions, and black belts are part of the deployment strategy. Under this heading, the deploying entity should provide considerations for training. These considerations should be specific and usually are subject to the flavor of both the deployment entity and the supplier doing the training, if any. The training should not exclude any other individual whose scope of responsibility intersects with the training function. Considerations such as geographic location, timing, and scheduling should be discussed in advance and set on the annual calendar so that they can be readily available for replacements, changes, and dropouts.

4.8 Elements Critical to Sustain DFSS Deployment

In what follows, we present some of the thoughts and observations that were gained through our deployment experience of Six Sigma, in particular DFSS. The purpose is to determine factors toward keeping and expanding the momentum of DFSS deployment to be sustainable.

This book presents the DFSS methodology that exhibits the merge of many tools at both the conceptual and analytical levels and penetrates dimensions such as characterization, optimization, and validation by integrating tools, principles, and concepts. This vision of DFSS should be a core competency in a company's overall technology strategy to accomplish its goals. An evolutionary strategy that moves the deployment of DFSS method toward the ideal configuration is discussed. In the strategy, we have identified the critical elements, necessary decisions, and deployment concerns.

The literature suggests that more innovative methods fail immediately after initial deployment than at any other stage. Useful innovation attempts that are challenged by cultural change are not directly terminated, but allowed to fade slowly and silently. A major reason for the failure of technically viable innovations is the inability of management to commit to an integrated, effective, and cost-justified evolutionary program for sustainability that is consistent with the company's mission. The DFSS deployment parallels in many aspects the technical innovation challenges from a cultural perspective. The

DFSS initiatives are particularly vulnerable if they are too narrowly conceived, are built on only one major success mechanism, or do not align with the larger organizational objectives. The tentative top-down deployment approach has been working where the top leadership support should be a significant driver. However, this approach can be strengthened when built around mechanisms such as the superiority of DFSS as a design approach and the attractiveness of the methodologies to designers who want to become more proficient professionals.

While it is necessary to customize a deployment strategy, it should not be rigid. The strategy should be flexible enough to meet expected improvements. The deployment strategy itself should be DFSS-driven and robust to (withstand) anticipated changes. It should be insensitive to expected swings in the financial health of the company and should be attuned to the company's objectives.

The strategy should consistently build coherent linkages between DFSS and daily design business. For example, engineers and architects need to see how all the principles and tools fit together, complement one another, and build toward a coherent whole. DFSS needs to be seen initially as an important part, if not the central core, of an overall effort to increase technical flexibility.

4.9   DFSS Sustainability Factors

Many current design methods, some called "best practices," are effective if the design is at a low level and need to satisfy a minimum number of functional requirements, such as a component or a process. As the number of requirements increases, the efficiency of these methods decreases. In addition, they are hinged on heuristics and developed algorithms [e.g., design for assembly (DFA)] limiting their application across the different development phases.

The design process can be improved by constant deployment of the DFSS concepts and tools, which begins from a different premise, namely, the conception and abstraction or generalization. The design axioms and principles are central to the conception part of DFSS. As will be explained in Chap. 8, axioms are general principles or truths that can't be derived, except that there are no counterexamples or exceptions. Axioms constituted the foundations of many engineering disciplines such as thermodynamic laws, Newton's laws, and the concepts of force and energy. Axiomatic design provides the principles to develop a good design systematically and can overcome the need for customized approaches.

We believe that management should provide more leadership and an overall strategy for economically achieving product, process, and service in the integration of the DFSS approach within a design program management system. In a sustainability strategy, the following attributes would be persistent and pervasive features:

1. Continued improvement in the effectiveness of DFSS deployment by benchmarking other successful deployment elsewhere

2. Developing a deployment measurement system that track the critical-to-deployment requirements, detect failure modes, and implement corrective actions

3. Enhanced control (over time) over the company's objectives via selected DFSS projects that really move the needle

4. Extending involvement of all levels and functions

5. Embedding DFSS into the everyday operations of the company

The prospectus for sustaining success will improve if the strategy yields a consistent day-to-day emphasis of the following recommendations:

- Recognizing that DFSS represents a cultural change and a paradigm shift and allows the necessary time for the project's success

- Extending DFSS to key suppliers and moving these beyond the component level to subsystem and system levels

- Integrating the DFSS methodology as a superior design approach with the company's design program management system (PMS) and an alignment of the issues of funding, timing, and reviews

- Linking DFSS design to design for reliability (DFR) and allowing a broader understanding of tools that can be used in this regard

- Stressing the usefulness and soundness of the methodologies rather than stressing conformance to narrow design problem-solving protocols

- Sustaining managerial commitment to adopting appropriate, consistent, relevant, and continuing reward and recognition mechanism for black belts and green belts

- Using DFSS as a consistent, complete, fully justified, and usable program for reference or as a design guideline to support expansion to other new programs and projects

- Recognizing the changes that are needed to accommodate altering a designer's tasks from individualized projects to broader scope and highly interdependent team assignments

- Providing relevant, on-time training and opportunities for competency enhancement, the capacity to continue learning, and alignment of rewards with competency and experience

- A prioritizing mechanism for future projects that targets the location, size, complexity, involvement of other units, type of knowledge to be gained, potential for generalization, and replication or transferability and fit within the strategic plan

- Instituting an accompanying accounting and financial evaluation effort to cope with the scope of consideration of the impact of the project on both fronts—hard and soft savings—and moves resources toward the beginning of the design cycle in order to accommodate DFSS methodology

The DFSS methodology, theory, and application, formed by integrating conceptual methods, design principles and axioms, quality, and analytical methods, is very useful to the design community in both creative and incremental design situations. However, this vision needs to evolve with more deployment as companies leave the DFSS pilot phase. While the pilot can be considered narrow when compared with the whole vision, emphasis on the key concepts that need to be maintained should gear deployment toward success.

If the DFSS approach is to become pervasive as a central culture underlying a technology development strategy, it must be linked to larger company objectives. In general, the DFSS methodology should be linked to

1. The societal contribution of a company in terms of developing more reliable, efficient, and environmentally friendly products, processes, and services

2. The goals of the company, including profitability and sustainability in local and global markets

3. The explicit goals of management embodied in company mission statements, including characteristics such as greater design effectiveness, efficiency, cycle-time reduction, and responsiveness to customers

4. A greater capacity for the deploying company to adjust and respond to customers and competitive conditions

5. The satisfaction of managers, supervisors, and designers

A deployment strategy is needed to sustain the momentum achieved in the pilot phase to subsequent phases. The strategy should show how DFSS allows black belts and their teams to respond to a wide variety of externally induced challenges and that complete deployment of DFSS will fundamentally increase the yield of the company's operations and its ability to provide a wide variety of design responses. DFSS deployment should be a core competency of a company. DFSS will enhance the variety of quality of design entities and design processes. These two themes should be continually stressed in strategy presentations to more senior management. As deployment proceeds, the structures and processes used to support deployment will also need to evolve. Several factors need to be considered to build into the overall sustainability strategy. For example, the future strategy and plan for sustaining DFSS need to incorporate more modern learning theory on the usefulness of the technique for green belts and other members when they need the information.

Design for Six Sigma Project Algorithm

5.1 Introduction

The design project is the core of DFSS deployment and has to be executed consistently using a process, an algorithm, that lays out the DFSS principles, tools, and methods within the company development processes. This chapter is intended primarily to achieve this task in order to support the black belt (BB) and the BB's team and the functional champion in project execution. As such, the material presented herein should be viewed and used as a generic template, a design algorithm, with ample flexibility for customization to fit the deploying company's specific needs. We choose the word *algorithm* over the word *process* to emphasize the consistency and repeatability of the DFSS approach. The DFSS algorithm is a high-level perspective of an iterative team-oriented process to design and embody solutions that consistently exceed expectations at Six Sigma quality level or higher. This vision can be accomplished by integrating design best practices, reducing design vulnerabilities using design axioms, and permitting a balance between creativity and discipline with accountability and flexibility. This algorithm provides the roadmap with tools and administrative details required for a smooth and successful initial DFSS deployment experience.

The flowchart presented in Fig. 5.1 depicts the proposed DFSS algorithm. The algorithm objective is to develop design entities with unprecedented customer wants, needs (expectations), and delights for its total life at Six Sigma quality level. This algorithm is based on the integrated theoretical frameworks of this book.

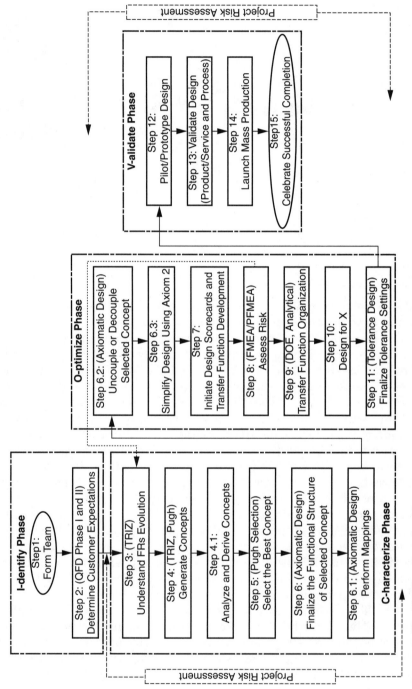

Figure 5.1 The DFSS project algorithm.

The DFSS algorithm has four phases:—I (identify), C (characterize), O (optimize), and V (validate) denoted ICOV as depicted in Fig. 5.1. The objective of this chapter is to mold DFSS principles, tools, and phases in a comprehensive implementable sequence in a manner that consolidates the advantages of both algorithmic and principle-based approaches and link the algorithm to the design entity total life cycle. The life cycle of a design entity, whether a product, a service, or a process, is usually formed and modeled using phases and tollgates. A "phase" represents a set of design activities of the project and is usually bounded by an entrance tollgate and an exit tollgate. A "tollgate" represents a milestone in the life cycle and has some formal meaning defined by the development process cascaded to the design team and recognized by management and other stakeholders. The life cycle of a designed entity, whether a product, a service, or a process, starts with some form of idea generation, whether in free invention format or using more disciplined creativity such as those surfaced by R&D departments. This is usually followed by several sequential activities. In the life cycle, the design process is followed by manufacturing or production activities followed by service, and aftermarket support. The design stages (phases) are described in Chap. 1.

In this algorithm, we emphasize the DFSS cross-functional team. A well-developed team has the potential to design winning Six Sigma–level solutions. Winning is contagious as successful design teams foster other DFSS teams. The growing synergy, which arises from ever-increasing numbers of successful teams, accelerates deployment throughout the company. The payback for small, upfront investments in team performance can be enormous. Continuous vigilance on the part of the black belt at improving and measuring team performance throughout the project life cycle will be rewarded with ever-increasing capability and commitment to deliver winning design solutions.

As depicted in Fig. 5.1, the process proposed here requires information and data to correctly formulate the objective of the design project. Correct formulation of the objective of the design project ranges from changes or modifications (incremental design) to very new design (creative design). In the algorithm presented here, the project option must be assessed on or before the conclusion of step 2 in a manner suggested by Fig. 5.2. Figure 5.3 is the step roadmap. If the results of the current step do not meet its objective, they might nevertheless prove useful if the objectives were wholly or partially changed. Accordingly, the degree of intensity in executing the different algorithm steps will vary. Occasional reference to either scenario will be highlighted when necessary.

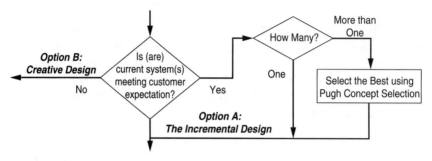

Figure 5.2 DFSS project tracks.

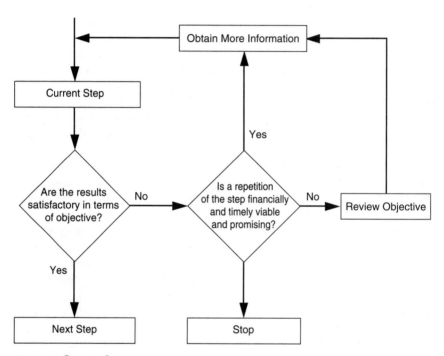

Figure 5.3 Step roadmap.

5.2   Form a Synergistic Design Team (DFSS Algorithm Step 1)

A cross-functional energetic design team is one of the ultimate means of any design effort. The success of design development activities depends on the performance of this team that is fully integrated with representation from internal and external (suppliers and customers) members. Special efforts may be necessary to create a multifunctional DFSS team that collaborates to achieve a shared project vision. Roles, responsibilities, membership, and resources are best defined upfront, collaboratively, by the teams. A key purpose of this step for the black belt is to establish the core team and get a good start, with clear direction from the champion and design owner. It is extremely important to "get it right" as early as possible to avoid costly downstream mistakes, problems, and delays.

Once the team is established, it is just as important for the black belt to maintain the team and continuously improve members' performance. This first step, therefore, is an ongoing effort throughout the DFSS ICOV cycle of the DFSS algorithm of planning, formulation, manufacturing, or production.

Design for Six Sigma teams usually share some common attributes such as corporate citizenship, passion for excellence in customer relations, systems engineering thinking with thorough knowledge about the design, and commitment for success.

The primary challenge for a design team is to learn and improve faster than their competitors. Lagging competitors must go faster to stay in the business race for customers. Leading competitors must go faster to stay on top. An energetic DFSS team should learn rapidly, not only about what needs to be done but also about how to do it— how to pervasively implement DFSS principles and tools with an adopted algorithm to achieve unprecedented customer satisfaction.

Learning without practicing is simply a gathering of information; it is not real learning. Real *learning* means gaining understanding, creating new thinking (breaking old paradigms), and understanding how to apply the new DFSS mental models in the big context—Six Sigma culture. No Six Sigma deploying company becomes world-class by simply knowing what is required, but rather by deploying and using the best contemporary DFSS methods on every project. Therefore, the team needs to project competitive performance through benchmarking of products and processes to help guide directions of change, use lessons learned to help identify areas for their improvement for scoping and selection, and use program and risk management best practices. The latter include finding and internalizing the best practices, developing deep expertise, and pervasively deploying the best practices throughout the project life cycle. This activity is key to achieving a winning rate of improvement by avoiding or eliminating risks. In addition, the team needs to apply design principles and systems thinking, specifically, thinking to capitalize on the total design (product, process, or service) in-depth knowledge.

5.3 Determine Customer Expectations (DFSS Algorithm Step 2)

The purpose of this step is to define and prioritize customer expectations and usage profiles, together with corporate, regulatory, and other internal company requirements. The emphasis here is placed on fostering deeper understanding of customers by enabling all design team members to learn by experiencing meaningful, direct engagements with customers.

Direct engagement with external and internal customers helps the DFSS team in interpreting customer satisfaction success parameters in increasing detail as the project progresses, thereby providing an ongoing reality check to help reduce expensive downstream design changes, scrap, and rework, thus avoiding the design hidden factory altogether. This direct engagement with customers will foster creativity and innovation, leading to unprecedented customer products.

The understanding of customer wants, needs, delights, and usage profiles; operating conditions; and environmental issues gives the DFSS team the information needed to design universal solutions. These overall attributes are called the "WHATs" array and will be referred to as such in discussion of the quality function deployment (QFD) methodology. The WHATs array is housed in the left room of the "house of quality" matrix of the QFD (see Chap. 7). Traditional design strategies tend to focus on the single aspect of eliminating dissatisfiers. In the customer domain, if the basic expectations are not satisfied, they become dissatisfiers. Six Sigma excellence mandates reaching beyond the market entry requirement of eliminating dissatisfiers to offering satisfiers and delighters to customers. The Kano model (Fig. 5.4), presented in Chap. 7, exhibits the relationships of dissatisfiers, satisfiers, and delighters to customer satisfaction and the importance of striving for unprecedented customer delight.

5.3.1 Research customer activities (DFSS algorithm step 2)

This is usually done by planning departments (product and process) or market research experts who should be represented in the DFSS team. The black belt and the DFSS team start by brainstorming all possible customer segments of the design. Use the affinity diagram method to group the brainstormed potential customer segments. The ultimate result is some grouping of markets, user types, or product/process applications types. From these groups, the DFSS team should work toward a list of clearly defined customer groups from which individuals can be selected.

Identify external and internal customers.   External customers might be service centers, independent sales/service organizations, regulatory agencies, and special societies. Merchants and, most importantly, the consumer (end user) should be included. The selection of external cus-

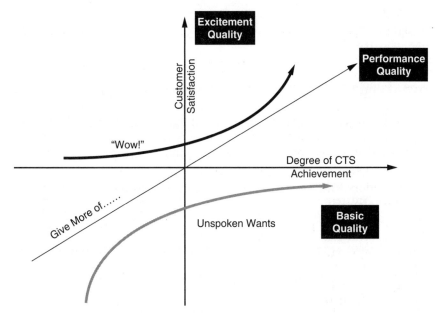

Figure 5.4   Kano model.

tomers must include existing and loyal customers and new conquest customers within the market segments. In the DFSS algorithm the objective is to design around key customer segments and try to include as many additional segments as possible. Internal customers might be in production and/or manufacturing, assembly, design services, display shops and other facilities, finance, employee relations, design groups, distribution organizations, and so on. Internal research might assist in selecting internal customer groups that would be most instrumental in identifying wants and needs in assembly and service operations.

5.3.2   Define the pursued (intended) ideal design from customer data (DFSS algorithm step 2)

The definition of "ideal" design is obtained by turning the knowledge gained from continuous monitoring of consumer trends, competitive benchmarking, and customer satisfiers and dissatisfiers into an initial definition of an ideal design. This will help identify areas for further research and allocate resources accordingly. The design should be described from the customer perspectives and should provide the first insight into what a good design may look like. This definition of customer-oriented ideal design will be detailed by concept methods such as TRIZ[*] (e.g., ideal final result) and axiomatic design[†] (e.g., axiom 1), which are good sources for evaluating consumer appeal and areas of likes or dislikes.

5.3.3 Understand the voice of the customer (DFSS algorithm step 2)

The identification of key customer design wants describes how the "voice of the customer" is collected and analyzed. A major step is listening to customers' capture wants and needs through focus groups, interviews, councils, field trials and observations, surveys, and so on. In addition, the team needs to analyze customer complaints and assign satisfaction performance ratings to design product and service attributes using a method called *quality function deployment* (QFD) (see Chap. 7). Market research is gathered in two ways: (1) through indirect information, obtained from surveys, questionnaires, competitive benchmarking and projections, consumer labs, trade journals, the media, and so on and (2) through direct customer engagement, including current, potential, and competitors' customers—from interviews, focus groups, customer councils, field observations and trials, and any other means appropriate.

Identify customer satisfaction attributes (DFSS algorithm step 2). *Attributes* are potential benefits that the customer could receive from the design and are characterized by qualitative and quantitative data. Each attribute is ranked according to its relative importance to the customer. This ranking is based on the customer's satisfaction with similar design entities featuring that attribute (incremental design case). A model recommended for data characterization was developed by Robert Klein [cited by Cohen (1995)]. Klein describes two ways to measure importance of customer wants and needs: direct method or by inference from other data. Attribute importance measured by the direct method is called "stated" importance. The method for inferring importance is conducted by measuring how strongly satisfaction with a specific attribute rates to overall design satisfaction. Attribute importance measured by this indirect method is called "revealed" importance. The Klein model uses both types of importance of each attribute to classify customer wants and needs into four quadrants (Fig. 5.5). This analysis identifies the key customer satisfaction attributes for further research studies.

---

[*]TRIZ is the Russian acronym for theory of inventive problem solving (TIPS), a systematic innovation technique developed by Genrich Altshuller of the former Soviet Union.

[†]A perspective design method developed by Professor N. P. Suh of MIT (Cambridge, Mass.). See Chap. 8.

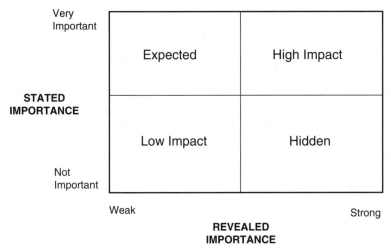

Figure 5.5 Klein model for customer satisfaction.

5.3.4 Categorize customer attributes into classes of wants, needs, and delights and map into critical-to-satisfaction (CTS) requirements: Phase 1 QFD (DFSS algorithm step 2)

The understanding of customer expectations (wants, needs), and delights by the team is a prerequisite to Six Sigma design development and is, therefore, the most important action in step 2. The fulfillment of these expectations and the provision of exciting delights will lead to satisfaction. This satisfaction will ultimately determine what products and services the customer will endorse and buy. In doing so, the DFSS team needs to identify constraints that limit the delivery of such satisfaction. Constraints present opportunities to exceed expectations and create delighters.

The identification of customer expectations is a vital step for the development of Six Sigma products and services that the customer will buy in preference to those of the leaping competitors. Noriaki Kano, a Japanese consultant, has developed a model relating design characteristics to customer satisfaction (Cohen 1995). This model (see Fig. 5.4) divides characteristics into three categories—dissatisfiers, satisfiers, and delighters—each of which affects customers differently. "Dissatisfiers" are also known as *basic, must-be,* or *expected* characteristics and can be defined as a characteristic which a customer takes for granted and causes dissatisfaction when it is missing. "Satisfiers" are also known as *performance, one-dimensional,* or *straight-line characteristics* and are defined as something the customer wants and expects; the more, the better. "Delighters" are features that exceed competitive offerings in creating unexpected, pleasant surprises. Not all customer satisfaction attributes are of equal importance. Some are more important to customers than others in subtly different ways. For example, dissatisfiers may not matter when they are met but may subtract from overall design satisfaction when they are not delivered.

The DFSS team should conduct a customer evaluation study. This is hard to do in creative design situations. Customer evaluation is conducted to assess how well the current or proposed design delivers on the needs and desires. The most frequently used method for this evaluation is to ask the customer (e.g., clinic or survey) how well the design project is meeting each customer's expectations. In order to beat the competition, the team must also understand the evaluation and performance of their toughest competition. In the planning matrix of the quality function deployment (QFD) method (Fig. 5.6), the team has the opportunity to grasp and compare, side by side, how well the current, proposed, or competitive design solutions are delivering on customer needs.

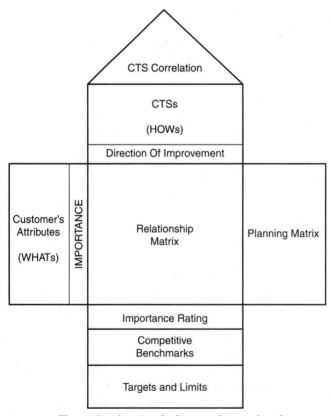

Figure 5.6 The quality function deployment house of quality.

The objective of the planning matrix evaluation is to broaden the team's strategic choices for setting customer performance goals. For example, armed with meaningful customer desires, the team could direct their efforts at either the strengths or weaknesses of best-in-class competitors, if any. In another choice, the team might explore other innovative avenues to gain competitive advantages.

5.3.5 Refine and prioritize customer wants, needs, and delights (DFSS algorithm step 2)

The objective of this step is to refine and prioritize customer wants, needs, and delights. The array of customer attributes should include all customer and regulatory requirements, together with social and environmental expectations. It is necessary to understand requirement and prioritization similarities and differences in order to understand what can be standardized (universally) and what needs to be tailored (locally).

Customer attributes and social, environmental, and other company wants can be refined in a matrix format for each identified market segment. These wants are also called the *WHATs* in the QFD literature. The *customer importance rating** is the main driver for assigning priorities from both the customer and the corporate perspectives, as obtained through direct or indirect engagement forms with the customer (see Sec. 5.3.3, subsection on identifying customer satisfaction attributes.)

Identify CTS array as related to the list of wants and needs for prioritization (DFSS algorithm step 2). The critical-to-satisfaction (CTS) array is an array of design features derived by the DFSS team to answer the WHATs array. The CTS array is also called the "HOWs" array. Each initial WHAT needs operational definition. The objective is to determine a set of critical-to-satisfaction requirements (CTSs) with which WHATs can be materialized. The answering activity translates customer expectations into design criteria such as speed, torque, and time to delivery. For each WHAT, there should be one or more HOWs describing a means of attaining customer satisfaction. For example, a "cool car" can be achieved through body style (different and new), seat design, legroom, lower noise, harshness, and vibration requirements. At this stage only overall requirements that can be measured and controlled need to be determined. These substitute for customer needs and expectations and are traditionally known as *substitute quality characteristics*. In this book, we will adopt the "critical to" terminology aligning with Six Sigma.

---

*Also known as *customer desirability index* (CDI). See Chap. 7 for more details.

Relationships between technical CTS and WHATs arrays are often used to prioritize customer wants and needs by filling the relationship matrix of QFD. For each customer want, the DFSS team has to assign a value that reflects the extent to which the defined CTS contributes to meeting the WHATs. This value, along with the importance index of the customer attribute, establishes the contribution of the CTSs to the overall customer satisfaction and can be used for prioritization.

The analysis of the relationships of customer wants and CTSs allows a comparison to other indirect information, which should be understood before prioritization can be finalized. The new information from the planning matrix in the QFD must be contrasted with the available design information (if any) to ensure that reasons for modification are understood. External customers on the DFSS team should be consulted to validate such changes and modifications. When customers interact with the team, delights often surfaced which neither would have independently conceived. Another source of delighters may emerge from team creativity as some features have the unintended result of becoming delights in the eyes of customers. Any design feature that fills a latent or hidden need is a delight, and with time, becomes a want. There are many means to create innovative delights by tools such as brainstorming, axiomatic design (Chap. 8), and TRIZ (Chap. 9). Delighters can be sought in areas of weakness, competition benchmarking, and technical, social, and strategic innovation.

5.3.6 Translating CTSs to functional requirements (FRs) (DFSS algorithm step 2)

The purpose of this step is to define a Six Sigma design in terms of customer expectations, benchmark projections, institutional knowledge, and interface management with other systems, and to translate this information into CTSs and later into technical functional requirements targets and specifications. This will facilitate the physical structure generation as proposed by the axiomatic design (Chap. 8) method. In addition, this step will provide a starting basis for the logical questions employed to define the physical structures of design.

A major reason for customer dissatisfaction is that the design specifications do not adequately link to customer use of the product or service. Often the specification is written after the design is completed. It may also be a copy of outdated specifications. This reality may be attributed to the current planned design practices that do not allocate activities and resources in areas of importance to customers and that waste resources by spending too much time in activities that provide marginal value, a gap that is nicely filled by the DFSS project algorithm. The approach is to spend time upfront understanding customer expectations and delights together with corporate and regulatory wants. This understanding is then translated into functional requirements (FRs) with design specifications (tolerances), which then cascade to all levels of design hierarchy. The power of first gaining complete understanding of requirements and then translating them into specifications is highlighted by Pugh (1991). This notion is also the basis of the strategy commonly associated with quality function deployment (QFD).

5.3.7   Map CTSs into functional requirements (FRs) (DFSS algorithm step 2)

The first formal mapping, in a QFD format of customer requirements to design characteristics was done in 1972 by the Kobe Shipyard of Mitsubishi Heavy Industries. This start led to the evolutionary development of the four phases of QFD. QFD phase 1 translates the customer needs and expectations into the CTSs. Subsequently, the CTSs must be converted into design actions. This conversion is completed by constructing QFD phase 2, a new house of quality, on which the WHATs are the CTSs and their target values from Fig. 5.6 (phase 1, house of quality). The HOWs and HOW MUCHs of each matrix are progressively deployed as WHATs on the charts or matrices that represent the next phase of the design development cycle. This conversion of HOWs to WHATs is continued from design planning to production planning.

While we recognize the mapping conducted in each of the four phases of QFD, we propose limiting the QFD exercise to only phases 1 and 2. We believe that the zigzagging method of axiomatic design is more powerful when armed with design axioms and vulnerability reduction techniques. Therefore, in the DFSS algorithm, we propose the following mappings as depicted in Fig. 5.7:

- Perform QFD phase 1 by mapping customer attributes to critical-to-satisfaction (CTS) requirements (step 2 of the DFSS algorithm).

- Perform QFD phase 2 by mapping CTS to functional requirements (FRs) (step 2 of the DFSS algorithm).

- Perform zigzag mapping of axiomatic design between the functional requirements (FRs) and design parameters (DPs) (step 6 of the DFSS algorithm).

- Perform zigzag mapping of axiomatic design between the design parameters (DPs) and the process variables (PVs) (step 6 of the DFSS algorithm).

The first mapping begins by considering the high-level customer attributes for the design. These are the true attributes, which define

104

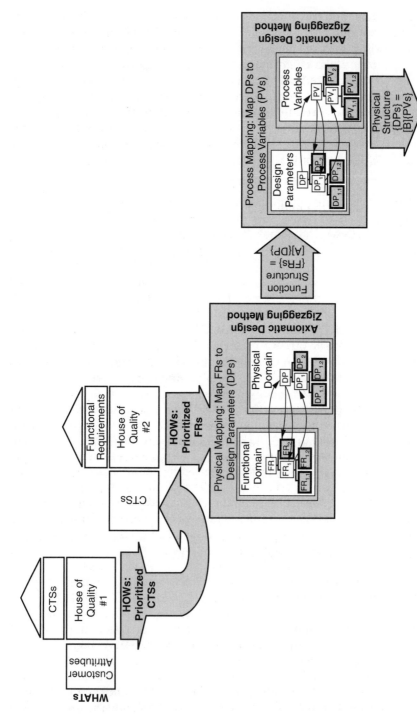

Figure 5.7 The design mappings in the DFSS algorithm.

what the customer would like if the design entity were ideal. This consideration of a product or service from a customer perspective must address the requirements from higher-level systems, internal customers (manufacturing/production, assembly, service, packaging, etc.), external customers, and regulatory legislation. True attributes are not directly operational in the world of the design teams. For this reason it is necessary to relate customer attributes to the CTSs and then to functional requirements that may be readily measured and, when properly targeted, will substitute or assure performance to the true quality of the attributes. The logic of a customer-to-design map is several levels deep, and a *tree diagram* is commonly used to create such logic (Cohen 1995).

In performing the mappings, the design team may start developing a testing matrix for validation and continue updating it as more details are achieved. They need to create tests that cover all customer attributes and eliminate unnecessary and redundant tests, specifically, a testing hidden factory.

5.3.8 Define FR specification target values and allowable variations (DFSS algorithm step 2)

Utilizing historical targets and variation provides an initial source of information in this step. Competitive benchmarking, usage profiles, and testing are useful tools to aid the DFSS team in understanding customer usage and competitive performance. It is also important to understand competition trends. The trend is vital because the team should set the design targets to beat what the competition will release, not what they have in the market now. On the basis of this information, the DFSS team selects the appropriate test target and allowable variation for each test. This selection is based on the team's understanding of the relationship matrix in the QFD so that the appropriate values may be chosen to satisfy design targets. Usually targets may be modified in light of customer studies. This involves verifying the target and variation with the actual customers. In some occasions, surrogates might be pieced together to measure customer reaction. In others, a meeting with internal customers may be necessary. Targets are tuned, and trade-off decisions are refined after assessing customer reaction. The preliminary specification may now be written. The DFSS team will select tests for the verification and in-process (ongoing) testing.

Step 2 actions are prerequisite to correctly proceed in the right path according to the design project classification, namely, incremental or creative. After mapping the classes of customer wants and needs to their corresponding CTSs, the cross-DFSS team needs to map the CTSs to FRs using the QFD methodology. The DFSS team then proceeds to check the availability of datum solutions that address the array of FRs.

The team will study the datum entities against the functional requirements generated by phase 2 QFD in order to check whether at least one solution, a design entity, exists that is the approximate physical translation of the functional requirements. If the answer is "Yes" (Fig. 5.2), then the selected entity should be within a slight variation from the pursued design. The team may declare the project as an incremental design problem and work toward improvements in order to satisfy the customer requirements, progressing from the datum design as a starting point. Adding, replacing, or eliminating design parameters (DPs) without altering the FRs is our definition of an "incremental design." The "creative design" includes alterations made in the incremental design case plus alteration made to the FR array. In the incremental design scenario, if more than one datum entity exists, the best entity could be selected using the Pugh concept selection method. In the absence of datum entities, the only option is the creative design, which requires more conceptualizing work and, therefore, more extensive deployment of TRIZ and axiomatic design methods.

In summary, the objective of specifying nominal and tolerances of the FRs and the DPs is to verify structure choices for functional solution entity elements and interfaces. A *structure* can be defined as an input-output or cause-and-effect relationship of functional elements. Mathematically, it can be captured by design mappings such as QFD and the zigzagging method of axiomatic design. Graphically, it is depicted in a "block" diagram that is composed from nodes connected by arrows depicting the relationships. A structure should capture all design elements within the scope and ensure correct flowdown to critical parameters. A structure is captured mathematically using mapping matrices, and matrices belonging to the same hierarchical level are clustered together. Hierarchy is built by the decomposing design into a number of simpler functional design matrices that collectively meet the high-level functional requirements identified in step 2 of the DFSS algorithm. Two structures are recognized in the DFSS algorithm:

- The physical structure between the functional requirements (FRs) and the design parameters (DPs)

- The process structure between the DPs and the process variables (PVs)

The *physical structure* is usually developed first to define the design concept. Preliminary concepts are best selected using the Pugh selection method. The preliminary work to verify structural choices should help the DFSS team get started on concept generation. The team needs to select the best solution entity element technologies in terms of design parameters (DPs) to meet or exceed requirements. Technology and structure choices are sometimes closely interlinked via the physical and process mappings when conducted following design axioms. New technologies (DPs) that can enable new structures and different technologies may suggest different mapping of functional requirements (FRs). The pursuit of linked technology and structure options may reveal new opportunities for delighters. Conversely, because axiom-driven structures often have very long lifespans, they need to be relatively insensitive to technology choices. An axiom-driven structure should enable reimplementation of new technologies without undue impact on either the structure or other mapping (portions) of the design. Therefore, to assure the insensitivity of the structure to future unknown technology, they need to be derived using design axioms. It is wise to examine the robustness of a structure against current, known technology and design alternatives. Structures need to be sturdy against customer use, misuse, and abuse; errors in requirements and specifications; unanticipated interactions with other portions of the solution entity; or process variations. The functional requirements should be verified over a range of operational parameters which exceed known requirements and specifications. This may require sensitivity studies in a form of a classical DOE (design of experiment) or Taguchi's parameter design. Determining sensitivity of design element performance due to changes in operating conditions (including local environment and solution entity interactions) over the expected operating range is an essential task for transfer function optimization within the DFSS algorithm (Sec. 5.10).

As preferred and optimized choices are made for physical structure including ideal and transfer functions (Chap. 6), the requirements cascade should be reverified to assure that the high-level requirements and Six Sigma specifications are met. A powerful approach to reverification of the requirements cascade is to compare the top-down mapping specification cascade with a bottom-up capability of achievement. The approach seeks to synthesize the functional requirements specifications established for the individual solution entity elements (components and subsystems) into specification for the next higher-level solution entity element. The synthesized specification is assessed against cascaded requirements and specifications for the higher-level solution entity element. The synthesis is repeated for each hierarchical level of design; components to subsystem, subsystem to system, and further specifications are evaluated against higher-level requirements and specifications established during the top-down mapping process identify specific regions which don't meet the Six-Sigma-level specification for further study.

5.3.9   Exiting step 2

What needs to be done, from the DFSS viewpoint, is to make sure that functional requirements (FRs) derived from the phase 2 QFD are optimized. Effectiveness and efficiency can be achieved by having the entire DFSS team plan the next design tasks in the DFSS algorithm upfront. Activities that should be considered include physical and process structure development, optimization and transfer function development, tests and designed experiments, validation, reliability methods, and other DFSS algorithm activities. The team needs to map the relationship of each activity in the DFSS algorithm to translated design (product, service, and process) characteristics. A matrix is created to compare the list of brainstormed design activities to the list of translated and implementable actions. Following QFD ratings, the relationships are designated as a "none" relationship, a "weak" relationship, a "medium" relationship, or a "strong" relationship. In addition, the design list of activities may be improved by adding steps to cover missing or weak areas of the matrix in the original list and deleting redundant activities that are already sufficiently covered by other planned design activities. The DFSS team may consider combining activities to gain efficiency.

Once the engineering activities are planned, workload and timing can be established and resources allocated. Recommended thoughts in this context include

- Use a project management approach such as the critical path method (CPM) in planning. The DFSS team needs to map out a sequence of planned events, determine which events must be done in series and/or in parallel, and identify the critical path. Project management software is ideal for such a task.

- Understand timing, resource constraints, and milestones. Workload cannot be established until the timing resource is understood. Constraints with respect to resources and budget also need to be assessed, as well as the requirements for various milestones in the development cycle. In crisp situations, the project evaluation and review technique (PERT) may be used to document the different activities of the project plan with latest and earliest start and finish times for each activity with no slippage on the projected completion date. A project Gantt chart with schedule, events, and DFSS team responsibilities should be detailed. Milestones are determined, agreed on, and synchronized with the company design development process. Milestones serve as communication "tollgates" to ensure that team review and update management about their progress prior to getting approval to proceed to the next step of the DFSS algorithm. Typically, milestone review distribution is usually dense toward project closure, a paradigm that should be changed when the DFSS algorithm is employed. These reviews should not be the only communication channel to the outside. Informal communication is found to be very beneficial to lessen the pressure of milestone deadlines and other internal reviews.

- Estimate workload associated with the DFSS algorithm activity. Design workload is now estimated with reference to the planned activities with the required timing.

- Allocate resources for the various activities.

5.4   Understand Functional Requirements Evolution (DFSS Algorithm Step 3)

Design solution entities evolution follows certain basic patterns of development. On the availability of historical data, evolutionary trends of the functional requirements (FR) performance can be plotted over time and have been found to evolve as an S curve (Fig. 5.8), a TRIZ concept. The *theory of inventive problem solving* (TIPS) (Russian acronym) / TRIZ) is a valuable methodology for gaining understanding and making projections about technical evolution (Alexander 1964; Altshuller 1988, 1991; Tsourikov 1993; Dovoino 1993).

This knowledge can be used by the DFSS team to predict logical next stages of development in the multigeneration plan of their design and to form opinions regarding the limitations of the current

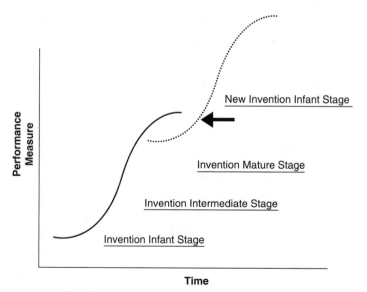

Figure 5.8   The S curve of evolution.

data, if any. The team needs to list historical breakthroughs in technology (i.e., technology mutation) and compare their design with generic design evolution. Therefore, they will relate technological breakthroughs with evolutionary improvements. Their study of the type of improvements will assist them in identifying the stage of development within their project. The relevant information can be accessed through literature and patent searches* together with benchmarking of best-in-class, competitive and noncompetitive companies. In addition, brainstorming of how generic evolutionary principles apply to the Six Sigma design elements is particularly useful. This activity involves the study of established technical paths of evolution to anticipate the future. In all cases, the compatibility of generic evolutionary principles to the project's applications needs to be evaluated.

5.5   Generate Concepts (DFSS Algorithm Step 4)

The DFSS team should develop a matrix to enable the evaluation of the alternative concepts against the FR array and other criteria selected by the team. The matrix provides structure and control to the process of analysis, generation, and evaluation of the project solutions. The (vertical) columns of the matrix are the criteria for evaluating these ideas or concepts in a visual and user-friendly fashion (e.g., schematics and sketches) while the (horizontal) rows are the criteria. The evaluation matrix will be used to justify that the best concept has been selected and to justify why certain solution entities are inferior and should be discarded. Conceptual entities should be detailed with sufficient clarity to ensure consistency in understanding by all team members. Additional clarity may be gained from word descriptions and modeling. All concepts should be presented at the same level of detail. Alternatives should be titled and numbered for ease of reference.

After the array of functional requirements (FRs) has been determined, different alternatives of solution entity are generated. These entities represent the physical translation of the functions defined in the functional domain. Alternatives are formed by the analysis and synthesis activities. *Synthesis* in this case means selecting a feasible structure where a function is physically mapped into possibly different entity. The techniques useful in idea generation and synthesis include analogy, brainstorming, combination, and evolution.

---

*TRIZ is very helpful in developing industry-specific patent database for future reference.

5.5.1 Derive and analyze concepts (DFSS algorithm step 4)

The key mechanisms for arriving at the best possible concept design or process solution entity are (1) axiomatic design, (2) TRIZ methodology, and (3) the method of "controlled convergence," which was developed by Dr. Stuart Pugh (1991) as part of his solution selection process.

Controlled convergence is a solution iterative selection process that allows alternate convergent (analytical) and divergent (synthetic) thinking to be experienced by the team. The method alternates between generation and selection activities (Fig. 5.9). We suggest the following enhancement to the controlled convergence method:

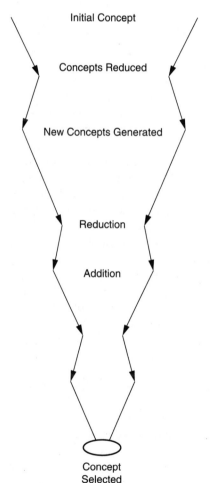

Figure 5.9 Controlled convergence method.

1. The "generation" activity can be enriched by the deployment of design axiom 1 and its entire derived theoretical framework, which calls for functional requirements independence. This deployment will be further enhanced by many TRIZ methodology concepts to resolve design vulnerabilities where applicable.

2. The "selection" activity can be enhanced by the deployment of axiom 2, which calls for design simplicity.

The controlled convergence method uses comparison of each alternative solution entity against a reference datum. Evaluation of a single solution entity is more subjective than objective. However, the method discourages promotion of ideas based on opinion and thus promotes objectivity. The controlled convergence method prevents adverse features and eliminates weak concepts, thereby facilitating the emergence of new concepts. It illuminates the best solution entity as the one most likely to meet the constraints and requirements of the customer as expressed by the design specification, and the one which is least vulnerable to immediate competition.

The development of the concepts through the combination of solution alternatives per functional requirement can be identified by a matrix technique called the *synthesis matrix*. In this matrix, the functional requirements (FRs) are listed in the rows and the solution alternatives (the design parameters) are laid down in the columns. At this step, the design parameters are usually known at a hierarchal level equivalent to components, subsystem, and subprocesses or in terms of physical effects (e.g., electric field). However, this knowledge is not detailed at this stage. The functional requirements need to be listed in the order of their hierarchy by the team, to the best of their knowledge at this step, and should be grouped according to their type of input (energy type, material type, information type).

The concepts are synthesized and generated from all possible feasible combinations of all possible design parameters (DPs) per functional requirement (FR) in the synthesis matrix. A feasible design is identified by connecting all possible solutions using arrows between the design parameters. The arrows can be connected only when the team is technically confident about the functional and production feasibility. For example, in Fig. 5.10, two concepts can be identified. Solutions for which the number of arrows is less than the number of rows are infeasible situations.

In conducting this exercise, the team will identify all possible feasible design solutions. In the next step, guided by their knowledge and DFSS algorithm, the team should concentrate only on promising solutions. The challenge here is to ensure that the physical-functional compatibility and other constraints are met and the appropriate flow

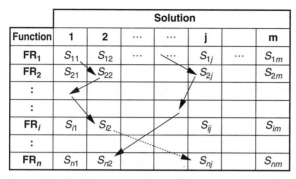

Function	\multicolumn{6}{c}{Solution}
	1	2	···	···	j		m
FR_1	S_{11}	S_{12}	···	····	S_{1j}	···	S_{1m}
FR_2	S_{21}	S_{22}			S_{2j}		S_{2m}
:							
:							
FR_i	S_{i1}	S_{i2}			S_{ij}		S_{im}
:							
FR_n	S_{n1}	S_{n2}			S_{nj}		S_{nm}

Table Legend

FR_i: Function indexed i in the functional structure

S_j: The physical solution of solution j

S_{ij}: The solution entity of group j that physically translate function i (e.g., hardware, software, field effect)

Figure 5.10 The design synthesis matrix.

of energy, material, and information is properly identified. These are the requirements to conceive sound design structure. A *structure* is a description of the design in the concerned mapping (see Sec. 5.3.8 and Chap. 8). The design is first identified in terms of its FRs and then progressively detailed in terms of its lower-level functional requirements and design parameters in the form of design matrices (mappings). This hierarchy is an output of the zigzagging method employment in the design structure detailing task. Normally, each functional requirement can be delivered by several possible DPs in a given structure hierarchical level within the structure. Therefore, the synthesis matrix exercise should be conducted at all levels of design structure.

Assume that we have a design array of n FRs, and FR_i is the ith row in the array. In addition, assume that an arbitrary functional requirement, say, FR_i (where $i = 1,2,3,...,n$) can be delivered physically by $j = 1,2,...,m_i$ design parameters (DPs). A synthesis matrix cell, say, S_{ij}, in the matrix is the design parameter indexed j, DP_j, of functional requirement indexed i, FR_i. The identification of all possible alternative solutions (DPs) per a functional requirement may be facilitated by the use of the morphological approaches of Zwicky (1984) and TRIZ methodology (Chap. 9).

Several feasible high-level and undetailed concepts are usually generated using the synthesis matrix. This generation of multiple concepts poses a selection problem, specifically, which concept to select for further detailing in the DFSS algorithm. The DFSS team must select the *best* concept using the Pugh concept selection method.*

---

*The concept selection problem was formulated by El-Haik and Yang (2000a, 2000b) as an optimization problem. They provided an integer programming formulation, both fuzzy and crisp, for the concept selection problem and employed design axioms as selection criteria.

5.6   Select the Best Concept (DFSS Algorithm Step 5)

In this step, the DFSS team produces the convergence on the best concept in iterative steps that are designed to be performed with DFSS discipline and rigor. The following sequence may be used to facilitate the convergence to the best concept by the DFSS team:

- List criteria on the rows of the Pugh matrix [functional requirements (FRs) array; constraints, regulatory and legal requirements from phase 2 QFD mapping]. These criteria should be measurable and defined with common understanding by all members of the team.

- List concepts on the columns of the Pugh matrix as obtained from the synthesis matrix.

- Choose a datum design with which all other concepts are to be compared from the alternative entities. The datum could be an existing baseline, as is the case of incremental design. In creative design situations, the datum could be any concept that the team may generate from the synthesis matrix. Evaluate concepts against the defined criteria. Use a numbering system rather than the traditional evaluation of plus (+) and minus (−). The datum will be the neutral element(s) of the numbering system chosen. For comparing each solution entity against the datum, rate either as plus (+), meaning better than the datum; or minus (−), meaning worse than the datum; or same (s), meaning same as the datum (see Fig. 5.11).

- Perform trade-off studies to generate alternatives using design axioms and TRIZ. Look at the negatives. What is needed in the design to reverse the negative (relative to the datum)? Will the improvement reverse one or more of the existing positives due to design *coupling?* If possible, introduce the modified solution entity into the matrix and retain the original solution entity in the matrix for reference purposes. Eliminate truly weak concepts from the matrix. This will reduce the matrix size. See if strong concepts begin to emerge from the matrix. If it appears that there is an overall uniformity of strength, this will indicate one of two conditions or a mixture of both. The criteria are ambiguous and, hence, subject to mixed interpretation by the DFSS team. Uniformity of one or more of the concepts suggests that they may be subsets of the others (i.e., they are not distinct). In this case, the matrix cannot distinguish where none exists.

- Having scored the concepts relative to the datum, sum the ranks across all criteria to get plus (+), minus (−), and (s) values. These scores must not be treated as absolute as they are for guidance only and as such must not be summed algebraically. Certain concepts will exhibit relative strengths, while others will demonstrate relative weaknesses.

Concepts FRs	 A	 B	 C	...	 K
FR_1	s	−2	+3		−1
FR_2	s	−2	−2		−2
:	s	−1	+1		+3
:	s	+3			+3
					−1
FR_m	s				
Total (−)					
Total (+)					

Figure 5.11 Pugh generation-selection matrix.

- Select the best concept with the maximum number of plus signs and minimum number of minus signs.

5.7   Finalize the Physical Structure of the Selected Concept (DFSS Algorithm Step 6)

The first step in design detailing is to develop the physical structure that determines the opportunity to capture the "maximum potential for customer satisfaction" defined in step 2. The purpose of the physical structural definition is to establish an enabler to subsequent concept and detail design efforts to realize this maximum potential. The axiomatic design method provides the zigzagging process as the means to define physical and process structures. The structure is captured mathematically using mapping matrices with matrices belonging to the same hierarchical level clustered together. Hierarchy is built by the decomposing design into a number of simpler functional design matrices that collectively meet the high-level functional requirements obtained from phase 2 QFD. The array of FRs should be checked for independence, that is, that they are different and distinct. For example, speed and torque are independent functional requirements, although they are constrained by physics. This requirement is needed because it forms a minimum array to design for that will have the potential to satisfy design requirements. Extra functional requirements may not be demanded by the customer and will result in either overdesign or poor value proposition to the customer. The collection of all design matrices obtained up to the detailing level that is satisfactory to the team forms the structure. The structure provides a means to track the chain of effects for design changes as they propagate across the design. The zigzagging process starts by using the minimum set of functional requirements that deliver the design tasks as defined by the customer and obtained from phase 2 QFD. The zigzagging process is guided by the creative and heuristic process of functional requirements definition through logical questions offered by the zigzagging method (Chap. 8). This structural definition is judged by following design axioms:

Axiom 1: The Independence Axiom   Maintain the independence of the functional requirements.

Axiom 2: The Information Axiom   Minimize the information content in a design.

After satisfying the independence axiom, design simplicity is pursued by minimizing the information contents per axiom 2. In this context, information content is defined as a measure of complexity and is related to the probability of successfully manufacturing (producing) the design as intended. Because of ignorance and other inhibitors, the exact deployment of design axiom might be concurrently infeasible as a result of technological and cost limitations. Under these circumstances, different degrees of conceptual vulnerabilities are established in the measures (criteria) related to the unsatisfied axiom. For example, a degree of *coupling* may be created because of axiom 1 violation and complexity due to violation of axiom 2 (Fig. 5.12).

A conceptually weak design may have limited potential to succeed in the use environment. The DFSS team should avoid or at least weaken coupling through creative selection of the DPs. Highly coupled concepts usually exhibit technical bottlenecks and conflicts, which on the positive thinking side offer opportunity source for innovation. The design and process failure mode–effect analysis (DPFMEA) exercise identifies the design vulnerabilities as failure causes (Chap. 11).

5.7.1   Perform mappings (DFSS algorithm step 6)

The mapping tool proposed here is the zigzagging process of axiomatic design. The primary intention of its use is to fulfill the minimum set of functions that in turn deliver the customer, corporate, and regulatory attributes. It is necessary to discuss and compare various solutions for the functional requirements identified in the physical structure as well as the methods of combining them to form a conceptual physical structure. This will decide both the feasibility and com-

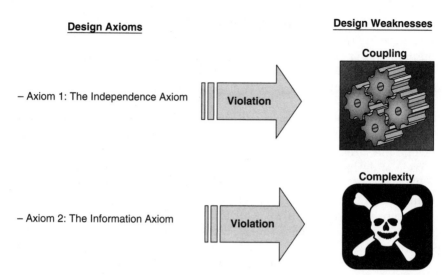

Figure 5.12 Vulnerabilities created by violation of the design axiom.

patibility of alternative solutions by narrowing down the conceptually feasible solutions to practically possible solutions of the preliminary structure.

Functional analysis and physical synthesis are the premier activities performed in the C (characterize) phase of the DFSS algorithm. At this stage of the project and after completing the detailing of the physical structure, the team should leap forward to perform the process mapping, namely, process design.

The team needs to seek big ideas for competitive advantage and customer delight, challenge conventional baseline physical structures with innovative ideas, capitalize on design parameter new technologies, and cast preliminary definition of functional requirements (FRs); design parameters and process variables; and minimize vulnerabilities. In so doing, the team seeks to uncouple and simplify both structures. In this step, the black belt should foster an environment in which "out of the box" thinking and brainstorming are encouraged utilizing the conceptual methods offered by the DFSS algorithm. When winnowing ideas, the black belt should foster a more structured, disciplined environment for the team as well as iterate back and forth between expansion and contraction of ideas.

5.7.2 Uncouple or decouple selected concepts (DFSS algorithm step 6)

The design process involves three mappings between four domains (Fig. 5.13). The first mapping involves the mapping between customer attributes and *critical-to-satisfaction* (CTS) metrics, followed by the second mapping from CTSs to the functional requirements (FRs). This mapping is very critical as it yields the definition of the high-level minimum set of functional requirements needed to accomplish the design value from the customer perspective. Once the minimum set of FRs are defined, the *physical mapping* should start. This mapping involves the FR domain and the codomain design parameter (DP). It represents the structuring activity and can be represented mathematically by design matrices (mapping) or, graphically, by block diagrams. The collection of design matrices forms the conceptual structure.

The *process mapping* is the last mapping and involves the DP domain and the process variables (PV) codomain. This mapping can be represented by matrices as well and provides the process structure needed to translate the DPs to PVs in manufacturing or production environments.

The zigzagging process of axiomatic design is a conceptual modeling technique that reveals how the design is *coupled* in the FRs. In the physical mapping, the FRs are the array of responses denoted as the array **y**.* Coupling of the FRs is a design weakness that negatively affects controllability and adjustability of the design entity. *Coupling* can be defined as the degree of lack of independence between the FRs, propagates over the design mappings, and limits the potential for Six Sigma design. Uncoupled designs not only are desirable in terms of controllability, quality, and robustness standpoints but also have a potential for high probability of producibility, namely, reduced defect per opportunity (DPO).

When a design matrix is square, where the number of FRs equals the number of DPs, and diagonal, the design is called *independent* or *uncoupled*. An uncoupled design is a one-to-one mapping.

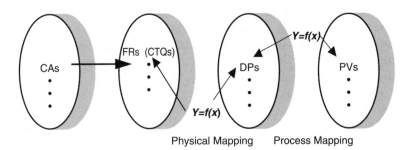

Physical Mapping   Process Mapping

Figure 5.13 The design process mappings.

---

*Mathematical bolded notation implies an array or a matrix.

In a *decoupled* design case, a design matrix of *lower/upper triangle* DPs is adjusted in some sequence conveyed by the matrix. Uncoupled and decoupled design entities possess conceptual robustness, where the DPs can be changed to affect intended requirements only without readjustment of any unintended functional requirements. Definitely, a coupled design results when the matrix has the number of requirements greater than the number of DPs. The coupled design may be uncoupled or decoupled by "smartly" adding extra DPs to the structure. Uncoupling or decoupling is an activity that is paced with structure detailing and can be dealt with using axiomatic design theorems and corollaries. Uncoupled and decoupled designs have higher potentials to achieve Six Sigma capability in all FRs than do coupled designs. Design for Six Sigma in the conceptual sense is defined as having an overall uncoupled or decoupled design by conducting the process mapping and physical mapping concurrently by the team.

5.7.3 Simplify design using axiom 2 (DFSS algorithm step 6)

After maintaining independence per axiom 1, the DFSS team should select the design with the least information content. The less information specified to manufacture or produce the design, the less complex it is; hence, information measures are measures of complexity. In general, "complexity" is defined as a quality of designed entities. Complexity in design has many facets, including the lack of transparency of the transfer functions between inputs and outputs in the physical structure, the relative difficulty of employed physical and transactional processes, and the relatively large number of assemblies, processes, and components involved (Phal and Beitz 1988). In Chap. 8, we explore different techniques to simplify the design. For now, suffice it to say that the number, variance, and correlation relationships of the design elements are components of design complexity.

5.8 Initiate Design Scorecards and Transfer Function Development (DFSS Algorithm Step 7)

The functional requirements in the physical structure can be further detailed by design *scorecards* and *transfer functions,* two unique concepts of the DFSS algorithm (Chap. 6). The transfer function is the means for dialing customer satisfaction and can be initially identified via the different design mappings. A transfer function is a relationship, preferably mathematical, in the concerned mapping linking controllable and uncontrollable factors. Transfer functions can be derived, empirically obtained from a DOE, or regressed using historical data.

In some cases, no closed mathematical formula can be obtained and the DFSS can resort to modeling. In the DFSS algorithm, there is a transfer function for every functional requirement, for every design parameter, for every process variable, and ultimately for every CTS and customer attribute (Fig. 5.14). Transfer functions are captured in the mappings **{FRs}** = [A] {DPs} and **{DPs}** = [B] {PVs} in Figs. 5.7

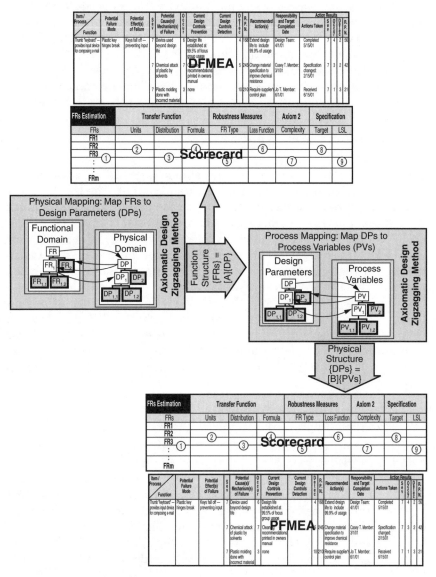

Figure 5.14   Design scorecards for transfer functions and FMEA.

and 5.13. The dependent variables in the transfer functions are optimized by either shifting their means, reducing their variations, or both. This can be achieved by adjusting their mapped-to-independent variables means and variance. This optimization propagates to the customer domain via the other high-level transfer functions in the design mappings resulting in increased customer satisfaction.

Design scorecard documents record and assess quantitatively the DFSS project progress, store the learning process, and exhibit all critical elements and performance of a design (CTSs, FRs, DPs, and PVs). Scorecards have many benefits, including showing gaps for improvements relative to customer attributes, documenting transfer functions and design optimization, predicting final results, enabling communication among all stakeholders in the project, and evaluating how well the design is supported by manufacturing and production processes at component and subassembly quality levels. We suggest documenting the transfer functions belonging to the same design hierarchy in the mapping in one scorecard, thus avoiding having one for each design element or requirement in the physical structure.

5.9   Assess Risk Using DFMEA/PFMEA (DFSS Algorithm Step 8)

An FMEA can be described as a systemized group of activities intended to

1. Recognize and evaluate the potential failures of a design and its effect.

2. Identify actions which could eliminate or reduce the chance of the potential failure occurring.

3. Document the process.

It is complementary to the design process to define positively what a design must do to satisfy the customer.

The failure mode–effect analysis (FMEA) [see AIAG (2001)] helps the DFSS team improve their project's product and process by asking "What can go wrong?" and "Where can variation come from?" Design and manufacturing or production, assembly, delivery, and other service processes are then revised to prevent occurrence of failure modes and to reduce variation. Specifically, the teams should study and completely understand physical and process structures as well as the suggested process mapping. This study should include past warranty experience, design and process functions, customer expectations and delights, functional requirements, drawings and specifications, and process steps. For each functional requirement (FR) and manufacturing/assembly process, the team asks "What can go wrong?" Possible design and process failure modes and sources of potential variation in manufacturing, assembly, delivery, and services processes should be determined. FMEA considerations include the variations in customer usage; potential causes of deterioration over useful design life; and potential process issues such as missed tags or steps, package and shipping concerns, and service misdiagnosis. The team should modify the design and processes to prevent "wrong things" from happening and involve the development of strategies to deal with different situations, the redesign of processes to reduce variation, and errorproofing (poka-yoke) of designs and processes. Efforts to anticipate failure modes and sources of variation are iterative. This action continues as the team strives to further improve the product design and processes.

We suggest using the FMEA concept to analyze systems and subsystems in the early concept and design stages. The focus is on potential failure modes associated with the functions of a system caused by the design. Design FMEA (DFMEA) is used to analyze designs before they are released to production. In the DFSS algorithm, a DFMEA should always be completed well in advance of a prototype build. The input to DFMEA is the array of FRs. The outputs are (1) list of actions to prevent causes or to detect failure modes and (2) history of actions taken and future activity.

Process FMEA (PFMEA) is used to analyze manufacturing, assembly, or any other processes. The focus is on process inputs. Software FMEA documents and addresses failure modes associated with software functions.

5.9.1 The Use of FMEA and its links in the DFSS algorithm (DFSS algorithm step 8)

Failure management using the FMEA is an iterative process that promotes system and concurrent engineering thinking. The most prominent failures are related to functional requirements. Since components, subsystems, and systems have their own FRs, the FMEA development should be paced with structure detailing.

FMEA can be easily linked to other tools in the DFSS algorithm, such as the P-diagram (process diagram; see Fig. 5.15), fishbone diagram, physical and process structures, transfer functions, process mappings, and DOE, both classical and robust design. Among these tools, the P-diagram deserves more attention as the newly introduced tool that was skipped in previous chapters. The P-diagram is a robust design tool used to summarize and capture inputs and outputs of a scoped design or process entity. It distinguishes between factors on which the DFSS team has control, the DPs at different hierarchal levels, and factors that they can't control or wish not to control because of technology or cost inhibitors, the "noise" factors. Noise factors cause design failures and do so not only through their mean effects but also via their interaction with the design parameters. A "failure" is the unplanned occurrence that causes the system or component not to deliver one or more of its FRs under the specified operating conditions. The noise factors can be categorized as environmental, unit-to-unit sources (manufacturing or production), coupling with other systems, customer usage, and deterioration (wear) factors. The P-diagram is a DFSS structure tool used to identify intended inputs and FRs for a system, the noise factors; the design parameters, transfer function including the *ideal function*; and failure modes. The P-diagram helps the DFSS team in assessing the causes, modes, effects, and controls of the failure and recommended actions.

The P-diagram depicted in Fig. 5.15 introduces another category of inputs: the signal array. The signal factors are usually exerted by the customer to excite the system causing the DPs to deliver the FRs in the physical structure or the PVs to deliver the DPs in the process structure. A nonzero signal array indicates the "dynamic" formulation in robustness design methodology; specifically, the FRs array follows the signal array and DPs are set to minimize any differences between the two arrays. The relationship between the physical structure, transfer function, and DFMEA is depicted in Fig. 5.16.[*]

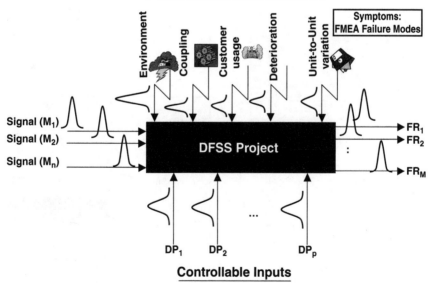

Figure 5.15 P-diagram.

---

[*]In Fig. 5.16, notice that the same relationship exists between the process structure or process mapping, the process transfer functions {**DPs**} = [**B**] {**PVs**}, and the PFMEA.

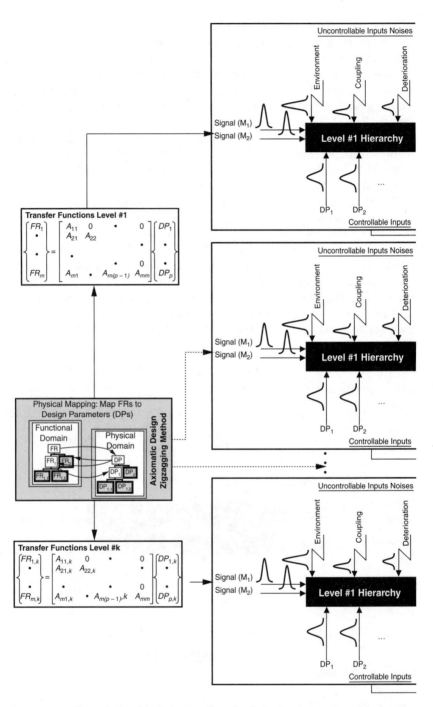

Figure 5.16 The relationship between the physical structures, transfer function array, and DFMEA.

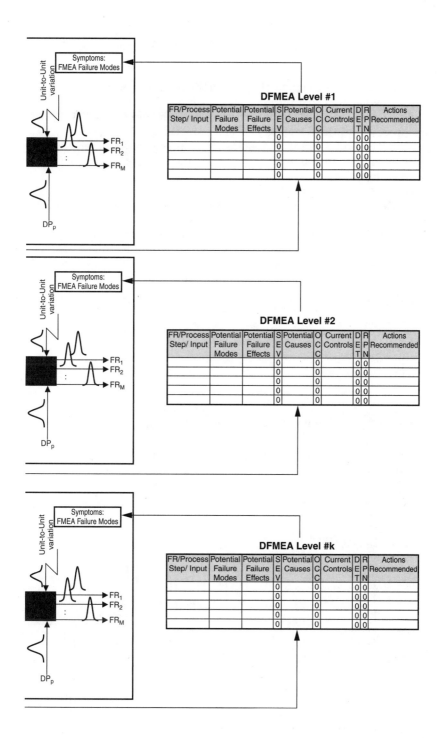

Let us illustrate the concepts in this step through the following example.

Example 5.1. The P-Diagram of the Press-Fit Process   A press fit is accomplished by forcing a shaft into a hole (in this study the pulley hub) that is slightly smaller than the shaft by relying on the materials' elasticity during the process and their tendency to maintain the old dimensions after the process. This will produce the grip that holds both parts together. The press fit is considered the simplest assembly method when sufficient hydraulic press power is available.

The press-fit joint consists of three components: the shaft, the pulley, and the pulley hub (Fig. 5.17). The shaft outside diameter is larger than the pulley hub inside diameter. Design books recommend some *allowance* for press fit over a specific range. In general, it is defined as the minimum clearance space between mating parts. Interference is a negative allowance. This interference creates the holding mechanism that joins the pulley to the shaft. Tolerance, on the other hand, is the permissible variation in the size of a part.

The P-diagram of the process as depicted in Fig. 5.16 includes the following components:

- *Customer intent*—what the customer wants
- *Signal*—the translation of the customer's intent to measurable engineering terms
- *Perceived result.*—what the customer gets
- *Response*—the translation of the customer's perceived result into measurable performance data in engineering terms

Transfer function and ideal function.   The *transfer function* is a relationship that relates all inputs to a given system to the outputs. In

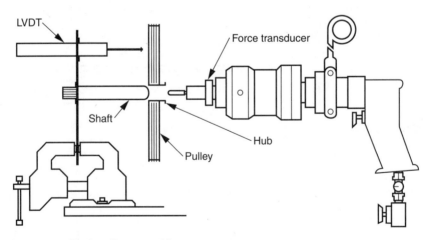

Figure 5.17 Shaft-pulley assembly.

terms of the P-diagram, this can be depicted as FR = f (signal factors, noise factors, design parameters) in the physical mapping and DP = f (signal factors, noise factors, process variables) in the process mapping. The zigzagging process helps the team identify the partial transfer function FR (DP) = f (signal factors, design parameters) in the physical structure. The P-diagram enables the teams to complement this function with the noise factors. At constant DP or PV settings and zero-noise-factor effects, the relationship FR or DP = f (signal factors) is called the "ideal function" in robust design methodology. The ideal function is a description of "how the system works if it performs its intended function perfectly." In design terms, the ideal function is a mathematical description of the energy transformation relationship between the signal and the response. The reason for studying the ideal function is to have a physics-based mathematical model of the system under consideration before testing. This allows the team to evaluate various control factor levels in spite of the presence of noise factors.

The ideal function of the press-fit process is

$$F_p = Z_0 \Delta_r^2 \qquad (5.1)$$

where F_p = press force (a measurable DP in the process structure)
Z_0 = joint material stiffness coefficient (accounts for pulley geometry and material properties)
Δ_r = relative interference (signal) , calculated as

$$\Delta_r = \frac{\Delta}{\mathrm{OD}_{shaft}} = \frac{\mathrm{OD}_{shaft} - \mathrm{ID}_{pulley}}{\mathrm{OD}_{shaft}} = \text{signal (M)} \qquad (5.2)$$

In other words, we present the process with some relative interference between the hub and the shaft, the signal, and using the process, we get a joint force holding both components together. The transfer function can be written as shown in Eq. (5.3). The exact mathematical transfer function relationship can be found empirically through a DOE as no equation is readily available in the literature to the author's knowledge.

$$F_p = f(\Delta_r^2; \qquad \text{signal} \qquad (5.3)$$

$$\left.\begin{array}{l} \text{Perpendicularity shaft to end} \\ \text{perpendicularity threads to end,} \\ \text{coaxiality of shaft to threads,} \\ \text{concentricity shaft to end,} \\ \text{molding deformation of hub,} \\ \text{gun aging noise factors;} \end{array}\right\} \quad \text{noise factors}$$

$$\left.\begin{array}{l}\text{shaft oiling,}\\ \text{hub coating,}\\ \text{relative surface finish,}\\ \text{hub material,}\\ \text{chamfer lead-in,}\\ \text{hydraulic flow rate,}\\ \text{pulley design}\end{array}\right\} \quad \text{process variables (PVs)}$$

(see also Fig. 5.18).

The DFSS algorithms strives for comprehensiveness development by recognizing that current vulnerable design entities are a result of the inefficient traditional design practices coupled with the inherent conceptual limitation of the design process itself, which is partially analytical in the early phases. Such obstacles create design vulnerabilities, which can be eliminated or reduced by the employment of the DFSS concepts and tools. Our research in design theory suggests maximizing the leverage of tool and concept integration as much as possible within the DFSS algorithm. To achieve such integration, it is necessary to highlight conceptual and theoretical relationships and to devise a workable format; hence the DFSS algorithm.

Figure 5.19 builds on the concepts presented here by linking FMEA to other methods such as the physical structure of axiomatic design, robustness noise factors, and next-higher (hierarchy)-level FMEA.

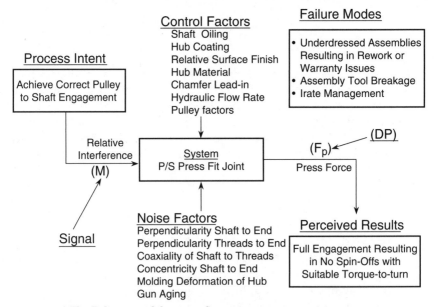

Figure 5.18 The P-diagram of the press-fit process.

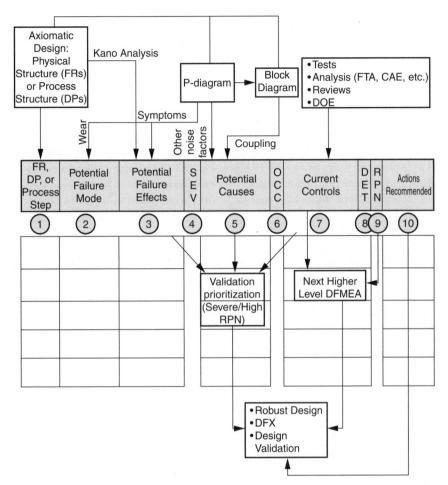

Figure 5.19 The relationship between FMEA, axiomatic design, and robustness tools (DFX = Design for X, where X = manufacturability, reliability, environment, assembly testing, service, aesthetics, packaging/shipping, etc.).

5.10   Transfer Function Optimization (DFSS Algorithm Step 9)

The objective of the DFSS team is to develop their project assignment, which functions as intended under a wide range of conditions for the duration of its life. In this context, the purpose of the optimization step in the DFSS algorithm is to minimize variation from the transfer function through exposing product and process functions to representative sources of variation, the noise factors. A transfer function in a form of a mathematical or an analytical model is developed to predict the optimum combination of design parameters or process variables and their target settings in the respective structures. This activity enables the simultaneous evaluation of many design parameters and process variables for their improvement potential. This step facilitates efficiency in project development by stressing the development of sound measurement strategies that are based on the measurement of FRs or the DPs. The transfer function optimization step may be conducted analytically or empirically. In either case, it involves a systematic way to anticipate downstream sources of product and process noise factors. This approach should take maximum advantage of the cost and quality performance leverage, which exists for preventing problems during the early stages of product and process design.

Optimization and robustness are analogous to each other in the context of the DFSS algorithm. *Robustness* means that the team must seek to identify the best expression of a design (product, service, and process) that is the lowest total cost solution to the customer-driven Six Sigma design specification. DFSS means generating concepts that are compatible with the optimization activity to produce Six Sigma level of quality. To analyze a system's robustness and adjustability, unique metrics such as the Z value, Taguchi's signal-to-noise ratios, and loss functions are available, which make it possible for the team to use mathematical analysis via Taylor series expansion or by employing powerful experimental methods to optimize a design's insensitivity to sources of variation. Because of the unavailability of transfer functions in the majority of the cases, we will concentrate on the parameter design as an empirical method to develop such transfer functions rather than the analytical method in this section and in Chaps. 13 to 15. Analytical methods will be discussed in more details in Chap. 6.

In the later case, parameter design of the robustness approach may be used. Parameter design is a systematic activity that extracts the best functional requirements performance from design concepts under development and produces performance that is minimally affected by noise factors. The Six Sigma design will provide functional performance with the least amount of sensitivity to uncontrollable sources of variation. The creative combination of noise factors enables the team to systematically anticipate the effects of uncontrollable sources of variation on functional requirements performance while simultaneously determining which design parameters will be useful in minimizing noise factor effects.

5.10.1 Finalize the physical structure (DFSS algorithm step 9)

The purpose of describing the design in terms of its input signals and its output responses (FRs) is to structure the development of the optimization strategy. The description of the physical structure provides a strategic summary of the level of system optimization taking place and the measurement approaches on which the optimization will be based. The characterization of operational transfer functions will identify a P-diagram for the functional elements in the structure. What is missing is the noise factors identification, specifically, which major sources of variation influence the solution entity selected at all levels of modularity.

5.10.2   Use the transfer function to identify design parameters for optimization (DFSS algorithm step 9)

The array of FRs is transferred into an array of transfer functions. The transfer function array is very informative. Using the transfer function, we need to optimize (shift the mean to target and reduce the variability) for all the FRs in the design. However, this optimization is not arbitrary. The selection for the DPs per an FR that will be used for optimization depends on the physical mapping from a coupling perspective. If the design is uncoupled, that is, if there is one-to-one mapping between FRs and DPs, then each FR can be optimized separately via its respective DP. Hence, we will have parameter design optimization studies equal to the number of FRs. If the design is decoupled, the design optimization routine has to follow the coupling sequence revealed by the respective design matrix in the structure. In coupled designs, the selection of DPs to be included in the study depends on the potential control required relative to the cost. The selection of design parameters will be done in a manner that will enable target values to be varied during experiments with no major impact on design cost. The greater the number of potential design parameters that are identified in the mapping, the greater the opportunity for optimization of function in the presence of noise.

A key philosophy of the DFSS algorithm is that during the optimization phase, inexpensive parameters can be identified and studied, and can be combined in a way that will result in performance that is insensitive to noise. This objective is sought at the Six Sigma quality level. The team's task is to determine the combined best design parameter settings that have been judged by the team to have the potential to improve the design. By analytically or experimentally varying the parameter target levels, a region of nonlinearity can be identified. This area of nonlinearity is the most robust setting for the parameter under study. Consider two levels or means of a design parameter (DP), level 1 (DP') and level 2 (DP''), having the same variance and distribution (Fig. 5.20). It is obvious that level 2 produces less variation in the functional requirement (FR) than level 1. Level 2 will also produce a lower-quality loss similar to the scenario on the right of Fig. 5.21. The design produced by level 2 is more robust than that produced by level 1. When the distance between the specification limits is 6 times the standard deviation, a Six Sigma optimized FR is achieved. When all design FRs are released at this level, a Six Sigma design is achieved.

The objective of parameter design is to suppress, as far as possible, the effect of noise by exploring the levels of the factors that make the design insensitive to them. Parameter design is concerned with the product or process functional requirement to provide this function at the lowest overall cost and targeted quality level under the variability produced by the noise factors.

5.10.3   Identify noise factors (DFSS algorithm step 9)

Noise factors cause the response to deviate from the intended target, which is specified by the signal factor value. Noise factors can be classified into three general categories:

■ Unit-to-unit sources (manufacturing/production and supplier variation) such as dimensional, assembly-related, or material property variation

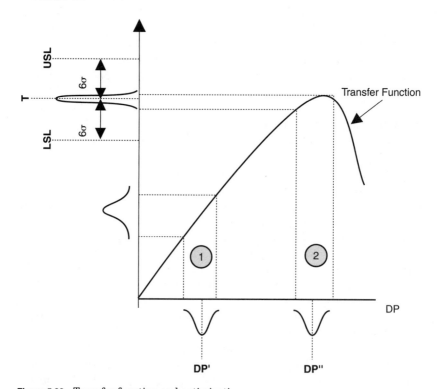

Figure 5.20   Transfer function and optimization.

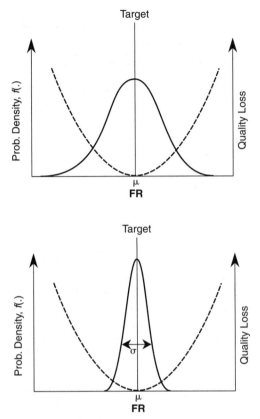

Figure 5.21 The quality loss function.

- External sources (usage and environment), such as temperature, user use, misuse, and abuse, and loading-related variation
- Deterioration sources (wearout) such as material fatigue or aging and wear, abrasion, and the general effects of usage over time

The noise factors affect the FRs at different segments in the life cycle (see Fig. 5.22). As a result, they can cause dramatic reduction in design robustness and reliability. Early-life failures can be attributed to manufacturing or production variability. The unit-to-unit noise causes failure in the field when the design is subjected to external noise. The random failure rate that characterizes most of the design life is attributed to external noise. Deterioration noise is active at the end of life. Therefore, a design is said to be robust (and reliable) when it is insensitive (impervious) to the effect of noise factors, even though the sources themselves have not been eliminated (Folkes and Creveling 1995).

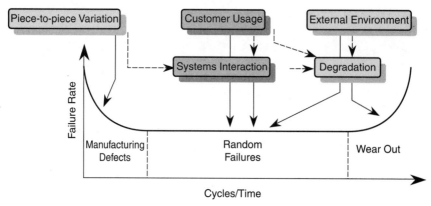

Figure 5.22 Effect of noise factors during the system life cycle.

In a robust design study, factors that are uncontrollable in use (or which are not practical to control) are selected to produce a testing condition during the experiment to find the transfer function. The objective is to produce variation in the functional response (experimental data set) that would be similar to the effect that would be experienced in actual use of the design. Simulating the effects of all the noise factors is not practical and is not necessary. The key requirement in the selection and combination of these noise factors is to select a few important factors at points that cover the spectral range and intensity of actual noises. Such selected noises are called "surrogate" noises. The rationale for this simplification approach is that the full spectral continuum of real-world noises should not cause variations very different from a small set of discrete choices positioned across the real-world spectrum.

5.10.4 Plan the optimization experiments (DFSS algorithm step 9)

The objective of this step is to coordinate all knowledge about the project under development into a comprehensive experimentation and data collection plan. The plan should be designed to maximize research and development efficiency through the application of testing arrays, design responses such as FRs, the loss functions, signal-to-noise ratios, and statistical data analysis. The team is encouraged to experimentally explore as many design parameters as feasible to investigate the functional requirements potential of the design or technology concept that is being applied within the design. Transferability of the improved FRs to the customer environment will be maximized because of the application of the noise factor test strategy during data collection.

Data from the optimization experiment will be used to generate the transfer function to be used for optimization and to improve design robustness to the Six Sigma level. The validity of this function and the resulting conclusions will be influenced by the experimental and statistical assumptions made by the DFSS team. What assumptions can be made regarding the existence of interactions between design parameters? Is there an assumption that the variance of the response remains constant for all levels within the transfer function, or is it assumed that the variance in the response is due to the effect of the noise factors? What assumptions can be made (if any) regarding the underlying distribution of the experimental data? What assumptions can be made regarding the effect of nuisance factors (other than the experimental noise factors) on the variance of the response? Are the optimum factor combinations predicted by the transfer function optimal? What assumptions can be made about the transferability of the results beyond the experimental environment, and what would substantiate these assumptions?

In the dynamic robustness formulation, the DFSS team should decide on noise strategy and signal range, and develop a design parameter strategy. Design parameters are specified freely by the team. If the experiment is exploratory, it is suggested to set levels at extreme values of the feasible operating range. Two levels will be appropriate for screening purposes, but more information on nonlinear effects will require a three-level strategy. The quality metrics that will be used in the analysis are FRs, loss functions of the FRs, and the signal-to-noise ratio. Depending on the FR, there are two broad forms of ratio are available. Static forms apply where the FR has a fixed value. Dynamic forms apply where the FR operates over a range of signal input values.

5.10.5 Collect data and analyze results
(DFSS algorithm step 9)

The individual values of the appropriate metric are calculated using the data from each experimental run. The purpose of determining the metric is to characterize the ability of DPs to reduce the variability of the FRs over a specified dynamic range. In a dynamic experiment, the individual values for transfer function sensitivity coefficients are calculated using the same data from each experimental run. The purpose of determining the sensitivity values is to characterize the ability of design parameters (DPs) to change the average value of the FRs across a specified dynamic range. The resulting sensitivity performance of a system is obtained by the best-fit curve.

DP level effects are calculated by averaging the metric to correspond to the individual levels as depicted by the orthogonal array diagram.

The importance of DPs for decreasing sensitivity is determined by comparing the gain in the metric ratio from level to level for each factor, comparing relative performance gains between each design parameter, and then selecting which ones produce the largest gains. The level for each DP with the optimized metric ratio is selected as the parameter's best target value. All of these best levels will be selected to produce the best parameter target combination. The same analysis and selection process is used to determine DPs, which can be best used to adjust the mean performance. These factors may be the same ones that have been chosen on the basis of metric improvement, or they may be factors that do not affect the optimization of the metric. DPs that don't contribute to improvements in the metric transfer function are set to their most economical values.

The DFSS team needs to run confirmation tests of optimum design combinations and verify assumptions, and perform a test with samples configured at the combined optimum design level and calculate the representative metric performance.

Compare the transfer function values to the predicted optimum values. If the actual performance is within the interval of performance that was predicted, then the predictive model validity is confirmed. There is a good chance at this point that the optimum results experienced in the confirmation run will translate to the usage environment. If the confirmation test values fall outside the interval, the team should reassess the original assumptions for this experiment since, in all likelihood, other conditions are operating which are not accounted for in the model.

A successful experiment will lead the team to clarify whether new technical information has been uncovered that will greatly improve the physical structure. The team will want to consider if other DP levels should now form the basis of a revised experimental plan. If the study failed to produce an improvement, the combination of noise factors that were in the original experiment may have overpowered the ability of the DPs to generate improved performance. A redesign of the DP strategy should be considered by the team. If improvement cannot be realized and the team has exhausted all possible DPs, there may be reason to conclude that the current concept being optimized will not be able to support the performance requirements for the design under development. This unfortunate situation usually happens because of the violation of design axioms and would justify the consideration and selection of a new concept or even a new physical structure.

5.11   Design for X (DFSS Algorithm Step 10)

The black belt should continually revise the DFSS team membership to reflect a concurrent approach in which both design and process members are key, equal team members. However, the individuals involved in the "design for X" (DFX, where X = manufacturability, reliability, environment, assembly, testing, service, aesthetics, packaging/shipping, etc.) merit special attention. X in DFX is made up of two parts: life-cycle processes (x) and performance measure (ability), i.e., X = x + ability. DFX is one of the most effective approach to implement concurrent engineering. For example, Design for Assembly (DFA) focuses on the assembly business process as part of production. The most prominent DFA algorithm is the Boothroyd-Dewhurst algorithm, which developed out of research on automatic feeding and automatic insertion.

DFX techniques are part of detail design where poka-yoke (error-proof) techniques can be applied when components are taking form and producibility issues are simultaneously considered [see Huang (1996)]. *Poka-yoke* is a technique for avoiding human error at work. The Japanese manufacturing engineer Shigeo Shingo developed the technique to achieve zero defects and came up with this term, which means "mistake proofing." A defect exists in either of two states; the defect either has already occurred, calling for defect detection, or is about to occur (i.e., is imminent), calling for defect prediction. Poka-yoke has three basic functions to prevent or reduce defects: shutdown, control, and warning. The technique starts by analyzing the process for potential problems, identifying parts by the characteristics of dimension, shape, and weight, detecting process deviation from nominal procedures and norms.

In design for reliability (DFR), not testing for reliability, the DFSS team needs to anticipate all that can go wrong and improve the reliability of the design by simplifying and reducing the number of and type of components (note the agreement with design axiom 2), standardizing the parts and material to reduce variability, considering design parameters to counteract environmental effects, minimizing damage from mishaps in shipping, service, and repair, employing robust processes that are insensitive to variation, and eliminating design vulnerabilities.

The design for maintainability (DFM) objective is to assure that the design will perform satisfactorily throughout its intended life with a minimum expenditure of budget and effort. DFM and DFR are related because minimizing maintenance can be achieved by improving reliability. An effective DFM minimizes (1) the downtime for maintenance, (2) user and technician maintenance time, (3) personnel injury resulting from maintenance tasks, (4) cost resulting from maintainability features, and (5) logistics requirements for replacement parts, backup units, and personnel. Maintenance actions can be preventive, corrective, or recycle and overhaul.

Design for environment (DFE) addresses environmental concerns in all stages of the DFSS algorithm as well as postproduction transport, consumption, maintenance, and repair. The aim is to minimize environmental impact, including strategic level of policy decision making and design development.

The team should take advantage of, and strive to design into, the existing capabilities of suppliers, internal plants, and assembly lines. It is cost-effective, at least for the near term. The idea is to create designs sufficiently robust to achieve Six Sigma design performance from current capability. Concurrent engineering enables this kind of upside-down thinking. Such concepts are applied in the DFSS algorithm to improve design for manufacturing, assembly, and service. These key "design for" activities are well known, including the Boothroyd-Dewhurst design for assembly, and a spectrum of design for manufacture and design for service methodologies. The major challenge is implementation. Time and resources need to be provided to carry out the "design for" activities.

A danger lurks in the DFX methodologies that can curtail or limit the pursuit of excellence. Time and resource constraints can tempt DFSS teams to accept the unacceptable on the premise that the shortfall can be corrected in one of the subsequent steps—the "second chance" syndrome. Just as wrong concepts cannot be recovered by brilliant detail design, bad first-instance detail designs cannot be recovered through failure mode analysis, optimization, or tolerance design.

5.12   Tolerance Design and Tolerancing (DFSS Algorithm Step 11)

The purpose of the tolerance design step is to assign tolerances to the part, assembly, or process, identified in the functional and physical structures, based on overall tolerable variation in FRs, the relative influence of different sources of variation on the whole, and the cost-benefit trade-offs. In this step, the DFSS team determines the allowable deviations in DPs values, tightening tolerances and upgrading materials only where necessary to meet the FRs. Where possible, tolerances may also be loosened.

In the DFSS algorithm, this step calls for thoughtful selection of design parameter tolerances and material upgrades that will be later cascaded to the process variables. Selection is based on the economics of customer satisfaction, the cost of manufacturing and production, and the relative contribution of sources of FR variation to the whole design project. When this is done, the cost of the design is balanced with the quality of the design within the context of satisfying customer demands. By determining which tolerances have the greatest impact on FR variation, only a few tolerances need to be tightened, and often, many can be relaxed at a savings. The quality loss function is the basis for these decisions. The proposed process also identifies key characteristics where functional criteria are met, but where further variability reduction will result in corresponding customer benefits.

When tolerances are not well understood, the tendency is to over-specify with tight dimensional tolerances to ensure functionality and thereby incur cost penalties. Traditionally, specification processes are not always respected as credible. Hence, manufacturing and production individuals are tempted to make up their own rules. Joint efforts between design and process in the team help improve understanding of the physical aspects of tolerance and thus result in tolerances that are cross-functional and better balanced. This understanding will be great-ly enhanced by the previous steps in the DFSS algorithm and by con-tinuous employment of design axioms, QFD, the zigzagging process, and other tools. The goal in the optimization step (Sec. 5.9) was to find combinations of dimensions that inherently reduced FR variation. Typically, further reduction in tolerances is necessary to meet the FR Six Sigma targets. This can be accomplished best by the tolerance design step.

Tolerance design can be conducted analytically on the basis of the val-idated transfer function obtained in Sec. 5.9 or empirically via testing. In either case, the inputs of this step are twofold—the DFSS team should have a good understanding of the product and process require-ments and their translation into product and process specifications using the QFD. The going-in (initial) position in the DFSS algorithm is to initially use tolerances that are as wide as possible for cost consider-ations, then to optimize the function of the design and process through a combination of suitable design parameters (DPs). Following this, it is necessary to identify those customer-related FRs that are not met through parameter design optimization methods. Tightening tolerances and upgrading materials and other parameters will usually be required to meet Six Sigma functional requirement targets. Systematic applica-tion of DFSS principles and tools such as QFD allows the identification of customer-sensitive characteristics and the development of target val-ues for these characteristics to meet customer expectations. It is vital that these characteristics be traced down to lowest-level mappings, and that appropriate targets and ranges be developed.

Decisions regarding tolerance reduction are based on the quadratic loss function, which suggests that loss to society is proportional to the square of the deviation of a design characteristic (such as a dimension) from its target value. The cost of being "out of specification" for the critical performance criteria must be estimated.

5.13   Pilot and Prototyping Design (DFSS Algorithm Step 12)

The objective of this step is to verify that the Six Sigma optimized design performs at a level that is consistent with the specifications. To accomplish this, a single, productionlike prototype is created in this step, which provides a framework for the systematic identification and efficient resolution of problems. In the product design scenario, this step also includes the verification of certain service-related and production/manufacturing-related support activities, such as maintenance procedures, packaging, dunnage, and shipping systems.

The DFSS team should identify processes, parts, and assemblies for prototype testing as they note which are carryovers and which are the results of new designs or Six Sigma optimization. The team proceeds to define test hierarchy (e.g., component, subsystem, and system) use specification, and FMEA information to identify tests normally performed to verify FR performance. The following are the major steps in this prototyping step:

1. Begin developing a total test matrix.

2. Define testing acceptance criteria.

3. Evaluate measurement systems (e.g., gauge R&R).

4. Develop a detailed testing schedule.

5. Establish a production resemblance plan.

6. Assemble a total prototype test plan.

7. Get test plan approval.

8. Order parts and build prototypes; procure prototype and production parts and assemblies, as necessary in the physical structure, to support testing based on step 4 (above).

9. Conduct the test.

10. Evaluate and analyze the results.

11. Verify service procedures and industrial engineering time studies. This activity verifies that the design work performed during the previous steps provides effective service for all systems and key subsystems and components. Design and process review of the project and supporting processes documentation will verify the design.

With these steps, it may be useful to calculate the prototype capability and compare this with internal capability of similar and baseline designs. Design prototypes in the category of incremental design made on the same production lines will assume the baseline process capability. The prototype sample of high-volume creative design and those outsourced to suppliers will use the applicable high-volume prototype sample testing (e.g., normality and pooled variance hypothesis testing are generally used). Low-volume creative design prototype samples can be estimated using χ^2 (chi-square) distribution.

5.14   Validate Deign (DFSS Algorithm Step 13)

This step provides a rapid, smooth confirmation of the design structures including manufacturing, assembly, and production capability, with the need for minimal refinements. Design and process specifications are completed and released before the start of this confirmation step of the DFSS algorithm. This step covers the installation of the equipment and the manufacture and evaluation of a significant production sample, which is produced under conditions of mass production. Production validation testing is used to verify that the process is producing a design that meets the FRs derived from the customer attributes. A functional evaluation of the design by the customer and DFSS team provides additional verification that design intent has been achieved.

Prior to installation, production processes should have already been optimized in the presence of potential sources of production and customer usage "noise" and have demonstrated potential capability both in house and at the supplier location. These activities have involved significant simulated production runs and form the basis for a low-risk installation at the intended facilities.

The DFSS team should identify the unanticipated installation trial concerns to ensure that the corrective actions are implemented prior to launch. Manufacturing and/or production capability is assessed and the process is exercised in a manner that would be expected for normal production operation.

The team needs to visit the training programs with focus on the transfer of knowledge from DFSS team to the on-site production personnel, including standby labor. This transfer of knowledge enables the operation individuals to add and improve on the control plan as they add their individual skills to the production planning.

The DFSS team membership should be enhanced to include more skills and expertise necessary to confirm production capability and to initiate an effective launch. Much of the earlier planning work done for quality control and maintainability of the process will be executed during the launch phase of production. The new team members who will participate in the launch now have to prepare themselves by reviewing the process documentation and control plans. This documentation, combined with supporting discussion by the new team members, will provide new members with the knowledge that the team has accumulated and will align them with the team's operating philosophy. New members for the team should include machine suppliers, purchased parts suppliers, maintenance, production, quality, and customers, both internal and external.

The location and availability of process history and control documentation should be organized according to the prioritization of operator needs. Process instructions, operator instruction aids, the process control plan, and reaction plans should be readily accessible in the operator workstation area. Design and manufacturing experts should be available and should recognize their major roles to enable efficient and effective decision making and identification of actions to resolve issues and to maximize transfer of their knowledge to operations personnel.

Based on the gauging R&R (control plan) procedures developed by the team during DFSS algorithm step 12, operators, machine setup personnel, and maintenance personnel should conduct the short-term versions of these procedures for all process gauge systems, including off-line systems. The performance of these gauges needs to be quantified subsequent to machine and gauge installation, and prior to process potential studies which will depend on these measurement systems.

Quality support personnel should be available to assist with these studies. This will reinforce the learning that needs to take place on the part of the operators and related maintenance personnel who have to utilize system-related gauge systems. As a reminder, the gauge capability studies will need to quantify the reproducibility, accuracy, stability, and repeatability. Ideally, overall gauge capability should be in the neighborhood of not more than 10 percent of the characteristic tolerance but less than 30 percent. The target for gauge system improvement should always be zero error. This issue becomes more critical as the gauging systems become a vital element of continuous process improvement. The ultimate level of acceptability will be a point of team review and agreement.

5.15 Launch Mass Production (DFSS Algorithm Step 14)

Implementation of the DFSS project with upfront launch planning, combined with just-in-time operator training, enables a smooth launch and rapid ramp to production at full speed. This step is also the final confirmation of deliverables, enabling the DFSS team learning derived from understanding variations in performance from expectations. Launch and mass production is an opportunity to confirm that the process of mass production has not changed design functional requirements. The program team must check this before the customer receives design. Launch and mass production also begin the transition from off-line to on-line quality control and daily management. With it comes the responsibility to implement poka-yoke techniques and DFX with on-line quality methods to continuously improve at a rate faster than that of competitors.

As results to targets become available, the team benefits from a cause-effect comparison of processes used in the design development process and the results obtained. Were the results as expected? Were the results better or worse than expected? Why? What practices should be utilized more in the development of future products? What do we learn as a result? Take every opportunity to learn, and share lessons learned with future program teams. Launch and mass production begin the return on investment for all the time and effort spent in upfront quality actions. The team will

- Develop and initiate a launch support plan for (1) manufacturing and (2) maintenance and service.
- Build a launch concern reaction strategy.
- Support a marketing launch plan.
- Implement a production launch plan.
- Implement a mass production confirmation–capability evaluation.
- Improve product and process at a rate faster than that of competitors.
- Use the disciplined procedures for collecting, synthesizing, and transferring information into the corporate design books.

5.16 Project Risk Management

This concept of the DFSS algorithm is overreaching across the whole DFSS algorithm. In effect, most of the DFSS algorithm tools are risk management tools, each with its own perspectives. Therefore, this section was postponed to the end of the chapter.

Many DFSS teams may start marching forward focused on methodical DFSS, but often without great acknowledgment of risks, in pursuit their project. They may feel confident about what the future may bring and their ability to manage resources in a way that leads to a fruitful and productive project closure. Events that they cannot fully foresee (e.g., market conditions at the time of conclusion) may change the picture completely. As a black belt, all that is expected of you and the rest of the team is the ability to weigh the potential losses and gains in order to make the best possible judgment to preserve that project and the effort that went into it. Events that change the team's perception of future events may occur unexpectedly, and often suddenly. These events arise from risk, that is, from the possibility that the future may be different from what they expect. These possibilities may bring bad or good results, generating threats of losses or presenting opportunities for gains.

In the context of a DFSS project, the discipline of risk management has been devoted to addressing loss from poor scoping, conceptual design vulnerabilities, lack of a business case, or errors in understanding customer needs. Risk management embraces both the upside and downside of risk. It seeks to counter all losses, from unfortunate business and technical judgments, and seize opportunities for gains through design innovation and growth.

When you choose to take a risk, it is wise to aim to lessen any negative impact and increase any positive impact that might happen. Risk is associated with events whose chances of occurrence are greater than zero (probable), but less than one (certain). Where change occurs, risk arises. For any company to succeed in its Six Sigma endeavor, its internal world must change; it cannot stand still, or expect to remain perfectly stable. Change brings risk, but without the change and its accompanied risk there could be no progress. DFSS is about progress; thus it is about change, and the team needs to learn to manage and tolerate a level of unpredictable risk.

Risk management is about visualizing other scenarios of the future and having the respective plans to accept (low risk), mitigate, prevent, or outsource risk. The most straightforward answer to the question "Why manage risk?" is "Manage risk to reduce potential losses." Risk management counters downside risks by reducing the probability and magnitude of losses (uncertainty); and recovery from these losses.

In the DFSS algorithm, the technical risk is usually handled using the FMEA. The noise factor strategy should be linked to transfer function optimization. Testing plans should be synchronized to reflect noise factor treatment which spans a range of options such as poka-yoke, feedforward, and backward controls (if applicable), and robustness. Project scoping risk can be reduced by deriving business and customer-driven projects.

DFSS Transfer Function and Scorecards

6.1 Introduction

Functional analysis and physical synthesis are the premier activities performed in the characterization (C) phase of the DFSS algorithm. In the following sections, we distinguish both of these activities, offer conceptual framework for their use, and link these concepts to the rest of the DFSS algorithm. The analysis part of the DFSS algorithm is covered in Chaps. 7 to 18. This chapter focuses on design synthesis. In addition to design synthesis, the other two major entities that are presented here are the transfer function and the design scorecard. A *transfer function* is the means of optimization and design detailing and usually documented in the scorecard. It is treated as a living entity within the DFSS algorithm that passes through its own life-cycle stages. A transfer function is first identified using the zigzagging method and then detailed by derivation, modeling, or experimentation. The prime uses are optimization and validation. Design transfer functions belonging to the same hierarchical level in the design structures should be recorded in that hierarchical level scorecard. A *scorecard* is used to record and optimize the transfer functions. The transfer function concept is integrated with other DFSS concepts and tools such as synthesis block diagram, robust design P-diagram, and design mapping within the rigor of design scorecards.

This chapter is linked in every way possible to all related DFSS algorithm concepts and tools. This link is particularly critical to Chap. 5, and occasional reference will be made where value can be gained.

6.2   Design Analysis

Axiomatic design provides the zigzagging method as the means to define the process and physical structures. A *structure* can be defined as an input-output or cause-and-effect relationship of functional elements. Graphically, it is depicted in a "block" diagram that is composed from nodes connected by arrows depicting the relationships. A structure should capture all design elements within the scope and ensure correct flowdown to critical parameters. A structure is captured mathematically using mapping matrices with matrices belonging to the same hierarchical level clustered together. Hierarchy is built by the decomposing design into a number of simpler functional design matrices that collectively meet the high-level functional requirements identified in step 2 of the DFSS algorithm. The collection of design matrices forms the conceptual physical or process structure. A structure provides a means to track the chain of effects for design changes as they propagate across the design. The decomposition starts by the identification of a minimum set of functional requirements that deliver the design tasks as defined by the customer and obtained from phase 2 QFD. The decomposition is guided by both the design axioms and the creative and heuristic process of function definition through logical questions as offered by the zigzagging method (Chap. 7). The efficient use of design principles can gear the synthesis and analysis activities to vulnerability-free solutions. The deployment of design axioms seems to be promising for two reasons: (1) history tells us that knowledge that is based on axioms will continue to evolve as long as the axioms are maintained, and (2) it strengthens the emphasis placed by robust design on the functional structure of the design. The deployment of design axioms achieves two main objectives: to uncouple or decouple design and to reduce design complexity (see Chaps. 5 and 7). Both objectives will eliminate or reduce conceptual design vulnerabilities.

6.3   DFSS Design Synthesis

The design mappings (matrices) conducted in the DFSS algorithm (Chap. 7) are conceptual representations at different hierarchical levels. These matrices don't stand alone, and a complete solution entity for the design project needs to be synthesized in the physical structure and later in the process structure. The ultimate goal of design mapping is to facilitate design detailing when the mathematical relationships are identified in the form of transfer functions. Design mapping, a design analysis step, should be conducted first prior to design synthesis activities.

A detailed transfer function is useful not only for design optimization but also for further design analysis and synthesis activities, in particular, design structure. For example, the functional requirement (FR) of a given subsystem or component can be the input signal of another component delivering another FR, and so on. These relationships create the design hierarchy. Uncoupled hierarchy is desirable to accomplish the high-level FRs of the design and can be examined using a functional block diagram. The *block diagram* is a graphical representation of the design mappings and noise factors effects. An example is depicted in Fig. 6.1. In process/service-oriented Six Sigma projects, the block diagram is actually a process map. This section provides the foundation for diagramming by linking the physical structure components using mathematical formulation and Taguchi robustness concepts (e.g., noise versus design factors) and tools (e.g., P-diagram; see Sec. 5.9.1).

6.3.1 P-Diagram

The robustness approach is based on a revolution in quality improvement because of the application of certain statistical methods to optimization by Dr. G. Taguchi. The principle idea is that statistical testing of products should be carried out at the design stage, in order to make the product robust against variations in the manufacturing and usage environments.

Using this methodology, quality is measured by statistical variability, such as standard deviation or mean-square error rather than percentage of defects or other traditional tolerance-based criteria. The objective is to keep performance on target value while minimizing the variability. *Robustness* means that a design performs its function as intended under all operating conditions throughout its intended life. Noise factors are the undesirable and uncontrollable factors that cause the FRs to deviate from target values. Noise factors adversely affect quality. However, it is generally impossible or too expensive to eliminate noise factors. Instead, through robust design, the effect of noise factors can be reduced.

Robust design is aimed at reducing the losses due to variation of performance from the target value based on the quality loss function, signal-to-noise (S/N) ratio, optimization, and experimentation. The design output is usually categorized into a desired part containing a useful portion of FR and extraneous or undesired portion. In the dynamic case, the desired portion is called "signal" and the undesired segment is called "error." Usually, both are added together to form the total output achieved. The primary objective of robust design is to reduce the effect of noise as much as possible.

The P-diagram (Fig. 6.2) of robust design represents all the elements of synthesis activity of the DFSS algorithm. The components of the P-diagram are:

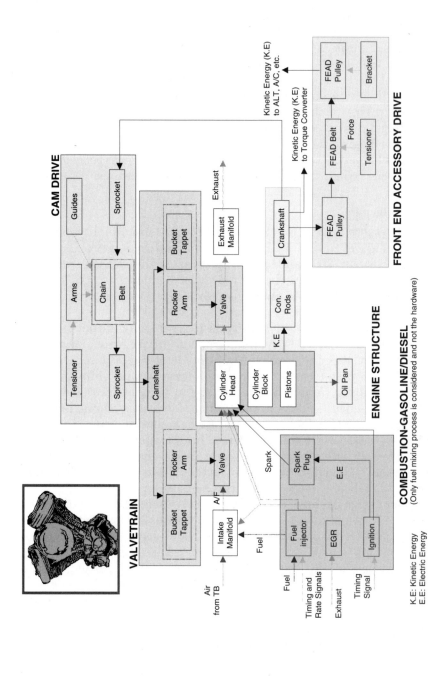

CAM DRIVE

VALVETRAIN

ENGINE STRUCTURE

FRONT END ACCESSORY DRIVE

COMBUSTION-GASOLINE/DIESEL
(Only fuel mixing process is considered and not the hardware)

K.E: Kinetic Energy
E.E: Electric Energy

Figure 6.1 Automotive engine block diagram.

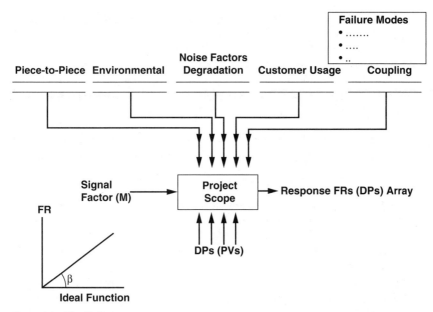

Figure 6.2 The P-diagram.

- *Signal (M).*   This is the translation of the customer's intent to measurable technical terms. A signal can be categorized as energy, information, or material.

- *Response.*   This is the translation of the customer's perceived result into measurable functional requirements (FRs) or design parameters (DPs). A FR (DP) can be categorized as energy, information, or material.

- *Design parameters (DPs).*   DPs are characteristics that are inherent to a specific product or process and are specified by the team.

- *Noise factors (N).*   Noise factors impact the performance of a product or process but are either uncontrollable or too expensive to control. Noise factors are selected according to their ability to affect the fundamental transfer functions. The main purpose of noise is to create variability in response. Noise factors can be categorized as piece-to-piece (or service-to-service) variation; coupling with other components, subsystems, or systems; customer usage; environmental factors; and degradation.

- *Ideal function.*   The ideal function is a description of "how the system works if it performs its intended function perfectly." In dynamic systems, the ideal function is a mathematical description of the energy transformation relationship between the signal and the response. The reason for studying the ideal function is to have a physics-based mathematical model of the system under consideration before testing. This allows the DFSS team to evaluate various DP levels in spite of the presence of noise factors. Ideal function is written as FR = βM, where β is sensitivity with respect to the signal. This sensitivity could be constant or a function of design and noise factors [i.e., $\beta(DP_1,\ldots,DP_p;N_1,N_2,\ldots,N_L)$].

In the following sections, we will assume that transfer functions can be approximated by a polynomial additive model with some modeling and experimental error, that is, $FR_i \cong \sum_{j=1}^{P} A_{ij} DP_j + \sum_{k=1}^{K} \beta_{ik} M_k + \text{error}$ (noise factors), where the A_{ij} and β_{ik} are the sensitivities of the FR (or any response) with respect to design parameters and signal factors, respectively, P is the number of DPs, and K is the number of signals. This approximation is valid in any infinitesimal local area or volume of the design space. The noise terms will be further modeled as an error term to represent the difference between the actual transfer function and the predicted one. The additivity is extremely desired in the DPs and all design mappings. As the magnitude of the error term reduces, the transfer function additivity increases as it implies less coupling and interaction. In additive transfer functions, the significance of a DP is relatively independent from the effect of other DPs, indicating uncoupled design per DFSS axiom 1. Physical solution entities that are designed following axiom 1 will have an additive transfer function that can be optimized easily, thus reducing the DFSS project cycle time. From an analysis standpoint, this additivity is needed in order to employ statistical analysis techniques like parameter design, design of experiment (DOE), and regression. Nonlinearity is usually captured in the respective sensitivities.

6.3.2 Block diagram and synthesis

Let $\{FR_1, FR_2, FR_3\}$ and $\{f_1, f_2, f_3\}$ be the set Y', the set of FRs; and F', the set of hypothesized or proven transfer functions, respectively. Each f_i can be written in the form $f_i(M_i, DPs_i)$, $i = 1,2,3$, where M is the signal (revisit Sec. 5.9.1). The mapping $f(M,DP)$ will be assumed to be additive. In addition, assume that the three mappings are complete and constitute a design project. The objective of the synthesis or block diagramming activity is to piece together the solution entity identified for each function in order to have a graphical representation of the design. This requires the identification of the operational relationships as depicted by the transfer functions (this step is a design analysis step) and the precedence or logical relationships in the design hierarchy that govern the P-diagram and the transfer functions input/output (this step is a design synthesis step). Inputs are classified by Phal and Beitz (1988) as information, material, or energy.

1. *Pure series synthesis.* *Hierarchy* in design mapping means that a lower-level functional entity receives its signal from the output of a higher-level entity. At a given level, there are some "parent" functions, which in effect provide the signal to all other functions. In our case, the parent function is the function to the far left and has the precedence in the dynamic flow. This is pure series functional hierarchy (Fig. 6.3). This pattern of hierarchy may be modeled by utilizing the mathematical concept of composite mapping (operation $\bigcirc$). Let f_1, f_2, f_3 be three physical mappings from FRs to F as introduced in Chaps. 5 and 7; then

$$f_1 \bigcirc f_2 : \text{FR} \rightarrow F : f_1 \rightarrow f_2 \, (f_1 \, (M_1, \text{DP}_1), \, \text{DP}_2)$$

$$f_2 \bigcirc f_3 : \text{FR} \rightarrow F : f_2 \rightarrow f_3 \, (f_2 \, (M_2, \text{DP}_2), \, \text{DP}_3)$$

where $f_1(M_1, \text{DP}_1) = M_2$ and $f_2(M_2, \text{DP}_2) = M_3$. Assume that the ideal function has the linear form $\text{FR}_i = \beta_i M_i, i = 1,2,3$. Then the transfer function equation of the pure series physical structure without the noise factors effect is given by

$$\{\text{FR}_3\} = \begin{Bmatrix} \beta_1\beta_2\beta_3 \\ \beta_2\beta_3 A_{11} \\ \beta_3 A_{22} \\ A_{33} \end{Bmatrix}' \begin{Bmatrix} M_1 \\ \text{DP}_1 \\ \text{DP}_2 \\ \text{DP}_3 \end{Bmatrix} \qquad (6.1)$$

where $A_{ii}, i = 1,2,3$ is the sensitivity $\partial\text{FR}_i/\partial\text{DP}_i$. When the noise factor effects are added, the transfer equation can be written as

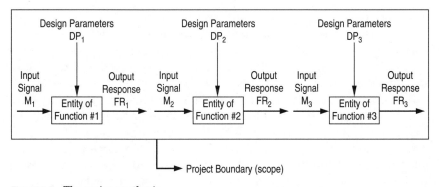

Figure 6.3 The series synthesis.

$$\{FR_3\} = \underbrace{\beta_1\beta_2\beta_3 M_1}_{\text{ideal function}} + \underbrace{\left.\begin{Bmatrix} \beta_2\beta_3 A_{11} \\ \beta_3 A_{22} \\ A_{33} \end{Bmatrix}\right.'\begin{Bmatrix} DP_1 \\ DP_2 \\ DP_3 \end{Bmatrix}}_{\text{design mapping}} + \varepsilon \text{ (noise factors)} \qquad (6.2)$$

In this mathematical form, FR_3 is called a *dependent variable,* while M, DP_1, DP_2, and DP_3 are independent variables. Note that we made no assumption about the sensitivity coefficients, A_{ij} values and it can be a linear or a nonlinear function of the DPs. The global parent function is function 1. Function 2 is the parent of function 3, the lower level function in the chain. The equation represents a redundant design where P, the number of DPs, is greater than m, the number of FRs. This redundant design may be changed to ideal design by fixing two DPs (the least sensitive DPs) and using the most sensitive DP as a robustness parameter.

2. *The pure parallel synthesis.* In the pure parallel arrangement (Fig. 6.4), we have the same input signal across the same hierarchical functional entities. For the case of three entities, the transfer function equation is given by

$$\begin{Bmatrix} FR_1 \\ FR_2 \\ FR_3 \end{Bmatrix} = \begin{bmatrix} A_{11} & 0 & 0 & \beta_1 & 0 & 0 \\ 0 & A_{22} & 0 & 0 & \beta_2 & 0 \\ 0 & 0 & A_{33} & 0 & 0 & \beta_3 \end{bmatrix} \begin{Bmatrix} DP_1 \\ DP_2 \\ DP_3 \\ M_1 \\ M_2 \\ M_3 \end{Bmatrix} \qquad (6.3)$$

$$= \underbrace{\begin{bmatrix} \beta_1 & 0 & 0 \\ 0 & \beta_2 & 0 \\ 0 & 0 & \beta_3 \end{bmatrix} \begin{Bmatrix} M_1 \\ M_2 \\ M_3 \end{Bmatrix}}_{\text{ideal functions}} + \underbrace{\begin{bmatrix} A_{11} & 0 & 0 \\ 0 & A_{22} & 0 \\ 0 & 0 & A_{33} \end{bmatrix} \begin{Bmatrix} DP_1 \\ DP_2 \\ DP_3 \end{Bmatrix}}_{\text{design mapping}} + \begin{matrix} \text{error} \\ \text{(noise factors)} \end{matrix}$$

with the constraints: $M = \sum_{i=1}^{3} M_i$ and $FR \le \sum_{i=1}^{3} FR_i$.

The pure parallel structure is an uncoupled design. We should expect a structure that is a combination of pure series and pure parallel arrangement to be a redundant design that may be reduced to achieve uncoupled design if certain DPs can be fixed.

3. *In-between synthesis.* At a certain level of hierarchy, we may have the following arrangement of Fig. 6.5. We refer to this arrangement as the "in-between" hierarchy since it is neither pure series nor pure parallel. This is due to the lack of an input relationship between the higher-level functions (functions 1 and 2). Both of these functions are global parents, thus violating the requirement of a pure series system.

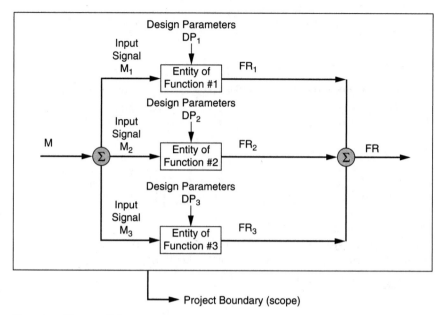

Figure 6.4 The parallel system.

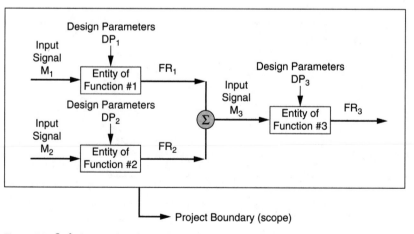

Figure 6.5 In-between structure.

In this case we have $f_3 \circ (f_1, f_2) = f_3 (\sum_{i=1}^{2} f_i (M, DP_i), DP_3) = f_3 (M_3, DP_3)$. The transfer function is given as

$$
\begin{Bmatrix} FR_1 \\ FR_2 \\ FR_3 \end{Bmatrix} = \begin{bmatrix} A_{11} & 0 & 0 & \beta_1 & 0 \\ 0 & A_{22} & 0 & 0 & \beta_2 \\ 0 & 0 & A_{33} & \beta_1\beta_3 & \beta_2\beta_3 \end{bmatrix} \begin{Bmatrix} DP_1 \\ DP_2 \\ DP_3 \\ M_1 \\ M_2 \end{Bmatrix} + \text{error (noise factors)} \quad (6.4)
$$

$$
= \begin{bmatrix} A_{11} & 0 & 0 \\ 0 & A_{22} & 0 \\ 0 & 0 & A_{33} \end{bmatrix} \begin{Bmatrix} DP_1 \\ DP_2 \\ DP_3 \end{Bmatrix} + \begin{bmatrix} \beta_1 & 0 \\ 0 & \beta_2 \\ \beta_1\beta_3 & \beta_2\beta_3 \end{bmatrix} \begin{Bmatrix} M_1 \\ M_2 \end{Bmatrix} + \text{error (noise factors)} \quad (6.5)
$$

with the constraint $M_3 = FR_1 + FR_2$.

4. *Coupled synthesis.* A coupled structure results when two or more FRs share at least one DP. Fig. 6.6 depicts a typical case. The transfer function can be written as

$$
\begin{Bmatrix} FR_1 \\ FR_2 \\ FR_3 \end{Bmatrix} = \underbrace{[\beta_1\beta_3 \quad \beta_2\beta_3]}_{\text{ideal function}} \begin{Bmatrix} M_1 \\ M_2 \end{Bmatrix} + \underbrace{\begin{bmatrix} \beta_3 A_{11} & 0 \\ \beta_3 A_{21} & 0 \\ 0 & 0 \end{bmatrix}}_{\text{design mapping}} \begin{Bmatrix} DP_1 \\ DP_3 \end{Bmatrix} + \text{error (noise factors)} \quad (6.6)
$$

Coupling here occurs because the number of design parameters p is less than the number of functional requirements m ($p = 2$, $m = 3$). Note that the design matrix is not ideal and that coupling resolution requires adding another DP (see Chap. 7 for more details).

6.3.3 Synthesis steps

The following synthesis steps can be used in both structures; however, the physical structure is used for illustration:

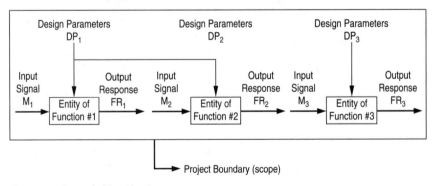

Figure 6.6 A coupled synthesis.

1. Obtain the high-level FRs from phase 2 QFD.

2. Define system boundaries from the project scope.

3. Conduct the zigzagging method to the lowest possible structure level and identify the transfer functions in every level. The lowest level represents the very standard design parameters (DPs) or process variables (PVs) (Chap. 7). For example, in product DFSS projects, dimensions, surface finish, and material properties are at the lowest level. On the process side, machine speed and feed rate are the corresponding-level PVs. On the service side, forms and fields are considered lowest levels. In many instances, the lowest hierarchical level of the structure is owned by a vendor or a group of vendors due to some outsourcing policy. A representation from the vendors should be added to the team as necessary.

4. Define the respective hierarchical levels of design mappings.

5. Within a level, for each mapping, and for each FR, classify the mapped-to DPs as

   - Signal *M,* and whether this is energy, information, or material
   - Other DPs and whether these are energy, information, or material

6. Plot the ideal function FR = $f(M)$ at a given constant DPs and absence of noise factors.

7. Plot the P-diagram of every FR using step 5.

8. Add the noise factors to all P-diagrams. *Noise factors* are uncontrollable factors that inhibit or affect FR delivery and generate soft and hard failure modes. The conceptual relation between scorecards, structures, and FMEA is depicted in Fig. 6.7. The noise factors are generally categorized as piece-to-piece variation (e.g., manufacturing), changes in dimension or strength over time (e.g., wear and fatigue), customer usage and duty cycle, external environment (climate, road conditions, etc.), and coupling (e.g., interaction with neighboring subsystems).

9. Aggregate the chains of P-diagrams in every hierarchical level into an overall structure using the precedence relationships in Sec. 6.3.2.

6.4 Design Scorecards and Transfer Function Development

The transfer functions in the physical and process structures are usually captured in *design scorecards,* which document and assess quantitatively the DFSS project progress, store the learning process, and show all critical elements of a design (CTSs, FRs, DPs, PVs) and their performance. Their benefits include documenting transfer functions

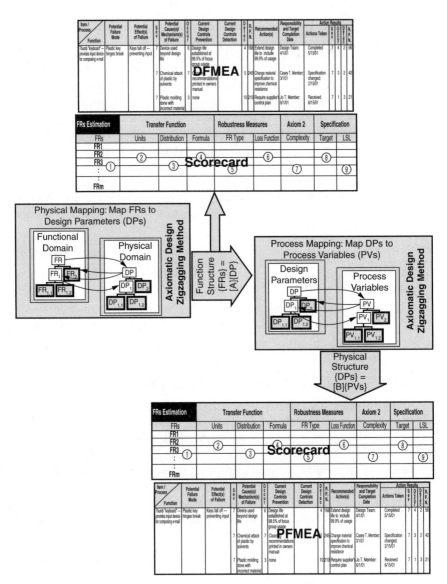

Figure 6.7 Design scorecards for transfer functions and FMEA.

and design optimization, predicting final results, enabling communication among all stakeholders in the project, and evaluating how well the design is supported by manufacturing and production processes.

The set of transfer functions of a given design are the means for dialing customer satisfaction and can be initially identified via the zigzagging method. A *transfer function* is a relationship that includes the concerned mapping linking, ideal function, and uncontrollable factors effect. Transfer functions can be mathematically derived, empirically obtained from a design of experiment or regression by using historical data. In several cases, no closed mathematical formula can be obtained and the DFSS team can resort to mathematical and/or simulation modeling. In the DFSS algorithm, there should be a transfer function (Fig. 6.8):

- For every functional requirement (FR) depicting the mapping {**FRs**} = [**A**] {**DPs**}

- For every design parameter (DP) depicting the mapping {**DPs**} = [**B**] {**PVs**}

- For every critical-to-satisfaction (CTS) requirement and ultimately for every customer attribute

Note that the transfer functions at a given hierarchy level are recorded in one scorecard. CTSs and customer attributes are each considered one level, in the DFSS algorithm, in their respective QFDs.

6.4.1 Transfer function development

Transfer functions are living entities in the DFSS algorithm. The life cycle of a transfer function in the DFSS algorithm passes into the following sequential stages (stages 1 to 6 are experienced in the DFSS project):

1. *Identification.* This is obtained by conducting the zigzagging method between the design domains.

2. *Uncoupling or decoupling.* This is effected by fixing, adding, replacing, and subtracting some of the independent variables in the codomain to satisfy design axioms.

3. *Detailing.* This is achieved by finding the cause-and-effect, preferably mathematical relationship, between the all variables after the "uncoupling/decoupling" stage in the concerned mapping domains. Detailing involves validating both the assumed relationship and the sensitivities of all independent variables.

4. *Optimization.* After detailing, the dependent variables in the transfer functions are optimized by either shifting their means or reducing their variation or both in the optimize (O) phase of the DFSS algorithm. This can be achieved by adjusting their mapped-to independent variables means and variance. This optimization propagates to the customer domain via the established relationships in the design mappings resulting in increased satisfaction.

5. *Validation.* The transfer function is validated in both structures.

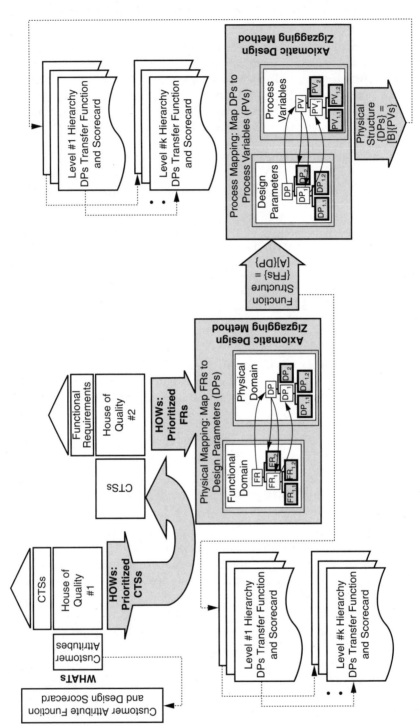

Figure 6.8 Transfer function and design scorecards in the DFSS algorithm.

6. *Maintenance.*   This is achieved by controlling all significant independent variables, after optimization, either in house or outside.

7. *Disposal.*   The design transfer functions are disposed or reach entitlement in delivering high-level FRs either when new customer attributes that can't be satisfied with the current design emerged or when the mean or controlled variance of the FRs are no longer acceptable by the customer. This stage is usually followed by the evolution of new transfer functions to satisfy the emerged needs.

8. *Evolution of a new transfer function.*   Per TRIZ, an evolution usually follows certain basic patterns of development. Evolutionary trends of the performance level of the functional requirements (FRs) of a certain design can be plotted over time and have been found to evolve in a manner that resembles an S curve (Fig. 6.9).

The following are the possible sources of "detailed" transfer functions:

1. Direct documented knowledge such as equations derived from laws of physics (e.g., force = mass × acceleration or voltage = current × resistance, profit = price − cost, interest formulas)

2. Derived transfer functions based on the direct documented knowledge as related to the project scope. For example, a DFSS project scope is to design a building elevator controller formed from many electric circuits. The team can rely on electric circuit analysis, which is based on direct documented knowledge of transfer functions such as

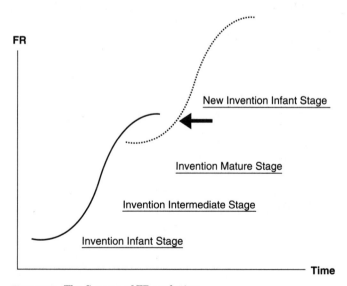

Figure 6.9   The S curve of FR evolution.

Ohm laws and digital logic circuit design and to derive their project FRs. The transfer functions obtained through this source is very dependent on the team's understanding and competency with their design and the discipline of knowledge it represents (engineering, social, economic, etc.). For example, the circuit in Fig. 6.10 [which is used as an example by Kapur and Lamberson (1977)] represents a series of three components where the output from the transformer goes to the amplifier and then to a phase-shifting synchronizer at the output of the amplifier with angle θ. We would like to have a transfer function of the output functional requirement V_0. Using electric circuit theory, we obtain $V_0 = V_2 \cos(\theta) + V_1 NK \sin(\theta)$ using Kirchhoff's laws, where N is the turns ratio of the transformer, K is the amplification multiplier, and V_1, V_2 are the input voltages of the transformer and the synchronizer, respectively.

3. Mathematical modeling using FR derivatives or sensitivities $(\partial FR_i/\partial DP_j)$, $(\partial FR_i/\partial M_k)$ with either the physical entity itself (prototype parts),the datum design, or a credible mathematical model. Sensitivities determine how an FR varies about a point in the design space. A design space is formed from [DPs, signal (M)]. A specific point in the design space triplet is usually referred to as a *design point, level,* or *setting.* The derivatives are estimated by the gradient at the design point. The gradient is determined by perturbing a design point axis, say, a DP, by a predetermined amount ΔDP and measuring the resulting perturbation in FR, ΔFR. The gradient is the ratio $\Delta FR/\Delta DP$ (see Fig. 6.11); that is, $\partial FR_i/\partial DP_j \cong \Delta FR_i/\Delta DP_j$. Therefore, the modeled transfer function of the FR is an approximation and can be written as

$$FR \cong \sum_{j=1}^{P} \frac{\Delta FR}{\Delta DP_j} DP_j + \sum_{k=1}^{K} \frac{\Delta FR}{\Delta M_k} M_k + \text{error (noise factors)} \quad (6.7)$$

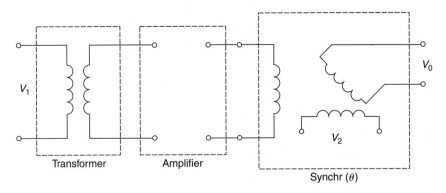

Figure 6.10   An electric circuit. [*From Kapur and Lamberson (1977).*]

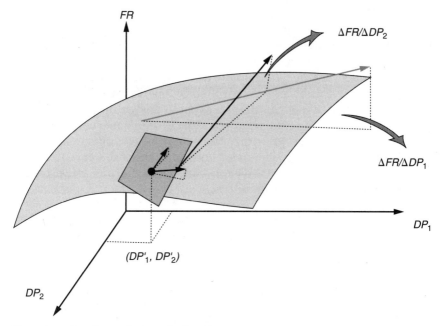

Figure 6.11 Gradients of a transfer function.

Notice that the transfer function in more than one dimension is either a surface (two variables) or a volume (more than two variables). In this method, we vary the parameters one at a time by an infinitesimal amount and observe ΔFR. Gradients of other FRs are collected simultaneously. This approach creates an incremental area or volume (formed from the infinitesimal variation in the respective parameters) around the design point of interest, where the transfer function will be valid. Extrapolation is not valid anywhere else for any FR.

This analysis identifies sensitive DPs that affect the mean of an FR, usually referred to as the *adjustment* parameters. An FR adjustment parameter will have a relatively large gradient magnitude compared to the other parameter gradients. DPs that affect the variance of an FR are called *variability* parameters. In the absence of historical data, the variance of an FR can be estimated using Taylor series expansion as

$$\sigma_{\text{FR}}^2 \cong \sum_{j=1}^{P+L+K} \underbrace{\left(\frac{\Delta\text{FR}}{\Delta x_i} \bigg|_{x\,=\,\mu} \right)^2}_{\text{sensitivities}} \underbrace{\sigma_j^2}_{\text{variance}} \tag{6.8}$$

where x stands for all the variables affecting the FR (signal, DPs, and noise) and L is the number of noise factors. The variance of the variables can be estimated from historical data or assessed. The worst-case scenario is usually used. Notice that the sensitivities are designed in by the team as they decide the DPs and the physical structure while the parameter variables are controlled by operations. This equation stresses the fact that a Six Sigma–capable design needs the contribution of both design and operations. The design experts in the team can use their expertise to modify the sensitivities and the operation members, the variances. We view this equation as the equation that captures the concurrent engineering concept in mathematical terms. The mean of the FR is estimated as

$$\mu_{FR} \cong \sum_{j=1}^{P+L+K} \left(\frac{\Delta FR}{\Delta x_j} \bigg|_{x_j = \mu_j} \right) \mu_j \tag{6.9}$$

In the absence of data, we usually tend to assume each x_j as normally distributed with mean μ_j and variance σ_j, and the variation around the nominal values, μ_j, is independent among the DPs within the x_j values.

4. Design of experiment (DOE) is another source of transfer function. In many perspectives, a DOE in another form of sensitivity analysis. DOE analysis runs the inputs throughout their completely experimental ranges, not through incremental area or volume. A DOE can be run using physical entities or mathematical and simulation models [e.g., Monte Carlo, CAD/CAM (computer-aided design/manufacturing), Simul8, SigmaFlow, Witness]. The predictive equation in MINITAB analysis is in essence a transfer function. The black belt may take the derivative of the predictive equation to estimate sensitivities. In simulation, the team needs to define DPs, signal, and noise factor distributions and parameters. The simulation model then samples from these distributions a number of runs and forms an output distribution. The statistical parameters of this distribution, such as the mean and variance, are then estimated. Afterward, the statistical inference analysis can be used. For example, assuming the FR normality, we can use the following Z value to calculate the DPM:

$$Z_{FR} = \frac{USL_{FR} - \mu_{FR}}{\sigma_{FR}}$$

(where USL = upper specification limit). Special DOE techniques are more appropriate such as response surface method (Chap. 17) and parameter design (Chaps. 14 and 15).

5. Regression transfer equations are obtained when the FRs are regressed over all input variables of interest. Multiple regressions coupled with multivariate analysis of variance (MANOVA) and Covariance (MANCOVA) are typically used.

6.4.2 Transfer function and optimization

*Transfer functions** are fundamental design knowledge to be treated as living documents in the design guides and best practices within Six Sigma deployment and outside the deployment initial reach. Transfer functions are usually recorded and optimized in design scorecards. Some of the transfer functions are readily available from existing knowledge. Others will require some intellectual (e.g., derivation) and monetary capital to obtain.

A transfer function is the means for optimization in the DFSS algorithm. *Optimization* is a design activity where we shift the mean to target and reduce the variability for all the responses in the DFSS project scope in the respective structure. However, this optimization is not arbitrary. The selection of DPs for an FR that will be used for optimization depends on the physical mapping and the design type from coupling perspective. If the design is uncoupled, that is, if there is a one-to-one mapping between FRs and DPs, then each FR can be optimized separately via its respective DP. Hence, we will have optimization studies equal to the number of FRs. If the design is decoupled, the optimization routine must follow the coupling sequence revealed by the design matrices in the structure (Chap. 7). In coupled scenarios, the selection of DPs to be included in the study depends on the potential control needed and affordable cost. The selection should be done in a manner that will enable target values to be varied during experiments with no major impact on design cost. The greater the number of potential design parameters that are identified, the greater the opportunity for optimization of function in the presence of noise.

A key philosophy of robust design is that during the optimization phase, *inexpensive parameters* can be identified and studied, and can be combined in a way that will result in performance that is insensitive to noise. The team's task is to determine the combined best settings (parameter targets) for each design parameter, which have been judged by the design team to have potential to improve the system. By varying the parameter target levels in the transfer function (design point), a region of nonlinearity can be identified. This area of nonlinearity is the most optimized setting for the parameter under study (Fig. 6.12).

---

*We are assuming continuity and existence of first-order derivatives in our discussions of transfer functions.

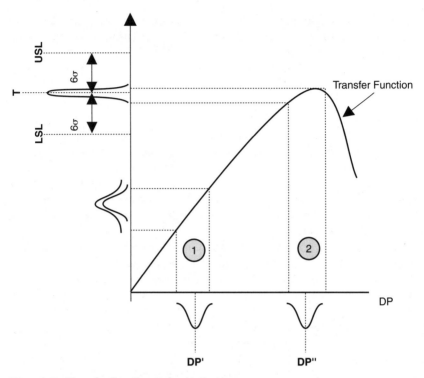

Figure 6.12 Transfer function and optimization.

Consider two levels or means of a design parameter (DP), level 1 (DP') and level 2 (DP"), having the same variance and distribution. It is obvious that level 2 produces less variation in the FR than level 1. Level 2 will also produce a lower quality loss (see Chaps. 13 to 15) similar to the scenario at the bottom of Fig. 6.13. The design produced by level 2 is more robust than that produced by level 1. When the distance between the specification limits is 6 times the standard deviation, a Six Sigma optimized FR is achieved. When all design FRs are released at this level, a Six Sigma design is obtained.

The black belt and the rest of the DFSS team needs to detail the transfer functions in context of the DFSS algorithm if they want to optimize, validate, and predict the performance of their project scope in the use environment.

However, the team should be cautious about the predictability of some of the transfer functions due to the stochastic effects of noise factors in the use environment that are particularly hard to predict. The team should explore all knowledge to obtain the transfer functions desired, including drawings. For example, stackup based on tolerances may contain descriptions of functionality that is based on lumped-

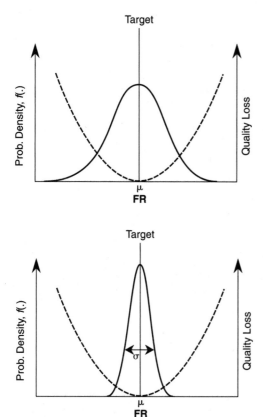

Figure 6.13 The quality loss function.

mass model and geometry. Transfer functions other than direct docu-mented knowledge are approximate and for the purposes of the DFSS algorithm are valid until proved wrong or disposed of by the evolution of other adopted technologies.

6.4.3 DFS scorecard development

The scorecard identifies which DPs contribute most to variation and mean in the response of the transfer function and the optimized design point. Tightening tolerances may be appropriate for these parameters. The proposed scorecard is depicted in Fig. 6.14. The team has the free-dom to customize the scorecard by adding more requirements.

Prior to discussing the entries of the scorecards we will make the fol-lowing remarks:

- The team should remember that a scorecard is related to hierarchical structure (see Fig. 6.8). There will be a number of scorecards equal to the number of hierarchical levels in the concerned structure.

DFSS Project Scorecard

Scope: Subsystem xxxx
Hierarchical Level: 1

Axiom 1 Measures (01)
Design Type: Coupled, Decoupled, Uncoupled
Reangularity = equation 6.4
Semangularity = equation 6.5

Design Mapping:
$$\begin{Bmatrix} FR1 \\ FR2 \\ FR3 \end{Bmatrix} = [\beta_1\ \beta_2\ \beta_3]\begin{Bmatrix} M_1 \\ M_2 \end{Bmatrix} + \begin{bmatrix} \beta_3 A_{11} & 0 \\ \beta_3 A_{21} & 0 \\ 0 & A_{31} \end{bmatrix}\begin{Bmatrix} DP_1 \\ DP_3 \end{Bmatrix} + \text{error (noise factors)}$$

Example — Ideal Function — Design Mapping (19)

FRs Estimation	Transfer Function			Robustness Measures		Axiom 2	Specification			Pred. FRs Capability			6σ Score			Actual	
FRs	Units	Distribution	Formula	FR Type	Loss Function	Complexity	Target	LSL	USL	μ	σ	Short/Long	z	z-shift	DPMO	μ	σ
FR1												L					
FR2 (2)			(4)		(6)		(8)		(10)		(12)	S	(14)		(16)		(18)
FR3 (1)		(3)		(5)		(7)		(9)		(11)		(13)		(15)		(17)	
⋮																	
⋮																	
FRm																	

DPs Estimation	Range			Sensitivity (Aij)	Specification		Axiom 2 Complexity	Pred. FRs Capability			6σ Score			Actual Capability		
	Units	Min	Max	Distribution	LSL	USL	Complexity	μ	σ	Short/Long	z	z-shift	DPMO	μ	σ	Short/Long
DP1																
DP2																
⋮																
⋮				(19)												
DPp																

DPs Estimation	Range			Sensitivity (Aij)	Specification		Axiom 2 Complexity	Pred. FRs Capability			6σ Score			Actual Capability		
	Units	Min	Max	Distribution	LSL	USL	Complexity	μ	σ	Short/Long	z	z-shift	DPMO	μ	σ	Short/Long
M1																
M2																
⋮				(20)												
⋮																

Noise Estimation	Noise Name	Units	Range		FRs Affected	Failure Mode
			Min	Max		
Customer Usage	N11					
	N12					
	⋮					
Piece-to-Piece	N21					
	N22		(21)			
	⋮					
Environment	N31				(22)	
	N32					
	⋮					
Coupling	N41					
	N42					
	⋮					
Degradation	N51					

Sensitivity Matrix

FRs \ DPs	DP1	DP2	..	..	..	DPp
FR1	A11	A12				
FR2	A21	A22				
FR3						
⋮						
⋮						
⋮						
⋮						
⋮			(19)			
⋮						
⋮						
FRm						Amp

Figure 6.14 Design scorecard.

- The scorecard is driven by the DFSS algorithm, and concepts such as coupling and loss function will be stressed.

- Noise factors and some times external signal factors can't be specified by the design team even when knowledge about them is readily available.

- We are using the physical structure for illustration purposes only, and the discussion here applies equally for process structure scorecards. Both structures set of scorecards are needed in the DFSS algorithm.

Notice that we numbered the entries with the scorecard for ease of reference in Fig. 6.14 (where DPMO is defects per million opportunities). After documenting the hierarchical level and scorecard scope in terms of the design mapping, the team needs to populate the scorecard with the following entries (where the listed numbers correspond to column numbers in Fig. 6.14):

0. Axiom 1 measures, which include type of design mapping in the structure addressed by the scorecard, *reangularity* (calculated) estimate using Eq. (6.4) and *semangularity* (calculated) estimate using Eq. (6.5). Both measures accurately estimate the degree of axiom 1 satisfaction, in particular, when nonlinear sensitivities exist. Entries of these two equations are documented in the sensitivity matrix in column 19 in Fig. 6.14.

1. List of all FRs within the scorecard scope as represented by the transfer function and design mapping.

2. Units used per the FRs; measurable and continuous FRs are expected.

3. The distribution of each FR is documented. Usually, "normal" is a safe assumption per the central-limit theorem (CLT) when the number of $(P + K + L)$ is greater than 5.

4. The transfer function equation per each FR is entered in this column. The team can make use of column 20 (Fig. 6.14) and the sensitivity matrix in column 19.

5. FR type according to robust design is documented to indicate the direction of optimization. Robust design optimization requires the use of one of four classifications of responses. These quality characteristics are classified by Dr. Taguchi as "the smaller, the better" (e.g., minimize vibration, reduce friction), "the larger, the better" (e.g., increase strength), "the [most] nominal, the best" (where keeping the product on a single performance objective is the main concern), and "dynamic" (where energy-related functional performance over a prescribed dynamic range of usage is the perspective).

6. The robust design loss function provides a better estimate of the monetary loss incurred as an FR deviates from its targeted performance value T. A *quality loss function* can be interpreted as a means to translate variation and target adjustment to monetary value. It allows the DFSS team to perform a detailed optimization of cost by relating design terminology to economical estimates. In its quadratic version, a quality loss is determined by first finding the functional limits* $T \pm \Delta FR$ for the concerned FR. The *functional limits* are the points at which the design would fail, producing unacceptable performance in approximately half of the customer applications. For further loss function literature, please refer to Chaps. 14 to 16. The expected loss function formulas by FR type as classified in Fig. 6.14 column 5 are as follows:

$$E[L(\text{FR},T)] = K[\sigma_{\text{FR}}^2 + (\mu_{\text{FR}} - T)^2] \quad \text{(nominal-the-best FR)}$$

$$(6.10)$$

$$E[L(\text{FR},T)] = K(\sigma_{\text{FR}}^2 + \mu_{\text{FR}}^2) \quad \text{(smaller-the-better FR)}$$

$$(6.11)$$

$$E[L(\text{FR},T)] = K\left[\frac{1}{\mu_{\text{FR}}^2} + \frac{3}{\mu_{\text{FR}}^4} \, \sigma_{\text{FR}}^2 \right] \quad (6.12)$$

The mean μ_{FR} and the variance σ_{FR}^2 of the FR are obtained from long-term historical data, short-term sampling, or using Eqs. (6.8) and (6.9).

$$[E[L(\text{FR},T)] = \int K\beta(\beta M - T)^2 \, g(\beta M) \, dM \quad \text{(dynamic)} \quad (6.13)$$

where g is the FR probability density function.

7. Calculate FR complexity in nats or bits per axiom 2. Depending on the FR distribution identified or assumed in Fig. 6.14 column 5, a complexity measure based on the entropy can be derived (see Sec. 6.5). For example, the complexity of a normally distributed FR is given by

$$h(\text{FR}_i) = \ln \sqrt{2\pi e \sigma_{\text{FR}i}^2} \quad (6.14)$$

where the FR variance is estimated from Eq. (6.8) in the absence of historical data. For other distributions, the DFSS team may

---

Functional limit or *customer tolerance* in robust design terminology is the design range in the axiomatic approach.

consult with their MBB for complexity-customized equations. A scorecard, say, s, complexity is the sum of entries in column 7:

$$h(\text{scorecard}_s) = \sum_{i=1}^{m_s} h(\text{FR}_i)$$

$$= \sum_{i=1}^{m_s} \ln \sqrt{2\pi e \sigma_{\text{FR}i}^2}$$

The sum of the complexity column in all hierarchal scorecards of a structure gives the structure complexity estimate as

$$h(\text{structure}) = \sum_{s=1}^{N} h(\text{scorecard}_s)$$

(where N = number of scorecards). The sum of the physical structure and the process structure complexities is the design complexity. Other forms of complexity are derived by El-Haik and Yang (1999).

The objective is to reduce complexity per axiom 2 by decreasing the variance of the FRs while achieving Six Sigma capability.

8. Enter the target of the FR as obtained from the phase 2 QFD.

9. Enter the lower FR specification limit (LSL) as obtained from the phase 2 QFD, if applicable.

10. Enter the upper FR specification limit (USL) as obtained from the phase 2 QFD, if applicable.

11. The predicted FR capability for the mean is entered here as estimated from Eq. (6.9), sampling, or historical long-term data.

12. The predicted FR capability for variance is entered here as estimated from Eq. (6.8), sampling, or historical long-term data.

13. Enter (L) for long and (S) for short if the sampling was used.

14. Calculate the Z value.

15. Use the historical shift if available, or estimate.

16. Use the tables to calculate the DPMO.

17. Record the historical mean of the FR, if available.

18. Record the standard deviation of the FR, if available.

19. Enter the sensitivities in the matrix as derived or obtained from the transfer function sources in Sec. 6.4.1.

20. Enter the ideal function sensitivity as derived or obtained from Sec. 6.4.1 transfer function detailing methods.

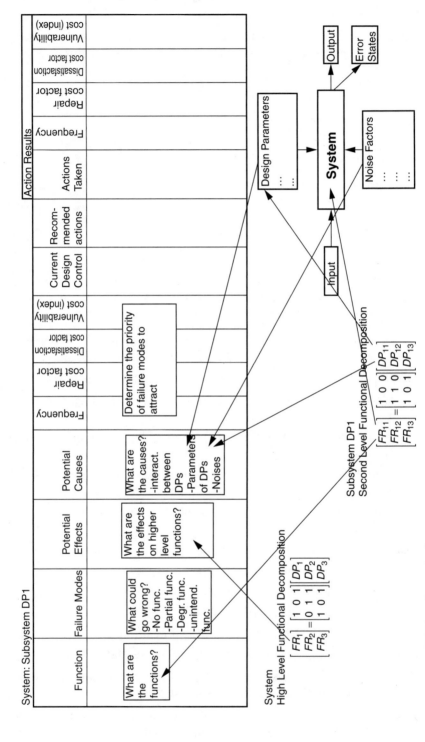

Figure 6.15 FMEA and transfer function.

21. List the noise factor units and ranges as brainstormed in the P-diagram or obtained from transfer function detailing activity. Against each noise factor, list the FRs affected from column 1. This column will act as a reminder for the DFSS team not to lose sight of the noise factors.

22. Enter the failure modes per the FRs affected in column 22.

Refer to Fig. 6.15.

Chapter

7

Quality Function Deployment (QFD)

7.1 Introduction

In the context of DFSS, QFD is best viewed as a planning tool that relates a list of delights, wants, and needs of customers to design technical functional requirements. With the application of QFD, possible relationships are explored between quality characteristics as expressed by customers and *substitute quality requirements* expressed in engineering terms (Cohen 1988, 1995). In the context of DFSS, we will call these requirements *critical-to* characteristics, which include subsets such as critical-to-quality (CTQ) and critical-to-delivery (CTD). In the QFD methodology, customers define the product using their own expressions, which rarely carry any significant technical terminology. The *voice of the customer* can be discounted into a list of needs used later as input to a relationship diagram, which is called QFD's *house of quality*.

The knowledge of customer needs is a "must" requirement in order for a company to maintain and increase its position in the market. Correct market predictions are of little value if the requirements cannot be incorporated into the design at the right time. Critical-to-innovation and critical-to-market characteristics are vital because companies that are first to introduce new concepts at Six Sigma (6σ) levels usually capture the largest share of the market. Wrestling market share away from a viable competitor is more difficult than it is for the first producer into a market. One major advantage of a QFD is the attainment of shortest development cycle, which is gained by companies with the ability and desire to satisfy customer expectation. The other significant advantage is improvement gained in the design family of the company, resulting in increased customer satisfaction.

The team should take the time required to understand customer wants and to plan the project more thoughtfully. Using the QFD, the DFSS team will be able to anticipate failures and avoid major downstream changes. Quality function deployment prevents downstream changes by an extensive planning effort at the beginning of the DFSS design or redesign project. The team will employ marketing and product planning inputs to deploy the customer expectations through design, process, and production planning and across all functional departments. This will assure resolution of issues, lean design, and focusing on those potential innovations (delighters) that are important to the customer.

Figure 7.1 shows that the company which is using QFD places more emphasis on responding to problems early in the design cycle. Intuitively, it incurs more time, cost, and energy to implement a design change at production launch than at the concept phase because more resources are required to resolve problems than to preclude their occurrence in the first place.

Quality function deployment (QFD) translates customer needs and expectations into appropriate design requirements. The intent of QFD is to incorporate the " voice of the customer" into all phases of the product development cycle, through production and into the marketplace. With QFD, quality is defined by the customer. Customers want products, processes, and services that throughout their lives meet customers' needs and expectations at a cost that represents value. The results of being customer-driven are total quality excellence, greater customer satisfaction, increased market share, and potential growth.

The real value of QFD is its ability to direct the application of other DFSS tools such as SPC and robustness to those entities that will have

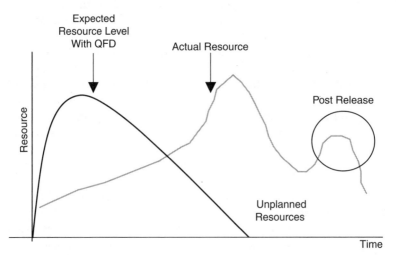

Figure 7.1 QFD effect on project resources.

the greatest impact on the ability to design and satisfy the needs of the customers, both internal and external. Quality function deployment is a *zooming* tool that identifies the significant design elements on which to focus design and improvement efforts and other resources. In the context of QFD, planning is key and is enhanced by reliable information in benchmarking and testing.

The objectives of this chapter are to

1. Provide the black belts, green belts, and other readers with the knowledge and skills they need to define quality function deployment.

2. Recognize and identify the four key elements of any QFD chart.

3. Have a basic understanding of the overall four phases of the QFD methodology.

4. Define the three quality features of the Kano model.

5. From the process standpoint, place QFD within the DFSS algorithm as highlighted in Chap. 5.

7.2   History of QFD

QFD was created by Mitsubishi Heavy Industry at Kobe Shipyards in the early 1970s. Stringent government regulations for military vessels coupled with the large capital outlay per ship forced Kobe Shipyard's management to commit to upstream quality assurance. The Kobe engineers drafted a matrix which relates all the government regulations, critical design requirements, and customer requirements to company technical controlled characteristics of how the company would achieve them. In addition, the matrix also depicted the relative importance of each entry, making it possible for important items to be identified and prioritized to receive a greater share of the available company resources.

Winning is contagious. Other companies adopted QFD in the mid-1970s. For example, the automotive industry applied the first QFD to the rust problem. Since then, QFD usage has grown as a well-rooted methodology into many American businesses. It has become so familiar because of its adopted commandment: "Design it right the first time."

7.3   QFD Benefits, Assumptions, and Realities

The major benefit of QFD is customer satisfaction. QFD gives the customers what they want, such as shorter development cycles, avoidance of a failures and redesign peaks (Fig. 7.1) during prelaunch, and

"know-how" knowledge as it relates to customer demand tha
served and transferred to the next design teams.

Certain assumptions are made before QFD can be implen
They include (1) forming a multidisciplinary DFSS team per st
the DFSS algorithm and (2) more time spent upstream understa
customer needs and expectations and defining the product or s
in greater detail.

There are many initial realistic concerns, which must be addr
in order to implement QFD successfully. For example, departn
represented in the team don't tend to talk to one another. In add
market research information that is not technically or design-foc
with QFD is more easily applied to incremental design than to brand
creative design. The traditional reality "problem prevention is not
rewarded as well as problem solving" will be faced initially by the
DFSS team. This reality will fade away as the team embarks on their
project using the rigor of the DFSS.

7.4   QFD Methodology Overview

Quality function deployment is accomplished by multidisciplinary DFSS
teams using a series of charts to deploy critical customer attributes
throughout the phases of design development. QFD is usually deployed
over four phases. The four phases are phase 1—CTS planning, phase
2—functional requirements, phase 3—design parameters planning,
and phase 4—process variables planning. Figure 7.2 shows these four

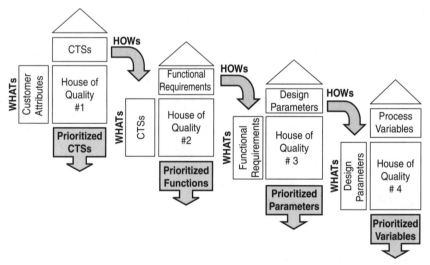

Figure 7.2   The four phases of QFD.

phases and their components. Each phase will be discussed in detail later in this chapter.

The position of QFD is best described by the block diagram in Fig. 7.3, modeled as a closed loop to reflect the ongoing design activities. The figure indicates the QFD iterative step-by-step process as represented by the loop of customer-QFD–physical solution entity. In this feedback loop, let A = customer needs, B = QFD analysis, C = desired designed entity (product/service/process), D = customer satisfaction gauge (e.g., surveys), E = other DFSS tools and concepts; then the gain C/A is given by $B/(1+BDE)$. Two general analysis activities occur in the loop on a continuous basis: the forward customer analysis activities of QFD (block B) and the backward customer analysis activities (block D). If the block product $BDE \gg 1$, then $C/A = (DE)^{-1}$. This means that our analytical capabilities should cope with our ability to synthesize concepts (Suh 1990).

QFD uses many techniques in an attempt to minimize and ease the task of handling large numbers of functional requirements that might be encountered. Applications in the range of 130 (engineering functions) $\times$ 100 (customer features) were recorded (Hauser and Clausing 1988). One typical grouping technique that may be used initially in a QFD study is the *affinity diagram*, which is a hierarchical grouping technique used to consolidate multiple unstructured ideas generated by the *voice of the customer.* It operates on the basis of intuitive similarities that may be detected from low-level standalone ideas (bottom) to arrangements of classes of ideas (up). This bundling of customer's features is a critical step. It requires a cross-functional team that has multiple capabilities such as the ability to brainstorm, evaluate, and revolutionize existing ideas in pursuit of identifying logical (not neces-

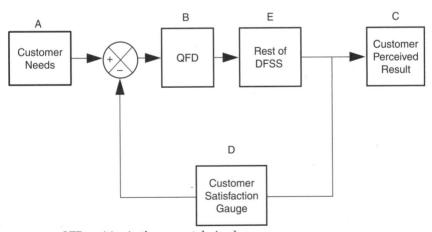

Figure 7.3   QFD position in the current design loop.

sarily optimum) groupings and hence, minimizing the overall list of needs into manageable classes.

Another technique is the *tree diagram,* which is a step beyond the affinity diagram. The *tree diagram* is used mainly to fill the gaps and cavities not detected previously in order to achieve a more completed structure leading to more ideas. Such expansion of ideas will allow the structure to grow but at the same time will provide more vision into the voice of the customer (Cohen 1988).

The "house of quality" (Fig. 7.4) is the relationship foundation of QFD. Employment of the house will result in improved communication, planning, and design activity. This benefit extends beyond the QFD team to the whole organization. Defined customer wants through QFD can be applied to many similar products and form the basis of a corporate memory on the subject of critical-to-satisfaction requirements (CTSs). As a direct result of the use of QFD, *customer intent* will become the driver of the design process as well as the catalyst for modification to design solution entities. The components that constitute the phase 1

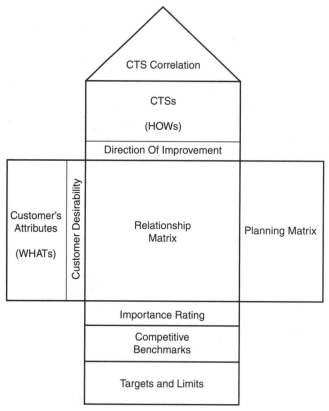

Figure 7.4 House of quality.

house of quality (Cohen 1988) are illustrated in Fig. 7.4 and described in Secs. 7.4.1 to 7.4.9.

7.4.1 Customer attributes (WHATs)

These are obtained from the *voice of customer* as represented by surveys, claim data, warranty, and promotion campaigns. Usually customers use fuzzy expressions in characterizing their needs with many dimensions to be satisfied simultaneously. *Affinity* and *tree diagrams* may be used to complete the list of needs. Most of these WHATs are very general ideas that require more detailed definition. For example, customers often say that they look for something "stylish" or "cool" when they purchase a product. "Coolness" may be a very desirable feature, but since it has different interpretations for different people, it cannot be acted on directly. Legal and safety requirements or other internal wants are considered extensions to the WHATs. The WHATs can be characterized using the Kano model (Sec. 7.5).

7.4.2 HOWs

The design features derived by the DFSS team to answer the WHATs are called the *HOWs*. Each initial WHAT requires operational definitions. The objective is to determine a set of critical-to-satisfaction (CTS) requirements with which WHATs can be materialized. The answering activity translates customer expectations into design criteria such as speed, torque, and time to delivery. For each WHAT, there should be one or more HOWs describing a means of attaining customer satisfaction. For example, a "cool car" can be achieved through a "stylish" body (different and new), seat design, legroom, lower noise, harshness, and vibration requirements. At this stage only overall requirements that can be measured and controlled need to be determined. These substitute for the customer needs and expectations and are traditionally known as *substitute quality characteristics*. In this book, we will adopt the critical-to terminology aligning with Six Sigma.

Teams should define the HOWs in a solution-neutral environment and not be restricted by listing specific parts and processes. Itemize just the means (the HOWs), from which the list of WHATs can be realized. The one-to-one relationships are not the *real world,* and

Direction of Improvement		
Maximize	↑	1.0
Target	○	0.0
Minimize	↓	−1.0

many HOWs will relate to many customer wants. In addition, each HOW will have some direction of goodness or improvement of the following:

The circle represents the nominal-the-best target case.

7.4.3 Relationship matrix

The process of relating "WHATs" to "HOWs" often becomes complicated by the absence of one-to-one relationships as some of the HOWs affect more than one WHAT. In many cases, they adversely affect one another. HOWs that could have an adverse effect on another customer want are important. For example, "cool" and "stylish" are two of the WHATs that a customer would want in a vehicle. The HOWs that support "cool" include lower noise, roominess, and seat design requirements. These HOWs will also have some effect on the "stylish" requirement as well. A relationship is created in the house of quality (HOQ) between the HOWs as columns and the WHATs in the rows. The relationship in every (WHAT, HOW) cell can be displayed by placing a symbol representing the cause-effect relationship strength in that cell. When employees at the Kobe Shipyards developed this matrix in 1972, they used the local horse racing symbols in their QFD as relationship matrix symbols; for instance, the double-centered circle means strong relationship, one circle means medium strength, and a triangle indicates a weak relationship. Symbols are used instead of direct numbers because they can be identified and interpreted easily and quickly. Different symbol notations have been floating around, and we found the following to be more common than others:

Standard 9-3-1		
Strong	●	9.0
Moderate	○	3.0
Weak	▽	1.0

After determining the strength of each (WHAT,HOW) cell, the DFSS team should take the time to review the relationship matrix. For example, blank rows or columns indicate gaps in either team's understanding or deficiency in fulfilling customer attributes. A blank row shows a need to develop a HOW for the WHAT in that row, indicating a potentially unsatisfied customer attribute. When a blank column exists, then one of the HOWs does not impact any of the WHATs. Delivering that HOW may require a new WHAT that has not been identified, or it might be a waste. The relationship matrix gives the DFSS team the opportunity to revisit their work, leading to better planning and therefore better results.

What is needed is a way to determine to what extent the CTS at the head of the column contributes to meeting customer attributes at the left of the row. This is a subjective weighing of the possible cause-effect relationships.

To rank-order the CTS and customer features, we multiply the numerical value of the symbol representing the relationship by the *customer desirability index.* This product, when summed over all the customer features in the WHATs array, provides a measure of the relative importance of each CTS to the DFSS team and is used as a planning index to allocate resources and efforts, comparing the strength, importance, and interactions of these various relationships. This importance rating is called the *technical importance rating.*

7.4.4   Importance ratings

Importance ratings are a relative measure indicating the importance of each WHAT or HOW to the design. In QFD, there are two importance ratings:

- *Customer desirability index.* This is obtained from the voice of the customer activities such as surveys and clinics, and is usually rated on the scale from 1 (not important) to 5 (extremely important) as follows:

Importance		
Extremely Important	◆	5.0
Very Important	◈	4.0
Somewhat Important	◇	3.0
A Little Import	◈	2.0
Not Important	◇	1.0

- *Technical importance ratings.* These are calculated as follows:

    1. By convention, each symbol in the relationship matrix receives a value representing the strength in the (WHAT,HOW) cell.
    2. These values are then multiplied by the customer desirability index, resulting in a numerical value for symbol in the matrix.
    3. The technical importance rating for each HOW can then be found by adding together the values of all the relationship symbols in each column.

The technical importance ratings have no physical interpretation, and their value lies in their ranking relative to one another. They are utilized to determine what HOWs are priority and should receive the most resource allocation. In doing so, the DFSS team should use the technical importance rating as a campus coupled with other factors such as difficulty, innovation, cost, reliability, and timing and all other measures in their project charter.

7.4.5   Planning matrix

This task includes comparisons of competitive performance and identification of a benchmark in the context of ability to meet specific customer needs. It is also used as a tool to set goals for improvement using a ratio of performance (goal rating/current rating). Hauser and Clausing (1988) view this matrix as a perceptual map in trying to answer the following question: How can we change the existing product or develop a *new* one to reflect customer intent, given that the customer is more biased toward certain features? The product of *customer value,* the *targeted improvement ratio* for the raw (feature), and the *sales point,* which is a measure of how the raw feature affects sales, will provide a weighted measure of the relative importance of this customer feature to be considered by the team.

7.4.6 HOWs correlation (the roof)

Each cell in the roof is a measure of the possible correlation of two different HOWs. The use of this information improves the team's ability to develop a systems perspective for the various HOWs under consideration.

Designing and manufacturing activities involve many trade-off decisions, due mainly to the violation of design axioms (Chap. 8). The correlation matrix is one of the more commonly used optional extensions over the original QFD developed by Kobe engineers. Traditionally, the major task of the correlation matrix is to make trade-off decisions by identifying the qualitative correlations between the various HOWs. This is a very important function in the QFD because HOWs are most often *coupled.* For example, a matrix contains "quality" and "cost." The design engineer is looking to decrease cost, but any improvement in this aspect will have a negative effect on the quality. This is called a *negative correlation* and must be identified so that a trade-off can be addressed. Trade-offs are usually accomplished by revising the long-term objectives (HOW MUCHs). These revisions are called *realistic objectives.* Using the negative correlation example discussed previously, in order to resolve the conflict between cost and quality, a cost objective would be changed to a realistic objective. In the correlation matrix, once again, symbols are used for ease of reference to indicate the different levels of correlation with the following scale:

Trade-Offs		
Synergy	+	1.0
Compromise	–	–1.0

In a coupled design scenario, both positive and negative interaction may result. If one HOW directly supports another HOW, a positive correlation is produced.

Correlations and coupling can be resolved only through conceptual methods such as TRIZ (Chap. 9) and axiomatic design (Chap. 8). Otherwise, a couple design results and trade-offs are inevitable, leading to compromised customer satisfaction with design physics.

Many of the coupling situations that occur are the result of a conflict between design intent and the laws of physics. Two DFSS tools, TRIZ and axiomatic design, are aimed at handling such conflicting requirements by providing principles and tools for resolution. In many cases, the laws of physics win mainly because of the ignorance of design teams. In several transactional DFSS projects, coupling situations may have to be resolved by high-level management because departmental and sectional functional lines are being crossed.

7.4.7 Targets or (HOW MUCH)

For every HOW shown on the relationship matrix, a HOW MUCH should be determined. The goal here is to quantify the customers' needs and expectations and create a target for the design team. The HOW MUCHs also create a basis for assessing success. For this reason, HOWs should be measurable. It is necessary to review the HOWs and develop a means of quantification. Target orientation to provide visual indication of target type is usually optional. In addition, the tolerance around targets needs to be identified according to the company marketing strategy and contrasting it with that of best-in-class competitors. This tolerance will be cascaded down using the axiomatic design method.

7.4.8 Competitive assessments or benchmarking

Competitive assessments are used to compare the competition's design with the team design. There are two types of competitive assessments:

- *Customer competitive assessment.* This is found to the right of the relationships matrix in the planning matrix. Voice-of-customer (VOC) activities (e.g., surveys) are used to rate the WHATs of the various designs in a particular segment of market.

- *Technical competitive assessment.* This is located at the bottom of the relationships matrix. It rates HOWs for the same competitor against HOWs from a technical perspective.

Both assessments should be aligned, and a conflict between them indicates a failure to understand the VOC by the team. In a case like this, the team needs to revisit the HOWs array and check their understanding and contrast that understanding with VOC data. Further research may be needed. The team may then add new HOWs that reflect the customer perceptions. Any unexpected items that violate conventional wisdom should be noted for future reference. Situations like this can be resolved only by having the DFSS team involved in the QFD, not only marketing people, comparing competitive designs. In this way, the team who is responsible for designing for customer attributes will interpret exactly what those wants are.

7.4.9   Other optional QFD chart extensions

The following items may be included for some QFD chart, which should be adapted to address the needs of the user such as *technology roadmaps,* to direct design toward the planned technology evolution and to prioritize resources, and service complaint and repair history.

7.5   Kano Model of Quality

In QFD, the "voice of the customer" activities such as market research, provide the array of WHATs that represent the customer attributes. Such WHATs are "spoken" by the customer and are called "performance quality" or "one-dimensional." However, more WHATs have to be addressed than just those directly spoken by the customer. As Fig. 7.5 shows, there are also "unspoken" WHATs. Unspoken WHATs are the basic features that customers automatically assume they will have in the design. Such WHATs are implied in the functional requirements of the design or assumed from historical experience. For example, customers automatically expect their lawnmowers to cut grass to the specified level, but they wouldn't discuss it on a survey unless they had trouble with one in the past. Unspoken wants have a "weird" property— they don't increase customer satisfaction. However, if they are not delivered, they have a strong negative effect on customer satisfaction.

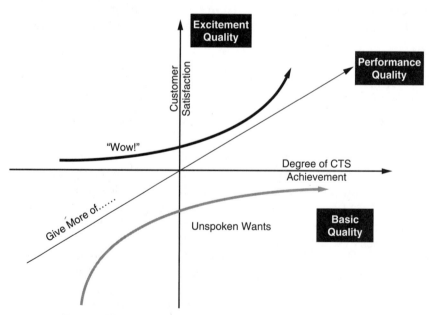

Figure 7.5   Kano model of customer attributes.

Another group of "unspoken" WHATs can be categorized as innovations or delighters. These pleasant surprises increase customer satisfaction in nonlinear fashion. For example, in the automotive industry, van owners were delighted by the second van side door and by baby seat anchor bolts.

Design features may change position on the Kano model over time. In the 1990s, the second side door in a caravan was a pleasant surprise for customers. Now, on most models, the second door is standard and expected to be installed without a specific request. The ideal DFSS project plan would include all three types of quality features: excitement quality (unspoken latent demands), performance quality (spoken and one-dimensional), and basic quality (unspoken or assumed).

7.6   The Four Phases of QFD

In the DFSS algorithm, including QFD, no single chart can accommodate all the information traded by the team to plan, control, and document their entire project development cycle. Targets and tolerances of the HOWs must be specified, team tasks must be clarified, and potential failures must be identified and countermeasures taken. In the DFSS algorithm, the QFD house-of-quality (HOQ) matrices have to be developed to plan the design and its production processes and controls and the procedures. Our experience indicates that an average QFD study will require many more charts than the four phases of QFD may propose.

The first QFD translates the customer needs and expectations into the CTSs and later into design actions. This conversion is completed by constructing a new relationship matrix (HOQ phase 2) on which WHATs are the CTSs and their target values from the previous matrix. The HOWs and HOW MUCHs of each matrix are progressively deployed as WHATs on the matrices that represent the next phase of the development cycle. This conversion of HOWs to WHATs is continued from design planning to process planning and finally to production planning.

This procedure should be continued until completion of production planning. As illustrated in Fig. 7.2, this entails deploying the customer requirements into CTSs, which are then deployed design parameters and then to process variables. At this point production requirements can be developed and QFD process is completed. Although only four charts are shown in the illustration, we suggest using the first phase of QFD and then proceeding with axiomatic design zigzagging process for cascading requirements. This will produce the hierarchy of the physical structure while employing design axioms.

QFD provides an efficient method to funnel the list of available options. Important, new, difficult, and high-risk HOWs are identified and moved to the next phase for further cascading and design detailing. The set of HOWs that do not require any special focus are not tracked to allow for the most effective use of team resources. This also ensures that HOWs critical to meeting the customer attributes receive the optimum allocation of time and resources as early as possible.

7.7   QFD Analysis

Completion of the first QFD house of quality may give the DFSS team a false impression that their job is completed. In reality, all their work to this point has been to create a tool that will guide future efforts toward deploying the VOC into the design. QFD matrix analysis in every phase will lead to the identification of design weaknesses, which must be dealt with as potential strength opportunities to be "best in class." A relatively simple procedure for analyzing the HOQ phase is provided below:

- *Blank or weak columns.*   HOWs that don't strongly relate to any customer attribute.

- *Blank or weak rows.*   Customer attributes that are not being strongly addressed by a HOW.

- *Conflicts.*   Technical competitive assessment that is in conflict with customer competitive assessment.

- *Significance.*   HOWs that relate to many customer attributes, safety/regulatory, and internal company requirements.

- *"Eye opener" opportunities.*   The team's company and competitors are doing poorly. The DFSS team should seize the opportunity to deliver on these sales points, which may be treated as delighters in the Kano model initially.

- *Benchmarking.*   Opportunities to incorporate the competitor's highly rated HOWs. The team should modify and incorporate using benchmarking and not resort to creation.

- *Deployment.*   Significant HOWs that need further deployment and work in phase 2, design parameters deployment.

7.8   QFD Example

The QFD example* is adapted with some alterations to illustrate the QFD diagnostics by a DFSS team.

---

*The example is contributed by Dave Roy, master black belt of Textron Inc.

7.8.1 QFD example highlights

The following are the highlights of the QFD example:

Project objectives. Design a global commercial process with Six Sigma performance.

Project problem statement

- Sales cycle time (lead generation to full customer setup) exceeds 182 business days. Internal and external customer specifications range from 1 to 72 business days.
- Only 54 percent of customer service requests are closed by the commitment date. The customers expect 100 percent of their service requests to be completed on time.
- Nonstandard commercial processes, none of which are Six Sigma–capable.

Business case

- There is no consistent, global process for selling to, setting up, and servicing accounts.
- Current sales and customer service information management systems do not enable measurement of accuracy and timeliness on a global basis.
- Enterprisewide customer care is a "must be" requirement—failure to improve the process threatens growth and retention of the portfolio.

Project goals

- Reduce prospecting cycle time from 16 to 5 business days.
- Reduce discovery cycle time from 34 to 10 business days.
- Reduce the deal-closing cycle time from 81 to 45 business days (net of all sales metrics customer wait time).
- Reduce setup cycle time from 51 to 12 business days.
- Increase the percentage of service requests closed by commitment date from 54 percent (1.6σ) to 99.97 percent (5.0σ).

7.8.2 QFD example steps

The basic QFD steps are described in the following paragraphs.

Step 1: Identify the WHATs and HOWs and their relationship. The DFSS team identifies customers and establishes customer wants, needs, delights, and usage profiles. Corporate, regulatory, and social requirements should also be identified. The value of this step is to greatly improve the understanding and appreciation DFSS team members have for customer, corporate, regulatory, and social requirements. The DFSS team, at this stage, should be expanded to include market research. A market research professional might help the black belt assume leadership during startup activities and perhaps later remain active participants as the team gains knowledge about customer engagement methods. The black belt should put plans in place to collaborate with identified organizations and/or employee relations to define tasks and plans in support of the project, and train team members in customer processes and forward-thinking methods such as brainstorming, visioning, and conceptualizing.

The DFSS team should focus on the key customers to optimize decisions around them and try to include as many additional customers as possible. The team should establish customer environmental conditions, customer usage, and operating conditions; study customer demographics and profiles; conduct customer performance evaluations; and understand the performance of the competition. In addition, the team should

- Establish a rough definition of an ideal service.

- Listen to the customer and capture wants and needs through interviews, focus groups, customer councils, field trials, field observations, and surveys.

- Analyze customer complaints and assign satisfaction performance ratings to attributes.

- Acquire and rank these ratings with the quality function deployment (QFD) process.

- Study all available information about the service including marketing plans.

- Create innovative ideas and delights, new wants by investigating improved functions and cost of ownership, and benchmarking the competition to improve weak areas.

- Create new delights by matching service functions with needs, experience, and customer beliefs. Innovate to avoid compromise for bottlenecks, conflicts, and constraints.

The following WHATS are used:

Direction of improvement
Available products
Professional staff
Flexible processes
Knowledgeable staff
Easy-to-use products
Speedy processes
Cost-effective products
Accuracy

Step 2: Identify the HOWs and the relationship matrix. The purpose of this step is to define a "good" product or process in terms of customer expectations, benchmark projections, institutional knowledge, and interface requirements, and to translate this information into CTS metrics. These will then be used to plan an effective and efficient DFSS project.

One of the major reasons for customer dissatisfaction and warranty costs is that the design specifications do not adequately reflect customer use of the product or process. Too many times the specification is written after the design is completed, or it is simply a reflection of an old specification that was also inadequate. In addition, poorly planned design commonly does not allocate activities or resources in areas of importance to customers and wastes engineering resources by spending too much time in activities that provide marginal value. Because missed customer requirements are not targeted or checked in the design process, procedures to handle field complaints for these items are likely to be incomplete. Spending time overdesigning and overtesting items not important to customers is futile. Similarly, not spending development time in areas important to customers is a missed opportunity, and significant warranty costs are sure to follow.

In DFSS, time is spent upfront understanding customer wants, needs, and delights together with corporate and regulatory requirements. This understanding is then translated into CTS requirements (CTSs), which then drive product and process design. The CTSs (HOWs) as well as the relationship matrix to the WHATs are given in the following table:

Importance to the customer
Meet time expectations
Know my business & offers
Save money/enhance productivity
Do it right the 1st time
Consultative
Know our products & processes
Talk to 1 person
Answer questions
Courteous
Adequate follow-up

A mapping begins by considering the high-level requirements for the product or process. These are the true CTSs which define what the customer would like if the product or process were ideal. This consideration of a product or process from a customer perspective must address the requirements from higher-level systems, internal customers (such as manufacturing, assembly, service, packaging, and safety), external customers, and regulatory legislation. Customer WHATs are not easily operational in the world of the black belt. For this reason it is necessary to relate true quality characteristics to CTSs—design characteristics that may be readily measured and, when properly targeted, will substitute or assure performance to the WHATs. This diagram, which relates true quality characteristics to substitute quality characteristics, is called a *relationship matrix,* the logic of which is several levels deep. A tree diagram, one of the new seven management tools, is commonly used to create the logic associated with the customer. The mapping of customer characteristics to CTS characteristics is extremely valuable when done by the DFSS team. A team typically begins with differing opinions and sharing stories and experiences when the logic is only a few levels deep. An experiment may even be conducted to better understand the relationships. When this experiment is completed, the entire team understands how product and process characteristics that are detailed on drawings relate to functions that are important to customers.

The full phase 1, 2, and 3 QFDs are given in Figs. 7.6 to 7.10. Our analysis below applies to phase 1. The reader is encouraged to apply such analysis on the other phases as well.

7.8.3   The HOWs importance calculation

Importance ratings are a relative comparison of the importance of each WHAT or HOW to the quality of the design. The 9-3-1 relationship matrix strength rating is used. These values are multiplied by the customer importance rating obtained from customer engagement activities (e.g., surveys), resulting in a numerical value. The HOWs importance rating is summed by adding all values of all relationships. For example, the first HOW of the Fig. 7.7 importance rating is calculated as $2.0 \times 3.0 + 4.0 \times 3.0 + 4.0 \times 3.0 + 4.0 \times 3.0 + 5.0 \times 9.0 + 5.0 \times 3.0 = 102$. Other HOW importance ratings can be calculated accordingly.

Phase 1 QFD diagnostics are described in the following paragraphs.

Weak WHATs.   The black belt needs to identify WHATs with only weak or no relationships. Such situations represent failure to address a customer attribute. When this occurs, the company should try to develop CTS(s) to address this WHAT. Sometimes the team may discover that present technology can't satisfy the WHAT. The DFSS team should resort to customer survey and assessment for review and further understanding.

No such WHAT exists in our example. The closest to this situation is "Available products" in row 1 and "Easy-to-use products" in row 5.

Direction of Improvement

Maximize	↑	1.0
Target	○	0.0
Minimize	↓	−1.0

	#	Importance to the Customer	Meet Time Expectations	Know My Business and Offers	Save Money/Enhance Probability	Do it Right the First Time	Consultative	Know Our Products and Processes	Talk to One Person	Answer Questions	Courteous	Adequate Follow-Up
			1	1	2	3	4	5	6	7	8	9	10
Direction of Improvement	1		↑	↑	↑	↑	↑	↑	↑	↑	↑	↑
Available Products	1	2.0	○		○		▽		○			
Professional Staff	2	3.0		▽		▽	○	●	▽		●	
Flexible Processes	3	4.0	○					●				
Knowledgeable Staff	4	4.0	○	●	○	●	●	●	●	●		▽
Easy-to-Use Products	5	4.0	○		○	○		▽	○			▽
Speedy Processes	6	5.0	●		●	○		○	○	○		▽
Cost-Effective Products	7	5.0	○	●	●	○		●	○			
Accuracy	8	5.0		●		●						
	9											

Figure 7.6 The WHATs, the HOWs, and the relationship matrix.

This was highlighted as the weakest WHAT but not weak enough to warrant the analysis above. However, the team is encouraged to strengthen this situation by a CTS with a strong relationship.

Weak HOWs. The team needs to look for blank or weak HOWs (in which all entries are inverted deltas). This situation occurs when CTSs are included that don't really reflect the customer attributes being addressed by the QFD. The black belt (BB) and the BB team may consider eliminating the CTSs from further deployment if it does not relate basic quality or performance attributes in the Kano model. The theme of DFSS is to be customer-driven and work on the right items; otherwise, we are creating a design "hidden factory."

In our example, The CTS "adequate follow-up" is weak (rated 13 on the importance rating scale). However, the WHAT "easy-to-use products" has no strong relationship with any CTSs, and eliminating "adequate follow-up" may weaken the delivery of this WHAT even further.

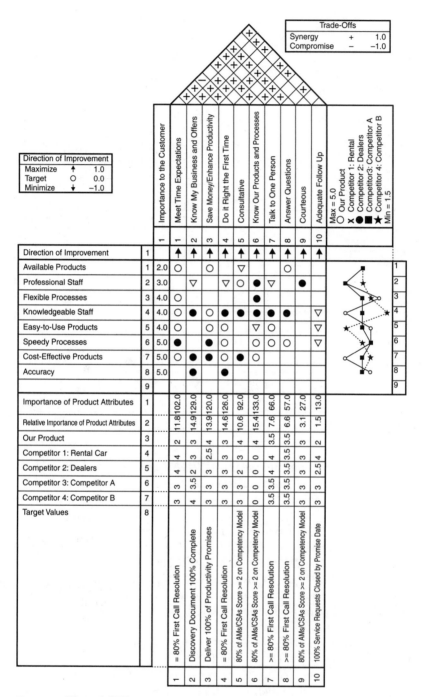

Figure 7.7   Phase 1 QFD.

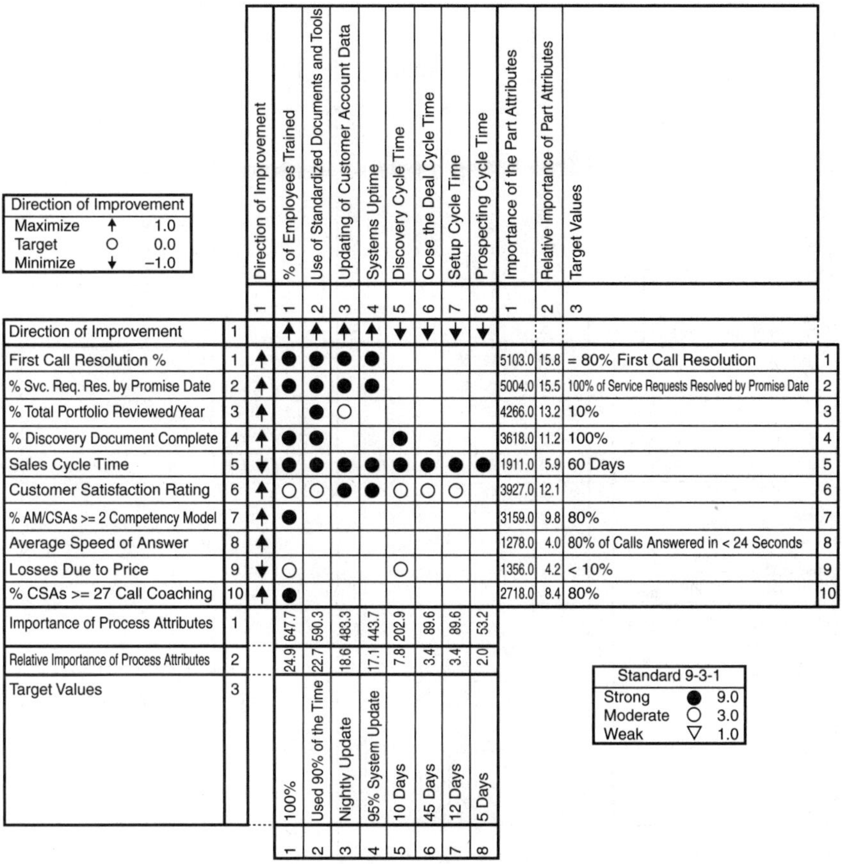

Direction of Improvement

Maximize ↑	1.0
Target O	0.0
Minimize ↓	−1.0

		Dir. of Improvement	% of Employees Trained	Use of Standardized Documents and Tools	Updating of Customer Account Data	Systems Uptime	Discovery Cycle Time	Close the Deal Cycle Time	Setup Cycle Time	Prospecting Cycle Time	Importance of the Part Attributes	Relative Importance of Part Attributes	Target Values	
		1	1	2	3	4	5	6	7	8	1	2	3	
Direction of Improvement	1		↑	↑	↑	↑	↓	↓	↓	↓				
First Call Resolution %	1	↑	●	●	●	●					5103.0	15.8	= 80% First Call Resolution	1
% Svc. Req. Res. by Promise Date	2	↑	●	●	●	●					5004.0	15.5	100% of Service Requests Resolved by Promise Date	2
% Total Portfolio Reviewed/Year	3	↑			●	O					4266.0	13.2	10%	3
% Discovery Document Complete	4	↑	●	●			●				3618.0	11.2	100%	4
Sales Cycle Time	5	↓	●	●	●	●	●	●	●	●	1911.0	5.9	60 Days	5
Customer Satisfaction Rating	6	↑	O	O		●	●	O	O	O	3927.0	12.1		6
% AM/CSAs >= 2 Competency Model	7	↑	●								3159.0	9.8	80%	7
Average Speed of Answer	8	↑									1278.0	4.0	80% of Calls Answered in < 24 Seconds	8
Losses Due to Price	9	↓	O				O				1356.0	4.2	< 10%	9
% CSAs >= 27 Call Coaching	10	↑	●								2718.0	8.4	80%	10
Importance of Process Attributes	1		647.7	590.3	483.3	443.7	202.9	89.6	89.6	53.2				
Relative Importance of Process Attributes	2		24.9	22.7	18.6	17.1	7.8	3.4	3.4	2.0				
Target Values	3		100%	Used 90% of the Time	Nightly Update	95% System Update	10 Days	45 Days	12 Days	5 Days				
			1	2	3	4	5	6	7	8				

Standard 9-3-1

Strong	●	9.0
Moderate	O	3.0
Weak	▽	1.0

Figure 7.8 Phase 2 QFD, third house of quality, process planning matrix.

Conflicts. The DFSS team needs to look for cases where technical benchmarking rates their product or service high but the customer assessment is low. Misconception of what the customer attributes is the major root cause. The team together with marketing can remedy the situation.

In our example, the "Cost-effective products," a WHAT, is addressed by many CTSs, including "Save money/enhance productivity." The customer rates our design as weak (rating 2), while the technical assessment is rated the highest (rating 4). Who is right? Conflicts may be a result of failure to understand the customer and must be resolved prior to further progress.

Strengths. By identifying the CTSs that contain the most "9" ratings, the DFSS team pinpoints which CTSs have the significant impact on the

Direction of Improvement

	Value
Maximize ↑	1.0
Target O	0.0
Minimize ↓	−1.0

		Dir. of Improvement (1)	% of Employees Trained (1)	Use of Standardized Documents and Tools (2)	Updating of Customer Account Data (3)	Systems Uptime (4)	Discovery Cycle Time (5)	Close the Deal Cycle Time (6)	Setup Cycle Time (7)	Prospecting Cycle Time (8)	Importance of the Part Attributes (1)	Relative Importance of Part Attributes (2)	Target Values (3)	
Direction of Improvement	1		↑	↑	↑	↑	↓	↓	↓	↓				
First Call Resolution %	1	↑	●	●	●	●					5103.0	15.8	= 80% First Call Resolution	1
% Svc. Req. Res. by Promise Date	2	↑	●	●	●	●					5004.0	15.5	100% of Service Requests Resolved by Promise Date	2
% Total Portfolio Reviewed/Year	3	↑		●	O						4266.0	13.2	10%	3
% Discovery Document Complete	4	↑	●	●			●				3618.0	11.2	100%	4
Sales Cycle Time	5	↓	●	●	●	●	●	●	●	●	1911.0	5.9	60 Days	5
Customer Satisfaction Rating	6	↑	O	O	●	●	O	O	O		3927.0	12.1		6
% AM/CSAs >= 2 Competency Mode	7	↑	●								3159.0	9.8	80%	7
Average Speed of Answer	8	↑									1278.0	4.0	80% of Calls Answered in < 24 Seconds	8
Losses Due to Price	9	↓	O				O				1356.0	4.2	< 10%	9
% CSAs >= 27 Call Coaching	10	↑	●								2718.0	8.4	80%	10
Importance of Process Attributes	1		647.7	590.3	483.3	443.7	202.9	89.6	89.6	53.2				
Relative Importance of Process Attributes	2		24.9	22.7	18.6	17.1	7.8	3.4	3.4	2.0				
Target Values	3		100%	Used 90% of the Time	Nightly Update	95% System Update	10 Days	45 Days	12 Days	5 Days				
			1	2	3	4	5	6	7	8				

Standard 9-3-1

Strong	●	9.0
Moderate	O	3.0
Weak	▽	1.0

Figure 7.9 Phase 3 QFD, third house of quality, process planning matrix.

total design. Change in these characteristics will greatly affect the design, and such effect propagates via the correlation matrix to other CTSs, causing positive and negative implications. The following CTSs are significant as implied by their importance ratings and number of "9" ratings in their relationships to WHATs: "Meet the expectations," "Know my business and offers," "Save money/enhance productivity," "Do it right the first time," and "Know our products and processes." Examining the correlation matrix (Fig. 7.10), we have positive correlation all over except in the cell "Do it right the first time" and "Meet time expectations."

Eye Openers. The DFSS team should look at customer attributes where

1. Their design as well as their competitors are performing poorly

Direction of Improvement		
Maximize	↑	1.0
Target	O	0.0
Minimize	↓	–1.0

		Direction of Improvement	Meet Time Expectations	Know My Business and Offers	Save Money/Enhance Productivity	Do it Right the First Time	Consultative	Know Our Products and Processes	Talk to One Person	Answer Questions	Courteous	Adequate Follow-Up	Importance of Product Attributes	Relative Importance of Product Attributes	Our Product	Competitor 1: Rental Car	Competitor 2: Dealers	Competitor 3: Competitor A	Competitor 4: Competitor B	Target Values	
		1	1	2	3	4	5	6	7	8	9	10	1	2	3	4	5	6	7	8	
Direction of Improvement	1		↑	↑	↑	↑	↑	↑	↑	↑	↑	↑									
Meet Time Expectations	1	↑		+		+		+	+	+		+	102.0	11.8	2	4	4	3	3	= 80% First Call Resolution	1
Know My Business and Offers	2	↑	+		+	+	+		+	+			129.0	14.9	3	3	2	3.5	4	Discovery Document 100% Complete	2
Save Money/Enhance Productivity	3	↑	+	+		+	+	+		+		+	120.0	13.9	4	2.5	3	3	3	Deliver 100% of Productivity Promises	3
Do it Right the First Time	4	↑	–	+				+	+	+			126.0	14.6	3	3	3	3	3	= 80% First Call Resolution	4
Consultative	5	↑		+	+			+				+	92.0	10.6	4	3	2	3	3	80% of AMs/CSAs Score >= 2 on Competency Model	5
Know Our Products and Processes	6	↑	+		+	+	+		+	+			133.0	15.4	3	0	0	0	0	80% of AMs/CSAs Score >= 2 on Competency Model	6
Talk to One Person	7	↑	+	+		+							66.0	7.6	3.5	4	4	3.5	3.5	>= 80% First Call Resolution	7
Answer Questions	8	↑	+	+		+	+	+	+				57.0	6.6	3.5	3.5	3.5	3.5	3.5	>= 80% First Call Resolution	8
Courteous	9	↑											27.0	3.1	3	3	3	3	3	80% of AMs/CSAs Score >= 2 on Competency Model	9
Adequate Follow-Up	10	↑		+	+		+	+		+			13.0	1.5	2	4	2.5	3	3	100% Service Requests Closed by Promise Date	10

		1	2	3	4	5	6	7	8	9	10
Importance of Product Attributes	1	102.0	129.0	120.0	126.0	92.0	133.0	66.0	57.0	27.0	13.0
Relative Importance of Product Attributes	2	11.8	14.9	13.9	14.6	10.6	15.4	7.6	6.6	3.1	1.5
Our Product	3	2	3	4	3	4	3	3.5	3.5	3	2
Competitor 1: Rental Car	4	4	3	2.5	3	3	0	4	3.5	3	4
Competitor 2: Dealers	5	4	2	3	3	2	0	4	3.5	3	2.5
Competitor 3: Competitor A	6	3	3.5	3	3	3	0	3.5	3.5	3	3
Competitor 4: Competitor B	7	3	4	3	3	3	0	3.5	3.5	3	3
Target Values	8	= 80% First Call Resolution	Discovery Document 100% Complete	Deliver 100% of Productivity Promises	= 80% First Call Resolution	80% of AMs/CSAs Score >= 2 on Competency Model	80% of AMs/CSAs Score >= 2 on Competency Model	>= 80% First Call Resolution	>= 80% First Call Resolution	80% of AMs/CSAs Score >= 2 on Competency Model	100% Service Requests Closed by Promise Date

Trade-Offs		
Synergy	+	1.0
Compromise	–	–1.0

Figure 7.10 Correlation matrix, product trade-offs.

2. The WHATs where they are performing compare poorly to those of their competitors for benchmarking

3. CTSs need further development in phase 2

We can pencil "flexible processes" in the first category and "accuracy" and "easy-to-use products" in the second category. The CTSs that deliver these WHATs should receive the greatest attention as they represent potential payoffs. Benchmarking represents the WHATs indicating areas in which the competitors are highly rated and makes incorporation of their design highly desirable. This saves design and research time.

The highest CTSs with the highest importance ratings are the most important. For example, "Know our products and processes" has the highest rating at 133. This rating is so high because it has three strong relationships to the WHATs. The degree of difficulty is medium (rating equal to 3) in the technical benchmarking. In addition, any CTS that has negative or strong relationships with this CTS in the correlation matrix should proceed to phase 2.

7.9 Summary

QFD is a planning tool used to translate customer needs and expectations into the appropriate design actions. This tool stresses problem prevention with emphasis on results in customer satisfaction, reduced design cycle time, optimum allocation of resources, and fewer changes. Together with other DFSS tools and concepts, it also makes it possible to release the designed entity at Six Sigma level. Since the customer defines quality, QFD develops customer and technical measures to identify areas for improvement.

Quality function deployment (QFD) translates customer needs and expectations into appropriate design requirements by incorporating the "voice of the customer" into all phases of the DFSS algorithm, through production and into the marketplace. In the context of DFSS, the real value of QFD is its ability to direct the application of other DFSS tools to those entities that will have the greatest impact on the team's ability to design their product, a service or a process that satisfies the needs and expectations of the customers, both internal and external.

The following items are a review of the different parts of the house of quality. The WHATs represent customer needs and expectations. The HOWs are critical-to-satisfaction (requirements) (CTSs) or substitute quality characteristics for customer requirements that the company can design and control. Relationships are identified between what the customer wants and how those wants are to be realized.

Qualitative correlations are identified between the various HOWs. Competitive assessment and importance ratings are developed as a basis for risk assessment when making decisions relative to trade-offs and compromises. Such trade-offs can be resolved with the employment of conceptual methods such as TRIZ and axiomatic design. Because of user preferences, especially in the automotive industry, the conventions, symbols, and even the shape of the house of quality have evolved with use. For example, the "roof" was added by Toyota, and Ford added the use of arrows to denote target orientation.

Axiomatic Design

8.1 Introduction

The theory and application of DFSS approach as defined in this book hinges on conceptual methods employed in the I (identify) phase. The axiomatic design approach, a perspective engineering design method, is a core method in this category.

Systematic research in engineering design began in Germany during the 1850s. More recent contributions in the field of engineering design include axiomatic design (Suh 1990), product design and development (Ulrich and Eppinger 1995), the mechanical design process (Ullman 1992), and Pugh's total design (Pugh 1991). These contributions demonstrate that research in engineering design is an active field that has spread from Germany to most industrialized nations around the world. To date, most research in engineering design theory has focused on design methods. As a result, a number of design methods are now being taught and practiced in both industry and academia. However, most of these methods overlooked the need to integrate quality methods in the concept phase so that only healthy (viable) concepts are conceived, evaluated, and launched with no or minimal vulnerabilities; hence, DFSS.

The current engineering practices exhibit many vulnerabilities leading to problematic quality issues in the designed entity that urgently call for the DFSS approach. These vulnerabilities can be categorized into the following groups:

- *Conceptual vulnerabilities.* These lead to lack of robustness at the conceptual level. This category is established in the designed systems (products/services/processes) due to violation of design guidelines and principles, in particular those promoted to axioms. In axiomatic design, a coupled system may result from violation of axiom 1 and system complexity due to the violation of axiom 2.

- *Operational vulnerabilities.* These lead to lack of robustness at the operational level, specifically, in the use environment and over the system life cycle, when the system is subjected to noise factors such as customer use or abuse, degradation, and piece-to-piece variation (see Chaps. 13 to 15).

The objective of this book is to develop a DFSS algorithm that provides solution methods to the two major categories of vulnerabilities listed above.

8.2   Why Axiomatic Design Is Needed

Design and its realization via manufacturing and production can be defined as sets of processes and activities that transform customers' wants into design solutions that are useful to society. These processes are carried over several phases starting from the concept phase. In the concept phase, conceiving, evaluating, and selecting *good* design solutions are tasks with enormous consequences. It is inevitable that the design and manufacturing organizations need to conceive healthy systems with no or minimal vulnerabilities in one development cycle.

Companies usually operate in two modes:

- *Fire prevention*—conceiving feasible and healthy conceptual entities with no or minimal conceptual vulnerabilities

- *Firefighting*—problem solving such that systems can live with minimal operational vulnerabilities

Unfortunately, the latter mode consumes the largest portion of the organization's human and nonhuman resources.

The crown of our DFSS theory is the methods of axiomatic design and robust design. These two methods in conjunction with the rest of this book provide a comprehensive DFSS approach that allows companies to work only in the first mode, which, in turn, opens the door to drastically improve business activities (products and processes) in a way that minimizes waste and resources while increasing customer satisfaction. It is a process that uses statistical techniques and conceptual methods to drive for results by supplementing means for decision making.

A design and its manufacturing and production process scenarios are continuously changing. Shorter life cycles and higher value of customer-oriented products are examples of present changes. We have reached the point where the product development time is rapidly shrinking. Therefore, design efficiency in terms of throughput and quality has become more significant than ever. This situation requires *healthy* design to be delivered to the customer on a continuous basis, which, in turn, requires efficient and systematic procedures to analyze, synthesize, and validate conceived concepts upfront. The activities of design must be based on general basic design principles, and not on accumulated empirical knowledge, simulation, and traditional engineering knowledge alone. A design process can be rapidly altered if the product follows some basic principles. If this approach can be extended to manufacturing and production, adaptation of novel products and future inexperienced creative design situations will become smoother and design organizations will gain the flexibility needed to accommodate changes quickly.

To stay competitive, the design industry needs to deliver high-quality products in a short time at the lowest cost. The impact of the early phases of design on both product and the manufacturing systems are discussed by Suh (1990, 2001). With increasing demands of shorter time to market, we encounter new products that lack the support of scientific knowledge and/or the presence of existing experience. It is no longer sufficient to rely solely on traditional knowledge. Concurrent engineering will facilitate somewhat in improving the situation, but only in designing the required incremental improvements of datum products and installed manufacturing systems. To design efficiently, design organizations need to support the practices of *synthesis* and *analysis* of new conceptual solution entities and base these activities on basic generic design principles. Basic principles do not substitute any other knowledge, nor do they replace the need to constantly learn, adopt, and implement new knowledge in the related disciplines. Deployment of basic principles complements the specific knowledge needed to develop products and manufacturing systems.

8.3 Design Axioms

Motivated by the absence of scientific design principles, Suh (1984, 1990, 1995, 1996, 1997, 2001) proposed the use of axiom as the scientific foundation of design. A design needs to satisfy the following two axioms along with many corollaries.

Axiom 1: The Independence Axiom. Maintain the independence of the functional requirements.

Axiom 2: The Information Axiom. Minimize the information content in a design.

After satisfying axiom 1, design simplicity is pursued by minimizing the information contents per axiom 2. In this context, information content is defined as a measure of complexity and is related to the probability of successfully manufacturing (producing) the design as intended.

8.4   The Independence Axiom (Axiom 1)

The design process involves three mappings between four domains (Fig. 8.1). The first mapping involves the mapping between *critical-to-satisfaction* (CTS) customer attributes and the functional requirements (FRs). A *functional requirement* is a solution-neutral, that is, design-parameter-independent, statement of what a system does and usually is expressed as a (verb, noun) pair or an active (verb, noun, phrase) triplet (e.g., carry load, conduct heat). In this mapping, in all elements of the physical structure there should be a team member responsible for the structure's life through the design algorithm stages. Functional specifications are established and propagated up regardless of the array of DPs used. The physical mapping is very critical as it yields the definition of the high-level minimum set of functional requirements needed to accomplish the design objective from the customer perspective. It can be performed by means of quality function deployment (QFD) (Chap. 7). Once the minimum set of FRs or CTSs are defined, the *physical mapping* may be started. This mapping involves the FR domain and the design parameter (DP) codomain. It represents the preliminary and detail design phases in the development cycle and can be represented by design matrices, hence the term "mapping" is used, as the high-level set of FRs are cascaded down to the lowest level of decomposition. The set of design matrices forms the conceptual physical structure that reveals coupling vulnerability and provides a means to track the chain of effects for design changes as they propagate across the structure.

The *process mapping* is the last mapping and involves the DP domain and the process variables (PV) codomain. This mapping can be represented by matrices as well and provides the process structure

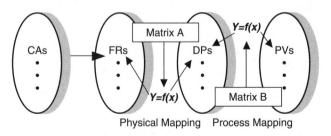

Figure 8.1 The design process mappings.

needed to translate the DPs into process variables (PVs) in manufacturing and production.

The equation $\mathbf{y} = \mathbf{f}(\mathbf{x})$ is used to reflect the relationship between domain (array $\mathbf{y}$) and the codomain (array $\mathbf{x}$) in the concerned mapping where the array $\{\mathbf{y}\}_{m \times 1}$ is the vector of requirements with m components, $\{\mathbf{x}\}_{p \times 1}$ is the vector of design parameters with p components, and $\mathbf{A}$ is the sensitivity matrix representing the physical mapping with $A_{ji} = \partial y_j/\partial x_i$. In the process mapping, matrix $\mathbf{B}$ represents the process mapping between the DPs and the PVs. The overall mapping is matrix $\mathbf{C} = \mathbf{A} \times \mathbf{B}$, the product of both matrices. The overall mapping (matrix $\mathbf{C}$) is what the customer will experience (Fig. 8.2). Excellence of conducting both mappings is necessary to gain customer satisfaction. This objective can't be achieved unless the design follows certain principles that allow repeatability of success, avoidance of failures, and moving faster toward the satisfaction that the customer desires.

Both mappings are the first interface of the DFSS team with the concept of *transfer function** in DFSS (Chap. 6). Initially, the DFSS team will identify the possible relationship between the two domains without being able to write the mapping mathematically. Later, the transfer functions can be derived from physics or identified empirically using regression and DOE. In almost all cases, modeling and simulation are required to approximate the transfer functions mathematically (see Chap. 6 for more details).

Per axiom 1, the ideal case is to have a one-to-one mapping so that a specific x can be adjusted to satisfy its corresponding y without

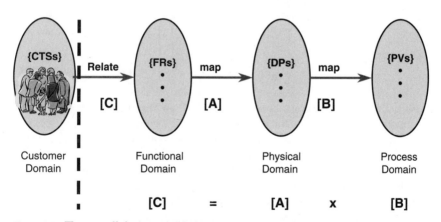

Figure 8.2   The overall design mappings.

---

*Note that the transfer function in this chapter is in matrix form with an FR array of size m.

affecting other requirements. This will give the DFSS team the advantage of vulnerability-free design, which allows a Six Sigma feasible design with unprecedented improvement flexibility. Axiom 2 states that the independent design that minimizes the information content is the *best*. Because of ignorance and other inhibitors, the exact deployment of design axiom on the redesign side might be infeasible because of technological and cost limitations. Under these circumstances, different degrees of conceptual vulnerabilities are established in the measures (criteria) related to the unsatisfied axiom (Fig. 8.3). For example, a degree of *coupling* may be created as a result of axiom 1 violation. A conceptually weak design may have limited chances to succeed in an operational vulnerability improvement phase.

When matrix **A** is a square diagonal matrix, that is, $m = p$ and $A_{ji} \neq 0$ when $i = j$ and 0 elsewhere, the design is called *uncoupled,* which means that each y can be adjusted or changed independent of the other y. An uncoupled design is a one-to-one mapping and is represented by

$$
\begin{Bmatrix} y_1 \\ \cdot \\ \cdot \\ y_m \end{Bmatrix} = \begin{bmatrix} A_{11} & 0 & \cdot & 0 \\ 0 & A_{22} & \cdot & \\ & \cdot & \cdot & 0 \\ 0 & \cdot & 0 & A_{mm} \end{bmatrix} \begin{Bmatrix} x_1 \\ \cdot \\ \cdot \\ x_m \end{Bmatrix}
\tag{8.1}
$$

In the decoupled design case, matrix **A** is a *lower/upper triangle* matrix, in which the maximum number of nonzero sensitivity coefficients equals $p(p - 1)/2$ and $A_{ij} \neq 0$ for $i = 1,j$ and $i = 1,p$. A decoupled design is represented by

$$
\begin{Bmatrix} y_1 \\ \cdot \\ \cdot \\ y_m \end{Bmatrix} = \begin{bmatrix} A_{11} & 0 & \cdot & 0 \\ A_{21} & A_{22} & & \\ & \cdot & \cdot & 0 \\ A_{m1} & A_{m2} & \cdot & A_{mm} \end{bmatrix} \begin{Bmatrix} x_1 \\ \cdot \\ \cdot \\ x_m \end{Bmatrix}
\tag{8.2}
$$

The decoupled design may be treated as uncoupled design when the **x** values are adjusted in some sequence conveyed by the matrix. Uncoupled and decoupled design entities possess conceptual robustness; that is, the **x** terms can be changed to affect other requirements in order to fit the customer attributes. Definitely, A coupled design results when the matrix has the number of requirements m greater than the number of **x** values p or when the physics is binding to such an extent that off-diagonal sensitivity elements are nonzero. The coupled design may be uncoupled or decoupled by "smartly" adding $m - p$ extra **x** terms to the problem formulation. A coupled design is represented by

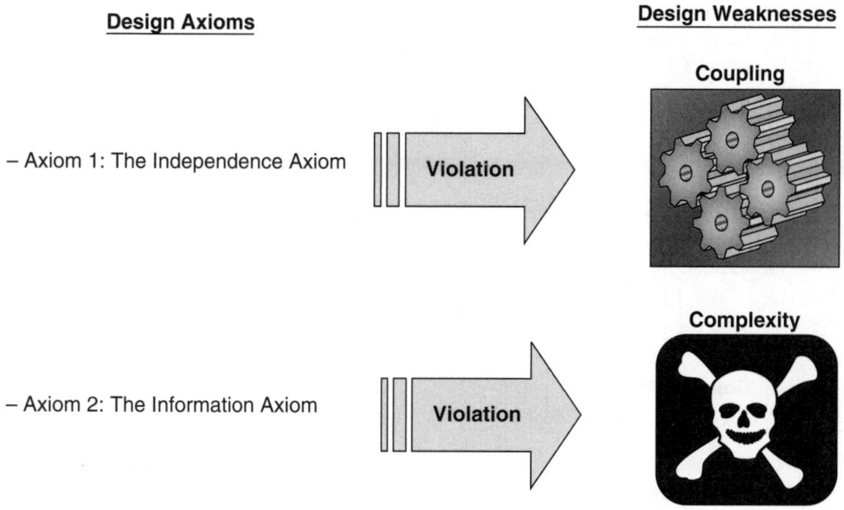

Design Axioms   **Design Weaknesses**

Coupling

– Axiom 1: The Independence Axiom   Violation

Complexity

– Axiom 2: The Information Axiom   Violation

Figure 8.3 Design vulnerabilities produced when the axioms are violated.

$$
\left\{ \begin{array}{c} y_1 \\ \cdot \\ \cdot \\ y_m \end{array} \right\} = \left[\begin{array}{cccc} A_{11} & A_{12} & \cdot & A_{1p} \\ A_{21} & A_{22} & & \cdot \\ \cdot & & \cdot & A_{(m-1)p} \\ A_{m1} & \cdot & A_{m(p-1)} & A_{mp} \end{array} \right] \left\{ \begin{array}{c} x_1 \\ \cdot \\ \cdot \\ x_m \end{array} \right\} \tag{8.3}
$$

An example of design categories is presented in Fig. 8.4, which displays two possible arrangements of the generic water faucet. The uncoupled architecture will have higher reliability and more customer satisfaction since the multiple adjustment of the two FRs can be done independently to fit customer demands.

The coupling of functional requirements is categorized as a design vulnerability. The DFSS team does not need to mix this concept with physical integration or consolidation of design parameters (DPs); that is, the one-to-one mapping for uncoupled design does not eliminate hosting more than one function in a component (see Sec. 10.2 for more details).

Uncoupling or decoupling of design matrices is the activity that follows their identification via "zigzagging" activity as soon as it is identified. Coupling occurs when a concept lacks certain design parameters or fails to meet its constraints. In performing the zigzagging process, the team will identify design constraints. *Constraints* are usually confused with functional requirements (FRs). Functional requirements represent what a design does; they are performance-related, and can be specified within some tolerance. Criteria such as cost, reliability,

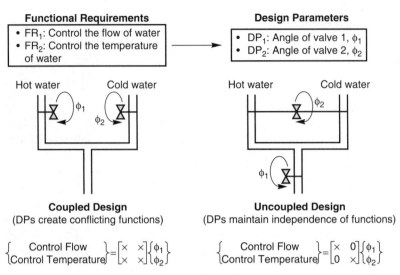

Functional Requirements
- FR$_1$: Control the flow of water
- FR$_2$: Control the temperature of water

Design Parameters
- DP$_1$: Angle of valve 1, ϕ_1
- DP$_2$: Angle of valve 2, ϕ_2

Coupled Design
(DPs create conflicting functions)

$$\begin{Bmatrix} \text{Control Flow} \\ \text{Control Temperature} \end{Bmatrix} = \begin{bmatrix} \times & \times \\ \times & \times \end{bmatrix} \begin{Bmatrix} \phi_1 \\ \phi_2 \end{Bmatrix}$$

Uncoupled Design
(DPs maintain independence of functions)

$$\begin{Bmatrix} \text{Control Flow} \\ \text{Control Temperature} \end{Bmatrix} = \begin{bmatrix} \times & 0 \\ 0 & \times \end{bmatrix} \begin{Bmatrix} \phi_1 \\ \phi_2 \end{Bmatrix}$$

Figure 8.4 Example of design coupling. [*From Swenson and Norlund (1996).*]

and environmental impact don't measure function directly and are not delivered by any particular DP; therefore they are called "constraints." A constraint is a property of a design, not something the design does. Typically, all elements—not only one element—in the design contribute to a constraint. We cannot add on a DP to improve the constraint. However, constraints are paced with the mapping in the zigzagging process; thus constraint modeling and transfer function modeling are both critical.

The importance of the design mappings has many perspectives. Chief among them is the revelation of both the transfer functions and coupling among the functional requirements, the domain in physical mapping, and the design parameters, the codomain. Knowledge of coupling is important because it gives the DFSS team clues of where to find solutions, make adjustments or changes, and how to maintain them over the long term with minimal negative effect.

The design matrices are obtained in a hierarchy when the *zigzagging method* is used (Suh 1990) as described in this chapter. At lower levels of hierarchy, sensitivities can be obtained mathematically as the FRs take the form of basic physical, mathematical, architectural, and engineering quantities. In some cases, the transfer functions are not readily available and some effort is needed to obtain them empirically or via modeling (e.g., CAE or discrete-event simulation). Lower levels represent the roots of the hierarchical tree where Six Sigma quality level can be achieved.

8.4.1   The zigzagging process

In the faucet example, the design is considered complete when the mapping from the functional domain to the physical domain is accomplished. However, in many design assignments of higher complexity, such as the transmission vane oil pump (Sec. 8.4.3), a process of cascading the high-level conceptual requirements is needed. The objective of this process is to decompose both the FRs and the DPs and the PVs for further detailing before manufacturing implementation. The process should be detailed such that it will enable the mapping from FRs to DPs in a certain decomposition level and from the DPs to the FRs of a further detailed level. The zigzagging process of axiomatic design does just that (Fig. 8.5). This process requires the decomposition in a solution neutral environment, where the DPs are chosen after the FRs are defined, and not vice versa. When the FRs are defined, we have to "zig" to the physical domain, and after proper DP selection, we have to "zag" to the functional domain for further decomposition. This process is in direct contrast to the traditional cascading processes, which utilizes only one domain, treating the design as the sum of functions or the sum of parts.

The process of zigzagging must continue until no further decomposition can be done. This is warranted, for example, when material properties or geometric dimensions are reached. Theoretically, the process can proceed to the physical and chemical structure of the design. The result of this process is the creation of the hierarchical tree, a physical structure, for the FRs and the DPs. This is the major

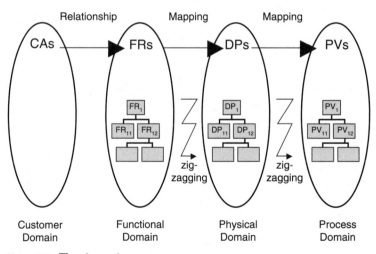

Figure 8.5   The zigzagging process.

output of this project from the technical side. Of course, the process should be conducted in the process domain via mapping the DPs to the PVs. The output is the process structure.

8.4.2 High-level zigzagging of automatic transmission

Automatic transmission is a power-transferring device that also can allow the engine to act as a braking device and can change the direction of the drive wheels (Brejcha 1982). The automatic transmission high-level FRs are to match engine speed and torque to the driver's needs and supply power to the driving wheels. There are many different transmission designs, but all are intended to provide power to the drive wheels at any engine speed without energy loss. Designing an automatic transmission to meet desired driver needs depends on many nontransmission aspects, including engine characteristics, which are determined by engine displacement, torque output, and operating specifications. A graphical curve, depicting torque versus engine speed, at constant throttle openings, determines the engine's useful torque. The transmission shift schedule can be correlated to the engine torque curve. It is also important to consider the type of transmission that the vehicle requires. A high-performance model may require lower ratios and firmer shifts to maximize the vehicle's ability to accelerate, meeting customer expectations. On the other hand, an economy model requires a more aggressive overdrive gear ratio. Luxury vehicles that carry heavy loads have still other requirements. Noise, vibration, and harshness (NVH) are another important consideration in automatic transmission design. The transmission itself must operate as smoothly and as quietly as possible, and the shift schedule should be designed to keep the engine from lugging or racing excessively.

Automatic transmission design may be approached from the axiomatic perspective. The following high-level FRs and their mapped-to DPs were identified by a group of transmission engineers:

- High-level FRs
  FR_1 = launch performance
  FR_2 = fuel economy
  FR_3 = high gear gradability
  FR_4 = provide engine isolation
  FR_5 = torque requirements
  FR_6 = absence of torque disturbances
- High-level DPs
  DP_1 = startup ratio

DP_2 = top gear ratio
DP_3 = structural characteristics
DP_4 = torsional damping
DP_5 = control strategy
DP_6 = torque converter

The design equation can be written as

$$
\begin{Bmatrix} FR_1 \\ FR_3 \\ FR_2 \\ FR_4 \\ FR_5 \\ FR_6 \end{Bmatrix}_{6 \times 1}
=
\begin{bmatrix}
A_{11} & 0 & A_{13} & 0 & 0 & 0 \\
0 & A_{22} & 0 & 0 & 0 & 0 \\
A_{31} & A_{32} & A_{33} & 0 & 0 & A_{36} \\
0 & 0 & A_{43} & A_{44} & 0 & 0 \\
A_{51} & 0 & A_{53} & 0 & A_{55} & 0 \\
0 & 0 & 0 & 0 & 0 & A_{66}
\end{bmatrix}_{6 \times 6}
\begin{bmatrix} DP_1 \\ DP_2 \\ DP_6 \\ DP_4 \\ DP_3 \\ DP_5 \end{bmatrix}_{6 \times 1}
\qquad (8.4)
$$

where A_{ij} is a nonzero entry (sensitivity). This equation exhibits two serious coupling issues. These coupling issues are that FR_1 (launch performance) and FR_2 (fuel economy) are coupled in DP_5 (control strategy) and DP_6 (torque converter). Otherwise, a decoupled design formulation results. It was concluded that these coupling issues can't be solved in the short term because of technological limitations. The only option is to reduce the degree of coupling. This formulation was further decomposed using the zigzagging process to yield the following FRs and DPs.

- Second-level FRs
  $FR_{1.1}$ = engine torque multiplication at launch
  $FR_{2.1}$ = engine torque multiplication at cruise
  $FR_{6.1}$ = engine transmission decoupling
  $FR_{6.2}$ = engine torque multiplication
  $FR_{4.1}$ = engine torque filtration
  $FR_{3.1}$ = engine power transmission
  $FR_{3.2}$ = holds oil
  $FR_{3.3}$ = support component
  $FR_{5.1}$ = torque multiplication scheduling
  $FR_{5.2}$ = shift quality control
  $FR_{5.3}$ = pump pressure control
  $FR_{5.4}$ = converter slip control
  $FR_{5.5}$ = component lubrication

- Second-level DPs
  $DP_{1.1}$ = low gear ratio
  $DP_{1.2}$ = final drive ratio
  $DP_{1.3}$ = transfer chain ratio
  $DP_{1.4}$ = torque converter ratio

$DP_{2.1}$ = high gear ratio
$DP_{2.2}$ = $DP_{1.2}$
$DP_{2.3}$ = $DP_{1.3}$
$DP_{2.4}$ = $DP_{1.4}$
$DP_{6.1}$ = fluid media
$DP_{6.2}$ = K factor
$DP_{4.1}$ = absorption mechanism
$DP_{4.2}$ = releasing mechanism
$DP_{4.3}$ = dissipation mechanism
$DP_{3.1}$ = case
$DP_{3.2}$ = internal components
$DP_{5.1}$ = apply/release friction elements
$DP_{5.2}$ = variator position
$DP_{5.3}$ = flow rate of primary pulley
$DP_{5.4}$ = pump excess discharge flow rate restriction
$DP_{5.5}$ = modulate bypass clutch
$DP_{5.6}$ = pump flow directing

The resultant design equation is given by Eq. (8.5). By definition, this design equation is redundant; some DPs need to be fixed. The criteria used to fix the extra DPs include cost, complexity, and variability optimization. This represents an opportunity for the designer to simplify the solution entity. Note that the logarithms of certain FRs were taken to achieve the additivity requirement of the axiomatic design formulation (Chap. 6). The deployment of axiom 1 enables Six Sigma targets for the design functional requirements. It presents a systematic approach for establishing the potential Six Sigma capability at the conceptual level in the designed system by reducing the coupling vulnerability between the functional requirements (FRs), represented by the array **y**. The equation **y** = **f(x)** is used where **y** is the array of functional requirements and **x** is the array of design parameters or process variables. It is by controlling the **x** terms that both the variation reduction and target adjustment objectives of **y** to Six Sigma level can be achieved.

8.4.3   Transmission vane oil pump

The transmission zigzagging exercise, in particular, $FR_{5.3}$ = pump pressure control, is continued at the pump level in this example (Brejcha 1982).

A hydraulic pump is a mechanism by which an external power source, namely, the engine, is used to apply force to a hydraulic media. Usually, the front pump drive is attached to the converter hub in automatic transmissions as depicted in Fig. 8.6. Figure 8.7 represents a side view, while Fig. 8.8 represents a top view of the vane pump (Brejcha 1982).

A hydraulic pump provides work when it transmits force and motion in terms of flow and pressure. In other words, the pump is the heart of automatic transmissions. Most currently used pumps are rotary type, which has the same mechanism of operation, where the hydraulic medium (also, fluid) is trapped in chambers that are cyclically

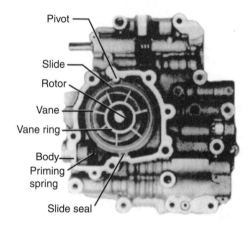

Figure 8.6 The pump schematic in automatic transmission [see Brejcha (1982)].

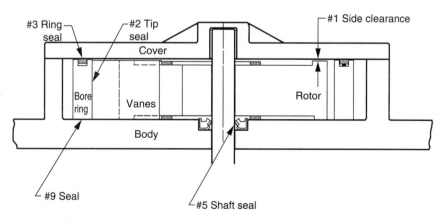

Figure 8.7   Pump side view.

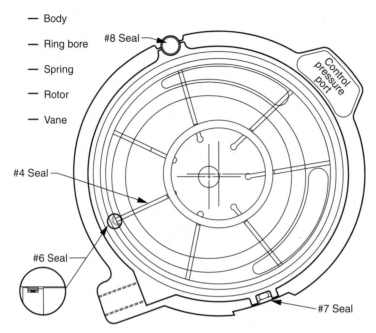

Figure 8.8 Pump top view.

expanding and collapsing. Expansion is needed at the pump inlet to draw fluid into the pump, while collapsing will occur at the outlet to force fluid into the system under pressure. The variable-vane pump is a rotary-type pump with variable capacity. The output will vary according to the requirements of the transmission to conserve power. The advantages are many. Chief among them are the ability to deliver a large capacity when the demand is high, especially at low speeds, and the minimal effort needed to drive at high speeds.

The mechanism of operation is as follows. When the priming spring moves the slide to the fully extended position, the slide and rotor are eccentric. As the rotor and vanes rotate within the slide, the expanding and contracting areas form suction (expanding) and pressure (collapsing) chambers. The hydraulic medium trapped between the vanes at the suction side is moved to the pressure side. A large quantity of fluid is moved from the pressure side back to the suction side as the slide moves toward the center (Fig. 8.6). A neutral condition (with no volume change) is created when concentricity is attained between the slide and rotor.

The function of the priming spring is to keep the slide in the fully expanded position such that full output can be commanded when the engine starts. Movement of the slide against the spring occurs when the pump pressure regulator valve reaches its predetermined value. At the design regulating point, the pressure regulator valve opens a port feed to the pump slide and results in a slide movement against the priming spring to cut back on volume delivery and maintain regulated pressure.

The pump physical structure using the zigzagging process. The FRs (array **y**) and design parameters (array **x**) must be decomposed into a hierarchy using the zigzagging process until a full structure in terms of design mappings is obtained. The DFSS team must zigzag between the domains to create such a structure. In the physical mapping, we first have to define the high-level FRs or functional requirements. In the pump case, there is one high-level requirement, y_1 = "convert external power to hydraulic power." This requirement is delivered by five design parameters: x_1 = "displacement mechanism," x_2 = "power source," x_3 = "inlet system," x_4 = "outlet system," x_5 = "hydraulic media," and x_6 = "external power coupling system." The mapping is depicted in Fig. 8.9, where x denotes a mapping or functional relationship. A P-diagram (Chap. 6) can be used to classify the mapping, in which x_2 is the signal and x_1, x_3, x_4, x_5, and x_6 are control design factors.

The level 1 mapping in the physical structure hierarchy represents a "zag" step. Not all of the design parameters will be zagged to the FR domain. The design parameters x_2 and x_5 will not be decomposed further as it is decided by other transmission requirements outside the scope of the pump. They can be treated as noise factors in this project. Figure 8.10 is mapping of x_1 = "displacement mechanism." We have four FRs (array **y**) with m, the number of FRs, equal to 4, and eight design parameters (array **x**) with $p = 8$, a redundant design since $p > m$. Once the zigzagging process is completed, the decoupling phase starts in all design mappings in the respective hierarchy.

8.5 Coupling Measures

The design categories have twofold importance: (1) they provide a sort of design classification scheme and (2) they strengthen the need to assess the degree of coupling in a given design entity. Renderle (1982) proposed the simultaneous use of *reangularity R* and *semangularity S* as coupling measures:

$$R = \prod_{\substack{j = 1, p - 1 \\ k = 1 + i, p}} \sqrt{\left[1 - \left(\sum_{i = 1}^{p} A_{kj} A_{kj} \right)^2 \middle/ \left(\sum_{k = 1}^{p} A_{kj}^2 \sum_{k = 1}^{p} A_{kj}^2 \right) \right]} \quad (8.6)$$

The array of FRs:
  y_1 = convert external power to hydraulic power

The array of DPs:
  x_1 = displacement mechanism
  x_2 = power source
  x_3 = inlet system
  x_4 = outlet system
  x_5 = hydraulic media
  x_6 = external power coupling system

The design mapping with ($m = 1, p = 6$) is given as

$$\{y_1\} = [A_{11}\ A_{12}\ A_{13}\ A_{14}\ A_{15}\ A_{16}] \begin{Bmatrix} x_1 \\ x_2 \\ x_3 \\ x_4 \\ x_5 \\ x_6 \end{Bmatrix}$$

Figure 8.9 The transmission pump level 1 mapping.

The array of FRs:
  $y_{1.1}$ = charge chamber
  $y_{1.2}$ = discharge chamber at uniform rate
  $y_{1.3}$ = does not allow slip flow to pass from outlet to inlet
  $y_{1.4}$ = provides displacement charge based on external hydraulic signal

The array of DPs:
  $x_{1.1}$ = expanding chamber
  $x_{1.2}$ = collapsing volumes
  $x_{1.3}$ = sealing device—geometry boundary between inlet and outlet
  $x_{1.4}$ = movable bore ring
  $x_{1.5}$ = bias spring
  $x_{1.6}$ = control pressure
  $x_{1.7}$ = rigid cover
  $x_{1.8}$ = rigid body

The design mapping with ($m = 4, p = 8$) is given as

$$\begin{Bmatrix} y_{1.1} \\ y_{1.2} \\ y_{1.3} \\ y_{1.4} \end{Bmatrix} = \begin{bmatrix} A_{11} & 0 & A_{13} & A_{14} & 0 & 0 & A_{17} & A_{18} \\ 0 & A_{22} & A_{23} & 0 & 0 & A_{26} & A_{27} & A_{28} \\ 0 & 0 & A_{33} & 0 & 0 & 0 & A_{37} & A_{38} \\ 0 & 0 & A_{43} & A_{44} & A_{45} & A_{46} & A_{47} & A_{48} \end{bmatrix} \begin{Bmatrix} x_{1.1} \\ x_{1.2} \\ x_{1.3} \\ x_{1.4} \\ x_{1.5} \\ x_{1.6} \\ x_{1.7} \\ x_{1.8} \end{Bmatrix}$$

Figure 8.10 The design matrix of x_1 = "displacement mechanism" (level 1.1).

$$S = \prod_{j=1}^{p} \left(\frac{|A_{jj}|}{\sqrt{\sum_{k=1}^{p} A_{kj}^2}} \right) \tag{8.7}$$

where A_{ij} are the elements of the design matrix (Suk 1990). Both of these measures are normalized by the magnitudes of the columns in the design matrix of the transfer functions. These measures can be understood in the context of vectors algebra where the arrays of DPs and FRs should be handled as vectors. Two vectors are orthogonal when the dot product between them is zero. *Reangularity R* is a measure of DP orthogonality in p-dimensional space; it is the absolute value of the product of sine function of all the angle pairs of the design matrix in the transfer function. R is maximum when the DPs are orthogonal. As the degree of coupling increases, R will decrease. This orthogonality measure can't assure axiom 1 satisfaction as the DPs can be orthogonal but not parallel to the FRs; that is, the one-to-one mapping can't be assured, hence *semangularity S*. This measure reflects the angular relationship between the corresponding axes of DPs and FRs. S is the product of the absolute values of the diagonal elements of the design matrix. When $S = 1$, the DPs parallel the FRs and uncoupled design is achieved. The different possibilities of design categories (according to axiom 1) in these two measures are given in Table 8.1.

8.5.1 The implication of coupling on design

The term "design" in the context of this book is not limited to product design. It should be extended to operations, manufacturing, or production—the processes by which the design entity is embodied. Axiom 1 is concerned with concept synthesis such that a healthy concept can be chosen. This axiom ensures that a potential for a Six Sigma capability in the design entity is established. This assurance should be made in both the physical (matrix **A**) and process (matrix **B**) mappings.

TABLE 8.1 Functional Independence Measures by Design Category

Design	Reangularity R	Semangularity S	Comments
Uncoupled	1	1	$R = S = 1$
Decoupled	<1	<1	$R = S$
Coupled	<1	<1	R could be greater or less than S

In the context of Sec. 8.4 and Fig. 8.2, we can express both mappings mathematically as

$$\{FR\}_{m \times 1} = [A]_{m \times p} \{DP\}_{p \times 1} \qquad (8.8)$$

$$\{DP\}_{p \times 1} = [B]_{p \times n} \{PV\}_{n \times 1} \qquad (8.9)$$

or equivalently

$$\{FR\}_{m \times 1} = [C] \{PV\}_{n \times 1} \qquad (8.10)$$

where A is the design mapping matrix, B is the process mapping matrix, and $[C]_{m \times n} = [A][B]$ is the overall design matrix. In either mapping, we seek to satisfy the independence axiom, axiom 1. Therefore the product matrix C should be diagonal, that is, uncoupled. The A and B matrices can be categorized from coupling perspective according to Eqs. (8.1) to (8.3). Accordingly, the different possibilities that can be taken by matrix C are given in Table 8.2. The following conclusions can be deduced:

- A decoupled design may be an upper or a lower triangular type of matrix depending on the formulation.
- For the overall design entity (product and process) to be totally uncoupled, both matrices should be uncoupled.

Uncoupled designs not only are desirable from controllability, quality, and robustness standpoints but also have potential for high probability of producibility, that is, reduced defect per opportunity (DPO). A decoupled design is the next choice when uncoupled design cannot be achieved; however, the revealed sequence of adjustment should be followed in executing the synthesis process of creative and incremental design situations. Uncoupled and decoupled designs have higher potentials to achieve Six Sigma capability in all FRs than do the coupled designs. Design for Six Sigma in the conceptual sense is defined

TABLE 8.2 Possibilities of Matrix [C]

[A]/[B]	＼	◿	▽	▨
＼	＼	◿	▽	▨
◿	◿	◿	▨	▨
▽	▽	▨	▽	▨
▨	▨	▨	▨	▨

Legend
▽ : Upper triangular matrix
◿ : Lower triangular matrix
＼ : Diagonal matrix
▨ : Coupled matrix (Upper, lower and diagonal)

as having an overall uncoupled or decoupled design by conducting the process mapping and physical mapping concurrently.

As expressed by Table 8.2, the following scenarios are observed:

- An overall *uncoupled design* is achieved only when both mappings are uncoupled.

- An overall *decoupled design* is achieved when

  Both mappings are decoupled, having similar triangular orientations.

  Either mapping is uncoupled while the other is decoupled.

- An overall *coupled design* is achieved when

  At least one mapping is coupled.

  Both mappings are decoupled with different triangular orientations.

With everything equal and left to chance, the odds are given by the probability distribution shown in Fig. 8.11. Design should not be left to chance, and the design community should endorse and adopt the DFSS approach.

In addition, Table 8.2 indicates where it is easier to implement a change for a problem solving, before or after release, without causing new problems or amplifying existing symptoms of the FRs. A design change in the form of a DFSS project can be more easily implemented and controlled in the case of uncoupled and decoupled designs than in the case of a coupled design. A DFSS project may be launched to target soft or hard design changes. Whether soft changes are solution-effective depends on the sensitivities, nominal and tolerance settings. Hard changes require alterations of the PV array, the DP array, or both, namely, a new design. Hard changes are usually followed by a soft-change phase for tuning and adjustment. In either case, the major cost is the expense of controlling the solution implementation.

Altering (hard changes) or adjusting (soft changes) the **x** values can be used to uncouple or decouple a system. Unfortunately, hard changes

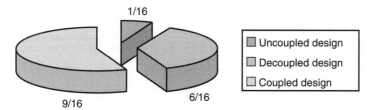

Figure 8.11 Probability of design.

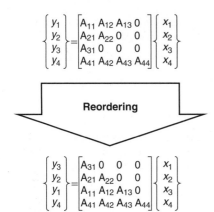

$$\begin{Bmatrix} y_1 \\ y_2 \\ y_3 \\ y_4 \end{Bmatrix} = \begin{bmatrix} A_{11} & A_{12} & A_{13} & 0 \\ A_{21} & A_{22} & 0 & 0 \\ A_{31} & 0 & 0 & 0 \\ A_{41} & A_{42} & A_{43} & A_{44} \end{bmatrix} \begin{Bmatrix} x_1 \\ x_2 \\ x_3 \\ x_4 \end{Bmatrix}$$

Reordering

Adjustment Sequence:		
DPs	Objective	Effects
x_1	y_3	y_1, y_2, y_4
x_2	y_2	y_1, y_4
x_3	y_1	y_4
x_4	y_4	–

$$\begin{Bmatrix} y_3 \\ y_2 \\ y_1 \\ y_4 \end{Bmatrix} = \begin{bmatrix} A_{31} & 0 & 0 & 0 \\ A_{21} & A_{22} & 0 & 0 \\ A_{11} & A_{12} & A_{13} & 0 \\ A_{41} & A_{42} & A_{43} & A_{44} \end{bmatrix} \begin{Bmatrix} x_1 \\ x_2 \\ x_3 \\ x_4 \end{Bmatrix}$$

Figure 8.12 Design matrices reordering.

are not always feasible because of incapable or outdated technology, cost of change, organizational culture, or other inhibitors. A company may choose to take an educated decision on keeping the coupled design entities but with reduced (minimized) degree of coupling among FRs. While the industry should recognize that this decision should be a short-term strategy, its adoption as a long-term strategy may result in loosing its competitive edge when coupling resolution rather than minimization is adopted by competition.

The hard changes that target decoupling are difficult and costly to implement after launch, a scenario that could have been avoided when the system was designed following DFSS. This is seldom the case, and companies may resort to soft changes first to improve their current systems. The aim of such a practice is usually problem solving of the immediate concerns. It is inevitable that the coupling be clearly characterized prior to proceeding to decouple the system. This may require rearrangement or reordering of the design matrices as described in Fig. 8.12.

8.5.2 Design sequencing

In this example, assume that we have a design equation [Eq. (8.11)]. The following design steps (sequence) may be concluded and should be used in the subsequent the design activities in the optimization and verification phases as follows:

1. Use DP_6 only to adjust and control FR_6.

2. Use DP_1 to adjust FR_1. Fix DP_1 value accordingly.

3. After fixing DP_1, use DP_3 to control FR_3. Fix DP_3 accordingly.

4. After fixing the values of DP_1 and DP_3, use DP_5 to adjust and control FR_5.

5. Use DP_2 only to adjust and control FR_2.

6. After fixing DP_2, use DP_4 to adjust FR_4.

$$\begin{Bmatrix} FR_1 \\ FR_2 \\ FR_3 \\ FR_4 \\ FR_5 \\ FR_6 \end{Bmatrix}_{6 \times 1} = \begin{bmatrix} A_{11} & 0 & 0 & 0 & 0 & 0 \\ 0 & A_{22} & 0 & 0 & 0 & 0 \\ A_{31} & 0 & A_{33} & 0 & 0 & A_{36} \\ 0 & A_{42} & 0 & A_{44} & 0 & 0 \\ A_{51} & 0 & A_{53} & 0 & A_{55} & 0 \\ 0 & 0 & 0 & 0 & 0 & A_{66} \end{bmatrix}_{6 \times 6} \begin{bmatrix} DP_1 \\ DP_2 \\ DP_3 \\ DP_4 \\ DP_5 \\ DP_6 \end{bmatrix}_{6 \times 1} \qquad (8.11)$$

where A is a nonzero sensitivity entry. Note that steps 1, 2, and 5 above can be performed simultaneously. This is not the case for the remaining rules. The greatest potential here to reduce the work at the optimization step is to utilize this step as a first perspective of the ideal functions definitions. Definition of the ideal function for each partitioned function is central to selecting and understanding good structural (solution entity) choices, as well as to setting the foundation for the rest of the DFSS algorithm. As activities progress through the development process, the definition of the ideal function will be clarified, refined, and finalized. The activities of analysis and synthesis require iterating rapidly with multiple forward and backward motion between design domains and within these domains. Iterative looping forward and backward is good practice in structure creation, as it is in most engineering activities.

A suggested sequence to be followed in this example with respect to the rest of the FR optimization studies is to use DP_1 only to optimize FR_1, use DP_3 only to optimize FR_3, use DP_4 only to optimize FR_4, and use DP_5 only to optimize FR_5. Any lower-level design parameters of DP_1 should not appear in the optimization of FR_3, for example. The selection of the DPs for optimization to avoid coupling, eliminate the interaction between control factors, and enforce an additive transfer function model.

8.5.3 Decoupling of coupled design (DFSS algorithm step 6)

There are many ways to decouple a design depending on the situation:

1. Make the size of array **y** equal to the size of array **x**: $m = p$. According to Theorem 2 in Suh (1990, p. 68), when a design is coupled because the number of FRs is greater than the number of the design parameters, it may be decoupled by the addition of new

parameters so that the number of FRs equals to the number of design parameters, if a subset of the design matrix containing $m \times m$ elements constitutes a triangular matrix (Fig. 8.13*).

2. Perform decoupling by utilizing the system's sensitivity. In this case, the designer is seeking parameters that have a minimal effect on FRs other than the targeted FR. This can be done by analyzing the magnitude of the off-diagonal elements in the design matrix by varying the **x** values over an extreme design range.

Methods 1 and 2 seek decoupling or uncoupling by adding, replacing, or changing the sensitivity of design parameters. These methods may greatly benefit from other axiomatic design theorems and corollaries (Suh 1990). In addition, a great solution synergy can be gained using TRIZ contradiction elimination principles (Chap. 9) to reduce or eliminate coupling vulnerability:

$$\begin{Bmatrix} \text{Functional requirement } (y_1) \\ \text{Functional requirement } (y_2) \end{Bmatrix} = \begin{bmatrix} A_{11} & A_{12} \\ A_{21} & A_{22} \end{bmatrix} \begin{Bmatrix} \text{design parameter (DP}_1) \\ \text{design parameter (DP}_2) \end{Bmatrix}$$

Assume the coupled design matrix above. The DFSS team can use TRIZ to make the sensitivities A_{12}, A_{21}, or both negligibly small by the right choice of the DPs using TRIZ Altschuller's contradiction matrix (Sec. 9.8).

TRIZ is based on principles extracted from international patents showing how persons have invented solutions for different categories of technical problems in the past. The principles are organized in problem categories for selective retrieval, and the methods include procedural algorithms. Because the principles are associated with similar problems successfully solved in the past the likelihood of success is enhanced. A simplified TRIZ process description is given in the following steps:

a. Convert the design problem statement into one of a conflict between two FRs considerations

$$\begin{Bmatrix} y_1 \\ y_2 \\ y_3 \end{Bmatrix} = \begin{bmatrix} \times & 0 \\ \times & \times \\ \times & \times \end{bmatrix} \begin{Bmatrix} x_1 \\ x_2 \end{Bmatrix} \quad \Rightarrow \quad \begin{Bmatrix} y_1 \\ y_2 \\ y_3 \end{Bmatrix} = \begin{bmatrix} \times & 0 & 0 \\ \times & \times & 0 \\ \times & \times & \times \end{bmatrix} \begin{Bmatrix} x_1 \\ x_2 \\ x_3 \end{Bmatrix}$$

Figure 8.13 Decoupling by adding extra design parameters.

---

*The entry $\times$ in the Fig. 8.13 design matrices is a shorthand notation for nonzero sensitivities $(\partial y_i / \partial x_i = A_{ji})$.

b. Match these two FRs considerations to any two of 39 generalize design requirements

c. Look up solution principles to the conflict of these two FRs using a Altschuller's TRIZ matrix

d. Convert this general solution principle into a working project solution

Decoupling methods 1 and 2 provide many opportunities in the case of new design. The degree of freedom in applying them will become limited in redesign situations with binding physical and financial constraints. Redesign scenarios that are classified as coupled call for another method that is based on tolerance optimization to reduce operational vulnerability.

3. Perform decoupling by tolerance optimization. Tolerances of the FRs have a strong role to play in decoupling a design. The FRs are always specified with some tolerances, $y_j \pm t_j$, $j = 1,...,m$, where t_j is the half-tolerance of FR_j and m is number of FRs in the array $\mathbf{y}$. Let's assume that we have a 2 × 2 coupled design with

$$\begin{Bmatrix} y_1 \\ y_2 \end{Bmatrix} = \begin{bmatrix} A_{11} & A_{12} \\ A_{21} & A_{22} \end{bmatrix} \begin{Bmatrix} x_1 \\ x_2 \end{Bmatrix}$$

In method 3, the issue is whether A_{12} or A_{21} can be neglected ($A_{12} = 0$ or $A_{21} = 0$) so that the design can be considered decoupled. If not, then method 3 is required. The transferred variation of y_1 is given by

$$\Delta y_1 = \frac{\partial y_1}{\partial x_1} \Delta x_1 + \frac{\partial y_1}{\partial x_2} \Delta x_2$$

On the basis of customer specification, we need to maintain $\Delta y_1 \leq t_j$; thus the change in the $FR(y_1)$ due to the changes in the design parameters is less than the tolerance specified by the customer. To achieve a decoupled design, we need to make A_{12} negligibly small, which translates into making $t_j \geq (\partial y_1/\partial x_2)\Delta x_2$, neglecting the off-diagonal element. This is the essence of Theorem 8 in Suh (1990, p. 122).

In summary, the decoupling or uncoupling actions (DFSS algorithm step 6) are

1. Start from high-level FRs (obtained from QFD phase 2 QFD).

2. Define high-level DPs.

3. Use the zigzagging process to map FRs to DPs to get the design matrices and physical structure.

4. Reorder and categorize design matrices at all levels as coupled, decoupled, or uncoupled.

5. Maintain independence of FRs at all levels of physical structure by employing the methods presented in this section.

6. Repeat steps 1 to 5 in the process mapping (the mapping from DPs to PVs).

8.5.4   The decoupling phase (DFSS algorithm 6.2)

This example is a continuation to Fig. 8.10 of Sec. 8.4.3, where the mapping of x_1 = "displacement mechanism" is depicted.

For sealing device 1, the rotor clearance to the chamber depicted in Fig. 8.14 will be used as a decoupling example. The zigzagging process up to the pump is summarized in Fig. 8.15. This device is selected in this example because sealing devices within the pump are not robust, resulting in low pump efficiency. Without the DFSS process, the pump manufacturer will resort to improving the robustness of the seal through an empirical experiment, an operational vulnerability improvement phase. This is depicted in Fig. 8.16. This may not be sufficient because the conceptual vulnerability of the seal is a coupled design. Without resolving the coupling, the best that can be done is a trade-off between the FR, y_1 = "minimize leak from high pressure to low pressure chamber" and the FR, y_2 = "lubricate running surfaces of the chamber," since both are delivered by one design parameter $x_{1.3.1}$ = "the tolerance between the vane and the rotor." The coupling occurs because the seal device 1 system is charged with two FRs and one design parameter; that is, the number of FRs ($m = 2$) is greater than the number of design parameters ($p = 1$). Clearly, another design parameter, say, $x_{1.3.2}$, needs to be introduced to resolve the coupling.

The array of FRs:
$y_{1.3.1.1}$ = minimize leak from high pressure to low pressure
$y_{1.3.2.1}$ = lubricate running surface of chamber

The array of DPs:
$x_{1.3.1.1}$ = close clearance

The design equation is given as
$$\begin{Bmatrix} y_{1.3.1.1} \\ y_{1.3.2.1} \end{Bmatrix} = \begin{bmatrix} A_{12} \\ A_{21} \end{bmatrix} \{x_{1.3.1.1}\}$$

Figure 8.14   The design matrix of $x_{1.3.1}$ = "ring rotor clearance to chamber" (level 1.3.1).

$x_{1.3.2.1}$: Displacement Mechanism

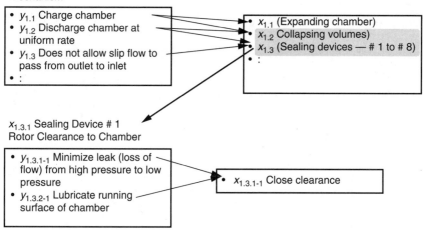

- $y_{1.1}$ Charge chamber
- $y_{1.2}$ Discharge chamber at uniform rate
- $y_{1.3}$ Does not allow slip flow to pass from outlet to inlet
- :

- $x_{1.1}$ (Expanding chamber)
- $x_{1.2}$ Collapsing volumes)
- $x_{1.3}$ (Sealing devices — # 1 to # 8)
- :

$x_{1.3.1}$ Sealing Device # 1
Rotor Clearance to Chamber

- $y_{1.3.1-1}$ Minimize leak (loss of flow) from high pressure to low pressure
- $y_{1.3.2-1}$ Lubricate running surface of chamber

- $x_{1.3.1-1}$ Close clearance

Figure 8.15 The zigzagging up to sealing device 1.

Effect of Coupling

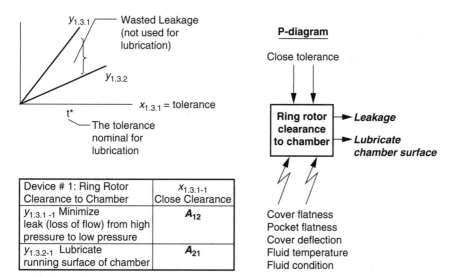

$y_{1.3.1}$ — Wasted Leakage (not used for lubrication)

$y_{1.3.2}$

$x_{1.3.1}$ = tolerance

t^*

The tolerance nominal for lubrication

Device # 1: Ring Rotor Clearance to Chamber	$x_{1.3.1-1}$ Close Clearance
$y_{1.3.1 -1}$ Minimize leak (loss of flow) from high pressure to low pressure	A_{12}
$y_{1.3.2-1}$ Lubricate running surface of chamber	A_{21}

P-diagram

Close tolerance

Ring rotor clearance to chamber → *Leakage*
→ *Lubricate chamber surface*

Cover flatness
Pocket flatness
Cover deflection
Fluid temperature
Fluid condition

Figure 8.16 The P-diagram of sealing device 1 without DFSS.

P-diagram

Device # 1: Ring Rotor Clearance to Chamber	$x_{1.3.1-1}$ Close Clearance	$x_{1.3.2-1}$ Material Coating
$y_{1.3.1-1}$ Minimize leak (loss of flow) from high pressure to low pressure	A_{11}	
$y_{1.3.2-1}$ Lubricate running surface of chamber	$A_{21} \longrightarrow 0$	A_{22}

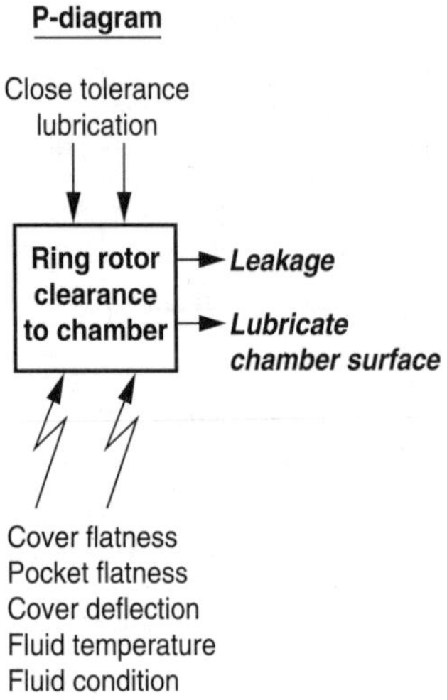

Close tolerance
lubrication

Ring rotor clearance to chamber → Leakage
→ Lubricate chamber surface

Cover flatness
Pocket flatness
Cover deflection
Fluid temperature
Fluid condition

Figure 8.17   The P-diagram of sealing device 1 with DFSS.

This parameter also should be *smartly* introduced, yielding uncoupled or at least decoupled designs. This parameter should deliver one of the FRs without adversely affecting the other FR. Using TRIZ, these characteristics fit the coating to be declared as $x_{1.3.2}$ in the sealing 1 mapping. Coating will help the lubrication by surface tensions of the hydraulic media keeping the surfaces wet and lubricated. Coating does not affect leakage, allowing the tolerance to be tightened to minimize the leakage. The resulting mapping and the P-diagram are given in Fig. 8.17.

8.6   The Implications of Axiom 2

Axiom 2 deals with design information content (complexity), which in essence is a function of the number of FRs and DPs (solution size) and their inherent variation. Shannon entropy can be used to quantify the information content. In a study on digital signal communication, Shannon (1948) defined a level of *complexity,* called the *entropy,* below which the signal can't be compressed. The principle of entropy was generalized to many disciplines and used as a measure of uncertainty. In the design context, Suh (1990) proposed information as a measure of complexity in axiom 2.

8.6.1   The complexity paradigm

Complexity in design has many facets, including the lack of transparency of the transfer functions between inputs and outputs in the functional structure, the relative difficulty of employed physical processes, and the relatively large number of assemblies and components involved (Phal and Beitz 1988). The term *complexity* is used in most of the literature in a pragmatic sense. It is easier to have an idea about complexity by shaping where it does exist and how it affects design rather than what it really means. Linguistically, *complexity* is defined as a quality of an object. In our case an object is a design entity, a product, a service, or a process. The object can have many interwoven elements, aspects, details, or attributes that make the whole object difficult to understand in a collective sense. Complexity is a universal quality that does exist, to some degree, in all objects. The degree of complexity varies according to the many explored and understandable phenomenon in the object. Ashby (1973) defines complexity as "the quantity of information required to describe the vital system." Simon (1981) defines a complex system as an object that "is made up of a large number of parts that interact in a non simple way." These definitions imply some level of communication between interrelated (coupled) elements of the complex system, which is translated to one of the major characteristics of a complex system, *hierarchy*. Simon (1981) illustrated that *hierarchy* has a broader dimension than the intuitive authority meaning. It reflects some level of communication or interaction between related entities. In a designed entity, the higher the level of interaction, the shorter is the relative spatial propinquity.

In his seminal paper, Weaver (1948) distinguished between two kinds of complexity: *disorganized* and *organized.* Systems of disorganized complexity are characterized by a huge number of variables. The effects of these variables and their interaction can be explained only by randomness and stochastic processes using statistical methods rather than by any analytical approach. The objective is to describe the system in an aggregate average sense. Statistical mechanics is a good example of a discipline that addresses this type of complexity. Analytical approaches work well in the case of *organized simplicity,* which is the extreme of the complexity spectrum at the lower end. Organized simplicity systems are characterized by a small number of significant variables that are tied together in deterministic relationships. Weak variables may exist but have little bearing in explaining the phenomena. (Refer to Sec. 3.7.)

The majority of design problems can't always be characterized as any of the two complexity extremes that have been discussed. It is safe to say that most problems often belong to a separate standalone category in between the two extremes called the *organized complexity.* This category of problem solutions utilizes statistical and analytical methods at different development stages. Design problems are more susceptible to analytical and conceptual approaches in the early stages and to statistical methods in the validation phases because of unanticipated factors called the *noise factors*. Organized complexity suggests the utilization of a new paradigm for simplification that makes use of information and complexity measures. (See Chap. 5.)

The amount of information generated or needed is a measure of the level of complexity involved in the design. In this product design context, complexity and information are related to the level of obedience of the manufacturing operations to the specifications required, *capability*. In addition, the selection of machining processes contributes to complexity, *compatibility*. Compatibility of machining to specification requirements may be considered as another ingredient of complexity. Compatibility is concerned with the related engineering and scientific knowledge. The selection of the wrong machine to attain a certain DP will increase the complexity encountered in delivering the right components. Compatibility is the essence of production and manufacturing engineering and material science, both of which are beyond the scope of this chapter.

Complexity in physical entities is related to the information required to meet each FR, which, in turn, is a function of the DP that it is mapped to. The ability to satisfy an FR is a function of machine capability since we can't always achieve the targeted performance. Thus, an FR should be met with the tolerance of the design range FR $\in [T \pm \Delta\text{FR}]$. The amount of complexity encountered in an FR is related to the probability of meeting its mapped-to DPs successfully. Since probability is related to complexity, we will explore the use of *entropy* information measures as a means of measuring complexity.

8.6.2   Entropy complexity measures

Shannon (1948) proved that, in a communication channel, the probability of transmission error increases with transmission rate only up to the channel capacity. Through his study of random processes, Shannon defined *entropy* as the level of *complexity* below which the signal can't be compressed. The introduction of the entropy principle was the origin of information theory. The original concept of entropy was introduced in the context of heat theory in the nineteenth century (Carnap 1977). Clausius used entropy as a measure of the disorganization of a system. The first fundamental form was developed by Boltzmann in 1896 during his work in the theory of ideal gases. He developed a connection between the macroscopic property of entropy and the microscopic state of a system. The Boltzmann relation between entropy and work is well known, and the concept of entropy is used in thermodynamics to supplement its second law.

Hartley (1928) introduced a logarithmic measure of information in the context of communication theory. Hartley, and later Shannon (1948), introduced their measure for the purpose of measuring information in terms of uncertainty. Hartley's information measure is essentially the logarithm of the cardinality or source alphabet size (see Definition 8.1), while Shannon formulated his measure in terms of probability theory. Both measures are information measures, and hence are measures of complexity. However, Shannon called his measure "entropy" because of the similarity to the mathematical form of that used in statistical mechanics.

Hartley's information measure (Hartley 1928) can be used to explore the concepts of information and uncertainty in a mathematical framework. Let X be a finite set with a cardinality $|X| = n$. For example, X can be the different processes in a transactional DFSS project, a flowchart, or a set of DPs in a product DFSS project. A sequence can be generated from set X by successive selection of its elements. Once a selection is made, all possible elements that might have been chosen are eliminated except for one. Before a selection is made, ambiguity is experienced. The level of ambiguity is proportional to the number of alternatives available. Once a selection is made, no ambiguity sustains. Thus, the amount of information obtained can be defined as the amount of ambiguity eliminated.

Hartley's information measure I is given by $I = \log_2 N$ (bits) where $N = n^s$ and s is the sequence of selection. The conclusion is that the amount of uncertainty needed to resolve a situation or the amount of complexity to be reduced in a design problem is equivalent to the potential information involved. A reduction of information of ΔI bits represents a reduction in complexity or uncertainty of ΔI bits.

Definition 8.1. A source of information is an ordered pair $\theta = (X,P)$ where $X = \{x_1, x_2,...,x_n\}$ is a finite set, known as a *source alphabet,* and P is a probability distribution on X. We denote the probability of x_i by p_i.

The elements of set X provide specific representations in a certain context. For example, it may represent the set of all possible tolerance intervals of a set of DPs. The association of set X with probabilities suggests the consideration of a random variable as a source of information. It conveys information about the variability of its behavior around some central tendency. Suppose that we select at random an arbitrary element of X, say, x_i with probability p_i. Before the sampling occurs, there is a certain amount of uncertainty associated with the outcome. However, an equivalent amount of information is gained about the source after sampling, and therefore uncertainty and information are related. If $X = \{x_1\}$, then there is no uncertainty and no information gained. At the other extreme, maximum uncertainty occurs when the alphabets carry equal probabilities of being chosen. In this situation, maximum information is gained by sampling. This amount of information reveals the maximum uncertainty that preceded the sampling process.

8.6.3 Shannon entropy

A function H satisfies the preceding three attributes if and only if it has the form

$$H_b\,(p_1, p_2,\ldots,p_m) = -\sum_{k=1}^{m} p_k \log_b p_k \qquad (8.12)$$

where $b > 1.^{*}$ If $p = 0$, then $p \log_b = 0$. The function H is called *b-ary entropy*.

Shannon entropy is a significant measure of information. When the probabilities are small, we are surprised by the event occurring. We are uncertain if rare events will happen, and thus their occurrences carry considerable amounts of information.[†] Therefore, we should expect H to decrease with an increase in probability. Shannon entropy is the expected value of function $\log_b 1/p$ according to distribution p. It is a measure of a discrete information source. Boltzmann entropy, on the other hand, may be used in the case of a continuous information source. It has an appealing mathematical form that may be considered the continuous analog to Shannon's entropy. Boltzmann information has an integral in which p_i is replaced with the probability density function, pdf $f(.)$. For some pdfs there are no closed-form integrals. One solution to this problem would be the use of a digitized version of Boltzmann entropy that has the form of Shannon's entropy. The discrete version is, however, an approximate solution (El-Haik 1996).

8.6.4 Boltzmann entropy

Boltzmann entropy $h(f)$ of a continuous random variable X with a density $f(x)$ is defined as

$$h_b(f) = -\int_{s} f(x) \log_b f(x)\, dx \qquad \text{(if an integral exists)} \quad (8.13)$$

where S is the support set, $S = \{x/f(x) \geq 0\}$, of the random variable.

Example 8.1 The *pitch diameter* (PD) of a spur gear is a basic design parameter of gear design on which most calculations are based. The *pitch circles*

---

*For $b=2(e)$, H has the units of bits (nats), respectively (1 nat $=1.44$ bits).

[†]Notice the link to outliers in a statistical random sample.

of a pair of mating gears are tangential to each other. However, the pitch diameters are hypothetical circles that can't be measured directly. The function of the gear is to "transfer speed." This function can be mapped to many design parameters that can be grouped basically into two subsets of DPs: a geometric subset and a material property (stiffness, hardness, etc.) subset. The diameter of the pitch circle PD follows a normal distribution as a result of manufacturing process variability: $PD \sim f(PD) = (1/\sqrt{2\pi\sigma^2})\, e^{-PD^2/2\sigma^2}$. Then, we have

$$h(f) = -\int f(PD)\ln f(PD)$$

$$= -\int f(PD)\left(\frac{-PD^2}{2\sigma^2} - \ln\sqrt{2\pi\sigma^2}\right)$$

$$= \frac{1}{2\sigma^2} E\,(PD^2) + \ln\sqrt{2\pi\sigma^2}$$

$$= \ln\sqrt{2\pi e\sigma^2}\ \text{nats} \tag{8.14}$$

This equation is depicted in Fig. 8.18.

In the case of a normal source of information, this example shows that information and complexity are both functions of variability. A reduction in the variance will reduce not only the probability of manufacturing nonconfirming gears but also the required information needed to manufacture the part. This is the power of design axioms.

Equation (8.14) states that, in the case of normal information source, a random variable, complexity and information are functions of variability.

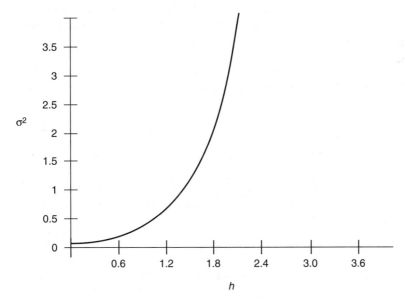

Figure 8.18   Variance as a function of complexity.

Reduction in the variance will reduce not only the probability of manufacturing nonconforming gears but also the required information needed to manufacture the part (Fig. 8.18). For the multivariate case, the joint density $\phi(DP)$ is given by

$$\frac{\exp\left[(-\tfrac{1}{2})(DP - M)'\,\Sigma^{-1}\,(DP - M)\right]}{\sqrt{(2\pi^p\,|\Sigma|)}}$$

or

$$\frac{\exp\left\{(-(\tfrac{1}{2})\sum_{i=1}^{p}\,[(DP_i - \mu_i)/\sigma_i]^2\right\}}{\sqrt{(2\pi)^p\,|\Sigma|}}$$

where $DP' = [DP_1,...,DP_p]$, $M' = [\mu_1,...,\mu_\pi]$, and

$$\Sigma = \begin{bmatrix} \sigma_1^2 & 0 & \cdot & 0 \\ 0 & \sigma_2^2 & \cdot & \cdot \\ \cdot & \cdot & \cdot & \cdot \\ 0 & \cdot & \cdot & \sigma_p^2 \end{bmatrix} \tag{8.15}$$

Then, complexity is given by[*]

$$h(DP_1,..., DP_p) = \ln \sqrt{(2\pi e)^p\,|\Sigma|} \ \text{nats} \tag{8.16}$$

For $p = 2$, we have

$$h(DP_1,..., DP_p) = \ln 2\pi e \sigma_1 \sigma_2 \ \text{nats} \tag{8.17}$$

Using the concept of Boltzmann entropy, we were able to identify variability as a source of complexity. However, variability is not the only source of complexity, as we shall see later. In fact, sensitivity adds to complexity as per the following theorem.

Theorem 8.1. The complexity of a design has two sources: variability and sensitivity. The total design complexity in the case of linear design is given by

$$h(\{\mathbf{FR}\}) = h(\{\mathbf{DP}\}) + \ln |\,[\mathbf{A}]\,| \tag{8.18}$$

where $|\,[\mathbf{A}]\,|$ is the determinant of the nonsingular design matrix $\mathbf{A}$.

Corollary 8.1 For process mapping (Fig. 8.2), in which $h(\{\mathbf{DP}\}) = h(\{\mathbf{PV}\}) + \ln |\,[\mathbf{B}]\,|$, then, by substitution in (8.18), the total design complexity is given by

$$h(\{\mathbf{FR}\}) = h(\{\mathbf{DP}\}) + \ln |\,[\mathbf{A}]\,| \tag{8.19}$$

$$= h(\{\mathbf{PV}\}) + \ln |\,[\mathbf{B}]\,| + \ln |\,[\mathbf{A}]\,|$$

---

[*]See El-Haik and Yang (1999) for proof.

$$= h(\{PV\}) + \ln | [B] [A] |$$

$$= h(\{PV\}) + \ln | [C] |$$

There are two components of complexity design: (1) that due to *variability* ($=h(DP)$) and (2) that due to *coupling vulnerability* ($=\ln|A|$). The term *coupling vulnerability* of Theorem 8.1 has a broader meaning than the numerical values of the sensitivity coefficients, the argument of the design matrix determinant. There are three ingredients in the coupling complexity component: mapping, sensitivity, and dimension. The *mapping* ingredient refers to the binary variable Z_{ij} denoting the mapping process between the functional domain and the physical domain and is defined as

$$Z_{ij} = \begin{cases} 1 & \text{if } FR_i \to DP_j \\ 0 & \text{elsewhere} \end{cases}$$

In other words, the mapping variable represents the position of the nonzero sensitivity coefficients in the design matrix **A**. The *sensitivity* ingredient refers to the magnitude and sign of nonzero $A_{ij} = \partial FR_i/\partial DP_j$ coefficients. The *dimension* ingredient refers to the size of the design problem: m, the number of FRs and p, the number of DPs in the squared design matrix. We view our interpretation of the complexity component due to vulnerability as the mathematical form of the Simon (1981) complexity definition.

The theme of Theorem 8.1 is that the designer experiences two complexity components in attaining an FR (in the physical mapping) or a DP (in the process mapping) if she or he does not know how its mapped-to variables vary (the variability component) and at what scale they vary (the vulnerability component). For an uncoupled design, the value of $|[A]|$ is the product of the diagonal elements, $|[A]| = \Pi_{i=1}^{p} A_{ii}$, and the complexity component due to sensitivity is $\sum_{i=1}^{p} \ln |A_{ii}|$. The total uncoupled design complexity (assuming that all DPs are normal information sources) is $\sum_{i=1}^{p} \ln \sqrt{(2\pi e \sigma_i A_{ii})}$ nats.

8.7 Summary

Axiomatic design (AD) is a *general principle* for design analysis and synthesis developed by Professor Nam P. Suh of MIT. It was developed to establish a scientific basis for design and to improve design activities by providing the design teams with a theoretical foundation based on logical and rational thought processes and tools. The objectives of axiomatic design are to establish a scientific basis for design activity to augment engineering knowledge and experience, provide a theoretical foundation based on logical and rational thought process-es and tools, minimize a random search or trial-and-error process, accelerate the product development process, and improve quality and reliability.

Axiomatic design is a simple and systematic tool for analyzing the merits of existing design and new concepts before further develop-ments while promoting thorough functional requirement analysis at the concept level in step 6 and beyond of the DFSS algorithm. Synergy is gained when axiomatic design is integrated with other design methodologies as needed, such as design optimization methods, TRIZ to identify appropriate design parameters to resolve coupling, and probabilistic design analysis.

Appendix: Historical Development of Axiomatic Design

Axiomatic design can be defined as a system design method. The process for system design is to identify the functional requirements or CTQs, determine the possible design parameters and process vari-ables, and then integrate them into a system. In this case, the ability to understand the relationship among the different requirements and parameters as represented in the design matrices and hierarchies is vital. Without this ability, the design process becomes a confusing assignment, which can ultimately lead to poor design. The basic assumption of the axiomatic approach to design is that there exists a fundamental set of good design principles that determines good design practice. The method was born out of the need to develop a disciplinary base for design. The work started in 1977, when Professor Nam Suh, the founder, was asked by Professor Herbert Richardson, MIT's Mechanical Engineering Department head, to establish a center for manufacturing research at MIT. Professor Suh started with a $40,000 fund and two part-time assistant professors and established the Laboratory for Manufacturing and Productivity (LMP). After some communication with the National Science Foundation (NSF), a pro-posal was put together for the method. By 1984, LMP had become a large and successful organization with substantial industrial funding. Several papers were published on the application, and the first book appeared in 1990. The first International Conference on Axiomatic Design (ICAD) was held in June 2000. Research on the axiomatic design method is beginning to manifest specific tracks. In the design-and-development process, a significant amount of valuable research has been developed (Sohlenius 1997; Hintersteiner 1999; Hintersteiner and Nain 1999; Nordlund et al. 1996; El-Haik and Yang 1999, 2000a, 2000b). Another track of development concentrates on concept synthesis and design weaknesses. For example, the sequence in which design changes can be made is discussed by Tate et al. (1998). The coupling weakness in design is discussed by Lee (1999).

The most significant contributions of the method are the zigzagging system architecture approach, the axioms, and the identification of design weaknesses. The method captures the requirements (the FR), components (DPs or PVs), and their relationships in design matrices. This information can be depicted in a variety of ways, including trees of design hierarchies and design matrices, flowcharts, and module junction structure diagrams (Kim et al. 1991; Suh 1996; Suh 1997, 2001).

Theory of Inventive Problem Solving (TRIZ)

9.1 Introduction

TRIZ (Teoriya Resheniya Izobreatatelskikh Zadatch) is the theory of inventive problem solving (TIPS) developed in the Soviet Union starting in the late 1940s. TRIZ has developed based on 1500+ person-years of research and study over many of the world's most successful solutions of problems from science and engineering, and systematic analysis of successful patents from around the world, as well as the study of the psychological aspects of human creativity (Darrell Mann 2002).

In the context of the DFSS algorithm, TRIZ can be used in concept generation and solving problems related to coupling vulnerability as discussed in Chap. 8. In essence, when two functional requirements are coupled, TRIZ may suggest different design parameters to uncouple the two, resulting in decoupled or uncoupled design.

Genrich S. Altshuller, the creator of TRIZ, initiated the investigation on invention and creativity in 1946. After initially reviewing 200,000 former Soviet Union patent abstracts, Altshuller selected 40,000 as representatives of inventive solutions. He separated the patents' different degrees of inventiveness into five levels, ranging from level 1, the lowest, to level 5, the highest. He found that almost all invention problems contain at least one contradiction; in this context a *contradiction* is defined as a situation in which an attempt to improve one feature of the system detracts from another feature. He found that the level of invention often depends on how well the contradiction is resolved.

Level 1: Apparent or conventional solution
32 percent; solution by methods well know
within specialty

Inventions at level 1 represent 32 percent of the patent inventions and employ obvious solutions drawn from only a few clear options. Actually level 1 inventions are not real inventions but narrow extensions or improvements of the existing systems, which are not substantially changed according to the application of invention. Usually a particular feature is enhanced or strengthened. Examples of level 1 invention include increasing the thickness of walls to allow for greater insulation in homes or increasing the distance between the front skis on a snowmobile for greater stability. These solutions may represent good engineering, but contradictions are not identified and resolved.

Level 2: Small invention inside paradigm 45
percent; improvement of an existing system,
usually with some compromise

Inventions at level 2 offer small improvements to an existing system by reducing a contradiction inherent in the system while still requiring obvious compromises. These solutions represent 45 percent of the inventions. A level 2 solution is usually found through a few hundred trial-and-error attempts and requires knowledge of only a single field of technology. The existing system is slightly changed and includes new features that lead to definite improvements. The new suspension system between the track drive and the frame of a snowmobile is a level 2 invention. The use of an adjustable steering column to increase the range of body types that can comfortably drive an automobile is another example at this level.

Level 3: Substantial invention inside
technology 18 percent; essential
improvement of an existing system

Inventions at level 3, which significantly improve the existing system, represent 18 percent of the patents. At this level, an invention contradiction is resolved with the existing system, often through the introduction of some entirely new element. This type of solution may involve a hundred ideas, tested by trial and error. Examples include replacing the standard transmission of a car with an automatic transmission, or placing a clutch drive on an electric drill. These inventions usually involve technology integral to other industries but not well known within the industry in which the invention problem arose. The resulting solution causes a paradigm shift within the industry. A level 3 invention is found outside an industry's range of accepted ideas and principles.

Level 4: Invention outside technology 4 percent; new generation of design using science not technology

Inventions at level 4 are found in science, not technology. Such breakthroughs represent about 4 percent of inventions. Tens of thousands of random trials are usually required for these solutions. Level 4 inventions usually lie outside the technology's normal paradigm and involve use of a completely different principle for the primary function. In level 4 solutions, the contradiction is eliminated because its existence is impossible within the new system. Thus, level 4 breakthroughs use physical effects and phenomena that had previously been virtually unknown within the area. A simple example involves using materials with thermal memory (shape-memory metals) for a key ring. Instead of taking a key on or off a steel ring by forcing the ring open, the ring is placed in hot water. The metal memory causes it to open for easy replacement of the key. At room temperature, the ring closes.

Level 5: Discovery 1 percent; major discovery and new science

Inventions at level 5 exist outside the confines of contemporary scientific knowledge. Such pioneering works represent less than 1 percent of inventions. These discoveries require lifetimes of dedication for they involve the investigation of tens of thousands of ideas. The type of solution occurs when a new phenomenon is discovered and applied to the invention problem. Level 5 inventions, such as lasers and transistors, create new systems and industries. Once a level 5 discovery becomes known, subsequent applications or inventions occur at one of the four lower levels. For example, the *laser* (light amplification by spontaneous emission of radiation), the technological wonder of the 1960s, is now used routinely as a lecturer's pointer and a land surveyor's measuring instrument.

Following extensive studies on inventions, other major findings of TRIZ include

1. Through inductive reasoning on millions of patents and inventions, TRIZ researchers found that most innovations are based on the applications of a very small number of inventive principles and strategies.

2. Outstanding innovations are often featured by complete resolution of contradictions, not merely a trade-off and compromise on contradictions.

3. Outstanding innovations are often featured by transforming wasteful or harmful elements in the system into useful resources.

4. Technological innovation trends are highly predictable.

9.1.1   What is TRIZ?

TRIZ is a combination of methods, tools, and a way of thinking (Darrel Mann 2002). The ultimate goal of TRIZ is to achieve absolute excellence in design and innovation. In order to achieve absolute excellence, TRIZ has five key philosophical elements. They are:

- *Ideality.* Ideality is the ultimate criterion for system excellence; this criterion is the maximization of the benefits provided by the system and minimization of the harmful effects and costs associated with the system.

- *Functionality.* Functionality is the fundamental building block of system analysis. It is used to build models showing how a system works, as well as how a system creates benefits, harmful effects, and costs.

- *Resource.* Maximum utilization of resource is one of the keys used to achieve maximum ideality.

- *Contradictions.* Contradiction is a common inhibitor for increasing functionality; removing contradiction usually greatly increases the functionality and raises the system to a totally new performance level.

- *Evolution.* The evolution trend of the development of technological systems is highly predictable, and can be used to guide further development.

Based on these five key philosophical elements, TRIZ developed a system of methods. This system of methods is a complete problem definition and solving process. It is a four-step process, consisting of problem definition, problem classification and tool selection, solution generation, and evaluation.

Problem definition.   This is a very important step in TRIZ. If you define the right problem and do it accurately, then that is 90 percent of the solution. The problem definition step includes the following tasks:

- *Project definition.*

- *Function analysis.*   This includes the function modeling of the system and analysis. This is the most important task in the "definition" step. TRIZ uses very sophisticated tools for function modeling and analysis.

- *Technological evolution analysis.*   This step looks into the relative maturity in technology development of all subsystems and parts. If a subsystem and/or part is technically "too" mature, it may reach its limit in performance and thus become a bottleneck for the whole system.

- *Ideal final result.* The ideal final result is the virtual limit of the system in TRIZ. It may never be achieved but provides us with an "ultimate dream" and will help us to think "out of the box."

Problem classification and tool selection. TRIZ has a wide array of tools for inventive problem solving; however, we must select the right tool for the right problem. In TRIZ, we must first classify the problem type and then select the tools accordingly.

Solution generation. In this step, we apply TRIZ tools to generate solutions for the problem. Because TRIZ has a rich array of tools, it is possible to generate many solutions.

Evaluation. In any engineering project, we need to evaluate the soundness of the new solution. TRIZ has its own evaluation approach. However, other non-TRIZ methods might also be used at this stage, such as axiomatic design and design vulnerability analysis.

In subsequent sections, we first discuss the philosophical aspects of TRIZ in order to lay a foundation for understanding. Then we discuss the four-step TRIZ problem definition and solving process, together with the tools used in TRIZ.

9.2 TRIZ Foundations

Ideality, functionality, contradictions, use of resources, and evolution are the pillars of TRIZ. These elements make TRIZ distinctively different from other innovation and problem-solving strategies. In this section, we describe all five elements.

9.2.1 Function modeling and functional analysis

Function modeling and functional analysis originated in value engineering (Miles 1961). A *function* is defined as the natural or characteristic action performed by a product or service. Usually, a product or service provides many functions. For example, an automobile provides customers with the ability to get from point *A* to point *B*, with comfortable riding environment, air conditioning, music, and so on.

Among all the functions, the most important function is the *main basic function,* defined as the primary purpose or the most important action performed by a product or service. The main basic function must always exist, although methods or designs to achieve it may vary. For example, for an automobile, "the ability to get from point *A* to point *B*" is a main basic function.

Besides the main basic function, there are other useful functions as well; we can call them *secondary useful functions*. There are several kinds of secondary useful functions:

1. *Secondary basic functions.* These are not main basic function, but customers definitely need them. For example, "providing a comfortable riding environment" is a "must have" for automobiles.

2. *Nonbasic but beneficial functions.* These functions provide customers with esteem value, comfort, and so on. For example, the paint finish in automobiles provides both basic and nonbasic functions; it protects the automobile from corrosion and rust, and creates a "sleek look" for the car.

Besides secondary useful functions, there are two other types of functions:

1. *Supporting function.* This function supports the main basic function or other useful function. It results from the specific design approach to achieve the main basic function or other useful functions. As the design approach to achieve the main basic function and other useful functions are changed, supporting functions may also change. There are at least two kinds of supporting functions: assisting functions and correcting functions.

   - *Assisting functions.* These functions assist other useful functions. For example, the engine suspension system provides the function of "locking the position of the engine in the automobile" to enable the engine to remain securely in place on the car while providing power.
   - *Correcting functions.* These functions correct the negative effects of another useful function. For example, the main basic function of the water pump in the automobile internal-combustion engine is to "circulate water in the engine system in order to cool the engine off"; it is a correcting function for the automobile engine. The main basic function of the engine is to provide power for the automobile, but the internal-combustion engine also creates negative effects, such as heat. A water pump's function is to correct this negative effect. If we change the design and use electricity as the power source of the automobile, the function of a water pump will be no longer needed.

2. *Harmful function.* This is an unwanted, negative function caused by the method used to achieve useful functions. For example, an internal-combustion engine not only provides power but also generates noise, heat, and pollution, and these are harmful functions.

In summary, the main basic function and secondary useful functions provide benefits for the customer. Supporting functions are useful, or at least they are not harmful, but they do not provide benefits directly to the customer and they incur costs. Harmful functions are not useful and provide no benefits at all.

Functional statement.   A function can be fully described by three elements: a subject, a verb, and an object. For example, for the automobile, its main basic function can be described as:

$$\underset{\text{(Subject)}}{\text{Car}} \quad \underset{\text{(Verb)}}{\text{moves}} \quad \underset{\text{(Object)}}{\text{people}}$$

For a toothbrush, its main basic function can be described as

$$\underset{\text{(Subject)}}{\text{Toothbrush}} \quad \underset{\text{(Verb)}}{\text{brushes}} \quad \underset{\text{(Object)}}{\text{teeth}}$$

Functional analysis diagram.   A functional analysis diagram is a graphical tool to describe and analyze functions. The following graph is a typical template for functional analysis diagram:

where the *subject* is the source of action, the *object* is the action receiver. Action is the *"verb"* in a functional statement, and it is represented by an arrow. In a technical system, the action is often being accomplished by applying some kind of *field*, such as a mechanical, electrical, or chemical field.   For example, the function "brush teeth" can be described by the following functional analysis diagram:

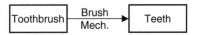

In the above diagram, "Mech" stands for "mechanical field." Clearly, brushing teeth is an application of one kind of mechanical field, force. In a functional analysis diagram, there are four types of actions, and they are represented by four types of arrows as illustrated in Fig. 9.1.

Example 9.1. Brushing Teeth   If we use a toothbrush correctly, and our teeth get cleaned properly, then we call this brushing action a "normal useful action." We can illustrate that by the following functional analysis diagram:

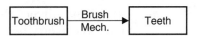

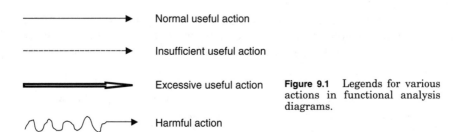

Figure 9.1   Legends for various actions in functional analysis diagrams.

However, if we use the toothbrush too gently and do not brush long enough, or we use a worn toothbrush, then our teeth will not get enough cleaning. In this case, we call it an "insufficient useful action" and we can express this by using the following functional analysis diagram:

If we use a very strong toothbrush, and brush our teeth with large force and strong strokes, then our gums will get hurt, and so will our teeth. We can use the following functional analysis diagram to describe this situation:

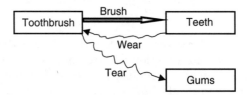

That is, the toothbrush delivers excessive brush action to teeth, and excessive brushing will deliver a harmful action, "tearing the gums," making them bleed; and the teeth also may deliver a harmful action, "wearing the toothbrush."

Functional modeling and analysis example.   Figure 9.2 is a schematic view of an overhead projector. Figure 9.3 is the functional modeling and analysis diagram for the whole system. In this example, there are many "chain actions"; that is, an "object" can be other object's subject. Then we have a sequence of "subject-action-object-action-" chains. Each chain describes a complete function. We can identify the following functions:

1. From "electricity to image in screen," that is, the function of "to project image in the film to screen," we can think that is the main basic function.

2. From "hand" to "focusing device" to "mirror," that is, the function of "to focus the image," which is a secondary basic function.

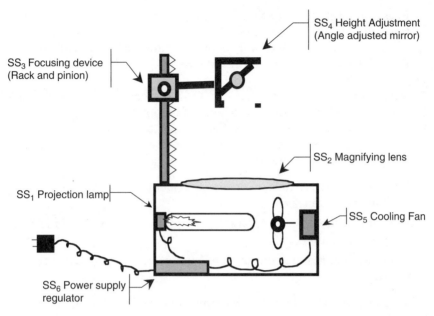

Figure 9.2 Overhead projector.

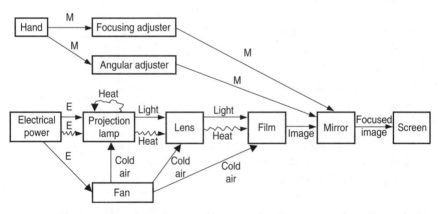

Figure 9.3 Functional analysis diagram for an overhead projector, where E stands for electrical field, and M stands for mechanical field.

3. From "hand" to "angular adjustment" to "mirror," that is, the function of "to project image to right position in screen," which is also a secondary basic function.

4. From "electricity" to "projection lamp" to "lens," and so on, which is a harmful function chain, without correction, that harmful function will damage the film and the device.

5. Because of the harmful function, we have to add a supporting function, namely, the chain from "electricity" to "fan" and end with lens
and film. This function is a correcting function to compensate for
the negative effect of harmful function.

Substance field functional models   In TRIZ methodology, the substance
field model is also a popular model for functional modeling and functional analysis. In the substance field model, there are three essential
elements. Figure 9.4 illustrates the template for the substance field
model: These three elements are:

Substance 1 (S1): Article, which is equivalent to "object" in functional
analysis diagram

Substance 2 (S2): Tool, which is equivalent to "subject" in functional
analysis diagram.

Field (F): Represents energy field between the interaction of S1
and S2.

We can give the following two simple examples of substance field
models:

- A vacuum clear cleans a carpet:
  S1—carpet (an article)
  S2—vacuum cleaner (a tool)
  F—cleaning (mechanical field)

- A person is painting a wall:
  S1—wall (an article)
  S2—person (a tool)
  F—painting (chemical field)

The substance field model may look complicated for TRIZ beginners.
However, it illustrates that "substances" and "fields" are essential
building blocks for any function.

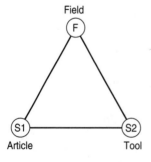

Field

Figure 9.4   Substance field model.

9.2.2 Resources

Maximum effective use of resources is very important in TRIZ. Also in TRIZ, we need to think of resources and make use of resources in creative ways.

For any product or process, its primary mission is to deliver functions. Because substances and fields are the basic building blocks of functions, they are important resources from TRIZ point of view. However, substances and fields alone are not sufficient to build and deliver functions, the important resources, space and time, are also needed. In the TRIZ point of view, information and knowledge base are also important resources.

We can segment resources into the following categories

1. Substance resources:
   - Raw materials and products
   - Waste
   - By-product
   - System elements
   - Substance from surrounding environments
   - Inexpensive substance
   - Harmful substance from the system
   - Altered substance from system

2. Field resources:
   - Energy in the system
   - Energy from the environment
   - Energy/field that can be built upon existing energy platforms
   - Energy/field that can be derived from system waste

3. Space resources:
   - Empty space
   - Space at interfaces of different systems
   - Space created by vertical arrangement
   - Space created by nesting arrangement
   - Space created by rearrangement of existing system elements

4. Time resources:
   - Prework period
   - Time slot created by efficient scheduling
   - Time slot created by parallel operation
   - Postwork period

5. Information/knowledge resources:
   - Knowledge on all available substances (material properties, transformations, etc.)

- Knowledge on all available fields (field properties, utilizations, etc.)
- Past knowledge
- Other people's knowledge
- Knowledge on operation

6. Functional resources:
   - Unutilized or underutilized existing system main functions
   - Unutilized or underutilized existing system secondary functions
   - Unutilized or underutilized existing system harmful functions

In TRIZ, it is more important to look into cheap, ready-to-use, abundant resources rather than expensive, hard-to-use, and scarce resources. Here is an example.

Example 9.2. Cultivating Fish in Farmland   The southeastern part of China is densely populated, so land is a scarce resource. Much of the land is used to plant rice. Agriculture experts suggest that farmland can be used to cultivate fish while the land is used to grow rice, because in rice paddies water is a free and ready resource, and fish waste can be used as a fertilizer for rice.

9.2.3 Ideality

Ideality is a measure of excellence. In TRIZ, ideality is defined by the following ratio:

$$\text{Ideality} = \frac{\sum \text{benefits}}{\sum \text{costs} + \sum \text{harm}} \qquad (9.1)$$

where $\sum$ benefits = sum of the values of system's useful functions. (Here the supporting functions are not considered to be useful functions, because they will not bring benefits to customers directly. We consider supporting functions are part of the costs to make the system work.)

$\sum$ costs = sum of the expenses for system's performance

$\sum$ harm = sum of "harms" created by harmful functions

In Eq. (9.1), a higher ratio indicates a higher ideality. When a new system is able to achieve a higher ratio than the old system, we consider it a real improvement. In TRIZ, there is a "law of increasing ideality," which states that the evolution of all technical system proceeds in the direction of increasing degree of ideality. The ideality of the system will increase in the following cases:

1. Increasing benefits.

2. Reducing costs

3. Reducing harms

4. Benefits increasing faster than costs and harms

In terms of TRIZ, any technical system or product is not a goal in itself. The real value of the product/system is in its useful functions. Therefore, the better system is the one that consumes fewer resources in both initial construction and maintenance. When the ratio becomes infinite, we call that the "Ideal final result" (IFR). Thus, the IFR system requires no material, consumes no energy and space, needs no maintenance, and will not break.

9.2.4 Contradiction

In the TRIZ standpoint, a challenging problem can be expressed as either a technical contradiction or physical contradiction.

Technical contradiction. A *technical contradiction* is a situation in which efforts to improve some technical attributes of a system will lead to deterioration of other technical attributes. For example, as a container becomes stronger, it becomes heavier, and faster automobile acceleration reduces fuel efficiency.

A technical contradiction is present when

- The useful action simultaneously causes a harmful action
- Introducing (intensification) of the useful action, or elimination or reduction of the harmful action causes deterioration or unacceptable complication of the system or one of its parts

A problem associated with a technical contradiction can be resolved by either finding a trade-off between the contradictory demands or overcoming the contradiction. Trade-off or compromise solutions do not eliminate the technical contradictions, but rather soften them, thus retaining the harmful (undesired) action or shortcoming in the system. Analysis of thousands of inventions by Altshuller resulted in formulation of typical technical contradictions, such as productivity versus accuracy, reliability versus complexity, and shape versus speed. It was discovered that despite the immense diversity of technological systems and even greater diversity of inventive problems, there are only about 1250 typical system contradictions. These contradictions can be expressed as a table of contradiction of 39 design parameters (see Table 9.1 and the chapter appendix).

From the TRIZ standpoint, overcoming a technical contradiction is very important because both attributes in the contradiction can be improved drastically and system performance will be raised to a whole

TABLE 9.1   Design Parameters

1	Weight of moving object	21	Power
2	Weight of nonmoving object	22	Waster of energy
3	Length of moving object	23	Waster of substance
4	Length of nonmoving object	24	Loss of information
5	Area of moving object	25	Waster of time
6	Area of nonmoving object	26	Amount of substance
7	Volume of moving object	27	Reliability
8	Volume of nonmoving object	28	Accuracy of measurement
9	Speed	29	Accuracy of manufacturing
10	Force	30	Harmful factors acting on object
11	Tension, pressure	31	Harmful side effects
12	Shape	32	Manufacturablity
13	Stability of object	33	Convenience of use
14	Strength	34	Repairability
15	Durability of moving object	35	Adaptability
16	Durability of nonmoving object	36	Complex of device
17	Temperature	37	Complexity of control
18	Brightness	38	Level of automation
19	Energy spent by moving object	39	Productivity
20	Energy spent by nonmoving object		

new level. TRIZ developed many tools for elimination of technical contradiction. These tools are discussed in Sec. 9.5.

Physical contradiction.   A physical contradiction is a situation in which a subject or an object has to be in two mutually exclusive physical states.

A physical contradiction has the typical pattern: "To perform function F_1, the element must have property P, but to perform function F_2, it must have property $-P$, or the opposite of P." For example, an automobile has to be light in weight (P) to have high fuel economy (F_1), but it also has to be heavy in weight ($-P$) in order to be stable in driving (F_2).

Example 9.3   When an electrotechnical wire is manufactured, it passes through a liquid enamel and then through a die which removes excess enamel and sizes the wire. The die must be hot to ensure reliable calibration. If the wire feed is interrupted for several minutes or more, the enamel in the hot die bakes and firm grips the wire. The process must then be halted to cut the wire and clean the die.
Physical contradiction.   The die should be both hot, for operation, and cold, to avoid baking the enamel.

In many cases a technical contradiction can also be formulated as a physical contradiction. Conventional design philosophy is based on compromise (trade-off). If a tool or object must be both hot and cold, it was usually made neither too hot nor too cold. Contrary to this approach, TRIZ offers several methods to overcome physical contradictions completely. These methods are discussed in Sec. 9.4.

9.2.5. S-curve and the evolution of a technical system

Based on the study of the evolution of many technical systems, TRIZ researchers have found that the "trends of evolution" of many technical systems are very similar and predictable. They found that many technical system will go through five stages in their evolution processes. These five stages are: pregnancy, infancy, growth, maturity, and decline. If we plot a time line in the horizontal axis (X-axis), and plot:

1. Performance
2. Level of inventiveness
3. Number of inventions (relating to the system)
4. Profitability

on the vertical axis (Y-axis), then we will get the four curves shown in Fig. 9.5. Because the shape of the first curve (performance versus evolution stages) (far left) has an S shape, it is also called the S-curve.

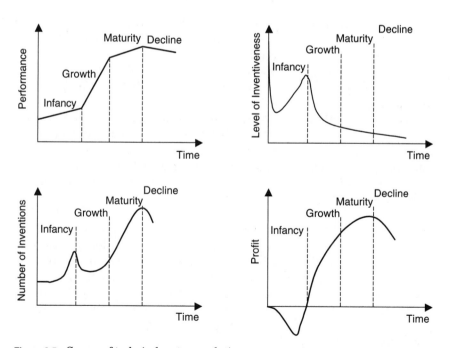

Figure 9.5 Curves of technical system evolution.

Pregnancy. For a technical system, its pregnancy stage is the time between an idea's inception and its birth. A new technological system emerges only after the following two conditions are satisfied:

- There is a need for the function of this system.
- There are means (technology) to deliver this function.

The development of the technical system, an airplane, can be used as an example. The need for the function of an airplane, that is, "to fly" was there a long time ago in many people's dreams and desires. However, the technical knowledge of aerodynamics and mechanics were not sufficient for the development of human flight until the 1800s. The technologies for the airplane were available since the development of glider flight in 1848 and the gasoline engine in 1859. It was the Wright brothers who successfully integrated both technologies in their aircraft in 1903, and a new technology got off the ground.

Infancy. The birth of a new technical system is the starting point of the infancy stage, and it is the first stage of an S-curve. The new system appears as a result of a high-level invention. Typically, the system is primitive, inefficient, unreliable, and has many unsolved problems. It does, however, provide some new functions, or the means to provide the function. System development at this stage is very slow, due to lack of human and financial resources. Many design questions and issues must be answered. For example, most people may not be convinced of the usefulness of the system, but a small number of enthusiasts who believe in the system's future continue to work toward its success.

In the infancy stage, the performance level is low and its improvement is slow (Fig.9.5, top left). The level of inventions is usually high, because the initial concept is often very inventive and patentable. It is usually level 3,4, or even 5 (top right). But the number of inventions in this system is usually low (bottom left), because the system is fairly new. The profit is usually negative (bottom right), because at this stage of the technical development the customers are usually few, but the expense is high.

Growth (rapid development). This stage begins when society realizes the value of the new system. By this time, many problems have been overcome, efficiency and performance have improved in the system, and people and organizations invest money in development of the new product or process. This accelerates the system's development, improving the results and, in turn, attracting greater investment. Thus, a positive "feedback" loop is established, which serves to further accelerate the system's evolution.

In the growth stage, the improvement of performance level is fast (Fig.9.5, top left), because of the rapid increase in the investment and the removal of many technical bottlenecks. The level of inventions is getting lower, because most inventions in this stage are dealing with incremental improvements. They are mostly level 1 or level 2 (top right), but the number of inventions is usually high (bottom left). The profit is usually growing fast (bottom right).

Maturity.   In this stage, system development slows as the initial system concept nears exhaustion of its potentials. Large amounts of money and labor may have been expended, however, the results are usually very marginal. At this stage, standards are established. Improvements occur through system optimization and trade-off. The performance of the system still grows but at a slower pace (Fig. 9.5, top left). The level of invention is usually low (top right) but the the number of inventions in the forms of industrial standards is quite high (bottom left). The profitability is usually dropping because of the saturation of the market and increased competition (bottom right).

Decline.   At this stage, the limits of technology have been reached and no fundamental improvement is available. The system may no longer be needed, because the function provided may be no longer needed. It is really important to start the next generation of technical systems long before the decline stage, in order to avoid the failure of the company. Figure 9.6 illustrate the S-curves of the succession of two generations of a technical system.

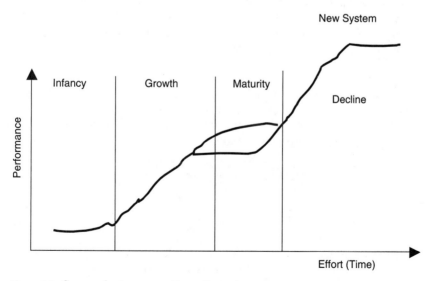

Figure 9.6   S-curve for two generations of a system

9.3   TRIZ Problem-Solving Process

TRIZ has a four-step problem-solving process: (1) problem definition, (2) problem classification and problem tool selection, (3) problem solution, and (4) solution evaluation. We shall describe each step in detail.

9.3.1   Problem definition

Problem definition is a very important step. The quality of the solution is highly dependent on problem definition.

The problem definition starts with several questions:

1. What is the problem?

2. What is the scope of the project?

3. What subsystem, system, and components are involved?

4. Do we have a current solution, and why is the current solution not good?

These are common questions to be asked in any engineering project. By answering these questions, we are able to define the scope of the project and focus on the right problem area.

Besides answering these common questions, several TRIZ methods are also very helpful in the problem definition stage.

Functional modeling and functional analysis.   After identifying the project scope, it is very helpful to establish the functional model of the subsystem involved in this project.   Functional modeling and analysis enables us to see the problem more clearly and precisely. We will recall the toothbrush example to illustrate how functional analysis can help the problem definition.

Example 9.4. Toothbrush Problem Revisited   Assume that we are a toothbrush manufacturer, and the current regular toothbrush is not satisfactory in performance, that is, teeth cannot be adequately cleaned. We can first draw the following functional diagram:

By analyzing the functional diagram, we may come up with the following possibilities:

1. The current lack of performance may be caused by "inadequate action," that is, the actual functional diagram is the following:

If that is the case, it belongs to the problem of "inadequate functional performance," and we can use the TRIZ standard solution technique to resolve this problem.

2. We may find the current functional model is too limiting, because the function statement "toothbrush brushes teeth" limits our solution to using the brush only and to using mechanical action only. We can develop the following alternative functional modeling:

The subject "toothbrush" is replaced by a more general "tooth cleaning device." The object "teeth" is changed to "dirt in teeth," which is more precise. The action "brush" is changed to a more general term "remove." Under this alternative functional modeling, many possible choices of "subjects" and "actions" can be open for selection. For example, we can use hydraulic action to clean teeth, or chemical action to clean teeth, we can even consider pretreatment of teeth to make them dirt free and so on. Clearly this alternative functional modeling opens the door for problem solving and innovation.

Ideality and ideal final result   After functional modeling and functional analysis, we can evaluate the ideality of the current system by using

$$\text{Ideality} = \frac{\Sigma \text{ benefits}}{\Sigma \text{ costs} + \Sigma \text{ harm}}$$

Ideal final result means the ultimate optimal solution for current system in which:

$$\Sigma \text{ benefits} \rightarrow \infty \quad \text{and} \quad \Sigma \text{ costs} + \Sigma \text{ harm} \rightarrow 0$$

By comparing the ideality of the current system with ideal final result, we can identify "where the system improvement should go" and "what aspects of system should be improved." This will definitely help the problem definition and identify "what problem should be solved."

S-curve analysis   It is very beneficial to evaluate the evolution stage of the current technical system involved in any TRIZ project. For example, if our current subsystem is at the growth stage, then we should focus our attention to gradual improvement. If our subsystem is near the maturity stage, then we will know that it is time to develop the next generation of this subsystem.

Contradiction analysis   By using the method described in Sec. 9.2.4, we can determine if there are any physical contradictions or technical contradictions in our current system. TRIZ has many methods to resolve contradictions.

9.3.2 Problem classification and tool selection

After we are finished with the problem definition, we should be able to classify the problem into the following categories. For each category there are many TRIZ methods available to resolve the problem.

Physical contradiction

   Methods. Physical contradiction resolution and separation principles.

Technical contradiction

   Methods. Inventive principles.

Imperfect functional structures.   This problem occurs when:

1. There are inadequate useful functions or lack of needed useful functions
2. There are excessive harmful functions

   Methods. Functional improvement methods and TRIZ standard solutions

Excessive complexity.   This problem occurs when the system is too complex and costly, and some of its functions can be eliminated or combined.

   Methods. Trimming and pruning.

System improvement.   This problem occurs when the current system is doing its job, but enhancement is needed to beat the competition.

   Method. Evolution of technological systems.

Develop useful functions.   This problem occurs when we can identify what useful functions are needed to improve the system but we do not know how to create these functions.

Methods. Physical, chemical, and geometric effects database.

9.3.3 Solution generation

After problem classification, there are usually many TRIZ methods available for solving the problem, so many alternative solutions could be found. These solutions will be evaluated in the next step.

9.3.4 Concept evaluation

There are many concept evaluation methods that can be used to evaluate and select the best solution. These methods are often not TRIZ related. The frequently used concept evaluation methods include Pugh concept selection, value engineering, and the axiomatic design method.

9.4   Physical Contradiction Resolution/Separation Principles

Usually, when we first encounter a contradiction, it often appears as a technical contradiction. After digging deeper into the problem, the fundamental cause of the technical contradiction is often a physical contradiction. A *physical contradiction* is a situation in which a subject or an object has to be in mutually exclusive physical state. A physical contradiction has the typical pattern: "To perform function F_1, the element must have property P, but to perform function F_2, it must have property –P, or the opposite of P."

9.4.1 Analyze the physical contradiction

In order to identify the physical contradiction that causes the technical contradiction, the following three steps are recommended to pre-analyze the conflict:

Step 1: Capture the functions involved in the conflict and establish the functional model for the contradiction

Step 2: Identify the physical contradiction. Physical contradiction often happens when a useful action and a harmful action coexist on the same object.

Step 3: Identify the zones of conflict (Domb 2003). There are two "zones" of conflict in a problem, spatial and temporal. In other words, the two "zones" are the location properties of conflict and the time properties of conflict. The identification of the zones can help to determine what separation principles can be used to resolve the physical contradiction.

We will use the following example to illustrate the above three steps.

Example 9.5* To seal ampoules containing medicine, flames from burners are applied to the neck of the ampoules to melt the glass. However, the flame may overheat the medicine in the ampoules, causing the medicine to decompose:

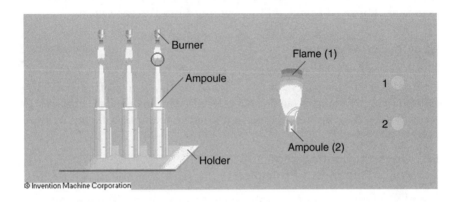

Step 1: Capture the functions involved in the conflict and establish the functional model for the conflict. Ampoules need to be sealed, but the drug should be kept intact. The flame will melt the glass and seal the ampoule, but it also overheats the drug. The functional model is the following:

Step 2: Identify the contradiction. Clearly, a useful action, "heat and seal ampoule," and a harmful action, "heat and decompose the drug" coexist on the ampoule. Therefore, the physical contradiction is

■ Ampoules need to be hot so they can be melted and sealed

■ Ampoules cannot be hot, or the drug would be decomposed.

Step 3: Identify the zones of conflict.

■ *Location property.* By examining the requirements to adequately seal the ampoule, it is very easy to find that the heat should only be applied to the tip of the ampoule. The bottom of the ampoule should never be heated in order to prevent drug decomposition.

■ *Time property.* In the current ampoule sealing process, the useful function, "heat and seal ampoule," and the harmful function "heat and decompose the drug" will happen simultaneously.

---

*The figures in Examples 9.5, 9.10, 9.15–9.17, 9.19–9.22, 9.24–9.29, and 9.44 are reprinted with permission from Invention Machine Corporation, www.invention-machine.com.

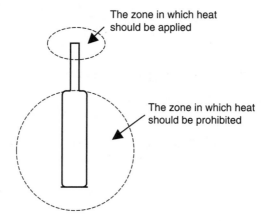

The zone in which heat
should be applied

The zone in which heat
should be prohibited

9.4.2   Separate the physical contradiction

After the identification of the physical contradiction, TRIZ has the following four approaches for resolving the contradiction. They are: separation in space, separation in time, separation between components, and separation between components and a set of components.

Approach 1: Separation in space.   Separation in space means: one part of an object has property P, while another part has an opposite property –P. By this separation the physical contradiction can be resolved. In order to accomplish the separation, we need to study the zones of conflict requirements. For example, in the "sealing glass ampoule" case, if we are able to keep the top zone of the ampoule hot and bottom zone of the ampoule cool, thus the physical contradiction can be resolved.

Example 9.6. Separation in Space

Problem.   Metallic surfaces are placed in metal salt solution (nickel, cobalt, chromium) for chemical coating. During the reduction reaction, metal from the solution precipitates onto the product surface. The higher the temperature, the faster the process, but the solution decomposes at a high temperature. As much as 75 percent of the chemicals settle on the bottom and walls of the container. Adding stabilizer is not effective, and conducting the process at a low temperature sharply decreases productivity.

Contradiction.   The process must be hot (from fast, effective coating) and cold (to efficiently utilize the metallic salt solution). Using the separation principle in space, it is apparent that only the areas around the part must be hot.

Solution.   The product is heated to a high temperature before it is immersed in a cold solution. In this case, the solution is hot where it is near the product, but cold elsewhere. One way to keep the product hot during coating is by applying an electric current for inductive heating during the coating process.

Example 9.7. Separation in Space (Fey and Revin 1997)

Problem.   A rotor link has to satisfy mutually exclusive requirements. It should be light (for a given shape) to reduce driving motors and/or to allow for larger payloads. However, light materials usually have reduced stiffness. On the other hand, it should be rigid to reduce end-of-arm deflection and the setting time in start/stop periods. However, rigid materials are usually heavy.

Contradiction.   The link must be light in order to carry large payloads, but it must be heavy for good rigidity.

Solution.   A simple analysis reveals that the end-of-link deflection is determined by the root segment of the link (the highest bending moment), while the inertia is determined mainly by the overhand segment. Thus the contradictory requirements are separated in space. The root section, which does not significantly influence the effective mass of the link but determines its stiffness, is made of a high Young's modulus material such as steel, while the overhang section, which does not noticeably contribute to stiffness but determines the effective mass, is made of something light (such as aluminum).

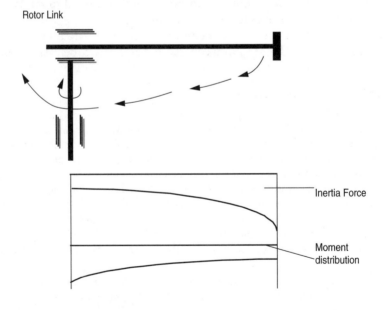

Rotor Link

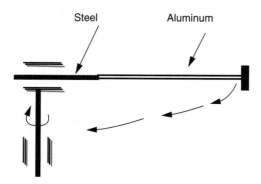

Steel   Aluminum

Approach 2: Separation in time. *Separation in time* means that at one time period an object has property P, and at another time period it has an opposite property −P.

In order to accomplish this, we need to study the time property of the conflict. If it is again a conflict of useful action versus harmful action, we need to identify the periods of both the useful action and the harmful action. We must then identify the time periods when the useful function has to be performed and harmful function eliminated. If we can separate these two periods completely, we may be able to eliminate this contradiction.

Example 9.8. Separation in Time (Teminko et al., 1998)

Problem. When an electrotechnical wire is manufactured, it passes through a liquid enamel and then through a die which removes excess enamel and sizes the wire. The die must be hot to ensure reliable calibration. If the wire feed is interrupted for several minutes or more, the enamel in the hot die bakes and firmly grips the wire. The process must then be halted to cut the wire and clean the die.

Contradiction. The die should be hot for operation and cold to avoid baking enamel. The separation-in-time principle suggests that the die should be hot when the wire is being drawn and cold when wire is not moving. Is there a way to automatically control heating of the die? While the wire is being drawn on the die, there is a significant force pulling the die in the direction of the wire pull and when the wire stops, there will be no pull.

Solution. The die can be fixed to a spring. When the wire moves, it pulls the die, which compresses the spring into the heating zone. The die is heated either by induction or by contact with the hot chamber walls. When the wire stops moving, the spring pushes the die back into the cold zone.

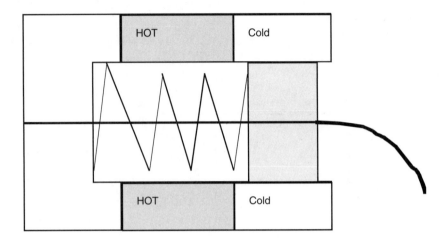

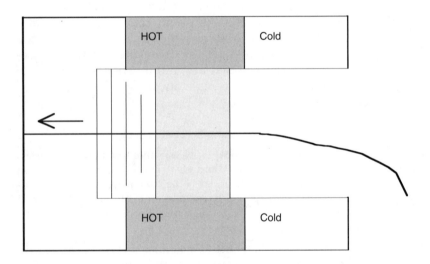

Example 9.9. Separation in Time

Problem.   Some buildings are supported by piles. The pile should have a sharp tip to facilitate the driving process. However, the sharp piles have reduced support capability. For better support capacity, the piles should have blunt ends. However, it is more difficult to drive a blunt-tipped pile.

Contradiction.   A pile should be sharp to facilitate the driving process but blunt to provide better support of the foundation.

Solution.   The situation clearly calls for the solution providing separation of contradictory properties in time. The pile is sharp during the driving process, and then its base is expanded, which could be realized by a small explosive charge.

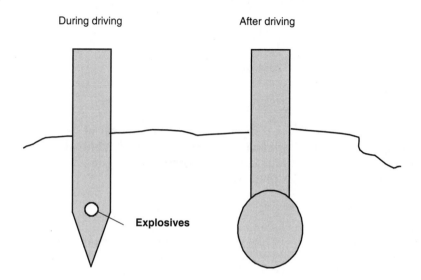

During driving   *After driving*

Explosives

Approach 3: Separation between the components. *Separation between the components* means that one component has property P, while another component is given an opposite property −P.

Sometimes we can limit the number of properties of the component involved in the conflict to one, and we introduce another component to have another property.

Example 9.10

Problem.   A conventional bench vise is designed to hold objects of regular shapes. To hold objects of irregular shapes, special jaws have to be installed. Fabrication of such jaws requires a time consuming and laborious process.

Contradiction.   The jaw must be rigid to clamp the part, and they must be flexible to accommodate themselves to the part's profile.

Solution.   It is proposed to use the principle of separation of opposite properties between the system and its components to make a jaw. Two flexible shells filled with a loose material are fixed to upper and lower vise jaws. The workpiece (parts) is placed between the shells and pressed. Once the shells envelope the workpiece, a vacuum pump is used to set the final shape, fastening the workpiece reliably. This design increases productivity and reduces the need for building a specialized fixture.

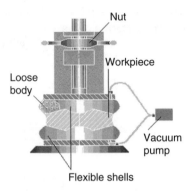

Flexible shells

Approach 4: Separation between components and the set of the components. This is an approach that has a set of components made, where every single component must have one property but the whole set of components will have another property.

Example 9.11

Problem. A remote-control unit (e.g., a TV set) is comprised of an infrared radiator, a capacitor, a power source (storage battery), and key-type control elements. The capacitor is permanently charged by the battery. When any of the panel keys are pressed, the capacitor discharges into the infrared radiator, which emits a controlling electromagnetic impulse. However, a high-value capacitor operating in charging or waiting mode has considerable leakage current. This is responsible for wasted discharge of the battery. A decrease in the capacitor value leads to a decrease in the range of the panel's action. An increase in the power element value causes an increase in the mass and cost of the panel.

Technical contradiction. We want to reduce leakage current by reducing the capacitor value, but the power of the emitted controlling impulse decreases. Obviously it is a typical technical contradiction since improving one parameter degrades another parameter.

Physical contradiction. In order to improve leakage current and avoid deterioration of the power, the capacitor should be in low value when it is charged and in high value when it discharges. So, the technical contradiction that relates to the whole system is changed into a physical contradiction that only relates to one component.

Solution. It is proposed to divide the energy flow while charging the capacitors and to combine the energy flow while exciting the infrared radiator. To do this, the capacitor is formed as a set of low-value capacitors combined in a battery so that their summed value is equal to the value of the panel's initial capacitor. The total summed leakage current of the battery of the capacitor is less than the leakage current of one high-value capacitor. The power of the emitted controlling pulse is kept the same as before.

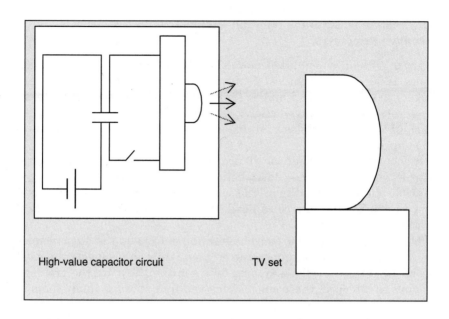

High-value capacitor circuit   TV set

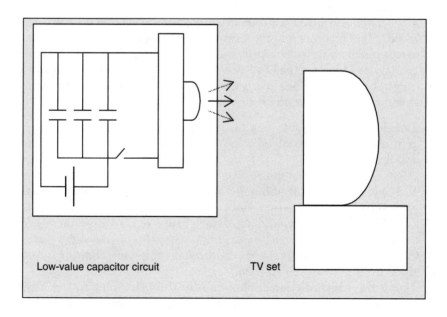

Low-value capacitor circuit   TV set

9.5   Technical Contradiction Elimination— Inventive Principles

Genrich Altshuller analyzed more than 40,000 patents and identified about 1250 typical technical contradictions. These contradictions are further expressed into a matrix of 39 × 39 "engineering parameters." To resolve these contradictions, Altshuller compiled 40 principles. Each of the 40 principles contains a few subprinciples, totaling up to 86 subprinciples.

It should be noted that the 40 principles are formulated in a general way. If, for example, the contradiction table recommends principle 30, "flexible shell and thin films," the solution of the problem relates somehow to change the degree of flexibility or adaptability of a technical system being modified.

The contradiction table (see chapter appendix) and the 40 principles do not offer a direct solution to the problem; they only suggest the most promising directions for searching for a solution. To solve the problem, one has to interpret these suggestions and find a way to apply them to a particular situation.

Usually people solve problems by analogical thinking. We try to relate the problem confronting us to some familiar standard class of problems (analogs) for which a solution exists. If we draw on the right analog, we arrive at a useful solution. Our knowledge of analogous problems is the result of educational, professional, and life experiences.

What if we encounter a problem analogous to the one we have never faced? This obvious question reveals the shortcomings of our standard approach to invention problems. So, the contradiction table and 40 principles offer us clues to the solution of the problems with which we are not familiar. When using the contradiction table and 40 principles, following this simple procedure will be helpful:

1. Decide which attribute has to be improved, and use one of the 39 parameters in the contradiction table to standardize or model this attribute.
2. Answer the following questions:
   a. How can this attribute be improved using the conventional means?
   b. Which attribute would be deteriorated if conventional means were used?
3. Select an attribute in the contradiction table (see chapter appendix) corresponding to step 2b.
4. Using the contradiction table, identify the principles in the intersection of the row (attributes improved) and column (attribute deteriorated) for overcoming the technical contradiction.

We list the 40 principles for reference:

Principle 1: Segmentation
- Divide an object into independent parts.
- Make an object easy to disassemble.
- Increase the degree of fragmentation (or segmentation) of an object.

Principle 2: Taking out.   Separate an "interfering" part (or property) from an object, or single out the only necessary part (or property) of an object.

Principle 3: Local quality
- Change an object's structure from uniform to nonuniform, or change an external environment (or external influence) from uniform to nonuniform.
- Make each part of an object function in conditions most suitable for its operation.
- Make each part of an object fulfill different and useful functions.

Principle 4: Asymmetry
- Change the shape of an object from symmetric to asymmetric.
- If an object is asymmetric, increase its degree of asymmetry.

Principle 5: Merging
- Bring closer together (or merge) identical or similar objects; assemble identical or similar parts to perform parallel operations.
- Make operations contiguous or parallel, and bring them together in time.

Principle 6: Universality.   Make a part or object perform multiple functions, to eliminate the need for other parts.

Principle 7: "Nested doll"
- Place each object, in turn, inside another, larger object.
- Make one part pass through a cavity in the other part.

Principle 8: Antiweight
- To compensate for the weight of an object, merge it with other objects that provide lift.
- To compensate for the weight of an object, make it interact with the environment (e.g., use aerodynamic, hydrodynamic, buoyancy, and other forces).

Principle 9: Preliminary antiaction
- If it will be necessary to perform an action with both harmful and useful effects, this action should be replaced later with antiactions to control harmful effects.

- Create stresses in an object that will oppose known undesirable working stresses later on.

Principle 10: Preliminary action

- Perform, before it is needed, the required modification of an object (either fully or partially).
- Prearrange objects in such a way that they can perform their intended actions expeditiously from the most convenient position.

Principle 11: Beforehand cushioning.   Prepare emergency means beforehand to compensate for the relatively low reliability of an object.

Principle 12: Equipotentiality.   In a potential field, limit position changes (e.g., change operating conditions to eliminate the need to raise or lower objects in a gravity field).

Principle 13: "The other way around"

- Invert the action(s) used to solve the problem (e.g., instead of cooling an object, heat it).
- Make movable parts (or the external environment) fixed, and fixed parts movable.
- Turn the object (or process) upside-down.

Principle 14: Spheroidality

- Instead of using rectilinear parts, surfaces, or forms, use curvilinear ones, moving from flat surfaces to spherical ones, or from parts shaped as a cube (parallelepiped) to ball-shaped structures.
- Use rollers, balls, spirals, and/or domes.
- Go from linear to rotary motion, using centrifugal force.

Principle 15: Dynamics

- Allow (or design) the characteristics of an object, external environment, or process to change to be optimal or to find an optimal operating condition.
- Divide an object into parts capable of movement relative to one another.
- If an object (or process) is rigid or inflexible, make it movable or adaptive.

Principle 16: Partial or excessive actions.   If 100 percent of an effect is hard to achieve using a given solution method, then, by using "slightly less" or "slightly more" of the same method, the problem may be considerably easier to solve.

Principle 17: Another dimension

- Move an object in two- or three-dimensional space.
- Use a multistory arrangement of objects instead of a single-story arrangement.

- Tilt or reorient the object, laying it on its side.
- Use "another side" of a given area.

Principle 18: Mechanical vibration

- Cause an object to oscillate or vibrate.
- Increase the object's frequency (even up to the ultrasonic level).
- Use an object's resonance frequency.
- Use piezoelectric vibrators instead of mechanical ones.
- Use combined ultrasonic and electromagnetic field oscillations.

Principle 19: Periodic action

- Instead of continuous action, use periodic or pulsating actions.
- If an action is already periodic, change the periodic magnitude or frequency.
- Use pauses between impulses to perform a different action.

Principle 20: Continuity of useful action

- Carry on work continuously; make all parts of an object work at full load, all the time.
- Eliminate all idle or intermittent actions or work.

Principle 21: Skipping.   Conduct a process, or certain stages (e.g., destructive, harmful, or hazardous operations), at high speed.

Principle 22: "Blessing in disguise"

- Use harmful factors (particularly, harmful effects of the environment or surroundings) to achieve a positive effect.
- Eliminate the primary harmful action by adding it to another harmful action to resolve the problem.
- Amplify a harmful factor to such a degree that it is no longer harmful.

Principle 23: Feedback

- Introduce feedback (referring back, cross-checking) to improve a process or action.
- If feedback is already used, change its magnitude or influence.

Principle 24: "Intermediary"

- Use an intermediate carrier article or intermediary process.
- Merge one object temporarily with another (which can be easily removed).

Principle 25: Self-service

- Make an object serve itself by performing auxiliary helpful functions.
- Use waste resources, energy, or substances.

Principle 26: Copying

- Instead of an unavailable, expensive, or fragile object, use simpler and inexpensive copies of it.
- Replace an object or process with its optical copies.
- If visible optical copies are already used, move to infrared or ultraviolet copies.

Principle 27: Cheap short-living. Replace an expensive object with a multitude of inexpensive objects, compromising certain qualities (e.g., service life).

Principle 28: Mechanical substitution

- Replace a mechanical means with a sensory (optical, acoustic, taste or smell) means.
- Use electric, magnetic, and electromagnetic fields to interact with the object.
- Change from static to movable fields, from unstructured fields to those having structure.
- Use fields in conjunction with field-activated (e.g., ferromagnetic) particles.

Principle 29: Pneumatics and hydraulics. Use gas and liquid parts of an object instead of solid parts (e.g., inflatable, liquid-filled, air-cushioned, hydrostatic, hydroreactive parts).

Principle 30: Flexible shells and thin films

- Use flexible shells and thin films instead of three-dimensional structures.
- Isolate the object from the external environment using flexible shells and thin films.

Principle 31: Porous materials

- Make an object porous or add porous elements (inserts, coatings, etc.).
- If an object is already porous, use the pores to introduce a useful substance or function.

Principle 32: Color changes

- Change the color of an object or its external environment.
- Change the transparency of an object or its external environment.

Principle 33: Homogeneity. Make objects interacting with a given object of the same material (or a material with identical properties).

Principle 34: Discarding and recovering

- Dispose of portions of an object that have fulfilled their function (discard by dissolving, evaporating, etc.) or modify them directly during operation.

- Conversely, restore consumable parts of an object directly during operation.

Principle 35: Parameter changes

- Change an object's physical state (e.g., to a gas, liquid, or solid).
- Change the concentration or consistency.
- Change the degree of flexibility.
- Change the temperature.

Principle 36: Phase transitions. Use phenomena occurring during phase transitions (e.g. volume changes, loss or absorption of heat).

Principle 37: Thermal expansion

- Use thermal expansion (or contraction) of materials.
- If thermal expansion is being used, use multiple materials with different coefficients of thermal expansion.

Principle 38: Strong oxidants

- Replace common air with oxygen-enriched air.
- Replace enriched air with pure oxygen.
- Expose air or oxygen to ionizing radiation.
- Use ozonized oxygen.
- Replace ozonized (or ionized) oxygen with ozone.

Principle 39: Inert atmosphere

- Replace a normal environment with an inert one.
- Add neutral parts, or inert additives to an object.

Principle 40: Composite materials. Change from uniform to composite (multiple) materials.

Example 9.12 Using 40 Principles and Contradiction Table to Improve Wrench Design   When we use a conventional wrench to take off an over-tightened or corroded nut (as shown in the picture), one of the problems is that the corners of the nut are getting concentrated load so they may wear out quickly. We can reduce the clearance between wrench and nut, but it will be difficult to fit in. Is there anything we can do to solve this problem (Darrel Mann 2002)?

It is clear that we want to reduce the space between wrench and nut to improve operation reliability, however, this led to the deterioration of operations. From a TRIZ standpoint, a technical contradiction is present when a useful action simultaneously causes a harmful action.

A problem associated with a technical contradiction can be resolved either by finding a trade-off between the contradictory demands, or by overcoming the contradiction. Trade-off or compromise solutions do not eliminate the technical contradictions, but rather soften them, thus retaining harmful (undesired) actions or shortcomings in the system. A trade-off solution in this example would be to make the clearance neither too big nor too small. An inventive solution for this problem can be found if the technical contradiction can be overcome completely. Forty principles and the contradiction table are important tools for overcoming contradictions.

1. *Build contradiction model.* Look into the problems and find a pair of contradictions. The contradiction should be described using 2 parameters of the 39 parameters for technical contradictions. In this problem, the contradiction is:

Things we want to improve: Reliability (parameter 27)

Things are getting worse: Ease of operation (parameter 33)

2. *Check contradiction table.* Locate the parameter to be improved in the row and the parameter to be deteriorated in the column in the contradiction matrix for inventive principles. The matrix offers the following principles 27, 17, and 40 (see the partial matrix show below).

What should be improved?	What is deteriorated?										
	25. Waste of time	26. Quantity of substance	27. Reliability	28. Measurement accuracy	29. Manufacturing precision	30. Harmful action at object	31. Harmful effect caused by the object	32. Ease of manufacture	33. Ease of operation	34. Ease of repair	35. Adaptation
25. Waste of time		35 38 18 16	10 30 4	24 34 28 32	24 26 28 18	35 18 34	35 22 18 39	35 28 34 4	4 28 10 34	32 1 10	35 28
26. Quantity of substance	35 38 18 16		18 3 28 40	3 2 28	33 30	35 33 29 31	3 35 40 39	29 1 35 27	35 29 10 25	2 32 10 25	15 3 29
27. Reliability	10 30 4	21 28 40 3		32 3 11 23	11 32 1	27 35 2 40	35 2 40 26		**27 17 40**	1 11	13 35 8 24
28. Measurement accuracy	24 34 28 32	2 6 32	5 11 1 23			28 24 22 26	3 33 39 10	6 35 25 18	1 13 17 34	1 32 13 11	13 35 2
29. Manufacturing precision	32 26 28 18	32 30	11 32 1			26 28 10 36	4 17 34 26		1 32 35 23	25 10	

3. *Interpret principles.* Read each principle and construct analogies between the concepts of principle and your situation, then create solutions to your problem. Principle 17 (another dimension) indicates that the wrench problem may be resolved to "move an object in two- or three-dimensional space" or "use a different side of the given area." From principle 27 (cheaper short living) and principle 40 (composition material), we may "replace an expensive object with multitude of inexpensive objects" and "change from uniform material to composite material."

4. *Resolve the problem.* The working surface of the wrench can be redesigned in nonuniform shape by applying principle 17 (see the illustration below). Principle 27 and 40 can be used together. The idea is to attach soft metal or plastic pads on the wrench working surfaces when tightening or undoing expensive nuts.

9.6 Functional Improvement Methods/TRIZ Standard Solutions

A function is the basic element for TRIZ analysis. Many problems in a technical system can be attributed to imperfect functional performances in part of the system. In the TRIZ point of view, at least three elements are needed to deliver a function. They are a subject, a field, and a object as illustrated here:

If any of the three elements is missing, then no function will be delivered. If any of these three elements is not working properly, then the function will not be delivered satisfactorily.

As we discussed in Sec. 9.2.1, sometimes some components in a system can also deliver harmful functions. Harmful functions will always reduce the ideality of the system and they are highly undesirable.

Functional improvement methods are the methods to improve the delivery of useful functions and to eliminate or contain harmful functions. Functional improvement methods are derived from TRIZ '76 standard solutions compiled by G.S. Altshuller and his associates between 1975 and 1985 (Domb et al. 1999). We will describe the functional improvement methods in the following two categories: the methods to improve useful functions and the methods to eliminate or contain harmful functions.

9.6.1 Methods to improve useful functions

If a useful function is not delivered or is not delivered properly, then there are two main reasons:

1. There are some element(s) missing in the subject-action-object model. The most frequently missing elements are the subject and action. In this case the function will not be delivered at all. To resolve this situation, we need to supply the missing elements in the subject-action-object model.

2. Some elements(s) in the subject-action-object model are not working properly, it could be the object, the field, or the subject, or a combination of them. In this case the useful function will not be adequately delivered. To resolve this situation, we need to improve the elements which are not working properly.

We will discuss several frequently used methods as follows.

Method 1: Fill the missing elements in a subject-action-object model

The most frequently missing elements in a subject-action-object model are subject and action, that is, the current situation is:

We need to find an action, or field, and an subject to complete the function model:

Example 9.13 A liquid contains vapor bubbles. The desired effect is to separate the bubbles from the liquid, however, that effect is not happening. The subject-action-object model for the actual situation is as follows:

With application of a centrifuge force, the vapor can be separated, then we need to add an action, or a field, which is "centrifuge force," and a subject, which is a rotation device, to make a complete subject-action-object model:

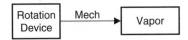

where "Mech" stands for mechanical field or mechanical action.

**Method 2: Add a subject and field to create
an adequate useful action**

There are cases where the original subject and field do exist, but they are not sufficient to create an adequate useful function:

Then we can add another subject and field to reinforce the effort, that is:

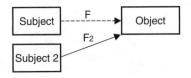

Example 9.14 Using only mechanical means to remove wallpaper is not efficient, but after spraying steam onto wallpaper, it will be much easier to remove.

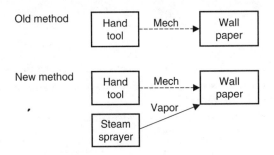

Method 3: Improve the object

In the case of inadequate useful action:

One of the common causes is that the object is not sensitive to the action or field. We can increase the sensitivity of the object to the field by altering the object by one of the following following ways:

- Replace the original object by a new substance
- Modify the substance of object
- Apply additive to outside object
- Apply additive to inside object
- Change the material structure or properties of object

That is, we change the original system to:

Example 9.15   Check for refrigerator leaks:

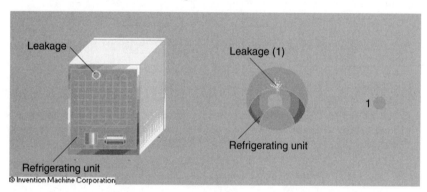

Initial situation.   There is a need to check a refrigeration system for leaks.
Current problem.   Normal techniques do not provide accurate detection and location of refrigerant leaks.
Analysis.   The human eye cannot see leakage (liquid flowing out), that is,

New design

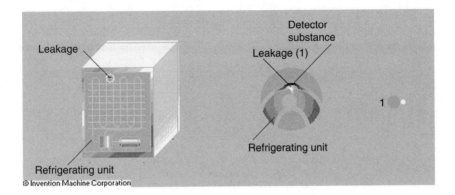

It is proposed to use a layer of detector substance on critical areas. The external surface is coated with heat-conductive paint (mixed with a detector substance). The paint swells and changes color to indicate the location of refrigerant leaks, that is,

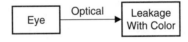

Examples 9.16 Measuring Surface Area

Old design

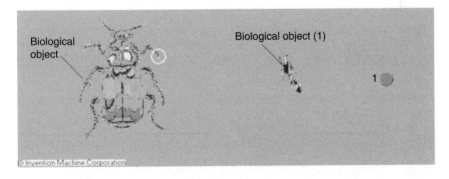

It is necessary to measure the surface area of biological specimens, such as insects. (*Disadvantage:* Small size and complex surface relief make measurement difficult.)

New design

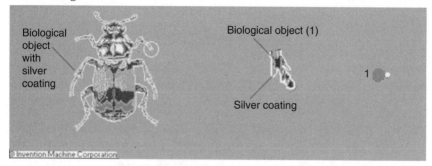

Biological object with silver coating

Biological object (1)

Silver coating

1

It is proposed to apply a thin coat of silver to the insect and measure its weight or amount. The coating is applied chemically in an acid solution of silver nitrate. The surface area of the insect is measured by the change in concentration of silver in solution (or the weight) after being coated.

Example 9.17   A Reliable Seal

Old design

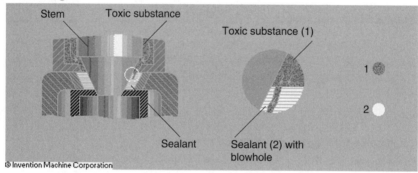

Stem   Toxic substance

Toxic substance (1)

1

2

Sealant   Sealant (2) with blowhole

A porous insert filled with sealant keeps a toxic agent from leaking by a stem. This is done using a tight fit of the insert against the stem surface. (*Disadvantage:* At high pressure, the toxic substance may press the seal away from the stem, forming a blowhole.)

New design

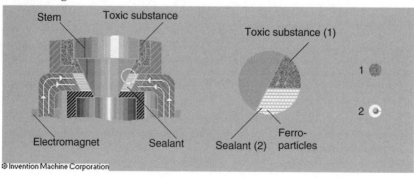

Stem   Toxic substance

Toxic substance (1)

1

2

Electromagnet   Sealant   Sealant (2)   Ferro-particles

It is proposed to introduce ferromagnetic particles into the sealant, pressing them (with a magnetic field) against the stem. The particles hold fast, increase sealant viscosity, and prevent formation of blowholes.

Method 4: Improve the field

In the case of inadequate useful action:

Another common cause is that the current field is not effective in delivering action to the object. We can try to change the field in one of the following ways:

- Change the direction of the field
- Change the intensity of the field
- Change the space structure of the field (uniform, nonuniform, and so on)
- Change the time structure of the field (inpulsive, accelerate, deaccelerate, back and forth, and so on)
- Apply a new substance between subject and object to alter the property of the field
- Add another field or fields to enhance the overall effects

That is, we change the original system to:

Example 9.18. Remote Electrical Welding   In the electronic manufacturing process, we need to weld thin wires in difficult-to-access areas. Electric current is applied in order to melt the wire:

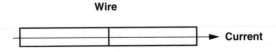

However, the wire won't melt unless the current is large; therefore, applying a high-electricity-resistant coating at the wire joint area is recommended to ensure a voltage drop near the joint sufficient to convert the electric field into a high-heat field at the wire joint area. This heat melts the wire and forms a bond, but very little voltage drop will occur at other parts of wire, leaving them intact, unaffected by the current.

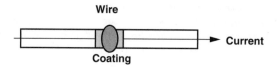

Wire

Coating

Current

Example 9.19 Change the Space Structure of the Field

Old design

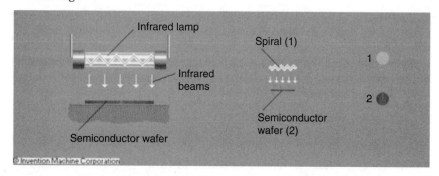

An infrared lamp heats a semiconductor wafer. (*Disadvantage:* The wafer heats unevenly; the edges cool down faster than the center portion.)

New design

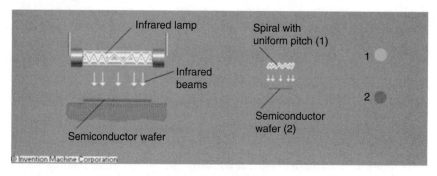

It is proposed to make the lamp heat uniformly. The lamp spiral heating element is made with a smaller winding pitch near the outer edges. This lamp produces an uneven heat flow which compensates for nonuniform cooling in the wafer.

9.6.2 Methods to eliminate or contain harmful functions

When a harmful function is delivered, we have:

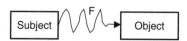

As we discussed earlier, all three elements are needed to deliver a function, including a harmful function.   Therefore, to eliminate a harmful function, we can try the following:

1. Block or disable the harmful action (field)
2. Destroy or disable the field of a harmful function
3. Draw harmful action (field) to another object
4. Add another field/fields to counterreact the harmful action (field)

We will discuss several frequently used methods as follows.

Method 1: Block or disable the harmful action (field)

We can block the harmful action (field) to the object by

- Inserting a substance to shield the object from harmful action, that is

- Inserting a substance between subject and object to alter the property of the harmful field, that is:

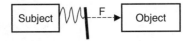

Example 9.20. Glass Ampoule Seal   In this problem, overheating of the medicine inside the ampoule is the harmful action. We will try to introduce an insulator to shield the ampoule bottom from the heat.

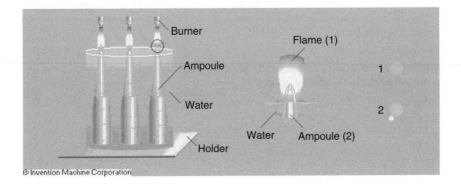

Specifically, we propose surrounding the ampoules with a water jacket. The water takes excess heat away from the ampoules to prevent overheating the drug.

Example 9.21. Bending Studded Pipe

Old design

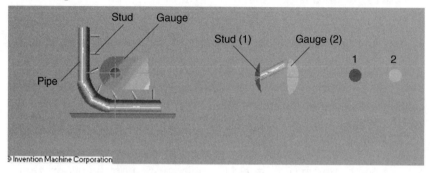

Studded pipe is bent over a mandrel to form it as needed. (*Disadvantage:* The mandrel damages the studs.)

New design

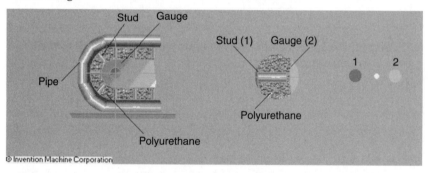

It is proposed to introduce an elastic (polyurethane) layer between the mandrel and the pipe. The elastic compresses without transmitting force from the mandrel to the studs. At the same time, it transmits enough compression force to bend the pipe wall between the studs. As a result, bending does not damage the studs.

Method 2: Add another field/fields to counterreact the harmful action (field)

Another field F_1 is added to counteract the harmful action, that is,

Example 9.22. Pollinating Plants

Old design

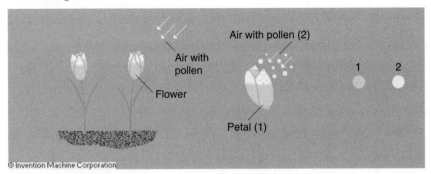

In artificial pollination, flowers are blown with a pollen-containing airstream. A disadvantage is that the airstream fails to open some flower petals because they are small and do not respond to the force of the wind.

New design

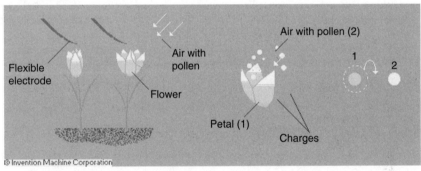

It is proposed to use an electric field. A flexible electrode is passed over the flowers, charging them. Then, an opposite-charge electrode is brought close to open the flower. At this point, pollen is applied with an airstream.

Method 3: Draw harmful action (field) to another object

Another object, object 1, is added to draw the harmful action to itself, that is:

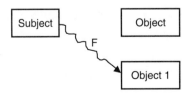

Example 9.23. Lightning Rod A lightning rod draws off thunderstorm-induced electrical shock from the building.

Example 9.24. Protecting Underground Cables

Old design

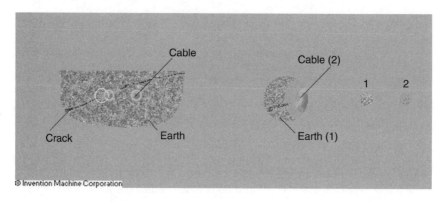

A cable is buried in the soil. [*Disadvantage:* Cracks in the earth (due to hard frost) can damage the cable.]

New design

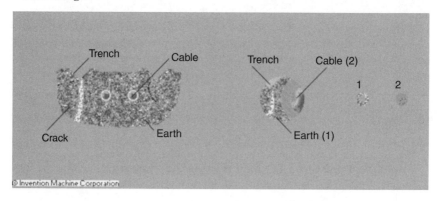

It is proposed to dig trenches in the earth parallel to the cable. These attract the frost-heaving cracks, keeping the cable undamaged.

Method 4: Trim or replace the subject of a harmful function

We can trim or replace the subject of the harmful function by one of the following ways in order to make the subject not create a harmful action (see Sec. 9.7):

- Simplify the system so the subject is eliminated (see Sec. 9.7).

- Replace the subject by another part of the system.
- Replace the subject by another substance.
- Switch on/off the magnetic influence on the subject.

Example 9.25. Grinding the Inner Surfaces of Parts   A ferromagnetic medium with abrasive particles moved by rotating magnetic field can be applied to polish the internal surface of a workpiece.

Old design

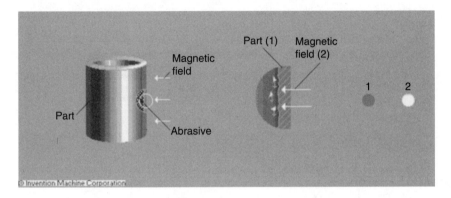

An external magnetic field moves ferromagnetic abrasive particles inside a part. The abrasive particles grind the part. (*Disadvantage:* The part walls weaken the magnetic field.)

New design

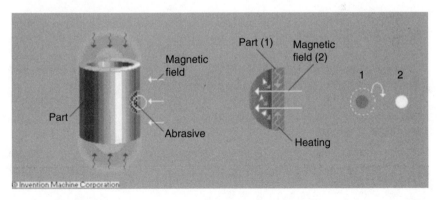

It is proposed to heat the part to the Curie point temperature. The part magnetic properties are removed to ensure that its metal does not weaken the magnetic field. The grinding intensity is increased by the stronger magnetic field and the higher part temperature.

Example 9.26. Restoring an Incandescent Filament For artificial lamp lighting to more closely resemble sunlight, it is necessary to increase the incandescent filament temperature. However, the higher the filament temperature, the higher the metal evaporation rate. Thus the filament quickly becomes thin and burns out.

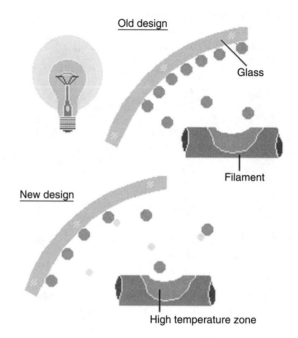

It is proposed to restore an incandescent filament during the lamp's operation by adding bromine into the lamp. Bromine interacts with the tungsten settled on the bulb and forms tungsten bromide. The compound evaporates and streams toward the high-temperature zone, where it decomposes and settles in the same place from which it had evaporated.

Method 5: Trim or replace the object of a harmful function

We can trim or replace the object of the harmful function by one of the following ways in order to make the object not sensitive to harmful action (field) (see Sec. 9.7):

- Simplify the system so the object is eliminated (Sec. 9.7).
- Replace the object by another part of the system.
- Replace the object by another object.
- Switch on/off the magnetic influence on the object.

Example 9.27. Tensometric Grid

Old design

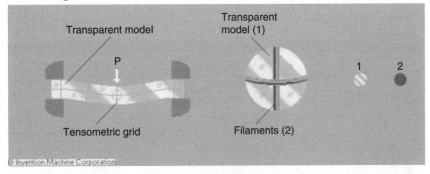

A *stress measurement grid* in a transparent model is made using filaments of material. (*Disadvantage:* The filament size and strength distort the stress measurement.)

New design

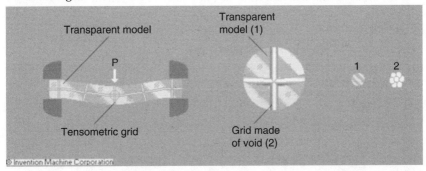

It is proposed to manufacture the grid from cylindrical microvoids. The microvoids (which are small in diameter) do not distort the model stress field, yet remain visible to accurately define the deflection of the model. The grid of voids is formed by etching away fine copper filaments embedded in the model.

Example 9.28. Removing Allergens

Old design

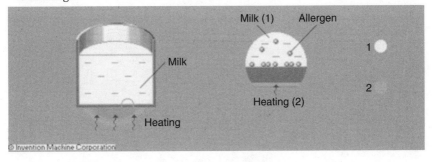

There is a need to reduce allergens in milk. To do this, the milk is boiled and then cooled, causing albumin to settle out (to be removed). (*Disadvantage:* Most globulin components, which have pronounced allergic properties, remain in the milk.)

New design

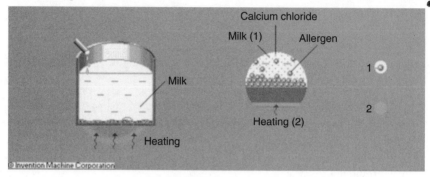

It is proposed that calcium chloride (0.03 to 0.1% concentration) be added to the milk (before treatment), to cause globulin fractions to settle out during treatment. As a result, milk allergens are reduced.

Example 9.29. Making Thin-Walled Pipes

Old design

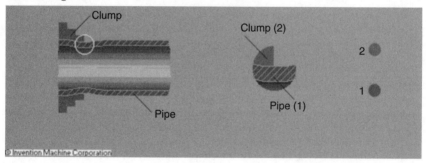

Thin-walled NiCr pipes are made by drawing operations. (*Disadvantage:* The pipes are easily deformed when clamped, machined, or transported.)

New design

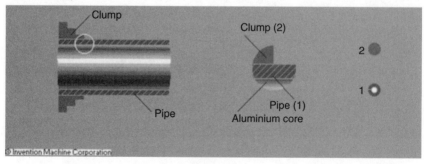

It is proposed to form an aluminum core inside the pipe to prevent it from deforming. When processing is finished, the core is removed by etching with an alkali reagent.

9.7 Complexity Reduction/Trimming

Trimming (or "pruning" or "part count reduction") is a set of systematic methods that can be used to eliminate redundant functions and parts, thus streamlining and simplifying the system. Redundant functions are functions that have little or no benefit to customers. Sometime redundant functions are hidden and we need to look hard to identify them. To identify redundant function, it is very helpful to draw a functional analysis diagram for the system and ask the question for each function in the model, is this function really needed?

Trimming can also be used to identify and eliminate redundant parts. Based on the fundamental ideas of ideality and the ideal final result, the best component is no component; the best utilization of resource is to use no resource. Reduction in redundant parts will reduce the cost and complexity of the system. Usually, the simpler the system, the fewer potential problems it will have.

What kind of parts should be trimmed? We should consider trimming the following kinds of parts:

- The parts that deliver no useful functions
- The parts that deliver many harmful function and few useful functions
- The parts that deliver functions that offer low value to customers
- The parts that have low utilization ratio

What kind of parts can be trimmed?

- The useful action of the part can be performed by another part in the system
- The useful action of the part can be performed by simply altering another part in the system
- The useful function of the part can be performed by the part that receives the action (self-service)
- The useful function of the part can be performed by a cheap substitute or a disposable substitute

Example 9.30. Using Trimming: Bulbless Lamps During the 1970s, the Soviet Union launched an unmanned lunar probe to the moon's surface to transmit TV pictures to the Earth. A projector using a lightbulb was designed to illuminate the lunar surface ahead of the vehicle. However, existing lightbulbs would not survive the impact of landing on the moon's surface. The most durable bulbs were ones used in military tanks, but even those bulbs would crack at the joint between the glass and screw base during tests. A new bulb design suitable for the application had to be developed. The situation was reported to the program leader, Dr. Babakin, who asked, "What is the purpose of the bulb?" The answer was obvious—to vacuum-seal around the filament. The moon's atmosphere, however, presents a perfect vacuum. Therefore, Babakin suggested lamps without bulbs. This is the case of identifying a redundant function.

9.8   Evolution of Technological Systems

Having researched hundreds of thousands of patents, Altshuller and his colleagues developed TRIZ, which concluded that evolution of technological systems is highly predictable and governed by several patterns:

- Increasing ideality
- Increasing complexity followed by simplification
- Nonuniform development of system elements
- Increasing the degree of dynamism

In our TRIZ problem solving process, if we encounter a problem that the current system is able to deliver useful functions and there is no apparent harmful function to eliminate and no contradiction to overcome, but the system is no longer competitive in the marketplace, there is an urgent need to enhance the performance of the system. We can then use the ideas in evolution of technological systems to generate improvement.

9.8.1   Increasing ideality

TRIZ states that the evolution of all technological system will proceed in the direction of increasing degree of ideality. Recall the definition of ideality:

$$Ideality = \frac{\Sigma \, benefits}{\Sigma \, costs + \Sigma \, harm}$$

In other words, the development of any technological system is always toward the direction of

1. Increasing benefits
2. Decreasing costs
3. Decreasing harm

There are many TRIZ techniques to increase ideality.

Techniques to increase benefits

Increasing the number of functions in a single system. This is featured by mono-bi-poly evolution. Mono means monosystem. A monosystem is defined as a single object having one function. Examples are as follows:

- A knife
- A one-barrel hunting rifle

Bi means bisystem. A bisystem is defined as the combination of two subsystems whose functions could be identical, similar, different, or even opposite. Here are some examples of bisystems:

- A two-barreled hunting rifle, which is a combination of two mono-barreled rifles.
- A straight/Phillips screw driver. One tip is straight and other is Phillips. It is a combination of two similar functions.
- Wristwatch calculator. It is a combination of a watch and calculator, a combination of different functions.
- A pencil with eraser. A combination of two opposite functions.

Poly means a polysystem. A polysystem is defined as the combination of three or more subsystems whose functions are identical, similar, different, or even opposite. The examples of a polysystem include multiple stage rocket, Swiss army knife and power tool sets, sound system (a combination of radio, CD player, tape player, etc.)

Mono-bi-poly evolution states that the development of a technological system is often featured by adding more and more functions, from a simple monosystem, to a bisystem, and to a poly system.

Increasing the magnitude and quality of a function. In the evolution of technological systems, not only are the number of functions likely to increase, but also the magnitude and quality of functions improve. For example, in the development of a firearm, the early generation of a firearm could only shoot at short distance, and was not very powerful, with very low frequency and poor accuracy. As the technology develops, it can shoot further and further, and it becomes more and more powerful, with higher and higher frequency and accuracy.

Example 9.31. OneCard System of a University The students of a university used to carry many cards, including a health club card, a library card, and a parking card. The university administration developed a OneCard system in which all the information is stored in one plastic card on the same magnetic strip. By using this card, a student can enter any facility, such as a health club or a library. This is the approach of increasing the number of useful functions.

Techniques to decrease the costs

Trimming

- Delete redundant functions
- Delete redundant parts
- Replace parts with cheap substitutes
- Combine several parts into a single part
- Perform part function by already existing substance resourses

Example 9.32. Self-Feeding Liquid A pump feeds colored liquid to a clearance between transparent plates in the roof of a greenhouse, forming a thin translucent sheet. The clearance thickness and opacity of the liquid sheet serve to control solar radiation entering the greenhouse. The function is adequate, but the hydraulic equipment increases the cost of operating the greenhouse.

To reduce cost, it is proposed to eliminate the hydraulic system. An expandable liquid is placed in the roof cavity. As the temperature inside the greenhouse rises, the liquid expands. The roof cavity becomes filled with fluid, preventing excessive solar energy from entering the greenhouse. The auxiliary function (the hydraulic equipment) is eliminated and the cost of the whole technological system is reduced.

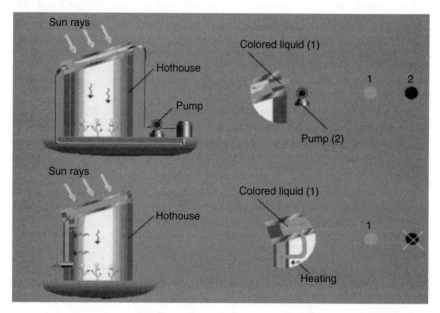

Reduce the number of supporting functions

- Trim supporting functions and related parts
- Simplify supporting functions and related parts
- Use free resources to deliver supporting functions

Techniques to decrease harms

1. Use functional improvement methods to reduce harm
2. Use trimming to reduce harmful functions
3. Use available resources to combat harmful functions

Example 9.33. Using a Pollutant (Waste) to Prevent Pollution   To prevent pollution, exhaust gas from a thermal power station is treated with alkaline chemicals. The alkaline slag is itself recovered from a coal-burning coal power station, where the slag had also been a source of pollution. In this case, the auxiliary function of preventing pollution is realized by using an existing resource, and a harmful function (pollution) is eliminated.

9.8.2   Increasing complexity followed by simplification

This evolution trend states that species of a technological system at an early stage of development will usually deliver only a very simple basic function with limited capacity, but as the technology matures, the system will add more functions with improved capacity in terms of function quality and quantity. The system will also add more components and become more complex, and further down the road of development, the system structure will improve, which means a reduced number of components, reduced complexity, and better use of space, material, and other resources, and the process will continue.

This trend is a close relative to the trend of increasing ideality. *Increasing ideality* as we described it is featured by "Do more or the same with less." The trend of increasing functionality and enhanced structures is featured by "Do more, spend more, and then consolidate, simplify, and spend less."

9.8.3   Nonuniform development of system elements

For a complex system with many components and subsystems, the evolution of the system will be also featured by nonuniform development of system elements:

- Each component or subsystem within a system may have its own S curve. Different components or subsystems usually evolve according to their own schedules. Likewise, different system components reach their inherent limits at various times.

- The components that reach their limits first are "holding back" the overall system. Such a component becomes the weak link in the design. An underdeveloped part is also a weak link.

- The formulation of weak links (contradiction) reveals the component(s) of the system that are holding back the development of the overall system. It seems obvious that the system can be improved by enhancing links (by eliminating the contradiction) that are constraining these components. A frequent mistake in system development, however, is the improvement of some strong element other than that which is limiting the system's development.

The following examples underscore the importance of focusing improvement on the weakest link in the system.

Example 9.34   Early airplanes were limited by poor aerodynamics. Yet for many years, rather than trying to improve the aerodynamics, engineers focused on increasing airplane engine power.

Example 9.35   A manufacturer of plastic car bumpers was producing scrap at twice the expected rate. All problem-solving efforts were directed at improving the manufacturing. Since the company president had been involved in formulation of material, changes to the formula were not seriously considered. Once, out of frustration, the organization purchased a commercial formulation. The manufacturing process became stable, and scrap production fell to one-tenth the target level.

9.8.4   Increasing the degree of dynamism

When a new technical system is developed, it is usually rigid and inflexible and able to provide only basic functions. The rigidity will hamper the performance of the system on different user environments. As the system develops further, the degree of dynamism will improve.

Example 9.36. Evolution of Lenses   Before 1950, the focal length of lens of a camera was fixed. Then the zoom lens was developed, and has improved constantly, with a focal length that can be varied at the mechanical level. One of the patents is presented here:

> U.S. Patent 4,958,179 (1990): *Camera with changeable focal length.* A variable-focal-length camera uses a lens system which has at least two different focal lengths: a relatively long focal length suitable for

a telephotographic mode and a relatively short focal length suitable for a wide-angle photographic mode. This system is equipped with a focal-length-hanging mechanism consisting of a rotatable focal-length-varying member driven by a motor, a lever-displacing member provided on the rotatable focal-length-varying member, a motor-switching lever engageable with the lever-displacing member for switching a motor switch, and a focal-length-changing member for varying the focal length of the taking lens on rotation of the rotatable changing member.

Since 1990, an optical fluid lens has been developed to replace the conventional solid lens in specific circumstances. Here is one of the patents applying this concept:

U.S. Patent 4,466,706 (1984): *Optical fluid lens.* A lens designed especially for applications requiring a large lens eliminates costly grinding and polishing operations. The lens embodies an adjustable chamber containing an optical fluid which can be pressurized in varying degrees by altering the size of the chamber. The curvatures of the resilient optical diaphragms at the ends of the chamber change in response to variations in the pressure of the fluid in the chamber to produce a lens of fixed or variable focal length.

9.9 Physical, chemical, and geometric effects database

Technical systems are designed and produced to deliver functions. As we discussed earlier, to deliver a function, we need at least three elements, a subject, an action, and an object. Subjects and objects are substances; actions are usually delivered by various fields. Therefore the knowledge base on the properties of substances and fields are very important in developing superior technical systems.

Many TRIZ software programs have huge databases on substances and fields properties, and their physical, chemical, and geometric effects. An example is the software developed by Invention Machine Corporation. This kind of database are very helpful in creating inventive solutions.

9.10   Comparison of Axiomatic Design and TRIZ

The following table summarizes the possible relations between axiomatic design (AD) (Chap. 8) and TRIZ design problem-solving tools. Seven corollaries and three theorems in AD (Suh 1990) are selected for comparison with TRIZ tools. Seven corollaries, which serve as the design rules, are derived from two axioms directly, so comparing these "lower-level design rules" with TRIZ tools is useful in order to understand these two methodologies. Only three theorems are selected because we do not think other theorems in AD can be linked with TRIZ. Mann (1999) gives the general comparisons of AD and TRIZ at the level of domain, mapping, hierarchies, and axioms.

Axiomatic design	TRIZ
Corollary 1: Decoupling of coupled design. Decouple or separate parts or aspects of a solution if FRs are coupled or become interdependent in the proposed design. This corollary states that functional independence must be ensured by decoupling if a proposed design couples the functional requirements. Functional decoupling may be achieved without physical separation. However, in many cases, such physical decomposition may be the best way of solving the coupling problem (Suh 1990).	*Contradiction* concept in TRIZ is similar to the functional coupling in AD. Overcoming contradiction in TRIZ means the removal of functional coupling in AD. There are two types of contradiction: technological contradiction and physical contradiction. A *technological contradiction* is derived from a *physical contradiction.* So, certain changes of the physical structure of a technological system guided by the "contradiction table" and the 40 "inventive principles" or "separation principles" are often required to remove contradiction.
Corollary 2: Minimization of FRs. Minimize the number of functional requirements and constraints. This corollary states that as the number of functional requirements and constraints increases, the system becomes more complex and thus the information content is increased. This corollary recommends that the designer strive for maximum simplicity in overall design or the utmost simplicity in physical and functional characteristics.	*Ideal final result* (IFR) philosophy corresponds to Corollary 2 in AD. IFR states that a system is a "fee" for realization of the required function and IFR will be realized if the system does not exist but the required function is performed. IFR helps an engineer focus on concepts that minimize requirements in substance, energy, and complexity of engineering product and process.
Corollary 3: Integration of physical parts. Integrate design features into a single physical process, device or system when FRs can be independently satisfied in the proposed solution. This corollary 3 states that the number of physical components should be reduced through integration of parts without coupling functional requirements. However, mere physical integration is not desirable if it results in an increase of information content or in a coupling of functional requirements.	*Evolution pattern 5:* Increased complexity followed by simplification. This pattern states that technological systems tend to develop first toward increased complexity (i.e., increased quantity and quality of system functions) and then toward simplification (where the same or better performance is provided by a less complex system). The term *mo-bi-poly* indicates that monofunction products evolve into bifunction or polyfunction products through integration of physical embodiments.

Axiomatic design	TRIZ

Corollary 4: Use of standardization.   Use standardization or interchangeable parts if the use of these parts is consistent with FRs and constraints. This corollary states a well-known design rule: Use standard parts, methods, operations and routine, manufacture, and assembly. Special parts should be minimized to decrease cost. Interchangeable parts allow for the reduction of inventory, as well as the simplification of manufacturing and service operations; that is, they reduce the information content.

No patterns, principles, or tools correspond to this corollary. TRIZ focuses its studies on inventive problem solving, so it pays less attention to the standardization and interchangeability of physical components.

Corollary 5: Use of symmetry.   Use symmetric shapes and/or arrangements if they are consistent with the FRs and constraints. It is self-evident that symmetric parts are easier to manufacture and easier to orient in assembly. Not only should the shape be symmetric wherever possible, but hole location and other features should be placed symmetrically to minimize the information required during manufacture and use. Symmetric parts promote symmetry in the manufacturing process.

Principle 4: Asymmetry (one of 40 inventive principles) in TRIZ is in opposition to corollary 5 in AD. TRIZ and AD propose opposite principles because AD theory states the general rules of engineering design, but TRIZ methodology concentrates its studies on the inventive problem-solving techniques. These techniques are derived from the patent database, which relates to novel methods and unique ideas.

Corollary 6: Greatest tolerance.   Specify the maximum allowable tolerance in stating functional requirements.

No corresponding tools are found in TRIZ. Corollary 6 is a general rule of design and has nothing to do with invention.

Corollary 7: Uncoupled design with less information.   Seek an uncoupled design that requires less information than coupled designs in satisfying a set of FRs. This corollary states that if a designer proposes an uncoupled design which has more information content than a coupled design, then the designer should return to the "drawing board" to develop another uncoupled or decoupled design having less information content than the coupled design.

The 40 inventive principles.   These principles provide the techniques to overcome contradictions.

Theorem 1: Coupling due to insufficient number of DPs.   When the number of DPs is less than the number of FRs, either a coupled design results or the FRs cannot be satisfied.

Substance-field analysis states that any properly functioning system can be modeled with a complete substance-field triangle and any deviation from a "complete" triangle, for example, missing one element, reflects the existence of a problem.

Axiomatic design	TRIZ
Theorem 2: Decoupling of coupled design. When a design is coupled because of the greater number of FRs than DPs ($m > n$), it may be decoupled by the addition of the design new DPs so as to make the number of FRs and DPs equal to each other, if a set of the design matrix containing $n \times n$ elements constitutes a triangular matrix.	*Building substance-field models,* class 1 of "76 standard solutions," shares the same idea with Theorem 2 in AD. This standard solution states that if a given object is unreceptive (or barely receptive) to required changes and the problem description does not include any restriction for introducing substances or fields, the problem can be solved by completing the substance-field model to introduce the missing element.
Theorem 5: Need for new design. When a given set of FRs is changed by the addition of a new FR, or substitution of one of the FRs by a new one, or by selection of a completely different set of FRs, the design solution given by the original DPs cannot satisfy the new set of FRs. Consequently, a new design solution must be sought.	*Enhancing substance-field model,* class 2 of "76 standard solutions," corresponds to Theorem 5. The addition of a new FR, or substitution of one of the FRs by a new one, means that the previous system is an inefficient substance-field model. In this case, enhancing substance-field model is required to improve the system functions.

9.10.1   A case study: Using TRIZ separation principles to resolve coupling

An *independence axiom* in AD implies that the design matrix be of a special form. The consequences of applying axiom 1 to the design matrix are as follows:

1. It is desirable to have a square matrix (i.e., $n = m$).
2. The matrix should be either diagonal or triangular.

In real design situations, we need to search for DPs that yield a diagonal or triangular design matrix. The degree of independence can be treated as the definition of tolerance. There are hierarchies in both the functional domain and the physical domain, and a zigzagging process between two domains in the design process. The domain process is most straightforward when the solution consists of uncoupled design at each level. When the design is uncoupled, we can deal with the individual FRs of a hierarchical level without considering other FRs of the same level and preceding hierarchical levels. When the design is coupled, we must consider the effect of a decision on other FRs and DPs. Therefore, the designer should try to find solutions by attempting to uncouple or decouple design in every level of the design hierarchy.

The problem is how to decouple a coupled design. It is obvious to modify a design matrix to be either diagonal or triangular. In practice, many coupled designs undergo changes and become decoupled through a trial-and-error process that is in opposition to TRIZ methodology. In

TRIZ methodology, a *coupled design* is defined as the existence of a contradiction. *Removal of dependency of coupling* means to overcome a technical or physical contradiction by applying inventive principles or separation principles. Thus, these principles can serve, with AD corollaries and theorems, as the guidelines of decoupling a coupled design.

The design process of the paper-handling mechanism (Sekimoto and Ukai 1994) illustrates how separation principles in TRIZ assist in satisfying axiom 1 in AD.

Paper-handling mechanism case study.   The function of the paper-handling mechanism used in an automatic teller machine (ATM) is to "isolate one bill from a pile of bills," which is the first FR of the system. Several physical structures can be used to realize this functional requirement, such as friction, vacuum, and leafing. The friction method is selected, and its mechanism is shown in Fig. 9.7.

However, this DP does not always work correctly because the friction varies under certain circumstances. If the friction force working on the top bill becomes excessive, two or more bills will be sent forward; if the force is too weak, the top bill may not be isolated. Therefore, we have to decompose the first-level functional requirement into two functional requirements: "Give a forward force to the first bill" and "Give a backward force to the second bill." To satisfy these two requirements, the new DP of this design is a pair of rollers rotating in the same direction as shown in Fig. 9.8. The friction coefficient of the upper roller is also greater than that of the lower roller. The design equation is

$$\begin{Bmatrix} FR_1 \\ FR_2 \end{Bmatrix} = \begin{bmatrix} A_{11} & A_{12} \\ A_{21} & A_{22} \end{bmatrix} \begin{Bmatrix} DP_1 \\ DP_2 \end{Bmatrix}$$

where FR_1 = give a forward force to the first bill
$\qquad FR_2$ = give a backward force to the second bill
$\qquad DP_1$ = upper roller
$\qquad DP_2$ = lower roller
$\qquad A_{11}$ = friction between upper roller and first bill
$\qquad A_{22}$ = friction between lower roller and second bill

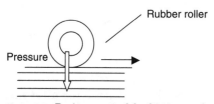

Figure 9.7   Basic concept of the friction mechanism.

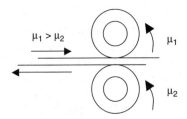

Figure 9.8 Basic concept of the paper isolation mechanism.

A_{12} and A_{21} represent the friction between two bills, so A_{12} is equal to A_{21}. Compared to A_{11} and A_{22}, A_{12} and A_{21} can be disregarded; thus two requirements can be satisfied independently.

The remaining questions are:

- What happens if three or more bills are inserted between the two rollers at the same time?
- What happens after the first bill is sent forward if the roller continues to rotate?
- What happens when the quality of the bill changes?

To answer these questions, the following four FRs must be defined:

FR_3 = slant the cross section of the piled bills to make isolation easy

FR_4 = pull out the isolate bill

FR_5 = adjust the friction force

FR_6 = decrease the forward force after one bill is gone

In AD theory, these six FRs are the minimum set of independent requirements that completely characterize the design objectives for the specific needs of the paper-handling mechanism. Six DPs in the physical domain are selected as follows, and the mechanism is illustrated in Fig. 9.8.

DP_1 = upper rollers

DP_2 = lower roller

DP_3 = wedge-shaped floor guide

DP_4 = carriage pinch rollers

DP_5 = press plate

DP_6 = cam

The function of the cam (DP_6) is to reduce the forward force after one bill is gone. However, when the cam turns, it also affects FR_1, FR_2, FR_3, and FR_5 because it changes the pressure and slope of the floor guide.

The design equation is as follows—clearly, this is the coupled design:

$$
\begin{Bmatrix} FR_1 \\ FR_2 \\ FR_3 \\ FR_4 \\ FR_5 \\ FR_6 \end{Bmatrix} =
\begin{bmatrix}
\times & 0 & 0 & 0 & 0 & x \\
0 & \times & 0 & 0 & 0 & x \\
0 & 0 & \times & 0 & 0 & x \\
0 & 0 & 0 & \times & 0 & 0 \\
0 & 0 & 0 & 0 & \times & x \\
0 & 0 & 0 & 0 & 0 & \times
\end{bmatrix}
\begin{Bmatrix} DP_1 \\ DP_2 \\ DP_3 \\ DP_4 \\ DP_5 \\ DP_6 \end{Bmatrix}
$$

However, from the TRIZ standpoint, FR_1 and FR_6 can be viewed as a technical contradiction because FR_1 requires a large forward force and FR_6 requires a small forward force. The technical contradiction can be overcome by applying the contradiction table and 40 inventive principles. However, if the technical contradiction can be transformed to a physical contradiction, the separation principles can be utilized to solve the problem.

In this case, FR_1 and FR_6 require the friction between the upper roller and the first bill to be both large and small. Physically, two factors control the friction force between the upper roller and the first bill: pressure and the friction coefficient. This means that the pressure, the friction coefficient, or both should be both large and small. Since FR_1 and FR_6 are not required at the same time, the pressure and friction coefficient should not be the same all the time. Therefore, the separation of opposite properties in time, one of the TRIZ separation principles, can be utilized to overcome the contradiction.

One design solution, making the pressure large and small, is given in Fig. 9.9. Another design alternative is illustrated in Fig. 9.10. A partial rubber roller is used to satisfy FR_1 and FR_6 because its friction

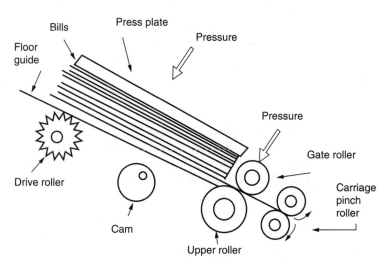

Figure 9.9 Design of paper isolation mechanism (solution 1).

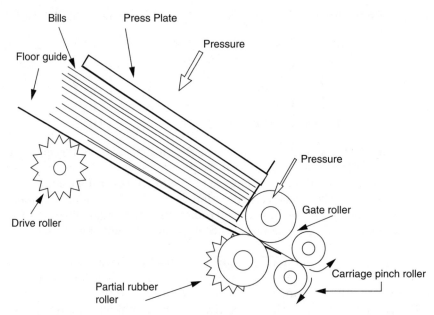

Figure 9.10 Design of paper isolation mechanism (solution 2).

coefficient is large at one time and small at another time when it turns. Thus, the technical contradiction is transformed to the physical one and the contradiction is overcome using TRIZ separation principles. In Fig. 9.10, two DPs are integrated into one part and five components are used to satisfy six functional requirements independently.

The design equation is

$$\begin{Bmatrix} FR_1 \\ FR_2 \\ FR_3 \\ FR_4 \\ FR_5 \\ FR_6 \end{Bmatrix} = \begin{bmatrix} \times & 0 & 0 & 0 & 0 & 0 \\ 0 & \times & 0 & 0 & 0 & 0 \\ 0 & 0 & \times & 0 & 0 & 0 \\ 0 & 0 & 0 & \times & 0 & 0 \\ 0 & 0 & 0 & 0 & \times & 0 \\ 0 & 0 & 0 & 0 & 0 & \times \end{bmatrix} \begin{Bmatrix} DP_1 \\ DP_2 \\ DP_3 \\ DP_4 \\ DP_5 \\ DP_6 \end{Bmatrix}$$

This is the uncoupled design. It is clear that the design solution in Fig. 9.10 is better because it is the uncoupled design and also has a simpler structure. Simple structure means that less information is needed and the structure is easy to produce.

Appendix: Contradiction Table of Inventive Principles

What is deteriorated ?

What should be improved?	1. Weight of movable object	2. Weight of fixed object	3. Length of movable object	4. Length of fixed object	5. Area of movable object	6. Area of fixed object	7. Volume of movable object	8. Volume of fixed object	9. Speed	10. Force	11. Stress, pressure	12. Shape	13. Object's composition stability
1. Weight of movable object			15 8 / 29 34		29 17 / 38 34		29 2 / 40 28		2 8 / 15 38	8 10 / 18 37	10 36 / 37 40	10 14 / 35 40	1 35 / 19 39
2. Weight of fixed object				10 1 / 29 35		35 30 / 13 2		5 35 / 14 2		8 10 / 19 35	13 29 / 10 18	13 10 / 29 14	26 39 / 1 40
3. Length of movable object	8 15 / 29 34				15 17 / 4		7 17 / 4 35		13 4 / 8	17 10 / 4	1 8 / 35	1 8 / 10 29	1 8 / 15 34
4. Length of fixed object		35 28 / 40 29				17 7 / 10 40		35 8 / 2 14		28 10	1 14 / 35	13 14 / 15 7	39 37 / 35
5. Area of movable object	2 17 / 29 4		14 15 / 18 4				7 14 / 17 4		29 30 / 4 34	19 30 / 35 2	10 15 / 36 28	5 34 / 29 4	11 2 / 13 39
6. Area of fixed object		30 2 / 14 18		26 7 / 9 39						1 18 / 35 36	10 15 / 36 37		2 38
7. Volume of movable object	2 26 / 29 40		1 7 / 35 4		1 7 / 4 17				29 4 / 38 34	15 35 / 36 37	6 35 / 36 37	1 15 / 29 4	28 10 / 1 39
8. Volume of fixed object		35 10 / 19 14	19 14	35 8 / 2 14						2 18 / 37	24 35	7 2 / 35	34 28 / 35 40
9. Speed	2 28 / 13 38		13 14 / 8		29 30 / 34		7 29 / 34			13 28 / 15 19	6 18 / 38 40	35 15 / 18 34	28 33 / 1 18
10. Force	8 1 / 37 18	18 13 / 1 28	17 19 / 9 36	28 10	19 10 / 15	1 18 / 36 37	15 9 / 12 37	2 36 / 18 37	13 28 / 15 12		18 21 / 11	10 35 / 40 34	35 10 / 21
11. Stress, pressure	10 36 / 37 40	13 29 / 10 18	35 10 / 36	35 1 / 36	10 15 / 36 28	10 15 / 36 37	6 35 / 10	35 24	6 35 / 36	36 35 / 21		35 4 / 15 10	35 33 / 2 40
12. Shape	8 10 / 29 40	15 10 / 26 3	29 34 / 5 4	13 14 / 10 7	5 34 / 4 10		14 4 / 15 22	7 2 / 35	35 15 / 34 18	35 10 / 37 40	34 15 / 10 14		33 1 / 18 4
13. Object's composition stability	21 35 / 2 39	26 39 / 1 40	13 15 / 1 28	37	2 11 / 13	39	28 10 / 19 39	34 28 / 35 40	33 15 / 28 18	10 35 / 21 16	2 35 / 40	22 1 / 18 4	
14. Strength	1 8 / 40 15	40 26 / 27 1	1 15 / 8 35	15 14 / 28 26	3 34 / 40 29	9 40 / 28	14 7	9 14 / 17 15	8 13 / 26 14	10 18 / 3 14	10 3 / 18 40	10 30 / 35 40	13 17 / 35
15. Duration of moving object's operation	19 5 / 34 31		2 19 / 9		3 17 / 19		10 2 / 19 30		3 35 / 5	19 2 / 16	19 3 / 27	14 26 / 28 25	13 3 / 35
16. Duration of fixed object's operation		6 27 / 19 16		1 40 / 35				35 34 / 38					39 3 / 35 23
17. Temperature	36 22 / 6 38	22 35 / 32	15 19 / 9	15 19 / 9	3 35 / 39 18	35 38	34 39 / 40 18	35 6 / 4	2 28 / 36 30	35 10 / 3 21	35 39 / 19 2	14 22 / 19 32	1 35 / 32
18. Illumination	19 1 / 32	2 35 / 32	19 32 / 16		19 32 / 26		2 13 / 10		10 13 / 19	26 19 / 6		32 30	32 3 / 27
19. Energy expense of movable object	12 18 / 28 31		12 28		15 19 / 25		35 13 / 18		8 15 / 35	16 26 / 21 2	23 14 / 25	12 2 / 29	19 13 / 17 24
20. Energy expense of fixed object		19 9 / 6 27							36 37				27 4 / 29 18

What is deteriorated ?

What should be improved?	14. Strength	15. Duration of moving object's operation	16. Duration of fixed object's operation	17. Temperature	18. Illumination	19. Energy expense of movable object	20. Energy expense of fixed object	21. Power	22. Waste of energy	23. Loss of substance	24. Loss of information	25. Waste of time	26. Quantity of substance
1. Weight of movable object	28 27 / 18 40	5 34 / 31 35		6 29 / 4 38	19 1 / 32	25 12 / 34 31		13 36 / 18 31	6 2 / 34 19	5 35 / 3 31	10 24 / 35	10 35 / 20 28	3 26 / 18 31
2. Weight of fixed object	28 2 / 10 27		2 27 / 19 6	28 19 / 32 22	35 19 / 35		18 19 / 28 1	15 19 / 18 22	18 19 / 28 15	5 8 / 13 30	10 15 / 35	10 20 / 35 26	19 6 / 18 26
3. Length of movable object	8 35 / 29 34	19		10 15 / 19	32	8 35 / 24		1 35	7 2 / 35 39	4 29 / 23 10	1 24	15 2 / 29	29 35
4. Length of fixed object	15 14 / 28 26		1 40 / 35	3 35 / 38 18	3 25			12 8	6 28	10 28 / 24 35	24 26	30 29 / 14	
5. Area of movable object	3 15 / 40 14	6 3		2 15 / 16	15 32 / 19 13	19 32		19 10 / 32 18	15 17 / 30 26	10 35 / 2 39	30 26	26 4	29 30 / 6 13
6. Area of fixed object	40		2 10 / 19 30	35 39 / 38				17 32	17 7 / 30	10 14 / 18 39	30 16	10 35 / 4 18	2 18 / 40 4
7. Volume of movable object	9 14 / 15 7	6 35 / 4		34 39 / 10 18	10 13 / 2	35		35 6 / 13 18	7 15 / 13 16	36 39 / 34 10	2 22	2 6 / 34 10	29 30 / 7
8. Volume of fixed object	9 14 / 17 15		35 34 / 38	35 6 / 4				30 6		10 39 / 35 34		35 16 / 32 18	35 3
9. Speed	8 3 / 26 14	3 19 / 35 5		28 30 / 36 2	10 13 / 19	8 15 / 35 38		19 35 / 38 2	14 20 / 19 35	10 13 / 28 38	13 26		10 19 / 29 38
10. Force	35 10 / 14 27	19 2		35 10 / 21		19 17 / 10	1 16 / 36 37	19 35 / 18 37	14 15	8 35 / 40 5		10 37 / 36	14 29 / 18 36
11. Stress, pressure	9 18 / 3 40	19 3 / 27		35 39 / 19 2		14 24 / 10 37		10 35 / 14	2 36 / 25	10 36 / 3 37		37 36 / 4	10 14 / 36
12. Shape	30 14 / 10 40	14 26 / 9 25		22 14 / 19 32	13 15 / 32	2 6 / 34 14		4 6 2	14	35 29 / 3 5		14 10 / 34 17	36 22
13. Object's composition	17 9 / 15	13 27 / 10 35	39 3 / 35 23	35 1 / 32	32 3 / 27 15	13 19	27 4 / 29 18	32 35 / 27 31	14 2 / 39 6	2 14 / 30 40		35 27	15 32 / 35
14. Strength		27 3 / 26		30 10 / 40	35 19 / 10	35		10 26 / 35 28	35	35 28 / 31 40		29 3 / 28 10	29 10 / 27
15. Duration of moving object's operation	27 3 / 10			19 35 / 39	2 19 / 4 35	28 6 / 35 18		19 10 / 35 38		28 27 / 3 18	10	20 10 / 28 18	3 35 / 10 40
16. Duration of fixed object's operation				19 18 / 36 40				16		27 16 / 18 38	10	28 20 / 10 16	3 35 / 31
17. Temperature	10 30 / 22 40	19 3 / 39	19 18 / 36 40			32 30 / 21 16	19 15 / 3 17	2 14 / 17 25	21 17 / 35 38	21 36 / 29 31		35 28 / 21 18	3 17 / 30 39
18. Illumination	35 19	2 19 / 6		32 / 35 19		32 1 / 19	32 35 / 1 15	32	19 16 / 1 6	13 1	1 6	19 1 / 26 17	1 19
19. Energy expense of movable object	5 19 / 9 35	28 35 / 6 18		19 24 / 3 14	2 15 / 19			6 19 / 37 18	12 22 / 15 24	35 24 / 18 5		35 38 / 19 18	34 23 / 16 18
20. Energy expense of fixed object	35			19 2 / 35 32						28 27 / 18 31			3 35 / 31

What is deteriorated ?

What should be improved?	27. Reliability	28. Measurement accuracy	29. Manufacturing precision	30. Harmful action at object	31. Harmful effect caused by the object	32. Ease of manufacture	33. Ease of operation	34. Ease of repair	35. Adaptation	36. Device complexity	37. Measurement or test complexity	38. Degree of automation	39. Productivity
1. Weight of movable object	3 11 1 27	28 27 35 26	28 35 26 18	22 21 18 27	22 35 31 39	27 28 1 36	35 3 2 24	2 27 28 11	29 5 15 8	26 30 36 34	28 29 26 32	26 35 18 19	35 3 24 37
2. Weight of fixed object	10 28 8 3	18 26 28	10 1 35 17	2 19 22 37	35 22 1 39	28 1 9	6 13 1 32	2 27 28 11	19 15 29	1 10 26 39	25 28 17 15	2 26 35	1 28 15 35
3. Length of movable object	10 14 29 40	28 32 4	10 28 29 37	1 15 17 24	17 15	1 29 17	15 29 35 4	1 28 10	14 15 1 16	1 19 26 24	35 1 26 24	17 24 26 16	14 4 28 29
4. Length of fixed object	15 29 28	32 28 3	2 32 10	1 18		15 17 27	2 25	3		1 35	1 26	26	30 14 7 26
5. Area of movable object	29 9	26 28 32 3	2 32	22 33 28 1	17 2 18 39	13 1 26 24	15 17 13 16	15 13 10 1	15 30	14 1 13	2 36 26 18	14 30 28 23	10 26 34 2
6. Area of fixed object	32 35 40 4	26 28 32 3	2 29 18 36	27 2 39 35	22 1 40	40 16	16 4	16	15 16	1 18 36	2 35 30 18	23	10 15 17 7
7. Volume of movable object	14 1 40 11	25 26 28	25 28 2 16	22 21 27 35	17 2 40 1	29 1 40	15 13 30 12	10	15 29	26 1	29 26 4	35 34 16 24	10 6 2 34
8. Volume of fixed object	2 35 16		35 10 25	34 39 19 27	30 18 35 4	35		1		1 31	2 17 26		35 37 10 2
9. Speed	11 35 27 28	28 32 1 24	10 28 32 25	1 28 35 23	2 24 35 21	35 13 8 1	32 28 13 12	34 2 28 27	15 10 26	10 28 4 34	3 34 27 16	10 18	
10. Force	3 35 13 21	35 10 23 24	28 29 37 36	1 35 40 18	13 3 36 24	15 37 18 1	1 28 3 25	15 1 11	15 17 18 20	26 35 10 18	36 37 10 19	2 35	3 28 35 37
11. Stress, pressure	10 13 19 35	6 28 25	3 35	22 2 37	2 33 27 18	1 35 16	11	2	35	19 1 35	2 36 37	35 24	10 14 35 37
12. Shape	10 40 16	28 32 1	32 30 40	22 1 2 35	35 1	1 32 17 28	32 15 26	2 13 1	1 15 29	16 29 1 28	15 13 39	15 1 32	17 26 34 10
13. Object's composition		13	18	35 24 18 30	35 40 27 39	35 19	32 35 30	2 15 10 16	35 30 34 2	2 35 22 26	35 22 39 23	1 8 35	23 35 40 3
14. Strength	11 3	3 27 16	3 27	18 35 37 1	15 35 22 2	11 3 10 32	32 40 28 2	27 11 3	15 3 2 3	2 13 28	27 3 15 40	15	29 35 10 14
15. Duration of moving object's operation	11 2 13	3	3 27 16 40	22 15 33 28	21 39 16 22	27 1 4	12 27	29 10 27	1 35 13	10 4 29 35	19 29 39 35	6 10	35 17 14 19
16. Duration of fixed object's operation	34 27 6 40	10 26 24		17 1 40 33	22	35 10	1	1	2		25 14 6 35	1	20 10 16 38
17. Temperature	19 35 3 10	32 19 24	24	22 33 35 2	22 35 2 24	26 27	26 27	4 10 16	2 18 27	2 17 16	3 27 35 31	26 2 19 16	15 28 35
18. Illumination		11 15 32	3 32	15 19	35 19 32 39	19 35 28 26	28 26 19	15 17 13 16	15 1 19	6 32 13	32 15	2 26 10	2 25 16
19. Energy expense of movable object	19 21 11 27	3 1 32		1 35 6 27	2 35 6	28 26 30	19 35	1 15 17 28	15 17 13 16	2 29 27 28	35 38	32 2	12 28 35
20. Energy expense of fixed object	10 36 23			10 2 22 37	19 22 18	1 4					19 35 16 25		1 6

What is deteriorated?

What should be improved?	1. Weight of movable object	2. Weight of fixed object	3. Length of movable object	4. Length of fixed object	5. Area of movable object	6. Area of fixed object	7. Volume of movable object	8. Volume of fixed object	9. Speed	10. Force	11. Stress, pressure	12. Shape	13. Object's composition stability
21. Power	8 36 38 31	19 26 17 27	1 10 35 37		19 38	17 32 13 38	35 6 38	30 6 25	15 35 2	26 2 36 35	22 10 35	29 14 2 40	35 32 15 31
22. Waste of energy	15 6 19 28	19 6 18 9	7 2 6 13	6 38 7	15 26 17 30	17 7	7 18 23	7		16 35 38	36 38		14 2 39 6
23. Loss of substance	35 6 23 40	35 6 22 32	14 29 10 39	10 28 24	35 2 10 31	10 18 39 31	1 29 30 36	3 39 18 31	10 13 28 38	14 15 18 40	3 36 37 10	29 35 3 5	2 14 30 40
24. Loss of information	10 24 35	10 35 5	1 26	26		30 26	30 16		2 22	26 32			
25. Waste of time	10 20 37 35	10 20 26 5	15 2 29	30 24 14 5	26 4 5 16	10 35 17 4	2 5 34 10	35 16 32 18		10 37 36 5	36 37 4	4 10 34 17	35 3 22 5
26. Quantity of substance	35 6 18 31	27 26 18 35	29 14 35 18		15 14 29	2 18 40 4	15 20 29		35 29 34 28	35 14 3	10 36 14 3	35 14	15 2 17 40
27. Reliability	3 8 10 40	3 10 8 28	15 9 14 4	15 29 28 11	17 10 14 16	32 35 40 4	3 10 14 24	2 35 24	21 35 11 28	8 28 10 3	10 24 35 19	35 1 16 11	
28. Measurement accuracy	32 35 26 28	28 35 25 26	28 26 5 16	32 28 3 16	26 28 32 3	26 28 32 3	32 13 6		28 13 32 24	32 2	6 28 32	6 28 32	32 35 13
29. Manufacturing precision	28 32 13 18	28 35 27 9	10 28 29 37	2 32 10	28 33 29 32	2 29 18 36	32 28 2	25 10 35	10 28 32	28 19 34 36	3 35	32 30 40	30 18 30 18
30. Harmful action at object	22 21 27 39	2 22 13 24	17 1 39 4	1 18	22 1 33 28	27 2 39 35	22 23 37 35	34 39 19 27	21 22 35 28	13 35 39 18	22 2 37	22 1 3 35	35 24
31. Harmful effect caused by the object	19 22 15 39	35 22 1 39	17 15 16 22		17 2 18 39	22 1 40	17 2 40	30 18 35 4	35 28 3 23	35 28 1 40	2 33 27 18	35 1	35 40 27 39
32. Ease of manufacture	28 29 15 16	1 27 36 13	1 29 13 17	15 17 27		16 40 26 12	1 40	35	13 29 1 40	35 12 8 1	35 19	1 28 1 37	11 13 1
33. Ease of operation	25 2 13 15	6 13 1 25	1 17 13 12		1 17 13 16	18 16 15 39	1 16 35 15	4 18 39 31	18 13 34	28 13 35	2 32 12	15 34 29 28	32 35 30
34. Ease of repair	2 27 35 11	2 27 35 11	1 28 10 25	3 18 31	15 13 32	16 25	25 2 35 11	1	34 9	1 11 10	13	1 13 2 4	2 35
35. Adaptation	1 6 15 8	19 15 29 16	35 1 29 2	1 35 16	35 30 29 7	15 16	15 35 29		35 10 14	15 17 20	35 16	15 37 1 8	35 30 14
36. Device complexity	26 30 34 36	2 26 35 39	1 19 26 24	26	14 1 13 16	6 36	34 26 6	1 16	34 10 28	26 16	19 1 35	29 13 28 15	2 22 17 19
37. Measurement or test complexity	27 26 28 13	6 13 28 1	16 17 26 24	26	2 13 18 17	2 39 30 16	29 1 4 16	2 18 26 31	3 4 16 35	36 28 40 19	35 36 37 32	27 13 1 39	11 22 39 30
38. Degree of automation	28 26 18 35	28 26 35 10	14 13 28 17	23	17 14 13		35 13 16		28 10	2 35	13 35	15 32 1 13	18 1
39. Productivity	35 26 24 37	28 27 15 3	18 4 28 38	30 14 26 7	10 26 34 31	10 35 17 7	2 6 34 10	35 37 10 2		28 15 10 36	10 37 14	10 10 34 40	35 3 22 39

What should be improved?	What is deteriorated ?

What should be improved?	14. Strength	15. Duration of moving object's operation	16. Duration of fixed object's operation	17. Temperature	18. Illumination	19. Energy expense of movable object	20. Energy expense of fixed object	21. Power	22. Waste of energy	23. Loss of substance	24. Loss of information	25. Waste of time	26. Quantity of substance
21. Power	26 10 28	19 35 10 38	16	2 14 17 25	16 6 19	16 6 19 37			10 35 38	28 27 18 38	10 19	35 20 10 6	4 34 19
22. Waste of energy	26			19 38 7	1 13 32 15			3 38		35 27 2 37	19 10	10 18 32 7	7 18 25
23. Loss of substance	35 28 31 40	28 27 3 18	27 16 18 38	21 36 39 31	1 6 13	35 18 24 5	28 27 12 31	28 27 18 38	35 27 2 31			15 18 35 10	6 3 10 24
24. Loss of information		10	10		19			10 19	19 10			24 26 28 32	24 28 35
25. Waste of time	29 3 28 18	20 10 28 18	28 20 10 16	35 29 21 18	1 19 26 17	35 38 19 18	1	35 20 10 6	10 5 18 32	35 18 10 39	24 26 28 32		35 38 18 16
26. Quantity of substance	14 35 34 10	3 35 10 40	3 35 31	3 17 39		34 29 16 18	3 35 31	35	7 18 25	6 3 10 24	24 28 35	35 38 18 16	
27. Reliability	11 28	2 35 3 25	34 27 6 40	3 35 10	11 32 13	21 17 27 19	36 23	21 11 26 31	10 11 35	10 35 29 39	10 28	10 30 4	21 28 40 3
28. Measurement accuracy	28 6 32	28 6 32	10 26 24	6 19 28 24	6 1 32	3 6 32		3 6 32	26 32 27	10 16 31 28		24 34 28 32	2 6 32
29. Manufacturing precision	3 27	3 27 40		19 26	3 32	32 2		32 2	13 32 2	35 31 10 24		32 26 28 18	32 30
30. Harmful action at object	18 35 37 1	22 15 33 28	17 1 40 33	22 33 35 2	1 19 32 13	1 24 6 27	10 2 22 37	19 22 31 2	21 22 35 2	33 22 19 40	22 10 2	35 18 34	35 33 29 31
31. Harmful effect caused by the object	15 35 22 2	15 22 33 31	21 39 16 22	22 35 2 24	19 24 39 32	2 35 6	19 22 18	2 35 18	21 35 22 2	10 1 34	10 21 29	1 22	3 24 39 1
32. Ease of manufacture	1 3 10 32	27 1 4	35 16	27 26 18	28 24 27 1	28 26 27 1	1 4	27 1 12 24	19 35	15 34 33	32 24 18 16	35 28 34 4	35 23 1 24
33. Ease of operation	32 40 3 28	29 3 8 25	1 16 25	26 27 13	13 17 1 24	1 13 24		35 34 2 10	2 19 13	28 32 2 24	4 10 27 22	4 28 10 34	12 35
34. Ease of repair	1 11 2 9	11 29 28 27	1	4 10	15 1 13	15 1 28 16		15 10 32 2	15 1 32 19	2 35 34 27		32 1 10 25	2 28 10 25
35. Adaptation	35 3 32 6	13 1 35	2 16	27 2 3 35	6 22 26 1	19 35 29 13		19 1 29	18 15 1	15 10 2 13		35 28	3 35 15
36. Device complexity	2 13 28	10 4 28 15		2 17	24 17 29 28	27 2 29 28		20 19 30 34	10 35 13 2	35 10 28 29		6 29	13 3 27 10
37. Measurement or test complexity	27 3 15 28	19 29 25 39	25 34 6 35	3 27 35 16	2 24 26	35 38	19 35 16	19 1 16 10	35 3 15 19	1 18 10 24	35 33 27 22	18 28 32 9	3 27 29 18
38. Degree of automation	25 13	6 9		26 2 19	8 32 19	2 32 13		28 2 27	23 28	35 10 18 5	35 33	24 28 35 30	35 13
39. Productivity	29 28 10 18	35 10 2 18	20 10 16 38	35 21 28 10	26 17 19 1	35 10 38 19	1	35 20 10	28 10 29 35	28 10 35 23	13 1 5 23		35 38

What is deteriorated ?

What should be improved?	27. Reliability	28. Measurement accuracy	29. Manufacturing precision	30. Harmful action at object	31. Harmful effect caused by the object	32. Ease of manufacture	33. Ease of operation	34. Ease of repair	35. Adaptation	36. Device complexity	37. Measurement or test complexity	38. Degree of automation	39. Productivity
21. Power	19 24 26 31	32 15 2	32 2	19 22 31 2	2 35 18	26 10 34	26 35 10	35 2 10 34	19 17 34	20 19 30 34	19 35 16	28 2 17	28 35 34
22. Waste of energy	11 10 35	32		21 22 35 2	21 35 2 22		35 32 1	2 19		7 23	35 3 15 23	2	28 10 29 35
23. Loss of substance	10 29 39 35	16 34 31 28	35 10 24 31	33 22 30 40	10 1 34 29	15 34 33	32 28 2 24	2 35 34 27	15 10 2	35 10 28 24	35 18 10 13	35 10 18	28 35 10 23
24. Loss of information	10 28 23			22 10 1	10 21 22	32	27 22				35 33	35	13 23 15
25. Waste of time	10 30 4	24 34 28 32	24 26 28 18	35 18 34	35 22 18 39	35 28 34 4	4 28 10 34	32 1 10	35 28	6 29	18 28 32 10	24 28 35 30	
26. Quantity of substance	18 3 28 40	3 2 28	33 30	35 33 29 31	3 35 40 39	29 1 35 27	35 29 10 25	2 32 10 25	15 3 29	3 13 27 10	3 27 29 18	8 35	13 29 3 27
27. Reliability		32 3 11 23	11 32 1	27 35 2 40	35 2 40 26		27 17 40	1 11	13 35 8 24	13 35 1	27 40 28	11 13 27	1 35 29 38
28. Measurement accuracy	5 11 1 23			28 24 22 26	3 33 39 10	6 35 25 18	1 13 17 34	1 32 13 11	13 35 2	27 35 10 34	26 24 32 28	28 2 10 34	10 34 28 32
29. Manufacturing precision	11 32 1			26 28 10 36	4 17 34 26		1 32 35 23	25 10		26 2 18		26 28 18 23	10 18 32 39
30. Harmful action at object	27 24 2 40	28 33 23 26	26 28 10 18			24 35 2	2 25 28 39	35 10 2	35 11 22 31	22 19 29 40	23 19 29 40	33 3 34	22 31 13 24
31. Harmful effect caused by the object	24 2 40 39	3 33 26	4 17 34 26							19 1 3 1	2 21 27 1	2	22 35 18 39
32. Ease of manufacture		1 35 12 18		24 2			2 5 13 16	35 1 11 9	2 13 15	27 26 1	6 28 11 1	8 28 1	35 1 10 28
33. Ease of operation	17 27 8 40	25 13 2 34	1 32 35 23	2 25 28 39		2 5 12		12 26 1 32	15 34 1 16	32 25 12 17		1 34 12 3	15 1 28
34. Ease of repair	11 10 1 16	10 2 13	25 10	35 10 2 16		1 35 11 10	1 12 26 15		7 14 16	35 1 13 11		34 35 7 13	1 32 10
35. Adaptation	35 13 8 24	35 5 1 10		35 11 32 31		1 13 31	15 34 1 16	1 16 7 4		15 29 37 28	1	27 34 35	35 28 6 37
36. Device complexity	13 35 1	2 26 10 34	26 24 32	22 19 29 40	19 1	27 26 1 13	27 926 24	1 13	29 15 28 37		15 10 37 28	15 1 24	12 17 28
37. Measurement or test complexity	27 40 28 8	26 24 32 28		22 19 29 28	2 21	5 28 11 29	2 5	12 26	1 15	15 10 37 28		34 21	35 18
38. Degree of automation	11 27 32	28 26 10 34	28 26 18 23	2 33	2	1 26 13	1 12 34 3	1 35 13	27 4 1 35	15 24 10	34 27 25		5 12 35 26
39. Productivity	1 35 10 38	1 10 34 28	32 1 18 10	22 35 13 24	35 22 18 39	35 28 2 24	1 28 7 19	1 32 10 25	1 35 28 37	12 17 28 24	35 18 27 2	5 12 35 26	

10

Design for X

10.1 Introduction

This chapter, for the most part, focuses on the product. This is attributed to the evolution of the Design for X family of tools in manufacturing industries. We will focus on only a vital few members of the DFX family. However, DFSS teams with transactional projects can still benefit from this chapter by drawing analogies between their processes and/or services and the topics presented here, in particular, Design for Serviceability (DFS) and Design for Life-Cycle Cost using activity-based costing with uncertainty. Many transactional DFSS teams found the concepts, tools, and approaches presented here very useful, acting in many ways as eye-openers by stimulating out-of-the-box thinking.

The concurrent engineering is a contemporary approach to DFSS. The black belt should continually revise the DFSS team membership to reflect the concurrent design, which means that both design and process members are key, equal team members. DFX techniques are part of detail design and are ideal approaches to improve life-cycle cost,* quality, increased design flexibility, and increased efficiency and productivity using the concurrent design concepts (Maskell 1991). Benefits are usually pinned as competitiveness measures, improved decision making, and enhanced operational efficiency. The letter "X" in DFX is made up of two parts: life-cycle processes x and performance measure (ability): $X = x +$ ability (Huang 1996). In product design, for example,

---

Life-cycle cost is the real cost of the design. It includes not only the original cost of manufacture but also the associated costs of defects, litigations, buybacks, distributions support, warranty, and the implementation cost of all employed DFX methods.

one of the first members of the DFX family is *Design for Assembly* (DFA). The DFX family is one of the most effective approaches to implement concurrent engineering. DFX focuses on vital business elements of concurrent engineering, maximizing the use of the limited resources available to the DFSS team. In DFA, the focus is placed on factors such as size, symmetry, weight, orientation, form features, and other factors related to the product as well as handling, gripping, insertion, and other factors related to the assembly process. In effect, DFA focuses on the assembly business process as part of production by studying these factors and their relationships to ease assembly.

The DFX family started with DFA but continues to increase in number as fostered by the need for better decision making upfront, in particular those related to manufacturing. Manufacturing and production issues are often ignored or omitted in early design steps. This oversight can't be generalized because of the early work of Matousek (1957), Neibel and Baldwin (1957), Pech (1973), and several workshops organized by CIRP (College Internationale de Recherches pour la Production) and WDK (Workshop Design-Konstrucktion). Other efforts started in the 1970s by a group of researchers in the United Kingdom and at University of Massachusetts and resulted in two different commercial DFA tools: that due to Boothroyd and Dewhurst (1983) and the Lucas DFA (Miles 1989). They employed worksheets, data, knowledge base, and systematic procedures to overcome limitations of design guidelines, differentiating themselves from the old practices. The DFA approach is considered a revolution in design for assembly.

The Boothroyd-Dewhurst DFA moved out of research of automatic feeding and insertion to broader industrial applications, including manual assembly, in particular, locomotive engine. This success led to the proliferation of a new array of DFX, expanding the family to Design for Manufacturability, Design for Reliability, Design for Maintainability, Design for Serviceability, Design for Inspectability, Design for Environmentality, Design for Recycability, and so on.

DFX tools collect and present factuals about both the design entity and its production processes, analyze all relationships between them, measure the CTQs of performance as depicted by the physical structure, generate alternatives by combining strengths and avoiding vulnerabilities, provide a redesign recommendation for improvement, provide if-then scenarios, and do all that with many iterations.

The objective of this chapter is to introduce the vital few members of the DFX family. It is up to the reader to seek more in-depth material using Table 10.1.

The DFSS team should take advantage of, and strive to design into, the existing capabilities of suppliers, internal plants, and assembly lines. It is cost-effective, at least for the near term. The idea is to cre-

TABLE 10.1   DFX Citation Table

X	DFX	Reference
Product or process		
Assembly	Boothroyd-Dewhurst DFA	O'Grady and Oh (1991)
	Lucas DFA	Sackett and Holbrook (1988)
	Hitachi AEM	Huang (1996)
Fabrication	Design for Dimension Control	Huang (1996)
	Hitachi MEM	
	Design for Manufacturing	Arimoto et al. (1993)
		Boothroyd et al. (1994)
Inspection and test	Design for Inspectability	Huang (1996)
	Design for Dimensional Control	
Material logistics	Design for Material Logistics	Foo et al. (1990)
Storage and distribution	Design for Storage and Distribution	Huang (1996)
Recycling and disposal flexibility	Design for Ease of Recycling	Beitz (1990)
	Variety reduction program	Suzue and Kohdate (1988)
Environmental repair	Design for Environmentality	Navichandra (1991)
	Design for Reliability and Maintainability	Gardner and Sheldon (1995)
Service		
Cost	Design for Whole Life Costs	Sheldon et al. (1990)
Service	Design for Serviceability	Gershenson and Ishii (1991)
Purchasing	Design for Profit	Mughal and Osborne (1995)
Sales and marketing	Design for Marketability	Zaccai (1994)
	QFD	This volume, Chap. 6
Use and operation	Design for Safety	Wang and Ruxton (1993)
	Design for Human Factors	Tayyari (1993)

ate designs sufficiently robust to achieve Six Sigma product performance from current capability. Concurrent engineering enables this kind of upside-down thinking. Such concepts are factored into the DFSS algorithm to improve design for manufacturing, improve design for assembly, and design for service. The key "design for" activities to be tackled by the team are as follows:

1. Use DFX as early as possible in the DFSS algorithm.

2. Start with DFA and Design for Variety for product projects and Design for Service for transactional projects.

3. From the findings of step 2, determine which DFX to use next. This is a function of DFSS team competence. Time and resources need to be provided to carry out the "design for" activities. The major challenge is implementation.

A danger lurks in the DFX methodologies that can curtail or limit the pursuit of excellence. Time and resource constraints can tempt DFSS teams to accept the unacceptable on the premise that the shortfall can be corrected in one of the subsequent steps—the second-chance syndrome. Just as wrong concepts cannot be recovered by brilliant detail design, bad first-instance detail designs cannot be recovered through failure-mode analysis, optimization, or tolerance.

10.2 Design for Manufacture and Assembly (DFMA)

DFM and DFA are systematic approaches within the DFX family that the DFSS team can use to carefully analyze each design parameter (DP) that can be defined as part or subassembly for manual or automated manufacture and assembly to gradually reduce waste. Waste, or *muda,* the Japanese term, may mean any of several things. It may mean products or features that have no function (do not add value) and those that should have been trimmed (reduced, streamlined) using the zigzagging method in the physical mapping. It may also mean proliferation of parts that can be eliminated using the zigzagging method in the process mapping as well. But the most leverage of DFX in the DFSS algorithm, beyond the design axioms, is attacking the following muda sources: (1) assembly directions that need several additional operations and (2) DPs with unnecessarily tight tolerances.

As a golden rule, the DFSS team should minimize the number of setups and stages through which a high-level DP (e.g., a part or subassembly) must pass through before it becomes a physical entity. This objective is now feasible and affordable because of the significant development in computer numerically controlled (CNC) machines with single-setup machining and multiaxis capabilities. The employment of CNC machines will reduce lead times, tooling, and setup costs while responding to customer demands with added flexibility. Single-setup machines are usually equipped with touch trigger probe measuring part position and orientation. By reducing expensive fixtures and setup times, CNC machines gradually become very attractive with typical savings of over 60 percent reduction in work in progress (WIP). However, the most significant advantages are the high-quality parts produced improving both the rolled throughput yield (RTY) (by minimizing the effect of hidden factories) and the overall defect per million occurrences (DPMO) by reducing scrap and rework.

Before embarking on the DFA and DFM tools, the team should

- Revisit the physical structure, the process structure of the DFSS algorithm (Chap. 5), as well as the marketing strategy. The team should be aware that the DFSS algorithm as a product design strategy is global and the process strategy is usually local, depending on the already existing manufacturing facilities.

- Review all processes involved in market analysis, customer attributes and the CTSs, and other requirements such as packaging and maintenance. Where clarification is sought, the team may develop necessary prototypes, models, experiments, and simulation to minimize risks. In doing so, the team should take advantage of available specifications, testing, cost-benefit analysis, and modeling to build the design.

- Analyze existing manufacturing and assembly functions, operations, and sequence concurrently using simulation tools to examine assembly and subassembly definitions of the product and find the best organization and production methods.

- Apply the most appropriate, rather than the latest, technology in the processes identified in the process structure.

- Follow the axiomatic design approach to create "modular" design, namely, standard physical entities in the form of components, parts, and subassemblies. Modular entities have many attractive pros (advantages), such as cost reduction, physical and process structures configuration ease, facilitation of engineering change implementation, more product derivatives, and higher quality and reliability.

- Design for minimum number of parts by using the idea of physical coupling, *not* the functional coupling, namely, multifunctional requirements parts with multiple DPs uncoupled in time or space. For example, consider the bottle-can opener in Fig. 10.1 (Suh 1990).

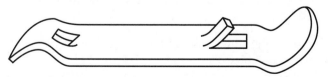

Figure 10.1 Bottle-can opener.

The functional requirements are

FR$_1$: open beverage bottle

FR$_2$: open beverage can

The DPs are

DP$_1$: beverage opener side

DP$_2$: can opener side

The design mapping is depicted in

$$\begin{Bmatrix} FR_1 \\ FR_2 \end{Bmatrix} = \begin{bmatrix} \times & 0 \\ 0 & \times \end{bmatrix} \begin{Bmatrix} DP_1 \\ DP_2 \end{Bmatrix}$$

By definition, the two functional requirements are independent or uncoupled per axiom 1 (Chap. 8). A simple device that satisfies these FRs can be made by stamping a sheetmetal as shown in Fig. 10.1. Note that a single device can be made without a functional coupling and hosted in the same part physically. *Functional coupling*, a design vulnerability, should not be confused with *physical coupling*. In addition, since the complexity of the product is reduced, it is also in line with axiom 2. The following steps are recommended:

- Choose the appropriate materials for fabrication ease.

- Apply the layered assembly principles and factors such as parts handling and feeding, orientation, identification, positioning, allowable tolerances, and mating.

- Use the appropriate DFM and DMA tools. Since DFM and DFA are interlinked, they can be used sequentially according to the roadmap in Fig. 10.2 as suggested by Huang (1996), who called the roadmap the "DFMA" approach.

10.2.1 The DFMA approach

With DFMA, significant improvement tends to arise from simplicity thinking, specifically reducing the number of standalone parts. The Boothroyd-Dewhurst DFA methodology gives the following three criteria against which each part must be examined as it is added to the assembly (Huang 1996):

1. During operation of the product, does the part move relative to all other parts already assembled?

2. Must the part be a different material than, or be isolated from, all other parts already assembled? Only fundamental reasons concerned with material properties are acceptable.

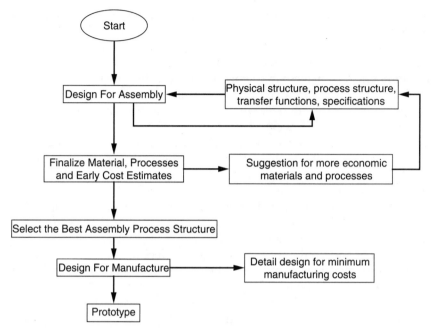

Figure 10.2   The DFMA steps (Huang 1996).

3. Must the part be separate from all other parts already assembled because the necessary assembly or disassembly of other separate parts would otherwise be impossible?

A "Yes" answer to any of these questions indicates that the part must be separate or using DFA terminology, a *critical* part. All parts that are not critical, can theoretically be removed or *physically coupled* with other critical parts. Therefore, theoretically, the number of critical parts is the minimum number of separate parts of the design.

Next, the DFSS team estimates the assembly time for the design and establishes its efficiency rating in terms of assembly difficulty. This task can be done when each part is checked to determine how it will be grasped, oriented, and inserted into the product. From this exercise, the design is rated and from this rating standard times are determined for all operations necessary to assemble the part. The DFA time standard is a classification of design features which affect the assembly process. The total assembly time can then be assessed, and using standard labor rates, the assembly cost and efficiency can be estimated. At this stage, manufacturing costs are not considered, but assembly time and efficiency provide benchmarks for new iterations. After all feasible simplification tasks are introduced, the next step is to analyze the manufacture of the individual parts. The objective of DFM within the DFMA is to enable the DFSS team to weigh alternatives, assess manufacturing cost, and make trade-offs between physical coupling (DPs consolidation) and increased manufacturing cost. The DFM approach provides experimental data for estimating cost of many processes. The DFSS team is encouraged to consult with the following studies where deemed appropriate: Dewhurst (1988) for injection molding, Dewhurst and Blum (1989) for die-cast parts, Zenger and Dewhurst (1988) for sheetmetal stamping, and Knight (1991) for powder metal parts.

The DFMA approach usually benefits from poka-yoke (errorproofing) techniques, which may be applied when components are taking form and manufacturing and assembly issues are considered simultaneously. *Poka-yoke* is a technique for avoiding human error at work. The Japanese manufacturing engineer Shigeo Shingo developed the technique to achieve zero defects and came up with this term, which means "errorproofing." A defect exists in either of two states: (1) it already has occurred, calling for defect detection, or (2) is about to occur, calling for defect prediction. Poka-yoke has three basic functions to use against defects: shutdown, control, and warning. The technique starts by analyzing the process for potential problems, identifying parts by the characteristics of dimension, shape, and weight, detecting processes deviating from nominal procedures and norms.

Example 10.1 In this exercise (Huang 1996) a motor-drive assembly must be designed to sense and control whether it is in position on two steel guiderails. The motor is fixed on a rigid base to enable the up-down movement over the rails and to support the motor system (Fig. 10.3). The motor and the measurement cylindrical sensor are wired to a power supply unit and control unit, respectively. The motor system is fully enclosed and has a removable cover for access to adjust position sensor when needed. The current design is given in Figs. 10.3 and 10.4.

The motor system is secured to the base with two screws. The sensor is held by a setscrew. To provide suitable friction and to guard against wear, the base is provided with two bushes. The end cover is secured by two end-plate screws, fastened to two standoffs, and screwed into the base. The end plate is fitted with a plastic bush for connecting wire passage. A box-shaped cover slides over the whole assembly from below the bases. The cover is held in place by four cover screws, two into the base and two into the end cover. Is this a good assembly design?

Solution   We need to take the following DFMA steps:

1. Study the current (datum) design and identify all parts and subassemblies. The proposed initial design is formed from 19 elements:
   a. Two purchased design subassemblies: the motor drive and the sensor.
   b. Eight other parts (end plate, cover, etc.)
   c. Nine screws

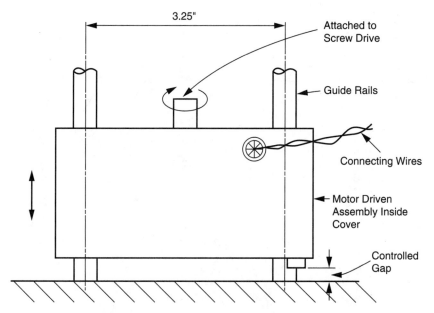

Figure 10.3 Motor-drive assembly front view (Huang 1996).

2. Apply the criteria given in Sec. 10.2.1 to every part to simplify the design and decide on the theoretical possible minimum number of parts. We need to simplify by achieving the minimum number of parts as follows:

   a. The motor and the sensor is a purchased and standard subassembly. Thus, no further analysis is required.

   b. The base is assembled into a fixture, and since there is no other part to assemble to, it is a "critical" part.

   c. The two bushes don't satisfy the criteria in Sec. 10.2.1. Theoretically, they can be assembled to the base or manufactured as the same material as end plate and combined with it.

   d. The setscrew, the four cover screws, and the end-plate screws are theoretically unnecessary. An integral fastening arrangement is usually possible.

   e. The two standoffs can be assembled to the base (don't meet the criteria).

   f. The end plate is a critical part for accessibility.

   If the motor and sensor subassemblies can be snapped or screwed to the base with a snapped-on plastic cover, only four separate items will be needed, representing a 79 percent reduction in parts because only four parts remain as the theoretically possible minimum: motor, sensor, base, and end plate.

3. Revisit all trimmed parts and check any practical, technical, or economic limitations for their removal. For example, some may argue that the two motor screws are needed to (a) secure the motor for higher fastening force or (b) hold the sensor because any other alternative will be

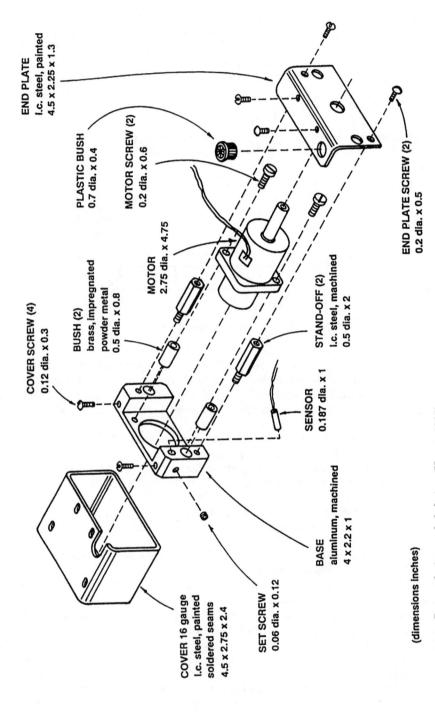

COVER SCREW (4)
0.12 dia. x 0.3

BUSH (2)
brass, impregnated
powder metal
0.5 dia. x 0.8

MOTOR
2.75 dia. x 4.75

PLASTIC BUSH
0.7 dia. x 0.4

MOTOR SCREW (2)
0.2 dia. x 0.6

END PLATE
I.c. steel, painted
4.5 x 2.25 x 1.3

END PLATE SCREW (2)
0.2 dia. x 0.5

STAND-OFF (2)
I.c. steel, machined
0.5 dia. x 2

SENSOR
0.187 dia. x 1

BASE
aluminum, machined
4 x 2.2 x 1

COVER 16 gauge
I.c. steel, painted
soldered seams
4.5 x 2.75 x 2.4

SET SCREW
0.06 dia. x 0.12

(dimensions inches)

Figure 10.4 Datum design exploded view (Huang 1996).

uneconomical because of its low volume. Other arguments may be that the two powder metal bushes may be unnecessary. In all cases, it is very difficult to justify the separate standoffs, the cover, and the six screws.

4. Estimate the assembly time and costs to account for savings in weighing assembly design alternatives. The DFMA database provides such estimates without having any detail drawings. Table 10.2 exhibits the result of the DFMA analysis with

    a. Total actual assembly time, T_1 = 163 s.

    b. Theoretical number of parts is 4, with an average of 3 s assembly time. Then, the total theoretical assembly time T_2 = 12 s.

    c. Calculate the datum assembly design η_1 efficiency using

$$\eta_1 = \frac{T_2}{T_1} \times 100\%$$

$$= \frac{12}{163} \times 100\%$$

$$= 7.362\% \qquad (10.1)$$

This is not an assembly-efficient design.

5. Redo step 4 for the optimum design (with minimum number of parts) after all practical, technical, and economic limitation considerations. Assume that the bushes are integral to the base, and the snap-on plastic cover replaces standoffs, cover, plastic bush, and six screws as shown in Fig. 10.5. These parts contribute 97.4 s in assembly time reduction, which amounts to $0.95 per hour assuming an hourly labor rate of $35. Other added improvements include using pilot point screws to fix the base, which was redesigned for self-alignment. The worksheet of the optimum design is given in Table 10.3.

TABLE 10.2   DFMA Worksheet for Datum Design

Item	Number	Theoretical part count	Assembly time, s	Assembly cost, U.S. cents
Base	1	1	3.5	2.9
Bush	2	0	12.3	10.2
Motor subassembly	1	1	9.5	7.9
Motor screw	2	0	21.0	17.5
Sensor subassembly	1	1	8.5	7.1
Setscrew	1	0	10.6	8.8
Standoff	2	0	16.0	13.3
End plate	1	1	8.4	7.0
End-plate screw	2	0	16.6	13.8
Plastic bush	1	0	3.5	2.9
Thread lead	—	—	5.0	4.2
Reorient	—	—	4.5	3.8
Cover	1	0	9.4	7.9
Cover screw	4	0	34.2	26.0
Totals	19	4	160.0	133.0

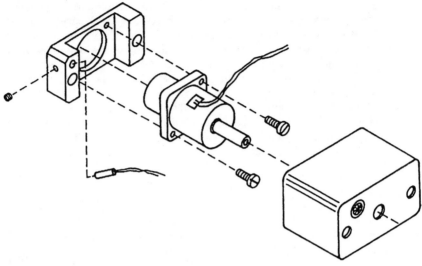

Figure 10.5 DFMA optimum design (Huang 1996).

TABLE 10.3   DFMA Worksheet for Optimum Design

Item	Number	Theoretical part count	Assembly time, s	Assembly cost, U.S. cents
Base	1	1	3.5	2.9
Motor subassembly	1	1	4.5	3.8
Motor screw	2	0	12.0	10.0
Sensor subassembly	1	1	8.5	7.1
Setscrew	1	0	8.5	7.1
Thread leads	—	—	5.0	4.2
Plastic cover	1	1	4.0	3.3
Totals	7	4	46.0	38.4

    a. Total actual assembly time $T_1 = 46$ s from the DFMA database.

    b. Theoretical number of parts is 4 with an average of 3 s assembly time. Then, the total theoretical assembly time $T_2 = 12$ s.

    c. Calculate the datum assembly design η_1 efficiency using

$$\eta_1 = \frac{T_2}{T_1} \times 100\%$$

$$= \frac{12}{46} \times 100\%$$

$$= 26.087\% \tag{10.2}$$

6. Calculate the parts cost savings as shown in Table 10.4. The saving = $35.44 − $21.73 = $13.71 in parts cost with a new fixed cover cost of $5000.

TABLE 10.4   Cost Differential Worksheet

Proposed design		Redesign	
Item	Cost, $	Item	Cost, $
Base (aluminum)	12.91	Base (nylon)	13.43
Bush (2)	2.40*	Motor screw (2)	0.20*
Motor screw (2)	0.20	Set screw	0.10*
Setscrew	0.10*	Plastic cover, including tooling	8.00
Standoff (2)	5.19		
End plate	5.89		
End-plate screw (2)	0.20*		
Plastic bush	0.1*		
Cover	8.05		
Cover screw (4)	0.40*		
Totals	35.44		21.73

*Purchased in quantity. Purchased motor and sensor subassemblies not included. *Redesign:* tooling costs for plastic cover = $5000.

7. Calculate the total savings in terms of both time (step 4) and parts reduction (step 6):

Total savings = savings from assembly time reduction + savings from

parts reduction

$$=\$0.95+\$13.71$$

$$=\$14.66 \tag{10.3}$$

The breakeven volume equals 342 total assemblies.

10.3   Design for Reliability (DFR)

Reliability is the probability that a physical entity delivers its functional requirements (FRs) for an intended period under defined operating conditions. The time can be measured in several ways. For example, time in service and mileage are both acceptable for automobiles, while the number of open-close cycles in switches is suitable for circuit breakers. The DFSS team should use DFR while limiting the life-cycle cost of the design.

The assessment of reliability usually involves testing and analysis of stress strength and environmental factors and should always include improper usage by the end user. A reliable design should anticipate all that can go wrong. We view DFR as a means to maintain and sustain Six Sigma capability over time. DFR adapts the law of probability to predict failure and adopts

1. Measures to reduce failure rates in the physical entity by employing design axioms and reliability science concurrently.

2. Techniques to calculate reliability of key parts and design ways to reduce or eliminate coupling and other design weaknesses.

3. *Derating*—using parts below their specified nominal values.

4. Design failure mode–effect analysis (DFEMA), which is used to search for alternative ways to correct failures. A "failure" is the unplanned occurrence that prevents the system or component from meeting its functional requirements under the specified operating conditions.

5. Robustness practices by making the design insensitive to all uncontrollable sources of variation (noise factors).

6. Redundancy, where necessary, which calls for a parallel system to back up an important part or subsystem in case it fails.

Reliability pertains to a wide spectrum of issues that include human errors, technical malfunctions, environmental factors, inadequate design practices, and material variability. The DFSS team can improve the reliability of the design by

- Minimizing damage from shipping, service, and repair
- Counteracting the environmental and degradation factors
- Reducing design complexity. (See El-Haik and Young 1999.)
- Maximizing the use of standard components
- Determining all root causes of defects, not symptoms, using DFMEA
- Controlling the significant and critical factors using SPC (statistical process control) where applicable
- Tracking all yield and defect rates from both in-house and external suppliers and developing strategies to address them

To minimize the probability of failure, it is first necessary to identify all possible modes of failure and the mechanism by which these failures occur. Detailed examination of DFR is developed after physical and process structure development, followed by prototyping; however, considerations regarding reliability should be taken into account in the conceptual phase when axiom 1 is employed. The team should take advantage of existing knowledge and experience of similar entities and any advanced modeling techniques that are available.

Failure avoidance, in particular when related to safety, is key. Various hazard analysis approaches are available. In general, these approaches start by highlighting hazardous elements and then proceed to identify all events that may transform these elements into hazardous conditions and their symptoms. The team then has to identify the corrective actions to eliminate or reduce these conditions. One of these approaches is called *fault-tree analysis* (FTA). FTA uses deductive logic gates to combine events that can produce the failure or the fault of interest (Sec. 11.3). Other tools that can be used in conjunction with FTA include DFMEA and PFMEA as well as the fishbone diagram.

10.4 Design for Maintainability

The objective of Design for Maintainability is to assure that the design will perform satisfactorily throughout its intended life with a minimum expenditure of budget and effort. Design for maintainability (DFM), Design for Serviceability (DFS), and Design for Reliability (DFR) are related because minimizing maintenance and facilitating service can be achieved by improving reliability. An effective DFM minimizes: (1) the downtime for maintenance, (2) user and technician maintenance time, (3) personnel injury resulting from maintenance tasks, (4) cost resulting from maintainability features, and (5) logistics requirements for replacement parts, backup units, and personnel. Maintenance actions can be preventive, corrective, or recycle and overhaul.

Design for Maintainability encompasses access and control, displays, fasteners, handles, labels, positioning and mounting, and testing. The DFSS team needs to follow these guidelines:

- Minimize the number of serviceable design parameters (DPs) with simple procedures and skills.

- Provide easy access to the serviceable DPs by placing them in serviceable locations. This will also enhance the visual inspection process for failure identification.

- Use common fasteners and attachment methods.

- Design for minimum hand tools.

- Provide for safety devices (guards, covers, switches, etc.)

- Design for minimum adjustment and make adjustable DPs accessible.

The DFSS team should devise the criteria for *repair* or *discard* decisions within the context of life-cycle costing. The major maintainability cost factors to consider include transportation, shipping, and handling; training of maintenance people; and repair logistics, which encompasses the design of service, production, distribution, and installation of repairable DPs (components and subassemblies).

The "repair" procedure should target

- Enhancing the field repair capability to react to emergency situations

- Improving current repair facilities to reflect the design changes

- Reducing cost using modularity and standard components
- Decreasing storage space

The "discard" procedure should consider

- Manufacturing cost
- Simplifying maintenance tasks (e.g., minimum skills, minimum tools, and standard attachment methods)
- Work site reliability: training technicians to avoid damaging the repair equipment
- Repair change adjustment to enable plug-in of new parts rather than field rework

10.5   Design for Serviceability

After the DFSS team finished DFR and DFMA exercises, the next step is to embark on Design for Serviceability, another member of the DFX family. *Design for Serviceability* (DFS) is the ability to diagnose, remove, replace, replenish, or repair any DP (component or subassembly) to original specifications with relative ease. Poor serviceability produces warranty costs, customer dissatisfaction, and lost sales and market share due to loss loyalty. The DFSS team may check their VOC (voice-of-the-customer) studies such as QFD for any voiced serviceability attributes. *Ease of serviceability* is a performance quality in the Kano analysis. The DFSS algorithm strives to have serviceability personnel involved in the early stages, as they are considered a customer segment. Many customers will benefit from DFS as applied in the DFSS algorithm, both internally and externally. For example, Fig. 10.6 depicts the automotive DFS customer segments. More customers indicate more benefit that can be gained that is usually more than the DFSS team realize initially.

The following consideration of DFS should be visited by the DFSS team:

1. Customer service attributes
2. Labor time
3. Parts cost
4. Safety
5. Diagnosis
6. Service simplification
7. Repair frequency and occurrence

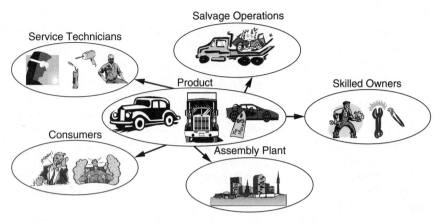

Figure 10.6   Automotive DFS customers.

8. Special tools

9. Failures caused by the service procedures

10.5.1   DFS guidelines

The DFS guidelines are

1. Reduce service functional requirements (FRs) by minimizing the need for service. This can be easily done in companies that track their product or service warranties. The DFSS team has the opportunity to make their DFS procedure data-driven by analyzing the possible failure rates of datum designs (incremental design scenarios) and rank them using Pareto analysis to address service requirements in prioritized sequence. DFX, axiomatic design, robustness, and DFR techniques can be used to improve the reliability. For example, DFMA improves reliability by reducing the number of parts; axiom 2 helps reduce design stiffness to reduce variation in the FRs (Chap. 7), which is the major cause of failure. In addition, axiomatic design helps generate ideas of physical coupling for DP consolidation, resulting in a smaller number of separate parts and thus enhanced reliability levels.

2. Identify customer service attributes and appropriate type of service required by any customer segment is the determinant of the DFS technique to be used. There are three types: standard operations, scheduled maintenance, and repairs. *Standard operations* consist of normal wear-and-tear items such as replenishing operating fluids. For standard operations, ease of service should be maximum and coupled with errorproofing (poka-yoke) techniques. In many industries, the end customer is usually the operator. *Scheduled maintenance* is usually

recommended for specific items in the customer manual, if any. In this category, customers expect less frequency and more ease. Under the pressure of minimum life-cycle cost, many companies are pushing the scheduled maintenance tasks to standard operations and "Do it yourself" procedures. A sound scheduled maintenance procedure should call for better reliability and durability, minimum tools (e.g., single standard fastener size), and easy removal paths. In *repair* service, ease of repair is key. This objective is usually challenged by limited accessibility space and design complexity. Repair service can be greatly enhanced by employing some sort of diagnostic system, repair kits, and modular design practices. Repair issues can take a spectrum of possible causes ranging from type 1 to type 2 errors in diagnostics systems, tools and parts logistics issues, and repair technicality.

3. Practice the DFS approach. If the serviceability requirements have not been serviced by now, the DFSS team is encouraged to use design mappings by employing the zigzagging method between serviceability FRs and its DPs. Once the team has identified all serviceability mapping, they can move to consider design alternatives. These alternatives may occasionally be inapplicable. In other cases, they may seem in conflict with one another. Nevertheless, the DFSS team should review the entire process to determine whether a Six Sigma–capable and rounded design is to be established in all requirements, including those related to serviceability. A *serviceability* set of FRs usually includes proper location, tools and parts standardization, protection from accelerated failure, ergonomics considerations, and diagnostic functions.

The DFSS team should generally perform the following steps to devise a sound DFS approach:

1. Review assumptions, serviceability customer CTSs and FRs from the QFD, serviceability types, customer segments, and Six Sigma targets.
2. Check datum designs and use the data available as a way to predict their design performance from datum (data) historical database(s). The team should also benchmark best-in-class competition to exceed customer satisfaction.
3. Identify types of service needed (e.g., standard operation, scheduled maintenance, or repair) and map them to appropriate customer segments.
4. Understand all service procedures in the company core books, including steps, sequence, and potential problems.
5. Estimate time of labor. Labor time is considered the foundation of serviceability quantification for warranty assessment purposes. It is the sum of repair recognition time, diagnostic time, logistic time, and actual repair time. The team should aim to beat the best-in-class labor time.

6. Minimize all service problematic areas by reviewing the customer concern tracking system (if any), determining and eliminating root causes, addressing the problem based on a prioritization scheme (e.g., Pareto analysis of warranty cost impact), searching for solutions in the literature and core books, and predicting future trends.
7. Determine solution approached in design from steps 1 to 6. The information extracted from the gathered data will lead to some formulation of a serviceability design strategy. Every separate component or critical part should be addressed for its unique serviceability requirements.
8. Introduce serviceability design parameters (DPs or solution) into the process structure. These can be categorized according to answers to the following questions:
   a. Orientation:
      (1) Do the parts have easy removal paths (sum of service steps)?
      (2) Do the service steps require re-orientation?
   b. Contamination:
      (1) Can the fluid, if any, be contained prior to or though service?
      (2) What is the possibility of contaminating parts during service?
   c. Access
      (1) *Assemblability.* Is it possible to group components for ease of service? Check the structure.
         (a) Is disassembly intuitive?
         (b) Can asymmetric components fit one way?
      (2) *Reachability.* Can the part be reached by hand? By tool? Can the part be removed from the assembly?
      (3) *Layerability.* Is the part in the assembly layer correlated to frequency of service?
      (4) *Real estate.* Possibility of moving or sizing parts for service space.
      (5) *Efficiency.* Unnecessary removal of parts which obstruct visibility or service.
      (6) *Diagnostics.* Can the part be accessed for diagnostics without disassembly?
      (7) *Service reliability.* Address potential damage of serviced or removed parts. Have all possibilities for parts minimization using DFMA been exhausted? Consider the use of standard parts (e.g., fasteners).
   d. Simplicity—Customer considerations:
      (1) *Tools.* Design for generic tools. Minimize use of specialized tools.
      (2) *Adjustment.* Reduce customer intervention through tuning and adjustment. Use robustness techniques.
      (3) *Poka-yoke.* Use color codes and very clear instructions.

10.5.2   Pressure recorder PCB (printed-circuit-board) replacement

This approach (Boothroyd and Dewhurst 1990, Huang 1996) has been used to study the service disassembly and reassembly processes by identifying all individual steps including part removal, tool acquisition, pickup and orientation, and insertion. The time standard in this procedure is the result of Abbatiello (1995) at the University of Rhode Island. An exploded view is given in Fig. 10.7. The worksheets in Tables 10.5 and 10.6 were developed to utilize the serviceability time database. The first step of the DFS approach is to complete the disassembly worksheet in Table 10.5. The DFSS team may disassemble the pressure

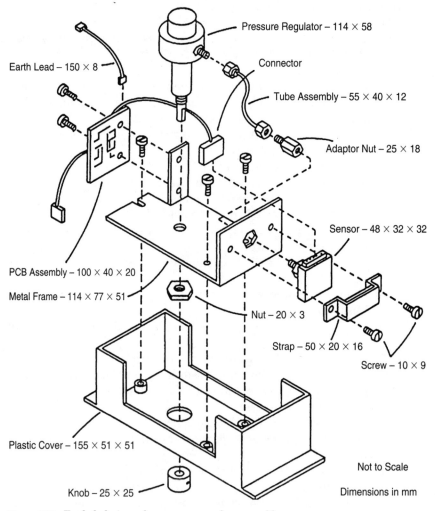

Pressure Regulator – 114 × 58

Earth Lead – 150 × 8

Connector

Tube Assembly – 55 × 40 × 12

Adaptor Nut – 25 × 18

Sensor – 48 × 32 × 32

PCB Assembly – 100 × 40 × 20

Metal Frame – 114 × 77 × 51

Nut – 20 × 3

Strap – 50 × 20 × 16

Screw – 10 × 9

Plastic Cover – 155 × 51 × 51

Not to Scale

Dimensions in mm

Knob – 25 × 25

Figure 10.7 Exploded view of pressure recorder assembly.

recorder to reach the PCB, the item requiring service. In the disassembly process, the team will access several disassembly locations and record all operations taken in the disassembly worksheet row by row.

Subassemblies are treated as parts when disassembly is not required for service; otherwise the disassembly operation recording will continue for removing them. Reference to the Abbatiello (1995) database is given in columns 3, 5, and 7 of Table 10.5. For example, the time in 4.2 s in column 4 is the average taken from hours of videotaped service work and includes a fraction of the time for tool replacement at the end of service. The estimated time for PCB disassembly T_d is 104.3 s. This time can be converted to labor cost by multiplying by the service labor hourly rate.*

The serviceability efficiency η is determined by parts necessity for removal or disassembly if they satisfy any of the following:

- The part or subassembly must be removed to isolate the service item(s).

- The part or subassembly removed contains the service item.

- The part or subassembly removed is a functional cover part enclosing the service item. For example, the plastic cover in the pressure recorder does not enclose the PCB; thus it is not considered a cover.

When a part or subassembly does not satisfy any of these requirements, it is not considered as a necessary part for disassembly. The sum in column 11 of Table 10.5 is the theoretical minimum justifiable and necessary number of disassembly operations N_m. In this example, only the removal of PCB is justified, $N_m = 1$.

The next step is to fill out the corresponding reassembly worksheet (see Table 10.6). The reassembly worksheet format is similar to the disassembly worksheet and requires reference to the insertion and fastening database.

The DFSS team noticed that the total removal time T_r equals 130.9 s and does not equal the total disassembly time.

On completion of both worksheets, the overall service efficiency of the service performed, replacing the PCB, can be calculated using the following steps:

1. Calculate the total time of service T_s as $T_s = T_d + T_r = 235.2$ s.
2. Determine the ideal service time based on the minimum amount of time required for all necessary operations, which include removal, set-aside, acquisition, and insertion. Several assumptions need to be made:

---

*Note that the division by 36 in column 10 of Table 10.5 is intended to convert dollars to cents and hours to seconds.

TABLE 10.5 Disassembly Worksheet

Assembly: pressure recorder

1	2	3	4	5	6	7	8	9	10	11	Labor rate, $/h, L = 30
ID number	Number of times operation is repeated	Four-digit tool acquisition code	Tool acquisition time, s	Four-digit item removal of operation code	Item removal or operation time, s	Four-digit item set-aside code	Item set-aside time, s	Operation time, s = (4) + (2) × [(6) + (8)]	Operation cost, cents = [(9) × L]/36	Number of service items, cover parts of functional connections	Service task performed
1	3	5700	4.2	1710	11.3	5500	1.4	42.3	35.25	0	Remove screws
2	1			5800	4.5			4.5	3.75	0	Reorientation
3	1	5700	4.2	4100	8			12.2	10.17	0	Loosen setscrew
4	1			1500	2.4	5500	1.4	3.8	3.167	0	Remove knob
5	1			1500	2.4	5500	1.4	3.8	3.167	0	Remove cover
6	1			5800	4.5			4.5	3.75	0	Reorient
7	1			4401	6.4			6.4	5.333	0	Unplug screw
8	2	5700	4.2	1700	8	5500	1.4	23	19.17	0	Remove screws
9	1			1500	2.4	5500	1.4	3.8	3.167	1	Remove PCB

104.3	86.92	1
T_d	C_d	N_m

Note: T_d = total operation time; C_d = total operation cost; N_m = total number of service items.

TABLE 10.6 The Reassembly Worksheet

Assembly: pressure recorder

1	2	3	4	5	6	7	8	9	10	
ID number	Number of times operation is repeated	Four-digit tool acquisition code	Tool acquisition time, s	Four-digit item acquisition code	Item acquisition time, s	Four-digit item insertion or operation code	Item insertion or operation time, s	Operation time, s = (4) + (2) × [(6) + (8)]	Operation cost, cents = (9) × L/36	Service task performed
1	1	5700	4.2	5601	3.4	0001	4.90	12.50	10.42	Add PCB
2	2			5600	1.4	0401	13.80	30.40	25.33	Fasten screw
3	1	5700	4.2			3000	4.40	8.60	7.167	Plug in sensor
4	1					5800	4.50	4.50	3.75	Reorient
5	1			5600	1.4	0001	4.90	6.30	5.25	Remove plastic cover
6	1			5600	1.4	0001	4.90	6.30	5.25	Remove knob
7	1					2700	8.00	8.00	6.667	Fasten setscrew
8	1	5700	4.2			5800	4.50	8.70	7.25	Reorient
9	3			5600	1.4	0401	13.80	45.60	38	Screw on cover

Labor rate, $/h, L = 30

T_r	C_r
130.9	109.1

a. All parts necessary for the service are placed within easy reach with no tools "ideally" required.

b. Following DFMA, the "$\frac{1}{3}$ DFS ideal design" rule of thumb is used, which states that in ideal design for assembly, approximately one in every three parts will need to be unsecured and later resecured by efficient methods such as snap-fit and release fastening.

Using these assumptions, the ideal service time for the parts that need no additional removal or insertion can be given by

$$t_{min} = \frac{3T_1 + T_2}{3} + \frac{2T_3 + T_4}{3} + T_5 + T_6 \qquad (10.4)$$

where T_1 = unsecured item removal time (= 2.4 s from database)
$ T_2$ = snap-fit item removal time (= 3.6 s from database)
$ T_3$ = unsecured item insertion time (= 3.8 s from database)
$ T_4$ = snap-fit item insertion time (= 2.2 s from database)
$ T_5$ = item acquisition time (= 4 s from database)
$ T_6$ = item set-aside time (= 1.4 s from database)

Therefore

$$t_{min} = \frac{3 \times 2.4 + 3.6}{3} + \frac{2 \times 3.8 + 2.2}{3} + 1.4 + 1.4$$

$$\cong 9 \text{ s} \qquad (10.5)$$

and with $N_m = 1$, the time-based efficiency is given by

$$\eta_{time} = \frac{t_{min} \times N_m}{T_s} \times 100\%$$

$$= \frac{9 \times 1}{235.2} \times 100\%$$

$$= 3.8\% \qquad (10.6)$$

The efficiency value is very low, leading us to conclude that the service procedure needs to be simplified using efficient disassembly methods and the assembly reconfigured so that the items requiring frequent service are conveniently accessed. In our example, considering PCB as a primary service structure, the assembly is reconfigured so that the board is on the outermost layer (Fig. 10.8). Using the same database values, the estimated time for both disassembly and reassembly T_s, equals 16.5 s. Hence, the new efficiency is

$$\eta_{\text{time}} = \frac{t_{\min} \times N_m}{T_s} \times 100\%$$

$$= \frac{9 \times 1}{16.5} \times 100\%$$

$$= 54.5\% \tag{10.7}$$

This DFS calculation approach can be extended to multiservice procedures, say, $i=1,2,...,k$. The overall time-based efficiency is a weighted

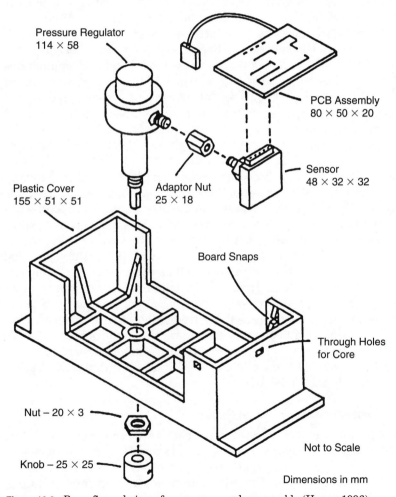

Pressure Regulator
114 × 58

PCB Assembly
80 × 50 × 20

Plastic Cover
155 × 51 × 51

Adaptor Nut
25 × 18

Sensor
48 × 32 × 32

Board Snaps

Through Holes
for Core

Nut – 20 × 3

Not to Scale

Knob – 25 × 25

Dimensions in mm

Figure 10.8   Reconfigured view of pressure recorder assembly (Huang 1996).

average of the procedures of interest by the failure frequencies f_i. This is given by

$$\eta_{\text{overall}} = \left(\cfrac{1}{\sum\limits_{i=1}^{k} f_i} \right)\left(\sum\limits_{i=1}^{k} f_i \eta_i \right) \tag{10.8}$$

10.6 Design for Environmentality

In an effort to meet the world's growing energy needs, the dependence on fossil fuels has become a necessary endeavor. Since the first oil crisis in 1973 and the Gulf War in 1991, the world's energy perspective has changed significantly. Since then many countries have attempted to reduce their dependence on oil by investigating alternative energy sources. More importantly, however, there has been an increased awareness concerning environmental pollution and efforts to reduce the effects of fossil fuel emissions. Global studies have concluded that increased fossil fuel consumption has led to increased carbon dioxide release, which in turn causes atmospheric heating. These theories, known as "the greenhouse theory" and "global warming," are both environmental concerns which have strongly affected the design and manufacturing industries. For example, increased legislation concerning automotive emission levels has driven the automotive industry to look for alternative fuel sources that would limit fossil fuel consumption while focusing on energy savings and lowering environmental impacts. Therefore, the motivation for environmentally friendly design is coming from the recognition that sustainable economic growth can occur without necessarily consuming the earth's resources. This trend opens the door for an evaluation of how the environment should be considered in design.

Design for Environmentality (DFE) (Myers 1984, Bussey 1998) addresses environmental concerns as well as postproduction transport, consumption, maintenance, and repair. The aim is to minimize environmental impact, including strategic level of policy decision making and design development. Since the introduction of DFE, one can view the environment as a customer! Therefore, the definition of defective design should encompass the designs that negatively impact the environment. As such, DFE usually comes with added initial cost, causing an increment of total life cost.

10.6.1 Technological and design issues

Most applied design technologies in the past have been developed by the U.S. Department of Energy demonstration projects. These technologies were often plagued with numerous kinds of errors. Common design principles were applied, and many solutions were found only by testing.

Environmentally friendly designs are still relatively expensive. For the most part, the technology gained since the early 1990s has proved itself able to contribute substantially to sustainability in many design applications. With companies concerned with short-term versus the long-term benefits, and until there is widespread use and mass marketing of these designs, commercial environmentally friendly designs will probably continue to be a conversation piece rather than a routine practice.

In addressing the question of whether DFE will be lucrative in a given DFSS project, it is imperative to consider designs that are optimal relative to other components in which they are used, specifically, datum technologies. Economic evaluation is required both for maximum economic benefit and to estimate what the expected dollar savings (or losses) will be. The major purpose of using economic analysis techniques is to consider environment concerns and profit concerns jointly in an attempt to reduce the use of nonrenewable energy and maximize recyclability. These techniques usually clarify the financial value of limiting nonrenewable energy use.

The actual financial value of any proposed solution can easily be evaluated according to the established economic criteria for the project. For example, solar economics deals with optimizing the trade-off between solar system ownership and operating costs and the future cost of the fuel saved by the solar system during its anticipated useful life. *Life-cycle cost* (LCC) is a term commonly used to describe a general method of economic evaluation by which all relevant costs over the life of a project are accounted for when determining the economic efficiency of a project. Life-cycle cost requires assessment of the following types of related costs:

1. System acquisition and installation costs (capital costs)

2. System replacement costs

3. Maintenance and repair costs

4. Operation cost (e.g., energy costs)

5. Salvage or resale value net of removal and disposal costs

A life-cycle costing approach can be implemented by applying any or all of the following evaluation techniques:

1. *Total life-cycle-cost* (TLCC) analysis, which sums the discounted value of all the equivalent costs over the time horizon.

2. *Net present worth* (NPW) analysis, which calculates the difference between the TLCC of a proposed project and its alternative as a dollar measure of the project's net profitability.

3. *Internal rate of return* (IRR) technique, which gives the percentage yield on an investment. (See Bussey 1998).

4. *Activity-based costing* (ABC) with or without uncertainty measures. (See Sec. 10.7.)

With its emphasis on costs, DFE is a suitable method for evaluating the economic feasibility of projects such as energy conservation or solar energy, which realize their benefits primarily through fuel cost avoidance.

10.6.2   DFE economic fundamentals and optimization

The value of a DFE investment, for the most part, is dictated by customer expectations. Each customer segment weighs differently factors such as extra initial cost, operating cost, energy savings, tax implications, payback, and the overall cash flow. Understanding the importance of each of these elements to the customer is necessary to find the correct balance of the life cycle cost. Common fundamentals that need to be considered in the economic evaluation of environmentally friendly design include the time value of money, opportunity costs, and the economic equivalence of neutral or negative environmentally friendly datum designs.

For example, in evaluating a solar project, a set of economic criteria must be developed to help facilitate the life-cycle economic analysis. To help clarify the DFE goals of the DFSS project, a differential analysis should be developed from a base which would be a 100 percent environment neutral datum design. For example, if we propose a solar energy system to operate vehicle accessories and to reduce fossil fuel consumption, we must analyze the base case of the conventional system in identical vehicle environments, microclimates, and so on, as those for which the solar system was being designed. To clarify the financial design goals for the proposed solution, the datum cost should be appraised. This will set the approximate value of the money that may be spent initially. By varying the proportion of this initial investment versus the reduction in annual energy costs, various rates of return will be obtained and used for comparison.

Several predesign tasks will play an integral part in determining the feasibility of the DFE approach to the DFSS project as presented in Table 10.7.

10.7   Design for Life-Cycle Cost (LCC): Activity-Based Costing with Uncertainty

Activity-based cost (ABC)* is a new and powerful method for estimating life-cycle design cost, in particular when coupled with uncertainty

---

*ABC has received this name because of its focus on activities performed in design practice.

TABLE 10.7   Questions to Ask in a Pre–Economic Evaluation of DFE

1. Amount of available funds for financing the project. Will the DFSS team have to ask for more budgets?

2. What is the minimum attractive rate of return (MARR)? What discount rate should be used to evaluate the project? What is investment timeframe?

3. What is the economic lifetime of the project?

4. Is the lifetime of the project the same as the customer's timeframe so that the investment will prove feasible?

5. What are the operating costs associated with the environmentally friendly design? For example, in the automotive industry, the escalation rate of fuel is a consideration. Will fuel inflate at a higher rate than the dollar?

6. Government incentives (federal/state/local). Will any incentives play a role in the overall economic evaluation?

provisions (Huang 1996). The method employs process action and sensitivity charts to identify and trace significant parameters affecting the LCC. The ABC method assumes that the design, whether a product, a service, or a process, *consumes activities.* This assumption differentiates ABC from conventional cost estimation methods that assume *resources consumption.* Let us use the example of a materials-handling step elimination in manufacturing. A conventional cost method may translate this cost reduction into reduction of direct labor with an equivalent amount. The ABC method translates this into elimination of activities to reduce materials-handling cost, a direct and precise translation. This assumption made ABC very attractive to modeling among other spectra of benefits that include superiority of cost tracing, separation of direct and indirect cost components, higher accuracy, and alignment to activity-based management systems. The ABC process is depicted in Fig. 10.9.

The ABC objective is to identify activities in the design life, and then assign reliable cost drivers and consumption intensities to the activities. Probability distributions are given to represent inherent cost uncertainty. Monte Carlo simulation and other discrete-event simulation techniques are then used to model uncertainty and to estimate the effect of uncertainty on cost. The uncertainty effect can be estimated by exercising a controlled virtual experiment within the simulation model. Different commercial simulation methods exist such as Crystal Ball (usually added to the Microsoft Excel spreadsheet) and SigmaFlow, derived from the Simul8 platform. SigmaFlow has additional Six Sigma analysis capabilities. Uncertainty can be modeled in many ways. We can model uncertainty based on historical data using probability theory, an incremental design scenario, or we can model uncertainty subjectively based on experience and fuzzy-set theory

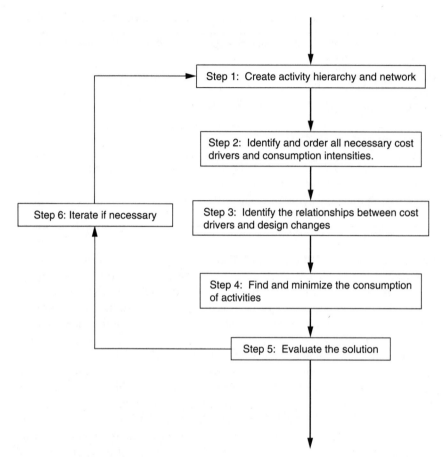

Figure 10.9 The ABC process (Huang 1996).

(Zadeh 1965), a creative design situation. In spreadsheet simulation models, the terms *assumption cell* and *forecast cell* are commonly used. The first is a source variable, while the latter is a response variable.

10.7.1 A product cost model

Consider the following simple example where a product cost is modeled as a sum function of direct labor and material costs (Huang 1996). According to our assumptions with respect to these two cost components, we would like to forecast the total cost (Fig. 10.10). Assumption cells are assigned uncertainty or probability distributions. A *direct labor* cell is distributed as a triangular distribution, while the *material* assumption cell is distributed elliptically. These uncertainty distributions are defined as the DFSS team find appropriate for various reasons; however, we found these assignments accurate in many applications:

1. Activity: machine cycle time
   - Constant for CNC machines
   - Uniform with parameters (min, max)
   - Triangular* with parameters (min, mode, max)

2. Activity: conveyor [also automated guided vehicle (AGV)] speed
   - Constant since little variation of actual conveyor speeds
   - Uniform with parameters (min, max)

3. Activity: travel time of AGV, forklift truck, etc.
   - Constant with fixed speed and distance as well as little blockage
   - Triangular with parameters (min, mode, max)
   - Lognormal[†] with parameters (mean, μ)

4. Activity: percent down [when repair time and/or time to (between) failure(s) is (are) unknown]
   - Must assume distributions (see activities 5 and 6, below) for TTR and TTF (or TBF) and parameters for one or the other. (We need both distributions, but only one set of parameters.)

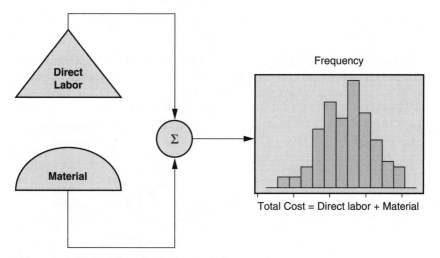

Figure 10.10 Direct labor simulation example.

Note: Mean = (min + mode + max)/3.

[†]The black belt should avoid this unless very long delays are possible.

- Use the following relation and solve for a single unknown: percent down/100 = MTTR/(MTTF + MTTR) = MTTR/MTBF.
- Works best when you use the black belt estimate MTTR and TTR distribution and use Erlang distribution* for TTF (or TBF) distribution with parameter based on solving the preceding equation for MTTF (or MTBF).

5. Activity: downtime—time to repair
- Triangular with parameters (min, mode, max)

6. Activity: downtime—time to (between) failure(s)
- Erlang with parameters (mean, $K^\dagger$)
- Exponential with parameter (mean)‡

7. Activity: scrap rate
- Binomial parameters: (percent success) with success = scrap part

8. Activity: interarrival time of parts
- Erlang parameters (mean, K)
- Lognormal§ with parameters (mean, σ)
- Exponential parameter (mean)

9. Activity: assignment of part type (or other discrete attributes) to parts
- Binomial parameters (percent success) success = first part type
- Discrete probability parameters¶ [percent A, percent $(A + B)$, 100 percent]

An iteration of the simulation routine provides random numbers from the assumption cells. These random numbers propagate through the sum equation in the model to calculate the value of the total cost, the *forecast* cell. When all iterations are performed, the calculated values of the forecast cell form a new statistical distribution. Six Sigma statistical analysis takes over, and the black belt can construct confidence intervals or test the hypothesis as if the data are collected from real random experiments.

Design for life-cycle cost involves choosing the best economical alternative as we did with other costing methods (Sec. 9.6.1). Design alternatives can be studied using ABC to select a cost optimum. However,

---

*Erlang distribution is a probability model.

$\dagger K$ = number of exponential distributions to sum (use $K = 3$ or $K = 2$ for best results).

‡Avoid when modeling with time between failures and when a very short (almost zero) duration is impossible.

§The black belt should avoid this unless very long delays are possible.

¶This is an extension of the binomial for three or more choices.

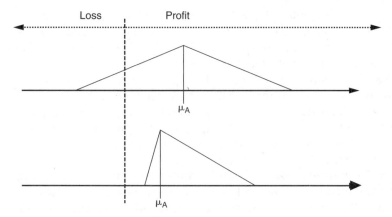

Figure 10.11 Possible ABC scenario.

because of the uncertainty factor, the picture is not crisp. Consider Fig. 10.11. Process structure or mapping of alternative *A* activities is more likely to give greater profit than alternative *B*; however, it involves the probability of loss. Which alternative to choose? The answer depends on the economic situation and risk management policy. A risk-averse team would choose alternative *A* because the expected profit is higher.

10.8 Summary

The DFX family provides systematic approaches for analyzing design from a spectrum of perspectives. It strengthens teamwork within the concurrent DFSS environment.

The Design for Manufacture and Assembly (DFMA) approach produces a considerable reduction in parts, resulting in simple and more reliable design with less assembly and lower manufacturing costs.

Design for Reliability (DFR) enables the DFSS team to gain insight into how and why a proposed design may fail and identifies aspects of design that may need to be improved. When the reliability issues are addressed at early stages of the DFSS algorithm, project cycle time will be reduced.

Per axiom 2, a simplified product can be achieved through the sequential application of DFMA followed by *Design for Serviceability* (DFS), which is the ability to diagnose, remove, replace, replenish, or repair any DP (component or subassembly), to original specifications, with relative ease. Poor serviceability produces warranty costs, customer dissatisfaction, and lost sales and market share due to loss loyalty.

In Design for Life-Cycle Cost, the activity-based cost (ABC) was introduced. ABC is a new and powerful method for estimating life-cycle design cost to help guide the DFSS team in decision making to achieve cost-efficient Six Sigma design in the presence of market and operations uncertainty. In effect, ABC is a zooming tool for cost, enabling the team to focus on top cost contributors.

Another DFX family member is Design for Maintainability. The objective of Design for Maintainability is to ensure that the design will perform satisfactorily throughout its intended life with a minimum expenditure of budget and effort. Design for Maintainability, DFS, and DFR are related because minimizing maintenance and facilitating service can be achieved by improving reliability.

Design for Environmentality (DFE) addresses environmental concerns as well as postproduction transport, consumption, maintenance, and repair. The aim is to minimize environmental impact, including strategic level of policy decision making and design development.

Failure Mode–Effect Analysis

11.1 Introduction

The *failure mode–effect analysis* (FMEA) helps DFSS team members improve product and process by asking "What can go wrong?" and "Where can variation come from?" Product design and manufacturing or production, assembly, delivery, and other service processes are then revised to prevent occurrence of failure modes and to reduce variation. Specifically, the team should study and completely understand physical and process structures as well as the suggested process mapping. Study should include past warranty experience, if any; customer wants, needs, and delights; performance requirements; drawings and specifications; and process mappings. For each functional requirement (FR) and manufacturing and assembly process the team needs to ask "What can go wrong?" They must determine possible design and process failure modes and sources of potential variation in manufacturing, assembly, delivery, and all service processes. Considerations include variations in customer usage; potential causes of deterioration over useful product life; and potential process issues such as missed tags or steps, shipping concerns, and service misdiagnosis. The team should modify product design and processes to prevent "wrong things" from happening and involve the development of strategies to deal with different situations, the redesign of processes to reduce variation, and errorproofing (poka-yoke) of designs and processes. Efforts to anticipate failure modes and sources of variation are iterative. This action continues as the team strives to further improve their design and its processes.

In the DFSS algorithm, various FMEA types will be experienced by the DFSS team. They are depicted in Fig. 11.1. We suggest using

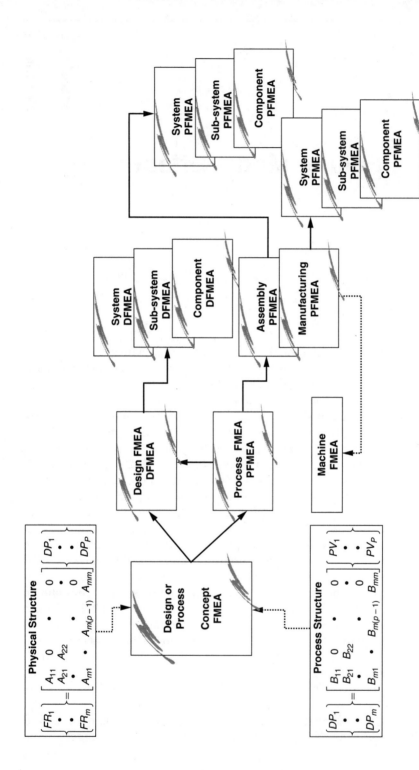

Figure 11.1 FMEA types.

concept FMEA to analyze systems and subsystems in the early concept and design stages. This focuses on potential failure modes associated with the functions of a system caused by the design. The concept FMEA helps the DFSS team review targets for the FRs, select optimum physical structure with minimum vulnerabilities, identify preliminary testing requirements, and determine if hardware system redundancy is required for reliability target settings. *Design FMEA* (DFMEA) is used to analyze designs before they are released to production. In the DFSS algorithm, a DFMEA should always be completed well in advance of a prototype build. The input to DFMEA is the array of functional requirements. The outputs are (1) list of actions to prevent causes or to detect failure modes and (2) history of actions taken and future activity. The DFMEA helps the DFSS team in

1. Estimating the effects on all customer segments

2. Assessing and selecting design alternatives

3. Developing an efficient validation phase within the DFSS algorithm

4. Inputting the needed information for Design for X (DFMA, DFS, DFR, DFE, etc.; see Chap. 10)

5. Prioritizing the list of corrective actions using strategies such as mitigation, transferring, ignoring, or preventing the failure modes

6. Identifying the potential special design parameters (DPs) in terms of failure

7. Documenting the findings for future reference

Process FMEA (PFMEA) is used to analyze manufacturing, assembly, or any other processes such as those identified as transactional DFSS projects. The focus is on process inputs. Software FMEA documents and addresses failure modes associated with software functions. The PFMEA is a valuable tool available to the concurrent DFSS team to help them in

1. Identifying potential manufacturing/assembly or production process causes in order to place controls on either increasing detection, reducing occurrence, or both

2. Prioritizing the list of corrective actions using strategies such as mitigation, transferring, ignoring, or preventing the failure modes

3. Documenting the results of their processes

4. Identifying the special potential process variables (PVs), from a failure standpoint, which need special controls

11.2 FMEA Fundamentals

An FMEA can be described as a systemic group of activities intended to

1. Recognize and evaluate the potential failures of a product or process and the effects of that failure
2. Identify actions which could eliminate or reduce the chance of the potential failure occurring
3. Document the entire process

It is complementary to the process of defining what a design or process must do to satisfy the customer (AIAG 2001).

In our case the process of "defining what a design or a process must do to satisfy the customer" is the DFSS algorithm. The DFSS team may visit existing datum FMEA, if applicable, for further enhancement and updating. In all cases, FMEA should be handled as a living document. The fundamentals of an FMEA inputs are depicted in Fig. 11.2 and the following list:

1. *Define scope, the FRs or DPs, and process steps.* For the DFSS team, this input column can be easily extracted from the physical and process structures or process mappings. However, we suggest doing the FMEA exercise for subsystems and components identified in the structures according to the revealed hierarchy resulting from the zigzagging method. At this point, it may be useful to translate the physical structure into a block diagram like the one depicted in Fig. 11.3 for an automotive engine. The block diagram is a pictorial translation of the structure related to the FMEA of interest. The block diagram strengthens the scope (boundary) of the FMEA in terms of what is included and excluded. In DFMEA, for example, potential failure modes include the delivery of "No" FR, partial and degraded FR delivery, over time, intermittent FR delivery, and unintended FR (not intended in the physical structure). The physical structure should help the DFSS team trace the coupling and with the help of the block diagram, pictorially classifies the coupling among the FRs in terms of energy, information, or material (Pahl and Beitz 1988).

2. *Identify potential failure modes.* Failure modes indicate the loss of at least one FR. The DFSS team should identify all potential failure modes by asking "In what way does the design fail to perform its FRs?" as identified in the physical structure. Failure modes are generally categorized as material, environment, people, equipment, methods, and so on. Failure modes have a hierarchy of their own, and a potential failure mode can be the cause or effect in a higher-level subsystem, causing failure in its FRs. A failure mode may, but not necessarily must,

FR, DP, or Process Step (1)	Potential Failure Mode (2)	Potential Failure Effects (3)	S E V (4)	Potential Causes (5)	O C C (6)	Current Controls (7)	D E T (8)	R P N (9)	Actions Recommended (10)
			0		0		0	0	
			0		0		0	0	
			0		0		0	0	
			0		0		0	0	
			0		0		0	0	

Figure 11.2 FMEA worksheet.

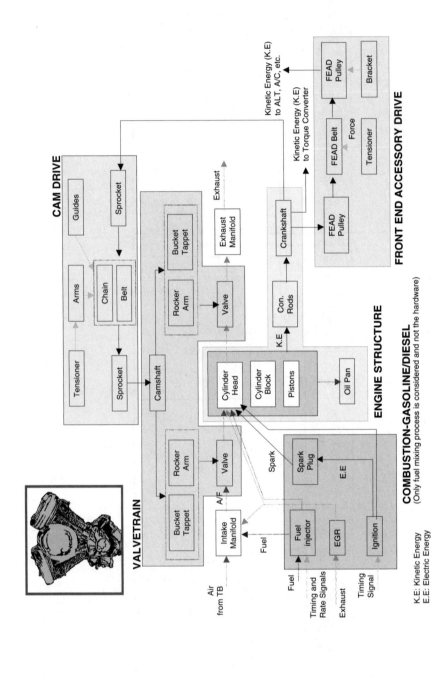

Figure 11.3 Engine block diagram.

occur. A potential failure mode may be studied from a previous baseline and current data, tests, fleet tests, and current FMEAs.

3. *Potential failure effects(s).* A *potential effect* is the consequence of the failure on other physical entities as experienced by the customer.

4. *Severity.* Severity is a subjective measure of "how bad" or "serious" the effect of the failure mode is. Usually severity is rated on a discrete scale from 1 (no effect) to 10 (hazardous effect). Severity ratings of 9 or higher indicate a potential special effect that needs more attention, and this typically is a safety or government regulation issue (Table 11.1). Severe effects are usually classified as "critical," "significant," or "control." "Critical" effects are usually a safety issue and

TABLE 11.1 Automotive Industry Action Group (AIAG) Severity Rating

Effect	Severity of effect defined	Rating
None	No effect	1
Very minor	Minor disruption to production line; a portion (<100%) of the product may have to be reworked on line but in station; fit/finish/squeak/rattle item does not conform; defect noticed by discriminating customers	2
Minor	Minor disruption to production line; a portion (<100%) of the product may have to be reworked on line but out of station; fit/finish/squeak/rattle item does not conform; defect noticed by average customers	3
Very low	Minor disruption to production line; product may have to be sorted and a portion (<100%) reworked; fit/finish/squeak/rattle item does not conform; defect noticed by most customers	4
Low	Minor disruption to production line; 100% of product may have to be reworked; vehicle/item operable, but some comfort/convenience item(s) operable at reduced level of performance; customer experiences some dissatisfaction	5
Moderate	Minor disruption to production line; a portion (<100%) may have to be scrapped (no sorting); vehicle/item operable, but some comfort/convenience item(s) inoperable; customers experience discomfort	6
High	Minor disruption to production line; product may have to be sorted and a portion (<100%) scrapped; vehicle operable, but at a reduced level of performance; customer dissatisfied	7
Very high	Major disruption to production line; 100% of product may have to be scrapped; vehicle/item inoperable, loss of primary function; customer very dissatisfied	8
Hazardous: with warning	May endanger operator; failure mode affects safe vehicle operation and/or involves noncompliance with government regulation; failure will occur with warning	9
Hazardous: without warning	May endanger operator; failure mode affects safe vehicle operation and/or involves noncompliance with government regulation; failure will occur without warning	10

require deeper study for all causes to the lowest level using, possibly, fault-tree analysis (FTA). "Significant" elements are important for the design itself. "Control" elements are regulated by the government for any public concern. A control plan is needed to mitigate the risks for the significant and critical elements. The team needs to develop proactive design recommendations. This information is carried to the PFMEA after causes have been generated in product projects. Process mappings are analogous to the block diagram in transactional DFSS projects.

5. *Potential causes.* Generally, these are the set of noise factors and the deficiencies designed in because of the violation of design axioms and best practices (e.g., inadequate assumptions). The study of the effect of noise factors helps the DFSS team identify the mechanism of failure. The analysis conducted by the DFSS team with the help of the block diagram allows identification of the interaction and coupling of their scoped project with the environment, with the customer, and within the DPs themselves. For each potential failure mode identified in column 2 of Fig. 11.2, the DFSS team needs to enter a cause in this column. See Sec. 11.3 for cause-and-effect tools linking FMEA columns 3 and 5. There are two basic causes: (1) the design is manufactured and assembled within specifications, (2) the design may include a deficiency or vulnerability that may cause unacceptable variation (misbuilds, errors, etc.), or (3) both.

6. *Occurrence. Occurrence* is the assessed cumulative subjective rating of the physical entities (parts/components or subsystems) as failures that could occur over the intended life of the design; in other words, it is the likelihood of the event "the cause occurs." FMEA usually assumes that if the cause occurs, so does the failure mode. On the basis of this assumption, occurrence is the likelihood of the failure mode also. Occurrence is rated on a scale of 1 (almost never) to 10 (almost certain), based on failure likelihood or probability, usually given in parts per million (ppm) defective. In addition to this subjective rating, a regression correlation models can be used. The occurrence rating is a ranking scale and does not reflect the actual likelihood. The actual likelihood or probability is based on the failure rate extracted from historical service or warranty data with the same parts or surrogate. See Table 11.2 for examples. In DFEMA, design controls help prevent or reduce the causes of failure modes, and the occurrence column will be revised accordingly.

7. *Current controls.* The objective of design controls is to identify and detect the design deficiencies and vulnerabilities as early as possible. Design controls are usually applied for *first-level failures.* A wide spectrum of controls is available, such as lab tests, project and design reviews, and design modeling (e.g., simulation, CAE). In the case of an

TABLE 11.2   Automotive Industry Action Group (AIAG) Occurrence Rating

Probability of failure	Occurrence	Rating
Very high—persistent failures	≥100 per 1000 vehicles/items (≥10%)	10
	50 per 1000 vehciles/items (5%)	9
High—frequent failures	20 per 1000 vehicles/items (2%)	8
	10 per 1000 vehicles/items (1%)	7
Moderate—occasional failures	5 per 1000 vehicles/items (0.5%)	6
	2 per 1000 vehicles/items (0.2%)	5
	1 per 1000 vehicles/items (0.1%)	4
Low—relatively few failures	0.5 per 1000 vehicles/items (0.05%)	3
	0.1 per 1000 vehicles/items (0.01%)	2
Remote—failure is unlikely	≤0.010 per 1000 vehicles/items (≤0.001%)	1

incremental DFSS project, the team should review relevant (similar failure modes and detection methods experienced on surrogate designs) historical information from the corporate memory such as lab tests, prototype tests, modeling studies, and fleet tests. In the case of creative design, the DFSS team needs to brainstorm new techniques for failure detection by asking: "In what means can they recognize the failure mode? In addition, how they can discover its occurrence?"

Design controls span a spectrum of different actions that include physical and process structure changes (without creating vulnerabilities), special controls, design guidelines, DOEs (design of experiments), design verification plans, durability, drawings, and modifications of standards, procedures, and best-practice guidelines.

8. *Detection. Detection* is a subjective rating corresponding to the likelihood that the detection method will detect the first-level failure of a potential failure mode. This rating is based on the effectiveness of the control system through related events in the design algorithm; hence, FMEA is a living document. The DFSS team should

- Assess the capability of each detection method and how early in the DFSS endeavor each method will be used.

- Review all detection methods in column 8 of Fig. 11.2 and condense the data on a detection rating.

- Rate the methods, selecting the lowest detection rating in case the methods tie.

See Table 11.3 for examples.

9. *Risk priority number* (RPN). This is the *product* of severity (column 4), occurrence (column 6) and detection (column 8) ratings. The range is between 1 and 1000. In addition to the product function, a weighted average of severity, detection, and occurrence is another method entertained, although on a small scale, to calculate RPN num-

TABLE 11.3   Automotive Industry Action Group (AIAG) Detection Rating

Detection	Likelihood of detection	Rating
Almost certain	Design control will almost certainly detect a potential cause/mechanism and subsequent failure mode	1
Very high	Very high chance design control will detect a potential cause/mechanism and subsequent failure mode	2
High	High chance design control will detect a potential cause/mechanism and subsequent failure mode	3
Moderately high	Moderately high chance design control will detect a potential cause/mechanism and subsequent failure mode	4
Moderate	Moderate chance design control will detect a potential cause/mechanism and subsequent failure mode	5
Low	Low chance design control will detect a potential cause/mechanism and subsequent failure mode	6
Very low	Very low chance design control will detect a potential cause/mechanism and subsequent failure mode	7
Remote	Remote chance design control will detect a potential cause/mechanism and subsequent failure mode	8
Very remote	Very remote chance design control will detect a potential cause/mechanism and subsequent failure mode	9
Absolute uncertainty	Design control will not and/or cannot detect a potential cause/mechanism and subsequent failure mode; or there is no design control	10

bers. RPN numbers are used to prioritize the potential failures. The severity, occurrence, and detection ratings are industry-specific and black belts should use their own company-adopted rating systems. Automotive industry ratings are summarized in Tables 11.1 to 11.3 (compiled AIAG ratings are shown in Table 11.4). The software FMEA is given in Table 11.5, while service FMEA ratings are given in Table 11.6.

10. *Actions recommended.* The DFSS team should select and manage recommended subsequent actions. That is where the risk of potential failures is high; an immediate control plan should be crafted to control the situation.

Over the course of the design project, the DFSS team should observe, learn, and update the FMEA as a dynamic living document. FMEA is not retrospective, but a rich source of information for corporate memory, including core design books. The DFSS team should document the FMEA and store it in a widely acceptable format in the company in both electronic and physical media.

11.3   Design FMEA (DFMEA)

The objective of DFMEA is to help the team design for Six Sigma by designing the failure modes out of their project. Ultimately, this objec-

TABLE 11.4   AIAG Compiled Ratings

Rating	Severity of effect	Likelihood of occurrence	Ability to detect
10	Hazardous without warning	Very high; failure is almost inevitable	Cannot detect
9	Hazardous with warning		Very remote chance of detection
8	Loss of primary function	High; repeated failures	Remote chance of detection
7	Reduced primary function performance		Very low chance of detection
6	Loss of secondary function	Moderate; occasional failures	Low chance of detection
5	Reduced secondary function performance		Moderate chance of detection
4	Minor defect noticed by most customers		Moderately high chance of detection
3	Minor defect noticed by some customers	Low; relatively few failures	High chance of detection
2	Minor defect noticed by discriminating customers		Very high chance of detection
1	No effect unlikely	Remote: failure is detection	Almost certain detection

tive will significantly improve the reliability of the design. Reliability, in this sense, can be defined simply as the quality of design (initially at Six Sigma level) over time.

The proactive use of DFMEA is a paradigm shift as this practice is seldom done or regarded as a formality. This attitude is very harmful as it indicates the ignorance of its significant benefits. Knowledge of the potential failure modes can be acquired from experience or discovered in the hands of the customer (field failures), or found in prototype testing. But the most leverage of the DFMEA is when the failure modes are proactively identified during the early stages of the project when it is still on paper.

The DFMEA exercise within the DFSS algorithm here is a function of the hierarchy identified in the physical structure. First, the DFSS team will exercise the DFMEA on the lowest hierarchical level (e.g., a component) and then estimate the effect of each failure mode at the next hierarchical level (e.g., a subsystem) and so on. The FMEA is a bottom-up approach, not a top-down one, and usually doesn't reveal all higher-level potential failures. However, this shortcoming is now fixed in the DFSS algorithm by utilizing the physical and process structures coupled with block diagrams as a remedy.

TABLE 11.5 The Software FMEA Rating

Rating	Severity of effect	Likelihood of occurrence	Detection
1	*Cosmetic error*—no loss in product functionality; includes incorrect documentation	1 per 100 unit-years (1/50M)	Requirements/design reviews
2	*Cosmetic error*—no loss in product functionality; includes incorrect documentation	1 per 10 unit-years (1/5M)	Requirements/design reviews
3	*Product performance reduction*—temporary; through timeout or system load the problem will "go away" after a period of time	1 per 1 unit-year (1/525K)	Code walkthroughs/unit testing
4	*Product performance reduction*—temporary; through timeout or system load the problem will "go away" after a period of time	1 per 1 unit-month (1/43K)	Code walkthroughs/unit testing
5	*Functional impairment/loss*—the problem will not resolve itself, but a "workaround" can temporarily bypass the problem area until fixed without losing operation	1 per week (1/10K)	System integration and test
6	*Functional impairment/loss*—the problem will not resolve itself, but a "workaround" can temporarily bypass the problem area until fixed without losing operation	1 per day (1/1440)	System integration and test
7	*Functional impairment/loss*—the problem will not resolve itself and no "workaround" can bypass the problem; functionality has been either impaired or lost but the product can still be used to some extent	1 per shift (1/480)	Installation and startup
8	*Functional impairment/loss*—the problem will not resolve itself and no "workaround" can bypass the problem; functionality has been either impaired or lost but the product can still be used to some extent	1 per hour (1/60)	Installation and startup
9	*Product halts/process taken down/reboot required*—the product is completely hung up, all functionality has been lost, and system reboot is required	1 per 10 min (1/10)	Detectable only once on line
10	*Product halts/process taken down/reboot required*—the product is completely hung up, all functionality has been lost, and system reboot is required	1+ per min (1/1)	Detectable only once on line

TABLE 11.6 Service FMEA Ratings

Severity		Occurrence		Detection	
Rating	Description	Rating	Description	Rating	Description
1	*Very low.* Unreasonable to expect this will be noticed in the process, or impact any process or productivity; nor negligible effect on product function. The customer will probably not notice the difference	1–2	*Remote*—probability of failure <0.015% of total (<150 ppm)	1–2	*Remote*—likelihood of defect being shipped is remote (<150 ppm)
2–3	*Low.* Very limited effect on local process, no downstream process impact; not noticeable to the system but slightly noticeable on product (subsystem and system)	3–5	*Low*—probability of failure from 0.015–0.375% of total (150–3750 ppm)	3–4	*Low*—likelihood of defect being shipped is low (151–750 ppm)
4–6	*Medium.* Effects will be throughout the process; may require unscheduled rework; may create minor damage to equipment; customer will notice immediately; effect on subsystem or product performance deterioration	6–7	*Moderate*—probability of failure from 0.375–1.5% of total (3751–15,000 ppm)	5–7	*Moderate*—likelihood of defect being shipped is moderate (751–15,000 ppm)
7–8	*High.* May cause serious disruptions to downstream process; major rework; equipment, tool, or fixture damage; effect on major product system but not on safety or government-regulated item	8–9	*High*-probability of failure from 1.5–7.5% of total (15,001–75,000 ppm)	8–9	*High*—likelihood of defect being shipped is high (15,001–75,000 ppm)
9–10	*Extreme.* Production shut down; injury or harm to process or assembly personnel; effect on product safety or involving noncompliance with government-regulated item	10	*Very high*-probability of failure >7.5% of total (>75,001 ppm)	10	*Very high*—likelihood of defect being shipped is very high (>100,000 ppm)

Before embarking on any FMEA exercise, we advise the black belt to book the FMEA series of meetings in advance, circulate all relevant information ahead of the FMEA meetings, clearly define objectives at the start of each meeting, adhere to effective roles, and communicate effectively.

The fundamental steps in the DFMEA to be taken by the DFSS team are

1. Constructing the project boundary (scope) as bounded by the physical structure. Components, subsystems, and systems are different hierarchical levels. The team will start at the lowest hierarchical level, the component level, and proceed upward. The relative information from the lower levels is inputted to the next higher level where appropriate in the respective columns.

2. Constructing the block diagram that fully describes coupling and interfaces at all levels of the hierarchy within the scope. Interfaces will include controlled inputs (the DPs) and uncontrolled input such as environmental factors, deterioration, manufacturing, and DFX methods.

3. Revisiting the physical structure at all hierarchical levels where the respective FRs are defined. The task here is to make sure that all DPs in the physical structure end up being hosted by some component or subsystem. The set of components constitute the bill of material of the project.

4. Identifying the potential failures for each hierarchical level in the structure. The team needs to identify all potential ways in which the design may fail. For each FR in the structure, the team will brainstorm the design failure modes. Failure modes describe how each hierarchical entity in the structure may initially fail prior to the end of its intended life. The potential design failure mode is the way in which a physical entity in the structure may fail to deliver its array of FRs.

5. Studying the failure causes and effects. The causes are generally categorized as weaknesses due to
   a. Design weakness because of axiom violation. In this type of failure causes, a component is manufactured and assembled to design specifications. Nevertheless, it still fails.
   b. Noise factors mean effects and their interaction with the DPs:
      (1) Unit-to-unit manufacturing and assembly vulnerabilities and deficiencies, such as a component not within specifications. Assembly errors, such as components manufactured to specifications but with attached assembly process error at the higher levels in the structure. In addition, material variation is fitted under this type.

(2) Environment and operator usage.

(3) Deterioration: wear over time.

The effect of a failure is a direct consequence of the failure mode on the next higher hierarchical level: the customer and regulations. Potential failure causes can be analyzed by tools such as fault-tree analysis (FTA), cause-and-effect diagram, and cause-and-effect matrix.

Two golden rules should be followed in cause identification: the team should start with modes with the highest severity rating and try to go beyond the first-, second-, or third-level cause.

6. Ranking of potential failure modes using the RPN numbers so that actions can be taken to address them. Each failure mode has been considered in terms of severity of the consequences, detection, and occurrence of its causes.

7. Classifying any special DPs as "critical" or "significant" characteristics that will require controls. When the failure mode is given a severity rating greater than specific critical rating, then a potential "critical" characteristic,* a DP, may exist. Critical characteristics usually affect safety or compliance to regulation. When a failure mode–cause combination has a severity rating in the some range below the "critical" threshold, then a potential "significant" characteristic, a DP, may exist. "Significant" implies significant to some CTSs in the QFD. Both types of classification are inputted to the PFMEA and are called "special" characteristics. Special characteristics require "special controls," requiring additional effort (administrative, measurement, overdesign, etc.) beyond the normal control. Robust design methodology is generally used to identify the "special" characteristics.

8. Deciding on "design controls" as the methods used to detect failure modes or causes. There are two types of controls:

   a. Those designed to prevent the occurrence of the cause or failure mechanism or failure mode and its effect. This control type also addresses reduction of the occurrence.

   b. Controls that address detection of cause, mechanism, or failure mode, by either analytical or physical methods, before the item is released to production.

9. Identification and management of corrective actions. On the basis of the RPN numbers, the team moves to decide on the corrective actions. The corrective-action strategy includes

---

*Manufacturing industry uses ∇, inverted delta, to indicate "critical" characteristics. The terms *critical* and *significant* were initially used in the automotive industry.

   a. Transferring the risk of failure to other systems outside the project scope
   b. Preventing failure altogether [e.g., design poka-yoke (error-proofing)]
   c. Mitigating risk of failure by
      (1) Reducing "severity" (altering or changing the DPs)
      (2) Reducing "occurrence" (decreasing complexity)
      (3) Increasing the "detection" capability (e.g., brainstorming sessions, concurrently, using top-down failure analysis such as FTA)
10. Review analysis, document, and update the DFMEA. The DFMEA is a living document and should be reviewed and managed on an ongoing basis. Steps 1 to 9 should be documented in the appropriate business publication media.

   The potential failure modes at any level can be brainstormed by leveraging existing knowledge such as engineering or architecture analysis, historical failure databases of similar design, possible designed-in errors, and physics of failures. For comprehensiveness and as a good practice, the black belt should instruct the DFSS team members to always maintain and update their specific list of failure modes.

   The understanding of safety-related and catastrophic failures can be enhanced by a fault-tree analysis (FTA), a top-down approach. FTA, like FMEA, helps the DFSS team answer the "What if?" questions. These tools deepen the understanding of the design team to their creation by identifying where and how failures may occur. In essence, FTA can be viewed as a mathematical model that graphically uses deductive logic gates (AND, OR, etc.) to combine events that can produce the failure or the fault of interest. The objective is to emphasize the lower-level faults that directly or indirectly contribute to high-level failures in the DFSS project structures. Facilitated by the structure's development, FTA needs to be performed as early as possible, in particular to safety-related failures as well as Design for Reliability (Chap. 10).

11.3.1 FTA example

In this example, the FTA will be applied to a vehicle headlamp. The electric circuit is very simple and includes the battery, the switch, the lamp itself, and the wire harness (Fig. 11.4). For simplicity, we will assume that the latter is reliable enough to be excluded from our study. We will also assume certain failure probabilities for some components. For a given time period, the probability of failure is the DPMO or the unreliability for the assigned distribution of failures (not necessarily normal). Such probabilities can be estimated from warranty and

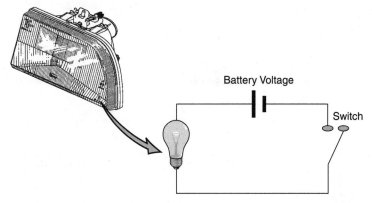

Figure 11.4 The vehicle headlamp circuit.

customer complaint databases that usually track failure rates. The probability of failure can be calculated by substituting the failure rate and the time of interest in the respective failure distribution. In this example, the following probabilities of failure will be assumed: $P_1 = 0.01$, $P_2 = 0.01$, $P_3 = 0.001$, and $P_5 = 0.02$. First, we need to define the high-level failure. In this case, it is the event "no light." The next step is to find the events that may cause such failure. We then proceed to identify three events that may cause the "no light" failure: "no power," "lamp failure," and "switch failure." Any failure event (or combination of these events) could cause the "no light" top failure; hence an OR logic gate is used. The FTA is given in Fig. 11.5.

From the theory of probability and assuming independence, we have

$$P_4 = P_1 + P_2 - P_1 \times P_2$$

$$= 0.0199 \tag{11.1}$$

$$P_6 = P_3 + P_4 + P_5 - P_3 \times P_4 - P_3 \times P_5 - P_4 \times P_5 + P_3 \times P_4 \times P_5$$

$$= 0.001 + 0.0199 + 0.02$$

$$- 0.001 \times 0.0199 - 0.001 \times 0.02 - 0.0199 \times 0.02$$

$$+ 0.001 \times 0.0199 \times 0.02$$

$$\cong 0.04046 \tag{11.2}$$

11.3.2 Cause-and-effect tools

The cause-and-effect diagram, also known as the "fishbone" or *Ishikawa* diagram, and the cause-effect matrix are two tools commonly

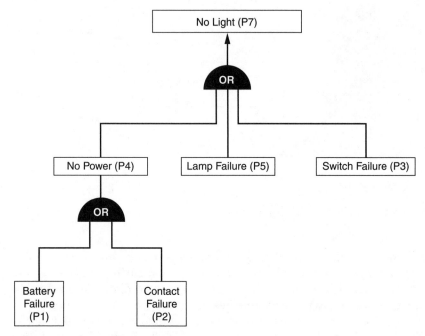

Figure 11.5   FTA of the vehicle headlamp.

used to help the DFSS team in their FMEA exercise. The cause-and-effect diagram classifies the various causes thought to affect the operation of the design, indicating with an arrow the cause-and-effect relation among them.

The diagram is formed from the causes that can result in the undesirable failure mode, which may have several causes. The causes are the independent variables, and the failure mode is the dependent variable. An example is depicted in Fig. 11.6. Failure of the assembly armplate of the press-fit pulley (Sec. 5.9.1) testing has generated scrap. Analysis of tested parts shows that the armplate that houses the torsion spring separates from the pivot tube, resulting in the disassembly.

The cause-and-effect matrix is another technique that can be used to identify failure causes. In the columns, the DFSS team can list the failure modes. The team then proceeds to rank each failure mode numerically using the RPN numbers. The team uses brainstorming to identify all potential causes that can impact the failure modes and list these along the left side of the matrix. It is useful practice to classify these causes as design weaknesses or noise factors by type (environment, wear, etc.). The team then rates, numerically, the effect of cause on each failure mode within the body of the matrix. This is based on the experience of the team and any available information. The team

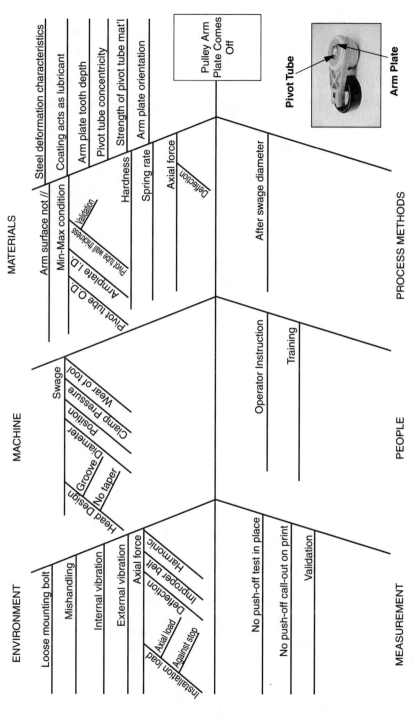

Figure 11.6 Pulley armplate disassembly failure mode.

then cross-multiply the rate of the effect by the RPN to total each row (cause). These totals are then used to analyze and prioritize where to focus the effort when creating the DFMEA. By grouping the causes according to their classification (coupling, environment, manufacturing, wear, etc.), the team will firm up several hypotheses about the strength of these types of causes to devise their attack strategy.

11.4   Process FMEA (PFMEA)

The activities in the PFMEA are similar to those of the DFMEA but with focus on process failures. The fundamental steps in the PFMEA to be taken by the DFSS team are

1. Constructing the project processes boundary (scope) as bounded by the process structure. The team can maximize its design quality by preventing all manufacturing, assembly, and production failures. Starting with the DFMEA at the lowest hierarchical level, the component level, the team should utilize the component process structure to map all the relevant processes. The team then proceeds to the next higher hierarchical, subsystem, level, finishes all subsystem PFMEAs, and proceeds upward.

2. Constructing the corresponding PFMEA process mapping as formed from step 1 above that fully describes coupling and interfaces at that level of the hierarchy within the process structure and scope. Interfaces will include controlled inputs (the PVs) and uncontrolled input noise such as manufacturing and assembly errors and variations. The team may start with macrolevel process mapping, but the maximum leverage is obtained at the micro mapping. Micromappings exhibit detailed operations, transportation, inspection stations, cycle times, and so on. The relative information from the lower levels is inputted to the next higher hierarchical PFMEA level, where appropriate, in the respective columns.

3. Revisiting the process structure at all hierarchical levels where the respective DPs are defined. The task here is to make sure that all PVs in the process structure end up being hosted by some process. The set of mutually exclusive processes constitute the production or manufacturing line of the project.

4. Identifying the potential failures for each hierarchical level in the process structure. Having gone through the corresponding hierarchical level DFMEA, the team needs to identify all potential ways in which the design may fail as a result of all process failures. For each design parameter in the structure, the team will brainstorm the process failure modes. Failure modes describe how each hierarchical entity in the structure may initially fail prior to the end

of its intended life. The potential process failure mode is the way in which a processed entity in the structure may fail to deliver its array of DPs.

5. Studying the failure causes and effects. The causes are generally categorized as weaknesses due to

   a. Process weakness because of axiom violation. In this cause of failure, a component is manufactured and assembled to process capability and design specifications. Nevertheless, it still can't be assembled or function as intended. Usually this will happen when the design was not conceived concurrently between design and manufacturing.

   b. Noise factors mean effects and their interaction with the PVs:

      (1) Manufacturing and assembly variation and deficiencies due mainly to incapable processes or material variation

      (2) Production environment and operator error

      (3) Machine deterioration: wear over time

   c. The effect of a failure is the direct consequence of the failure mode on the next higher hierarchical level processes, and ultimately the customer. Effects are usually noticed by the operator or the monitoring system at the concerned process or downstream from it. Potential failure causes can be analyzed at the process level by tools such as

      (1) Fault-tree analysis (FTA)

      (2) Cause-and-effect diagram

      (3) Cause-and-effect matrix

   d. Two golden rules should be followed in cause identification

      (1) The team should start with the modes with the highest severity ratings from the related DFMEA(s).

      (2) They should try to go beyond the first-level cause to second- or third-level causes. The team should ask the following process questions:

         (a) What incoming source of variations could cause this process to fail to deliver its array of DPs?

         (b) What could cause the process to fail, assuming that the incoming inputs are correct and to specifications?

         (c) If the process fails, what are the consequences on operator health and safety, machinery, the component itself, the next downstream processes, the customer, and regulations?

6. Ranking of potential process failure modes using the RPN numbers so that actions can be taken to address them. Each potential failure mode has been considered in terms of severity of its effect, detection likelihood, and occurrence of its causes.

7. Classifying any special PVs as "special" characteristics that will require controls such as "operator safety" characteristics as related to process parameters that do not affect the product but may impact safety or government regulations applicable to process operation. Another category of process "special" characteristics are the *high-impact characteristics,* which occur when out-of-specification tolerances severely affect operation or subsequent operations of the process itself but do not affect the component(s) or subsystem being processed. Both types of classification are inputted to the PFMEA and are called "special" characteristics.

8. Deciding on "process controls" as the methods to detect failure modes or the causes. There are two types of controls: (*a*) those designed to prevent the cause or failure mechanism or failure mode and its effect from occurring and (*b*) those addressing the detection of causes, or mechanisms, for corrective actions.

9. Identifying and managing of corrective actions. According to the RPN numbers, the team moves to decide on the corrective actions, as follows

   a. Transferring the risk of failure to other systems outside the project scope

   b. Preventing failure altogether [e.g., process poka-yoke (error-proofing)]

   c. Mitigating risk of failure by

   (1) Reducing "severity" (altering or changing the DPs)

   (2) Reducing "occurrence"

   (3) Increasing the "detection" capability (e.g., brainstorming sessions, concurrently using top-down failure analysis such as FTA)

   PFMEA should be conducted according to the process structure. It is useful to add the PFMEA and DFMEA processes to the design project management charts. The PERT or CPM approach is advisable. The black belt should schedule short meetings (less than 2 h) with clearly defined objectives. Intermittent objectives of an FMEA may include task time measurement system evaluation, process capability verifications, and conducting exploratory DOEs. These activities are resource- and time-consuming, introducing sources of variability to the DFSS project closure cycle time.

10. Review analysis, document, and update the PFMEA. The PFMEA is a living document and should be reviewed and managed on an ongoing basis. Steps 1 to 9 should be documented in the appropriate media.

An example of the time delivery distribution process PFMEA is depicted in Fig. 11.7.

Process Step/Input	Potential Failure Mode	Potential Failure Effects	SEV	Potential Causes	OCC	Current Controls	DET	RPN	Actions Recommended
1. Distribution	Warehouse shipping document does not print	Late delivery	10	Systems failure	1	Identified next day by customer service center	5	50	Open orders screen
2. Distribution	No transfer entered	Late delivery	10	Not authorized due to clerical error	1	Not controlled	1	10	Computer system includes load building
				Parameters in computer system are incorrect	5	Weekly procurement reports	5	250	Computer system includes load building
3. Distribution	Hub warehouse does not load correctly onto transfer carrier	Late delivery	10	Loading errors. Load not checked	1	Transfer report	5	50	Check load/ request for loading
				Phantom inventory (due to poor procedures & controls)	5	Transfer report	5	250	Daily inventory reconciliation
4. Distribution	Late delivery, spoke to customer	Late delivery	10	Matrix does not reflect realistic 'hub to spoke' transfer time	3	Transfer report	5	150	Validate matrix
			10	Poor performance by transfer carrier	5	Transfer report	5	250	Measure performance
5. Distribution	Late delivery spoke to customer	Late delivery	10	Wrong delivery information given to transfer carrier	1	Transfer report	5	50	Warehouse system creates bill of load
			10	Poor performance by carrier	5	Customer complaint	1	50	Monitor performance, change carrier
			10	Carrier is not aware of customer delivery date/ does not care	9	Carrier service contract, carrier reports	1	90	Record delivery date on bill of load for every shipment
			10	Wrong delivery information given to carrier	3	Not controlled	1	30	Warehouse system creates bill of load
Total			110		39		39	1230	

Figure 11.7 PFMEA of a distribution process.

11.5   Quality Systems and Control Plans

Control plans are the means to sustain any DFSS project findings. However, these plans are not effective if not implemented within a comprehensive quality operating system. A solid quality system can provide the means through which a DFSS project will sustain its long-term gains. Quality system certifications are becoming a customer requirement and a trend in many industries. The validate (V) phase of the ICOV DFSS algorithm requires that a solid quality system be employed in the DFSS project area.

The quality system objective is to achieve customer satisfaction by preventing nonconformity at all stages from design through service. A quality system is the Six Sigma–deploying company's agreed-on method of doing business. It is not to be confused with a set of documents that are meant to satisfy an outside auditing organization (i.e., ISO900x). In other words, a quality system represents the actions, not the written words, of a company. The elements of an effective quality system include quality mission statement, management reviews, company structure, planning, design control, data control, purchasing quality-related functions (e.g., supplier evaluation and incoming inspection), design product and process structure for traceability, process control, preventive maintenance, process monitoring and operator training, capability studies, measurement system analysis (MSA), audit functions, inspection and testing, service, statistical analysis, and standards.

Specifics from QS9000 as they apply to the DFSS project are found in QS9000 sec. 4.1.2.1: "define and document organizational freedom and authority to"

1. Initiate action to prevent the occurrence of any nonconformities relating to the product, process, and quality system.

2. Identify and record any problems relating to the product, process, and quality system.

3. Initiate, recommend, or provide solutions through designated channels.

4. Verify the implementation solutions.

11.5.1   Control methods

Automated or manual control methods are used for both design (service or product) and design processes. Control methods include tolerancing, errorproofing (poka-yoke), statistical process control (SPC)* charting

---

*Examples of SPC charting are X-bar and range (R) or X and moving range (MR) charts (manual or automatic), p & np charts (manual or automatic), and c & u charts (manual or automatic).

DFSS Team:

Date (Orig):

Date (Rev):

Current Control Plan

Process Step	Input	Output	Process Spec (LSL, USL, Target)	Cpk / Date (Sample Size)	Measurement System	%R&R or P/T	Current Control Method (from PFMEA)	Who?	Where?	When?	Reaction Plan?

Figure 11.8 Control plan worksheet.

with or without warning and trend signals applied to control the PVs or monitor the DPs, standard operating procedures (SOPs) for detection purposes, and short-term inspection actions. In applying these methods, the DFSS team should revisit operator training to ensure proper control functions and to extract historical long-term and short-term information.

Control plans are the living documents in the manufacturing, assembly, or production environment, which are used to document all process control methods as suggested by the FMEA or yielded by other DFSS algorithm steps such as optimization and robust design studies. The control plan is a written description of the systems for controlling parts and processes (or services). The control plan should be updated to reflect changes of controls on the basis of experience gained over time. A form is suggested in Fig. 11.8.

12

Fundamentals of Experimental Design

12.1 Introduction to Design of Experiments (DOE)

Design of experiments is also called *statistically designed experiments.* The purpose of the experiment and data analysis is to find the cause-and-effect relationship between the output and experimental factors in a process. The process model of DOE is illustrated in Fig. 12.1. This model is essentially the same as the P-diagram model discussed in Chap. 2.

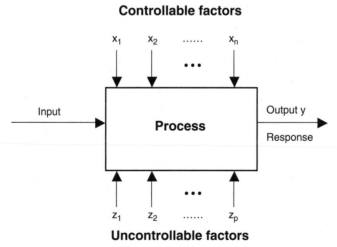

Controllable factors

x_1 x_2 $\cdots\cdots$ x_n

Input

Process

Output y

Response

z_1 z_2 $\cdots\cdots$ z_p

Uncontrollable factors

Figure 12.1 A process model.

Example 12.1. Agricultural Experiment Design of experiments was first developed as a research design tool to improve farm yields in the early 1930s. The output, or response, variable y in such an experiment was usually the yield of a certain farm crop. Controllable factors, such as $\mathbf{x} = (x_1, x_2,..., x_n)$, were usually the "farm variables," such as the amount of various fertilizers applied, watering pattern, and selection of seeds. Uncontrollable factors, such as $\mathbf{z} = (z_1, z_2,..., z_p)$, could be soil types, weather patterns, and so on. In early agricultural experiments, the experimenter would want to find the cause-and-effect relationship between the yield and controllable factors, specifically, how different types of fertilizers, application quantities, watering patterns, and types of seeds would influence crop yield.

In any DOE project, we will deliberately change those experimental factors and observe their effects on the output. The data obtained in the experiment will be used to fit empirical models relating output y with experimental factors. Mathematically, we are trying to find the following functional relationship:

$$y = f(x_1,x_2,...,x_n) + \varepsilon \qquad (12.1)$$

where ε is experimental error, or experimental variation. The existence of ε means that there may not be an exact functional relationship between y and $(x_1,x_2,...,x_n)$. This is because

1. Uncontrollable factors $(z_1,z_2,...,z_p)$ will influence the response y but are not accounted for in Eq. (12.1).

2. There are experimental and measurement errors on both y and $(x_1,x_2,...,x_n)$ in the experiment.

A DOE project will take many steps, described below.

Step 1: Project definition

This is the first but certainly not a trivial step. We need to identify the objective of the project and find the scope of the problem. For example, in a product design, we need to identify what we want to accomplish. Do we want to reduce defect? Do we want to improve the current product's performance? What is the performance? What is the project scope? Do we work on a subsystem or a component?

Step 2: Selection of response variable (output)

After project definition, we need to select the response variable y. In selecting this variable, the experimenter should determine if it could provide useful information about the process under study. Usually, the response variable is a key performance measure of the process. We would also want y to be

- A continuous variable, which would make data analysis much easier and meaningful
- A variable that can be easily and accurately measured

Step 3: Choice of factors, levels, and ranges

Actually, steps 2 and 3 can be done simultaneously. It is desirable to identify all the important factors which may significantly influence the response variable. Sometimes, the choice of factors is quite obvious, but in some cases a few very important factors are hidden.

There are two kinds of factors: the continuous factor and the discrete factor. A *continuous factor* can be expressed by continuous real numbers. For example, weight, speed, and price are continuous factors. A *discrete factor* is also called as *category variable,* or *attributes.* For example, type of machines, type of seed, and type of operating system are discrete factors.

In a DOE project, each experimental factor will be changed at least once; that is, each factor will have at least two settings. Otherwise, that factor will not be a variable but rather a fixed factor in the experiment. The numbers of settings of a factor in the experiment are called *levels.* For a continuous factor, these levels often correspond to different numerical values. For example, two levels of temperature could be given as 200 and 300°C. For continuous factors, the range of the variable is also important. If the range of variable is too small, then we may miss lots of useful information. If the range is too large, then the extreme values might give infeasible experimental runs. For a discrete variable, the number of levels is often equal to "the number of useful choices." For example, if the "type of machine" is the factor, then the number of levels depends on "How many types are there?" and "Which types do we want to test in this experiment?"

The choice of number of levels in the experiment also depends on time and cost considerations. The more levels we have in experimental factors, the more information we will get from the experiment, but there will be more experimental runs, leading to higher cost and longer time to finish the experiment.

Step 4: Select an experimental design

The type of experiment design to be selected will depend on the number of factors, the number of levels in each factor, and the total number of experimental runs that we can afford to complete.

In this chapter, we are considering primarily full factorial designs and fractional factorial designs. If the numbers of factors and levels are given, then a full factorial experiment will need more experimental runs, thus becoming more costly, but it will also provide more information about the process under study. The fractional factorial will need a smaller number of runs, thus costing less, but it will also provide less information about the process.

We will discuss how to choose a good experimental design in subsequent sections.

Step 5: Perform the experiment

When running the experiment, we must pay attention to the following:

- Check performance of gauges and/or measurement devices first.
- Check that all planned runs are feasible.
- Watch out for process drifts and shifts during the run.
- Avoid unplanned changes (e.g., swap operators at halfway point).
- Allow some time (and backup material) for unexpected events.
- Obtain buy-in from all parties involved.
- Preserve all the raw data
- Record everything that happens.
- Reset equipment to its original state after the experiment.

Step 6: Analysis of DOE data

Statistical methods will be used in data analysis. A major portion of this chapter discusses how to analyze the data from a statistically designed experiment.

From the analysis of experimental data, we are able to obtain the following results:

1. *Identification of significant and insignificant effects and interactions.* Not all the factors are the same in terms of their effects on the output. When you change the level of a factor, if its impact on the response is relatively small in comparison with inherited experimental variation due to uncontrollable factors and experimental error, then this factor might be insignificant. Otherwise, if a factor has a major impact on the response, then it might be a significant factor. Sometimes, two or more factors may interact, in which case their effects on the output will be complex. We will discuss interaction in subsequent sections. However, it is also possible that none of the experimental factors are found to be significant, in which case the experiment is inconclusive. This situation

may indicate that we may have missed important factors in the experiment. DOE data analysis can identify significant and insignificant factors by using analysis of variance.

2. *Ranking of relative importance of factor effects and interactions.* Analysis of variance (ANOVA) can identify the relative importance of each factor by giving a numerical score.

3. *Empirical mathematical model of response versus experimental factors.* DOE data analysis is able to provide an empirical mathematical model relating the output y to experimental factors. The form of the mathematical model could be linear or polynomial, plus interactions. DOE data analysis can also provide graphical presentations of the mathematical relationship between experimental factors and output, in the form of main-effects charts and interaction charts.

4. *Identification of best factor level settings and optimal output performance level.* If there is an ideal goal for the output, for example, if y is the yield in an agricultural experiment, then the ideal goal for y would be "the larger, the better." By using the mathematical model provided in paragraph 3, DOE data analysis is able to identify the best setting of experimental factors which will achieve the best possible result for the output.

Step 7: Conclusions and recommendations

Once the data analysis is completed, the experimenter can draw practical conclusions about the project. If the data analysis provides enough information, we might be able to recommend some changes to the process to improve its performance. Sometimes, the data analysis cannot provide enough information, in which case we may have to do more experiments.

When the analysis of the experiment is complete, we must verify whether the conclusions are good. These are called *confirmation runs*.

The interpretation and conclusions from an experiment may include a "best" setting to use to meet the goals of the experiment. Even if this "best" setting were included in the design, you should run it again as part of the confirmation runs to make sure that nothing has changed and that the response values are close to their predicted values.

In an industrial setting, it is very desirable to have a stable process. Therefore, one should run more than one test at the "best" settings. A minimum of three runs should be conducted. If the time between actually running the experiments and conducting the confirmation runs is more than a few hours, the experimenter must be careful to ensure that nothing else has changed since the original data collection.

If the confirmation runs don't produce the results you expected, then you need to

1. Check to see that nothing has changed since the original data collection.

2. Verify that you have the correct settings for the confirmation runs.

3. Revisit the model to verify the "best" settings from the analysis.

4. Verify that you had the correct predicted value for the confirmation runs.

If you don't find the answer after checking these four items, the model may not predict very well in the region that you decided was "best." However, you still will have learned something from the experiment, and you should use the information gained from this experiment to design another follow-up experiment.

12.2 Factorial Experiment

Most industrial experiments involve two or more experimental factors. In this case, factorial designs are the most frequently used designs. By a *factorial design,* we mean that all combinations of factor levels will be tested in the experiment. For example, we have two factors in the experiment, say, factors A and B. If A has a levels, B has b levels, then in a factorial experiment, we are going to test all ab combinations. In each combination, we may duplicate the experiment several times, say, n times. Then, there are n replicates in the experiment. If $n=1$, then we call it a *single replicate.* Therefore, for two factors, the total number of experimental observations is equal to abn.

Example 12.2 An article in *Industrial Quality Control* (1956, vol. 13, no. 1, pp. 5–8) describes an experiment to investigate the effect of the type of glass and the type of phosphor on the brightness of a television tube. The response variable is the current necessary (in microamps) to obtain a specific brightness level. The data are listed in Table 12.1.In this example, we can call the glass type *factor A,* so the number of A levels is $a = 2$; the phosphor type, *factor B,* with $b = 3$ as the number of B levels. The number of replicates in this example is $n = 3$. Total experimental observations = $2 \times 3 \times 3 = 18$ runs. Both glass and phosphor types are discrete factors. We would like to determine the following:

1. How the glass and phosphor types will affect the brightness of the TV, and whether these effects, if any, are significant, and whether there are any interactions.

2. Y is defined to be the current needed to achieve a certain brightness level, so the smaller current means higher efficiency. We would like to find a glass-phosphor combination that gives the best efficiency.

Data analysis for this example is given in Example 12.4.

TABLE 12.1   Data of Example
12.2

Glass	Phosphor type		
type	1	2	3
1	280	300	290
	290	310	285
	285	295	290
2	230	260	220
	235	240	225
	240	235	230

12.2.1   The layout of two-factor factorial experiments

In general, a two-factor factorial experiment has the arrangement shown in Table 12.2.

Each "cell" of Table 12.2 corresponds to a distinct factor level combination. In DOE terminology, it is called a "treatment."

12.2.2   Mathematical model

If we denote A as x_1 and B as x_2, then one possible mathematical model is

$$y = f_1(x_1) + f_2(x_2) + f_{12}(x_1, x_2) + \varepsilon \qquad (12.2)$$

Here $f_1(x_1)$ and $f_2(x_2)$ are the main effects of A and B, respectively, and $f_{12}(x_1, x_2)$ is the interaction of A and B.

In many statistics books, the following model is used

$$y_{ijk} = \mu + A_i + B_j + (AB)_{ij} + \varepsilon_{ijk} \qquad (12.3)$$

where
$$i = 1,2,\ldots,a$$
$$j = 1,2,\ldots,b$$
$$k = 1,2,\ldots,n$$
A_i = main effect of A at ith level
B_j = main effect of B at jth level
$(AB)_{ij}$ = interaction effect of A at ith level and B at jth level

12.2.3   What is interaction?

Let's look at Eq. (11.2) again; if there is no interaction, it will become

$$y = f_1(x_1) + f_2(x_2) + \varepsilon \qquad (12.4)$$

TABLE 12.2　General Arrangement for a Two-Factor Factorial Design

		Factor B			
		1	2	...	b
Factor A	1	$Y_{111}, Y_{112}, ..., Y_{11n}$	$Y_{121}, Y_{122}, ..., Y_{12n}$	...	
	2	$Y_{211}, Y_{212}, ..., Y_{21n}$	...		
	⋮	⋮			
	a				$Y_{ab1}, Y_{ab2}, ..., Y_{abn}$

where $f_1(x_1)$ is a function of x_1 alone and $f_1(x_2)$ is a function of x_2 alone; we call Eq. (12.4) the *additive model*. However, if the interaction effect is not equal to zero, then we do not have an additive model. Let's look at the following example.

Example 12.3　There are three different kinds of painkillers, say, A, B, and C. Taking each one of them will suppress the pain for a number of hours. If you take two different kinds of pills at the same time, then the results are as given in Table 12.3.

The values inside each table are the effective pain-suppressing hours. For example, in Table 12.3a, if you take one A pill alone, it can suppress the pain for 4 hours, if you take one B pill alone, it can suppress the pain for 2 hours, if you take one pill each, then it will suppress the pain for 6 hours. By simply plotting the response versus different factor level combination, we have the three interaction charts shown in Fig. 12.2.

Clearly, the effects of painkillers A and B are additive, since the effect of taking both A and B is equal to the summation of effects of taking A and B separately. The corresponding interaction chart is parallel. But for A and C or B and C, the effects are not additive. If the effect of taking both A and C together is more than the added effects of taking them separately, we call it *synergistic interaction*; if the effect of taking B and C together is less than the added effects of taking them separately, we call it *antisynergistic interaction*. In Fig. 12.2b and c, the corresponding interaction charts are not parallel.

12.2.4　Analysis of variance (ANOVA)

For any set of real experimental data, for example, the data from Table 12.2, the data most likely vary. (What would happen if all the data were the same?) Some variability of the data might be caused by changing of experimental factors and some might be due to unknown causes or experimental measurement errors. The ANOVA method attempts to accomplish the following:

1. Decompose the variation of your experimental data according to possible sources; the source could be the main effect, interaction, or experimental error.

TABLE 12.3 Interaction in a Factorial Experiment

(a)

		Number of B pills taken	
		0	1
Number of A pills taken	0	0	2
	1	4	6

(b)

		Number of C pills taken	
		0	1
Number of A pills taken	0	0	3
	1	4	10

(c)

		Number of C pills taken	
		0	1
Number of B pills taken	0	0	3
	1	2	3

2. Quantify the amount of variation due to each source.

3. Identify which main effects and interactions have significant effects on variation of data.

The first step of ANOVA is the "sum of squares" decomposition. Let's define

$$\overline{Y}_{i..} = \frac{\sum\limits_{j=1}^{b}\sum\limits_{k=1}^{n} y_{ijk}}{bn} \quad \text{(row average)}$$

$$\overline{y}_{.j.} = \frac{\sum\limits_{i=1}^{a}\sum\limits_{k=1}^{n} y_{ijk}}{an} \quad \text{(column average)}$$

$$\overline{y}_{ij.} = \frac{\sum\limits_{k=1}^{n} y_{ijk}}{n} \quad \text{(cell average)}$$

$$\overline{y}_{...} = \frac{\sum\limits_{i=1}^{a}\sum\limits_{j=1}^{b}\sum\limits_{k=1}^{n} y_{ijk}}{abn} \quad \text{(overall average)}$$

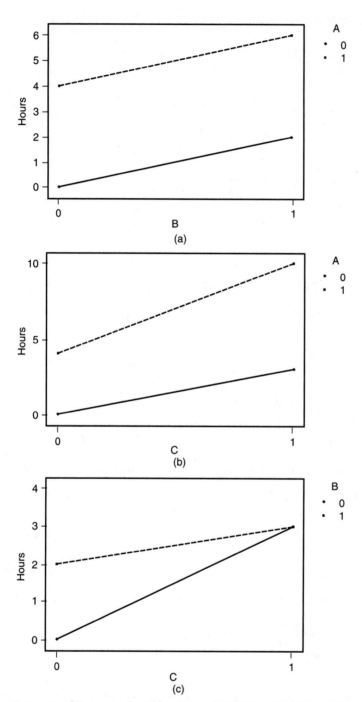

Figure 12.2 Interaction charts for (*a*) *A* and *B*, (*b*) *A* and *C*, (*c*) *B* and *C*.

It can be shown that

$$\sum_{i=1}^{a}\sum_{j=1}^{b}\sum_{k=1}^{n}(y_{ijk} - \bar{y}...)^2 = bn\sum_{i=1}^{a}(\bar{y}_{i..} - \bar{y}...)^2 + an\sum_{j=1}^{b}(\bar{y}_{.j.} - \bar{y}...)^2$$

$$+ n\sum_{i=1}^{a}\sum_{j=1}^{b}(\bar{y}_{ij.} - \bar{y}_{i..} - \bar{y}_{.j.} + \bar{y}...)^2$$

$$+ \sum_{i=1}^{a}\sum_{j=1}^{b}\sum_{k=1}^{n}(y_{ijk} - \bar{y}_{ij.})^2 \qquad (12.5)$$

or simply

$$SS_T = SS_A + SS_B + SS_{AB} + SS_E \qquad (12.6)$$

where SS_T is the *total sum of squares,* which is a measure of the total variation in the whole data set; SS_A is the *sum of squares* due to A, which is a measure of total variation caused by main effect of A; SS_B is the sum of squares due to B, which is a measure of total variation caused by main effect of B; SS_{AB} is the sum of squares due to AB, which is the measure of total variation due to AB interaction; and SS_E is the sum of squares due to error, which is the measure of total variation due to error.

In statistical notation, the number of degree of freedom associated with each sum of squares is as shown in Table 12.4.

Each sum of squares divided by its degree of freedom is a mean square. In analysis of variance, mean squares are used in the F test to see if the corresponding effect is statistically significant. The complete result of an analysis of variance is often listed in an ANOVA table, as shown in Table 12.5.

In the F test, the F_0 will be compared with F-critical values with the appropriate degree of freedom; if F_0 is larger than the critical value, then the corresponding effect is statistically significant. Many statistical software programs, such as MINITAB, are convenient for analyzing DOE data.

TABLE 12.4   Degree of Freedom for Two-Factor Factorial Design

Effect	Degree of freedom
A	$a - 1$
B	$b - 1$
AB interaction	$(a - 1)(b - 1)$
Error	$ab(n - 1)$
Total	$abn - 1$

TABLE 12.5   ANOVA Table

Source of variation	Sum of squares	Degree of freedom	Mean squares	F_0
A	SS_A	$a - 1$	$MS_A = \dfrac{SS_A}{a - 1}$	$F_0 = \dfrac{MS_A}{MS_E}$
B	SS_B	$b - 1$	$MS_B = \dfrac{SS_B}{b - 1}$	$F_0 = \dfrac{MS_B}{MS_E}$
AB	SS_{AB}	$(a - 1)(b - 1)$	$MS_{AB} = \dfrac{SS_{AB}}{(a - 1)(b - 1)}$	$F_0 = \dfrac{MS_{AB}}{MS_E}$
Error	SS_E	$ab(n - 1)$		
Total	SS_T	$abn - 1$		

Example 12.4. Data Analysis of Example 12.2 The data set of Example 12.2 was analyzed by MINITAB, and we have the following results:

```
Analysis of Variance for y, using Adjusted SS for Tests
Source          DF   Seq SS   Adj SS   Adj MS      F      P
Glass            1  14450.0  14450.0  14450.0  273.79  0.000
Phosphor         2    933.3    933.3    466.7    8.84  0.004
Glass*Phosphor   2    133.3    133.3     66.7    1.26  0.318
Error           12    633.3    633.3     52.8
Total           17  16150.0
```

How to use the ANOVA table. In Example 12.4, there are three effects: glass, phosphor, and glass-phosphor interaction. In some sense, the larger the sum of squares and the more variation is caused by that effect, the more important that effect is. In Example 12.4, the sum of square for glass is 14450.0, which is by far the largest. However, if different effects have different degrees of freedom, then the results might be skewed. The F ratio is a better measure of relative importance. In this example, the F ratio for glass is 14450.0; for phosphor, 8.84; and for glass-phosphor interaction, 1.26. So, clearly, glass is the most important factor. In DOE, we usually use the p value to determine whether an effect is statistically significant. The most commonly used criterion is to compare the p value with 0.05, or 5%, if p value is less than 0.05, then that effect is significant. In this example, the p value for glass is 0.000, and for phosphor, is 0.004, both are smaller than 0.05, so the main effects of both glass and phosphor are statistically significant. But for glass-phosphor interaction, the p value is 0.318, which is larger than 0.05, so this interaction is not significant.

From the interaction chart in Fig. 12.3, it is clear that two lines are very close to parallel, so there is very little interaction effect.

In Fig. 12.4 we can clearly see that glass type 2 gives a much lower current. For phosphor, type 3 gives the lowest current. Overall, for achieving the lowest possible current (maximum brightness), glass type 2 and phosphor type 3 should be used.

12.2.5   General full factorial experiments

The results and data analysis methods discussed above can be extended to the general case where there are a levels of factor A, b levels of factor B, c levels of factor C, and so on, arranged in a factorial experiment. There will be $abc\cdots n$ total number of trials if there are n replicates. Clearly, the number of trials needed to run the experiment will increase very rapidly with increase in the number of factors and the number of levels. In practical applications, we rarely use general full factorial experiments for more than two factors; two-level factorial experiments are the most popular experimental methods.

12.3   Two-Level Full Factorial Designs

The most popular experimental designs are *two-level factorial designs,* factorial designs in which all factors have exactly two levels. These designs are the most popular designs because

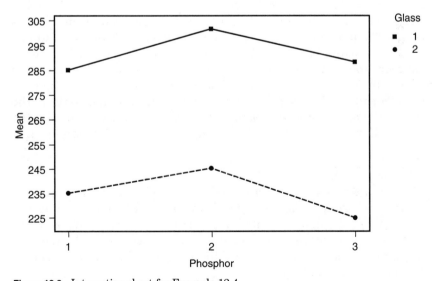

Figure 12.3 Interaction chart for Example 12.4.

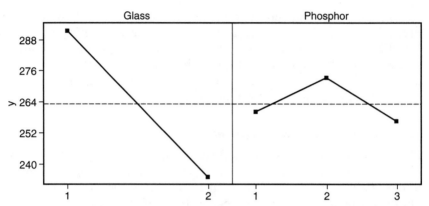

Figure 12.4   Main-effects chart for Example 12.4.

1. Two levels for each factor will lead to factorial designs with the least number of runs, so they will lead to the most economical experiments.

2. Two-level factorial designs will be ideal designs for screening experiments.

3. Two-level factorial designs are the basis of the fractional two-level factorial designs. They are the most flexible, efficient, and economical experimental designs. In practical applications of DOE, fractional factorial designs are the most frequently used designs.

A two-level full factorial design is also called a "2^k design," where k is the number of experimental factors and 2 means two levels. This is because the number of treatment combinations in a two-level full factorial of k factors is $2 \times 2 \cdots 2 = 2^k$. If there are n replicates at each treatment combination, then the total number of experimental trials is $2^k n$. Because there are only two levels for each factor, we call the level with low setting the *low level,* and the level with the high setting, the *high level.* For example, if a factor is temperature, with two levels, 100 and 200°C, then 100°C is the low level and 200°C is the high level.

12.3.1   Notation for two-level designs

The standard layout for a two-level design uses $+1$ and -1 notation to denote the "high level" and the "low level," respectively, for each factor. For example, the following matrix describes an experiment in which four trials (or runs) were conducted with each factor set to high or low during a run according to whether the matrix had a $+1$ or -1 set for the factor during that trial:

Run number	$A(X_1)$	$A(X_2)$
1	−1	−1
2	+1	−1
3	−1	+1
4	+1	+1

The use of +1 and −1 for the factor settings is called *coding* the data. This aids in the interpretation of coefficients fit to any experimental model.

Example 12.5 A router is needed to cut location notches on a printed-circuit board (PCB). The cutting process creates vibration on the board and causes the dimensional variation of the notch position. An experimental study is conducted to identify the cause of vibration. Two factors, bit size (A) and cutting speed (B), are thought to influence the vibration. Two bit sizes (1/16 and 1/8 in) and two speeds [40 and 90 rpm (r/min)] are selected, and four boards are cut at each set of conditions shown in Table 12.6. The response variable is the vibration measured on each test circuit board.

12.3.2 Layout of a general 2^k design

If the experiment had more than two factors, there would be an additional column in the matrix for each additional factor. The number of distinct experimental runs is $N = 2^k$. For example, if $k = 4$, then, $N = 2^4 = 16$.

Table 12.7 gives a *standard layout* for a 2^4 factorial experiment. The run number is sequenced by *standard order*, which is featured by a −1 1 − 1 1 ⋯ sequence for A, −1 −1 11 for B, four −1s and four 1s for C, and so on. In general, for a two-level full factorial with k factors, the first column starts with −1 and alternates in sign for all 2^k runs, and the second column starts with −1 repeated twice, and then alternates with 2 in a row of the opposite sign until all 2^k places are filled. The third column starts with −1 repeated 4 times, then 4 repeats of +1s

TABLE 12.6 Experiment Layout and Data for Example 12.5

Run number	A (bit size)	B (cutting speed)	Y [vibration (four replicates, $n = 4$)]			
			1	2	3	4
1	−1	−1	18.2	18.9	12.9	14.4
2	+1	−1	27.2	24.0	22.4	22.5
3	−1	+1	15.9	14.5	15.1	14.2
4	+1	+1	41.0	43.9	36.3	39.9

*Continued in Sec. 12.3.3.

TABLE 12.7   Experimental layout for a 2^4 Design

Run number	Factors				Response (with replicates)			Total (computed by adding responses for each row)
	A	B	C	D	1	...	n	
1	-1	-1	-1	-1				(1)
2	1	-1	-1	-1				a
3	-1	1	-1	-1				b
4	1	1	-1	-1				ab
5	-1	-1	1	-1				c
6	1	-1	1	-1				ac
7	-1	1	1	-1				bc
8	1	1	1	-1				abc
9	-1	-1	-1	1				d
10	1	-1	-1	1				ad
11	-1	1	-1	1				bd
12	1	1	-1	1				abd
13	-1	-1	1	1				cd
14	1	-1	1	1				acd
15	-1	1	1	1				bcd
16	1	1	1	1				$abcd$

and so on; the ith column starts with 2^{i-1} repeats of -1 followed by 2^{i-1} repeats of $+1$, and so on.

There could be n replicates; when $n = 1$, it is called *single replicate*. Each run can also be represented by the symbols in the last column of the table (Table 12.7), where the symbol depends on the corresponding levels of each factor; for example, for run 2, A is at high level (1); B, C, and D are at low level, so the symbol is a, meaning that only A is at high level. For run 15, B, C, and D are at high level, so we use bcd; for the first run, all factors are at low level, so we use high level (1) here, where (1) means all factors are at low level. In data analysis, we need to compute the *total* for each run, which is the sum of all replicates for that run. We often use those symbols to represent those totals.

12.3.3   Data analysis steps for two-level full factorial experiment

For a 2^k full factorial experiment, the numerical calculations for ANOVA, the main-effects chart, the interaction chart, and the mathematical model become easier, in comparison with general full factorial experiment. In the following paragraphs we give a step-by-step procedure for the entire data analysis.

Step 0: Preparation

Establish analysis matrix for the problem. The *analysis matrix* is a matrix that has not only all columns for factors but also the columns for all interactions. The interaction columns are obtained by multiplying the corresponding columns of factors involved. For example, in a 2^2 experiment, the analysis matrix is as follows, where the AB column is generated by multiplying the A and B columns:

Run number	A	B	AB
1	-1	-1	$(-1)^*(-1)=+1$
2	$+1$	-1	$(+1)^*(-1)=-1$
3	-1	$+1$	$(-1)^*(+1)=-1$
4	$+1$	$+1$	$(+1)^*(+1)=+1$

Attach experimental data on the analysis matrix. In Table 12.8 we use the data in Example 12.5 to illustrate this attachment.

Step 1: Compute contrasts. The vector of column coefficients multiplying the vector of *totals* computes a *contrast*. In Table 12.8, the column coefficients for A (second column) is $(-1, +1, -1, +1)$ and the vector of the total is $[(1), a, b, ab] = (64.4, 96.1, 59.7, 161.1)$. Therefore

$$\text{Contrast}_A = -(1) + a - b + ab = -64.4 + 96.1 - 59.7 + 161.1 = 133.1$$

Similarly

$$\text{Contrast}_B = -(1) - a + b + ab = -64.4 - 96.1 + 59.7 + 161.1 = 60.3$$

$$\text{Contrast}_{AB} = (1) - a - b + ab = 64.4 - 96.1 - 59.7 + 161.1 = 69.7$$

Contrasts are the basis for many subsequent calculations.

Step 2: Compute effects. *Effects* include both main effects and interaction effects. All effects are computed by the following formula:

TABLE 12.8   Analysis Matrix and Data for Example 11.5

Run number	Effects			Responses				Total
	A	B	AB	1	2	3	4	
1	-1	-1	$+1$	18.2	18.9	12.9	14.4	$(1) = 64.4$
2	$+1$	-1	-1	27.2	24.0	22.4	22.5	$a = 96.1$
3	-1	$+1$	-1	15.9	14.5	15.1	14.2	$b = 59.7$
4	$+1$	$+1$	1	41.0	43.9	36.3	39.9	$ab = 161.1$

$$\text{Effect} = \frac{\text{contrast}}{2^{k-1} \times n} = \frac{\text{contrast}}{N \times n \div 2} \qquad (12.7)$$

where N is the total number of runs. The definition for any main effect, for example, main effect of A, is

$$A = \bar{y}_{A+} - \bar{y}_{A-} \qquad (12.8)$$

which is the average response for A at high level minus the average of response for A at low level.

By Eq. (12.7)

$$A = \frac{\text{contrast}_A}{2^{k-1} \times n} = \frac{133.1}{2^{2-1} \times 4} = 16.63$$

Similarly

$$B = \frac{\text{contrast}_B}{2^{k-1} \times n} = \frac{60.3}{2 \times 4} = 7.54$$

$$AB = \frac{\text{contrast}_{AB}}{2^{k-1} \times n} = \frac{69.7}{2^{2-1} \times 4} = 8.71$$

Step 3: Compute sum of squares. *Sum of squares* (SS) is the basis for the analysis of variance computation; the formula for the sum of squares is

$$SS = \frac{\text{contrast}^2}{2^k \times n} = \frac{\text{contrast}^2}{N \times n} \qquad (12.9)$$

Therefore

$$SS_A = \frac{\text{contrast}_A^2}{2^2 \times n} = \frac{133.1^2}{4 \times 4} = 1107.22$$

$$SS_B = \frac{\text{contrast}_B^2}{2^2 \times n} = \frac{60.3^2}{4 \times 4} = 227.25$$

$$SS_{AB} = \frac{\text{contrast}_{AB}^2}{2^2 \times n} = \frac{69.7^2}{4 \times 4} = 303.6$$

To complete ANOVA, we also need SS_T and SS_E. In two-level factorial design

$$SS_T = \sum_{i=1}^{2} \sum_{j=1}^{k} \sum_{k=1}^{n} y_{ijk}^2 - \frac{y_{...}^2}{N \times n} \qquad (12.10)$$

where y_{ijk} is actually each individual response and $y_{...}$ is the sum of all individual responses.

In Example 12.5

$$SS_T = 18.2^2 + 18.9^2 + \cdots + 39.9^2$$

$$-\frac{(18.2 + 18.9 + \cdots + 39.9)^2}{16} = 1709.83$$

SS_E can be calculated by

$$SS_E = SS_T - SS_A - SS_B - SS_{AB}$$

In Example 12.5

$$SS_E = SS_T - SS_A - SS_B - SS_{AB} = 1709.83 - 1107.22 - 227.25 -$$
$$- 303.6 = 71.72$$

Step 4: Complete ANOVA table. The ANOVA table computation is the same as that of general factorial design. MINITAB or other statistical software can calculate the ANOVA table conveniently. In the example above, the ANOVA table computed by MINITAB is as follows:

```
Analysis of Variance for y, using Adjusted SS for Tests
Source   DF     Seq SS     Adj SS     Adj MS       F       P
A         1    1107.23    1107.23    1107.23  185.25   0.000
B         1     227.26     227.26     227.26   38.02   0.000
A*B       1     303.63     303.63     303.63   50.80   0.000
Error    12      71.72      71.72       5.98
Total    15    1709.83
```

Clearly, both main effects, A and B, as well as interaction AB, are all statistically significant.

Step 5: Plot main-effects and interaction charts for all significant effects. For any main effect, such as main-effect A, the main-effects plot is actually the plot of $\bar{y}_{A-}$ and $\bar{y}_{A+}$ versus the levels of A. The interaction chart is plotted by charting all combinations of $\bar{y}_{A-B-}$, $\bar{y}_{A-B+}$, $\bar{y}_{A+B-}$ and $\bar{y}_{A+B+}$. For Example 12.5, the main-effects plot and interaction chart are shown in Figs. 12.5 and 12.6, respectively.

Step 6: Establish a mathematical model. We can establish a regression model for the data. Here are the rules:

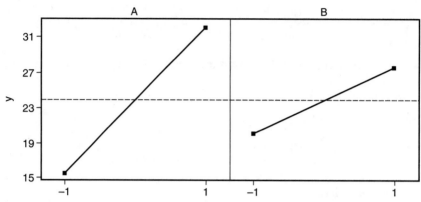

Figure 12.5 Main-effects chart of Example 12.5—LS (least-squares) means for *y*.

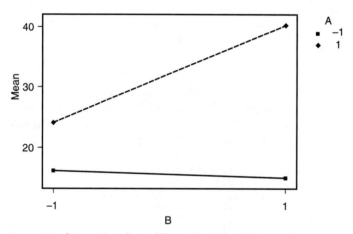

Figure 12.6 Interaction chart of Example 12.5—LS means for *y*.

1. Only significant effects are included in the model. For Example 12.5, since *A*, *B*, and *AB* are all significant, they are all included.

2. Usually, we use x_1 to express *A*, x_2 for *B*, x_3 for *C*, and so on and x_1x_2 for *AB* interaction, x_1x_3 for *AC* interaction, $x_1x_2x_3$ for *ABC* interaction, and so on.

The model for our example is

$$Y = \beta_0 + \beta_1 X_1 + \beta_2 X_2 + \beta_{12} X_1 X_2 + \text{experimental error} \qquad (12.11)$$

where β_0 = average of all responses and other β_i = effect/2. For example, $\beta_1 = A/2 = 16.63/2 = 8.31$, and for Example 12.5

$$Y = 23.8313 + 8.31x_1 + 3.77x_2 + 4.36x_1x_2$$

where x_1 and x_2 are coded values.

Step 7: Determine optimal settings. Depending on the objective of the problem, we can determine the optimum setting of the factor levels by examining the main-effects chart and interaction chart; if there is no interaction, the optimal setting can be determined by looking at one factor at a time. If there are interactions, then we have to look at the interaction chart. For the problem above, since AB interaction is significant, we have to find optimal by studying the AB interaction. From the interaction chart, if the vibration level is "the smaller, the better," then A at low level and B at high level will give the lowest possible vibrations.

12.3.4 2^3 factorial experiment

Consider the two-level, full factorial design for three factors, namely, the 2^3 design. This design has 8 runs. Graphically, we can represent the 2^3 design by the cube shown in Fig. 12.7. The arrows show the direction of increase of the factors. The numbers 1 through 8 at the corners of the design box reference the *standard order* of runs.

Example 12.6.   A 2^3 experiment   An experiment is conducted to determine the effects of three factors—holding pressure, booster pressure, and screw speed—on the part shrinkage in an injection-molding process. The experimental layout and results are given in Table 12.9.

Now we will carry out the step-by-step procedure to conduct data analysis, with the help of MINITAB. The analysis matrix is presented in Table 12.10. The block with the 1s and −1s is called the *analysis matrix*. The table formed by columns A, B, and C is called the *design matrix*.

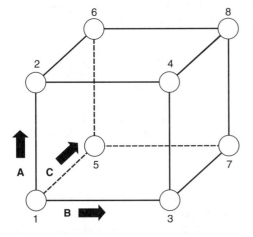

Figure 12.7 A 2^3 two-level, full factorial design; factors A,B,C.

TABLE 12.9   Experiment Layout and Data for Example 12.6

Run number	Factors			Response (part shrinkage)	
	A (booster pressure)	B (booster pressure)	C (screw speed)	1	2
1	−1	−1	−1	21.9	20.3
2	1	−1	−1	15.9	16.7
3	−1	1	−1	22.3	21.5
4	1	1	−1	17.1	17.5
5	−1	−1	1	16.8	15.4
6	1	−1	1	14.0	15.0
7	−1	1	1	27.6	27.4
8	1	1	1	24.0	22.6

TABLE 12.10   Analysis Matrix for a 2^3 Experiment

Analysis matrix								Response variables	
I	A	B	AB	C	AC	BC	ABC	1	2
+1	−1	−1	+1	−1	+1	+1	−1	21.9	20.3
+1	+1	−1	−1	−1	−1	+1	+1	15.9	16.7
+1	−1	+1	−1	−1	+1	−1	+1	22.3	21.5
+1	+1	+1	+1	−1	−1	−1	−1	17.1	17.5
+1	−1	−1	+1	+1	−1	−1	+1	16.8	15.4
+1	+1	−1	−1	+1	+1	−1	−1	14.0	15.0
+1	−1	+1	−1	+1	−1	+1	−1	27.6	27.4
+1	+1	+1	+1	+1	+1	+1	+1	24.0	22.6

In this problem, there are three main effects, A,B,C, and 3 two-factor interactions, AB, AC, and BC, and 1 three-factor interaction, ABC. By using MINITAB, we obtain the following ANOVA table; clearly, main effects A, B, and C are significant, and so are interactions BC and AC:

```
Analysis of Variance for Shrinkage, using Adjusted SS for Tests
Source    DF     Seq SS      Adj SS      Adj MS       F        P
A          1     57.760      57.760      57.760    103.14    0.000
B          1    121.000     121.000     121.000    216.07    0.000
C          1      5.760       5.760       5.760     10.29    0.012
A*B        1      1.440       1.440       1.440      2.57    0.147
A*C        1      3.240       3.240       3.240      5.79    0.043
B*C        1     84.640      84.640      84.640    151.14    0.000
A*B*C      1      1.960       1.960       1.960      3.50    0.098
Error      8      4.480       4.480       0.560
Total     15    280.280
```

MINITAB can plot the Pareto chart for effects, which gives very good ideas about the relative importance of each effect (see Fig. 12.8). For this example, the most dominant effects are B, BC, and A. (See main-effects plot in Fig. 12.9 and interaction chart in Fig. 12.10.)

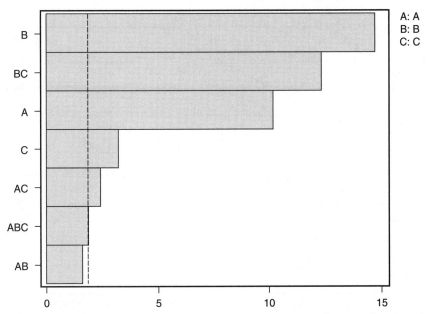

Figure 12.8 Pareto chart of the standardized effects for Example 12.6. (Response is shrinkage, alpha = 0.10.)

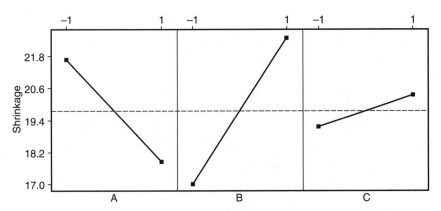

Figure 12.9 Main-effects chart (data means) for shrinkage for Example 12.6.

MINITAB can also generate a *cube plot* (see Fig. 12.11), which is very useful when three-factor interaction is significant.

In the injection-molding process, the smaller the part shrinkage and the less deformation of the part, the better quality the product will be. Since there are significant interactions, *AC* and *BC*, to ensure minimal shrinkage, we can find the optimal setting in the interaction chart, or cube plot: *A* at high level, *B* at low level, and *C* at high level.

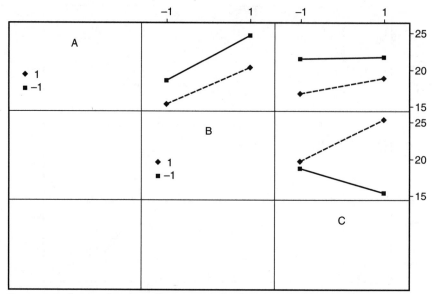

Figure 12.10   Interaction chart (data means) for shrinkage for Example 12.6.

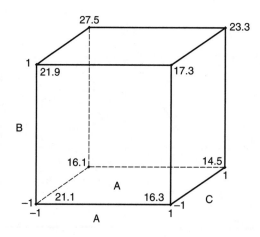

Figure 12.11 Cube plot (data means) for shrinkage for Example 12.6.

If all effects are significant, then the full model of this problem is

$$= \beta_0 + \beta_1 X_1 + \beta_2 X_2 + \beta_3 X_3$$

$$+ \beta_{12} X_1 X_2 + \beta_{23} X_2 X_3 + \beta_{13} X_1 X_3$$

$$+ \beta_{123} X_1 X_2 X_3 \qquad (12.12)$$

However, since AB and ABC are not significant, we need to estimate only the following reduced model:

$$y = \beta_0 + \beta_1 x_1 + \beta_2 x_2 + \beta_3 x_3 + \beta_{13} x_1 x_3 + \beta_{23} x_2 x_3 + \varepsilon \qquad (12.13)$$

From MINITAB, this is

$$Y = 19.75 - 1.90x_1 + 2.75x_2 + 0.60x_3 + 0.45x_1 x_3 + 2.30x_2 x_3$$

12.3.5 Full factorial designs in two levels

If there are k factors, each at two levels, a full factorial design has 2^k runs.

As shown in Table 12.11, when the number of factors is five or greater, a full factorial design requires a large number of runs and is not very efficient. A fractional factorial design or a Plackett-Burman design is a better choice for five or more factors.

12.4 Fractional Two-Level Factorial Design

As the number of factors k increase, the number of runs specified for a full factorial can quickly become very large. For example, when $k = 6$, then $2^6 = 64$. However, in this six-factor experiment, there are six main effects, say, $A,B,C,D,E,F,$ 15 two-factor interactions, $AB,AC,AD,AE,AF,BC,BD,BE,BF,CD,CE,CF,DE,DF,EF,$ 20 three-factors interactions, $ABC,ABD,...,$ 15 four-factor interactions, $ABCD,...,$ 6 five-factor interactions, and one six-factor interaction.

During many years of applications of factorial design, people have found that higher-order interaction effects (i.e., interaction effects involving three or more factors) are very seldom significant. In most experimental case studies, only some main effects and two-factor interactions are significant. However, in the 2^6 experiment above, out of 63 main effects and interactions, 42 of them are higher-order interactions, and only 21 of them are main-effects and two-factor interactions. As k increases, the overwhelming proportion of effects in the full factorials will be higher-order interactions. Since those effects are most likely to be insignificant, a lot of information in full factorial

TABLE 12.11 Number of Runs for a 2^k Full Factorial

Number of factors	Number of runs
2	4
3	8
4	16
5	32
6	64
7	128

experiments is wasted. In summary, for full factorial experiments, as the number of factors k increases, the number of runs will increase at an exponential rate that leads to extremely lengthy and costly experiments. On the other hand, as k increases, most of data obtained in the full factorial are used to estimate higher-order interactions, which are most likely to be insignificant.

Fractional factorial experiments are designed to greatly reduce the number of runs and to use the information from experimental data wisely. Fractional experiments run only a fraction of the runs of a full factorial; for two-level experiments, they use only $1/2$, $1/4$, $1/8$,... of runs from a full factorial. Fractional factorial experiments are designed to estimate only the main effects and two-level interactions, and not three-factor and other higher-order interactions.

12.4.1   A 2^{3-1} design (half of a 2^3)

Consider a two-level, full factorial design for three factors, namely, the 2^3 design. Suppose that the experimenters cannot afford to run all 2^3 = 8 treatment combinations, but they can afford four runs. If a subset of four runs is selected from the full factorial, then it is a 2^{3-1} design.

Now let us look at Table 12.12, where the original analysis matrix of a 2^3 design is divided into two portions.

In Table 12.12 we simply rearrange the rows such that the highest interaction, ABC contrast coefficients, are all $+1$s in the first four rows and all -1s in the second four rows. The second column in this table is called the *identity column,* or *I column,* because it is a column with all $+1$s.

If we select the first four runs as our experimental design, this is called a *fractional factorial* design with the defining relation $I = ABC$, where ABC is called the *generator.*

TABLE 12.12 2^{3-1} Design

Treatment combination	Factorial effects							
	I	A	B	C	AB	AC	BC	ABC
a	+1	+1	−1	−1	−1	−1	+1	+1
b	+1	−1	+1	−1	−1	+1	−1	+1
c	+1	−1	−1	+1	+1	−1	−1	+1
abc	+1	+1	+1	+1	+1	+1	+1	+1
ab	+1	+1	+1	−1	+1	−1	−1	−1
ac	+1	+1	−1	+1	−1	+1	−1	−1
bc	+1	−1	+1	+1	−1	−1	+1	−1
(1)	+1	−1	−1	−1	+1	+1	+1	−1

In Table 12.12 we can find that since all the contrast coefficients for *ABC* are +1s, we will not be able to estimate the effect of *ABC* at all. However, for other main effects and interactions, the first four runs have equal numbers of +1s and −1s, so we can calculate their effects. However, we can find that the contrast coefficients of *A* are identical as those of *BC*, and the contrast coefficients of *B* are exactly the same as those of *AC*, as well as *C* and *AB*. Since the effects are computed using the contrast coefficient, there is no way to distinguish the effects of *A* and *BC*, *B* and *AC*, and *C* and *AB*. For example, when we estimate the effect of *A*, we are really estimating the combined effect of *A* and *BC*. This mixup of main effects and interactions is called *aliases* or *confounding*.

All alias relationships can be found from the defining relation: $I = ABC$. If we simply multiply *A* on both sides of the equation, we get $AI = AABC$. Since multiplying identical columns will give an *I* column, this equation becomes $A = BC$. Similarly, we can get $B = AC$ and $C = AB$. This half-fraction based on $I = ABC$ is called the *principal fraction*.

If we use the second half of Table 12.12, the defining relationship will be $I = -ABC$. Because all *ABC* coefficients are equal to −1s, we can easily determine that $A = -BC$, $B = -AC$, and $C = -AB$. Therefore *A* is aliased with −*BC*, *B* is aliased with −*AC*, and *C* is aliased with −*AB*.

In summary, in the case of half-fractional two-level factorial experiments, we will completely lose the information about the highest order interaction effect and partially lose some information about lower-order interactions.

12.4.2   How to lay out a general half fractional 2^k design

The half-fractional 2^k design is also called 2^{k-1} design, because it has $N = 2^{k-1}$ runs.

Using the definition relationship to lay out the experiment, we describe the procedure to lay out 2^{k-1} design, and illustrate it with an example.

Step 1: Compute $N = 2^{k-1}$ and determine the number of runs. For Example 12.6, for $k = 4$, $N = 2^{k-1} = 2^3 = 8$.

Step 2: Create a table with N runs and lay out the first $k - 1$ factors in standard order. For example, for $k = 4$, the factors are *A, B, C,* and *D,* and the first $k - 1 = 3$ factors are *A, B,* and *C,* as shown in Table 12.13.

We will lay out the first three columns with *A, B,* and *C* in standard order.

TABLE 12.13 2^{3-1} Design

Run number	Factors			
	A	B	C	$D = ABC$
1	−1	−1	−1	−1
2	+1	−1	−1	+1
3	−1	+1	−1	+1
4	+1	+1	−1	−1
5	−1	−1	+1	+1
6	+1	−1	+1	−1
7	−1	+1	+1	−1
8 = N	+1	+1	+1	+1

Step 3: Use a defining relation to create the last column. In Example 12.6, if we use $I = ABCD$ as the defining relation, then for $D = ABC$, we can then get the D column by multiplying the coefficients of A,B,C columns in each row.

In step 3 above, $I = ABCD$, we can derive the following alias relationships: $A = BCD$, $B = ACD$, $C = ABD$, $D = ABC$; $AB = CD$, $AC = BD$, $AD = BC$.

Unlike the case in a 2^{3-1} design, the main effects are not aliased with two-factor interactions, but 2 two-factor interactions are aliased with each other. If we assume that three-factor interactions are not significant, then main effects can be estimated free of aliases. Although both 2^{3-1} and 2^{4-1} are half-fractional factorial designs, 2^{4-1} has less confounding than 2^{3-1}. This is because their resolutions are different.

12.4.3   Design resolution

Design *resolution* is defined as the length of the shortest word in the defining relation. For example, the defining relation of a 2^{3-1} is $I = ABC$, there are three letters in the defining relation, so it is a resolution III design. The defining relation of a 2^{4-1} is $I = ABCD$, and there are four letters in the defining relation, so it is a resolution IV design. Resolution describes the degree to which estimated main effects are aliased (or confounded) with estimated two-, three-, and higher-level interactions. Higher-resolution designs have less severe confounding, but require more runs.

A resolution IV design is "better" than a resolution III design because we have less severe confounding pattern in the former than the latter; higher-order interactions are less likely to be significant than low-order interactions. However, a higher-resolution design for the same number of factors will require more runs. In two-level fractional factorial experiments, the following three resolutions are most frequently used:

Resolution III designs. Main effects are confounded (aliased) with two-factor interactions.

Resolution IV designs. No main effects are aliased with two-factor interactions, but two-factor interactions are aliased with each other.

Resolution V designs. No main effects or two-factor interaction is aliased with any other main effects or two-factor interaction, but two-factor interactions are aliased with three-factor interactions.

12.4.4 ¼ fraction of 2^k design

When the number of factors k increases, 2^{k-1} will also require many runs. Then, a smaller fraction of factorial design is needed. A ¼ fraction of factorial design is also called a 2^{k-2} *design.*

For a 2^{k-1} design, there is one defining relationship, and each defining relationship is able to reduce the number of runs by half. For a 2^{k-2} design, two defining relationships are needed. If one P and one Q represent the generators chosen, then $I = P$ and $I = Q$ are called *generating relations* for the design. Also, because $I = P$ and $I = Q$, it follows that $I = PQ$. $I = P = Q = PQ$ is called the *complete defining relation.*

The 2^{6-2} design. In this design, there are six factors, say, $A, B, C, D, E,$ and F. For a 2^{6-1} design, the generator would be $I = ABCDEF,$ and we would have a resolution VI design. For a 2^{6-2} design, if we choose P and Q to have five letters, for example, $P = ABCDE, Q = ACDEF,$ then $PQ = BF,$ from $I = P = Q = PQ,$ in the complete defining relation, $I = ABCDE = ACDEF = BF,$ we will have only resolution II! In this case even the main effects are confounded, so clearly it is not good. If we choose P and Q to be four letters, for example, $P = ABCE, Q = BCDF,$ then $PQ = ADEF,$ and $I = ABCE = BCDF = ADEF,$ this is a resolution IV design. Clearly, it is also the highest resolution that a 2^{6-2} design can achieve.

We can now develop a procedure to lay out 2^{k-2} design, outlined in the following steps:

Step 1: Compute $N = 2^{k-2}$ and determine the number of runs. For the 2^{6-2} example, for $k = 6, N = 2^{k-2} = 2^4 = 16$.

Step 2: Create a table with N runs and lay out the first $k - 2$ factors in standard order. For example, for $k = 6,$ the factors are A,B,C,D,E,F.

The first $k - 2 = 4$ factors are A,B,C,D. We will lay out the first four columns with A,B,C,D in standard order (see Table 12.14).

Step 3: Use the defining relation to create the last two columns. In the 2^{6-2} example, if we use $I = ABCE$ as the defining relation, then $E = ABC$, and $I = BCDF$, then $F = BCD$.

Example 12.7. A fractional factorial design   The manager of a manufacturing company is concerned about the large number of errors in invoices. An investigation is conducted to determine the major sources of error. Historical data are retrieved from a company database that contains information on customers, type of product, size of shipment, and other variables.

The investigation group identified four factors relating to the shipment of product and defined two levels for each factor (see Table 12.15).

The group then set up a 2^{4-1} factorial to analyze the data, and the data from the last two quarters are studied and percentage errors in invoice are recorded (see Table 12.16).

For data analysis of two-level fractional factorial experiments, we can use the same step-by-step procedure for the two-level full factorial experiments, except that N should be the actual number of runs. By using MINITAB, we get the following results:

TABLE 12.14 2^{6-2} Design

Run number	A	B	C	D	E = ABC	F = BCD
				Factors		
1	−1	−1	−1	−1	−1	−1
2	1	−1	−1	−1	1	−1
3	−1	1	−1	−1	1	1
4	1	1	−1	−1	−1	1
5	−1	−1	1	−1	1	1
6	1	−1	1	−1	−1	1
7	−1	1	1	−1	−1	−1
8	1	1	1	−1	1	−1
9	−1	−1	−1	1	−1	1
10	1	−1	−1	1	1	1
11	−1	1	−1	1	1	−1
12	1	1	−1	1	−1	−1
13	−1	−1	1	1	1	−1
14	1	−1	1	1	−1	−1
15	−1	1	1	1	−1	1
16	1	1	1	1	1	1

TABLE 12.15   Factors and Levels for Example 12.7

Factor	Level	
Customer C	Minor (-1)	Major ($+1$)
Customer location L	Foreign (-1)	Domestic ($+$)
Type of product T	Commodity (-1)	Specialty ($+$)
Size of shipment S	Small ($+$)	Large ($+$)

TABLE 12.16   Experiment Layout and Data for Example 12.7

Factors				Percentage of error	
C	L	T	$S = CLT$	Quarter 1	Quarter 2
-1	-1	-1	-1	14	16
1	-1	-1	1	19	17
-1	1	-1	1	6	6
1	1	-1	-1	1.5	2.5
-1	-1	1	1	18	20
1	-1	1	-1	24	22
-1	1	1	-1	15	17
1	1	1	1	21	21

```
Estimated Effects and Coefficients for %Error (coded units)
Term        Effect         Coef       SE Coef      T        P
Constant                  15.000      0.2864     52.37    0.000
C            2.000         1.000      0.2864      3.49    0.008
L           -7.500        -3.750      0.2864    -13.09    0.000
T            9.500         4.750      0.2864     16.58    0.000
S            2.000         1.000      0.2864      3.49    0.008
C*L         -1.500        -0.750      0.2864     -2.62    0.031
C*T          2.500         1.250      0.2864      4.36    0.002
C*S          5.000         2.500      0.2864      8.73    0.000

Analysis of Variance for %Error, using Adjusted SS for Tests
Source     DF    Seq SS      Adj SS     Adj MS       F        P
C           1     16.00       16.00      16.00     12.19    0.008
L           1    225.00      225.00     225.00    171.43    0.000
T           1    361.00      361.00     361.00    275.05    0.000
S           1     16.00       16.00      16.00     12.19    0.008
C*L         1      9.00        9.00       9.00      6.86    0.031
C*T         1     25.00       25.00      25.00     19.05    0.002
C*S         1    100.00      100.00     100.00     76.19    0.000
Error       8     10.50       10.50       1.31
Total      15    762.50

Alias Structure
I + C*L*T*S
C + L*T*S
L + C*T*S
```

```
T + C*L*S
S + C*L*T
C*L + T*S
C*T + L*S
C*S + L*T
```

On the basis of the Pareto effect plot (Fig. 12.12), factors T (type of product) and L (location) as well as interaction CS (customer and size of shipment) are the top three effects. However, since CS and TL are aliased, the interaction CS could be the effect of TL as well. By using some common sense, the team thinks that TL (type of product and location of customer) interaction is more likely to have a significant effect.

If the type of product is found to be most significant, there are more invoice errors for specialty products. There are far fewer invoice errors for commodity products, especially for domestic customers, for whom only 5 percent of invoices contain errors. Location of customer is the second most significant factor, for there are a lot more invoice errors for foreign customers, even for commodity products. (A main-effects plot and an interaction chart for this example are given in Figs. 12.13 and 12.14, respectively.)

12.4.5   The general 2^{k-p} fractional factorial design

A 2^k fractional factorial design having 2^{k-p} runs is called a $1/2^p$ fraction of a 2^k design, or 2^{k-p} fractional factorial design. These designs need p

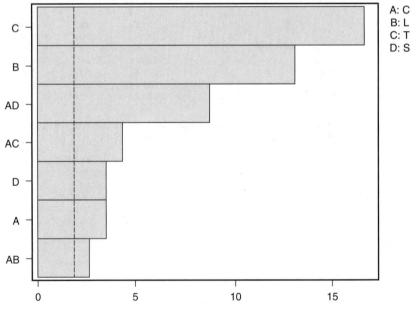

Figure 12.12   Pareto effect chart for Example 12.7. (Response is percent error, alpha = 0.10.)

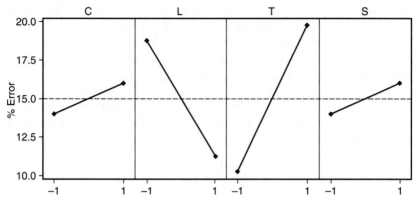

Figure 12.13 Main-effects chart, data means for percent error, for Example 12.7.

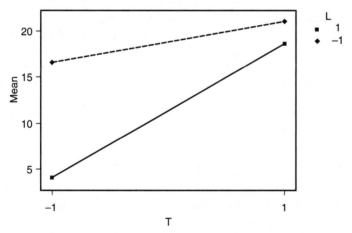

Figure 12.14 Interaction chart, data means for percent error, for Example 12.7.

independent generators. Selection of those p generators should result in a design with the highest possible resolutions. Montgomery (1997) lists many good 2^{k-p} fractional factorial designs.

Summary. 2^{k-p} designs are the workhorse of industrial experiments, because they can analyze many factors simultaneously with relative efficiency (i.e., with few experimental runs). The experiment design is also straightforward because each factor has only two settings.

Disadvantages.   The simplicity of these designs is also their major flaw. The underlying use of two-level factors is the belief that mathematical relationships between response and factors are basically *linear* in nature. This is seldom the case, and many variables are related to response in a nonlinear fashion.

Another problem of fractional designs is the implicit assumption that higher-order interactions do not matter, because sometimes they do. In this case, it is nearly impossible for fractional factorial experiments to detect higher-order interaction effects.

12.5   Three-Level Full Factorial Design

The three-level design is written as a 3^k factorial design. This means that k factors are considered, each at three levels. These are (usually) referred to as *low, intermediate,* and *high* levels, expressed numerically as 0, 1, and 2, respectively. One could have considered the digits -1, 0, and $+1$, but this may be confused with respect to the two-level designs since 0 is reserved for centerpoints. Therefore, we will use the 0,1,2 scheme. The three-level designs were proposed to model possible curvature in the response function and to handle the case of nominal factors at three levels. A third level for a continuous factor facilitates investigation of a quadratic relationship between the response and each factor.

Unfortunately, the three-level design is prohibitive in terms of the number of runs, and thus in terms of cost and effort.

12.5.1   The 3^2 design

This is the simplest three-level design. It has two factors, each at three levels. The nine treatment combinations for this type of design are depicted in Fig. 12.15.

A notation such as "20" means that factor A is at its high level (2) and factor B is at its low level (0).

12.5.2   The 3^3 design

This design consists of three factors, each at three levels. It can be expressed as a $3 \times 3 \times 3 = 3^3$ design. The model for such an experiment is

$$Y_{ijk} = \mu + A_i + B_j + AB_{ij} + C_k + AC_{ik} + BC_{jk} + ABC_{ijk} + \varepsilon_{ijk} \quad (12.14)$$

where each factor is included as a nominal factor rather than as a continuous variable. In such cases, main effects have 2 degrees of freedom, two-factor interactions have $2^2 = 4$ degrees of freedom, and

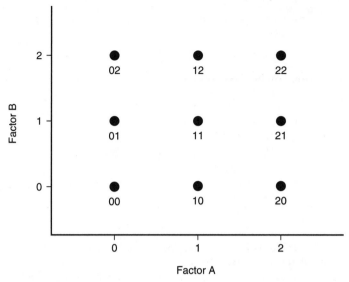

Figure 12.15 A 3^2 design.

k-factor interactions have 2^k degrees of freedom. The model contains 2 + 2 + 2 + 4 + 4 + 4 + 8 = 26 degrees of freedom. Note that if there is no replication, the fit is exact and there is no error term (the epsilon term) in the model. In this nonreplication case, if we assume that there are no three-factor interactions, then we can use these 8 degrees of freedom for error estimation.

In this model we see that $i = 1,2,3$, and similarly for j and k, making 27 treatments. These treatments may be tabulated as in Table 12.17, and the design can be depicted as Fig. 12.16.

Example 12.8. A Three-Factor Factorial Experiment A study was designed to evaluate the effect of wind speed and ambient temperature on scale accuracy. The accuracy is measured by loading a standard weight on the scale and recording the difference from the standard value. A 3×3 factorial experiment was planned. The data listed in Table 12.18 were obtained.

Data analysis of the 3^k design is the same as that of general full factorial design. By using MINITAB, we obtained the following results:

```
Analysis of Variance for Deviatio, using Adjusted SS for Tests
Source           DF    Seq SS    Adj SS    Adj MS      F      P
Wind              2     0.107     0.107     0.053   0.04  0.958
Temperat          2    50.487    50.487    25.243  20.42  0.000
Wind*Temperat     4     0.653     0.653     0.163   0.13  0.969
Error            18    22.253    22.253     1.236
Total            26    73.500
```

TABLE 12.17   The 3^3 Design

Factor B	Factor C	Factor A 0	Factor A 1	Factor A 2
0	0	000	100	200
0	1	001	101	201
0	2	002	102	202
1	0	010	110	210
1	1	011	111	211
1	2	012	112	212
2	0	020	120	220
2	1	021	121	221
2	2	022	122	222

From the ANOVA table, it is clear that the temperature is the only significant factor. Wind speed has very little effect on measurement deviation. Figure 12.17 shows that temperature influences the measurement deviation in a nonlinear fashion. An interaction chart is shown in Fig. 12.18.

12.5.3   Fractional factorial 3^k design

Fractional factorial 3^k experiments can be designed (Montgomery 1997). However, these designs cannot handle interactions very well, because they cannot give clear mathematical description of interaction. Therefore, fractional three-level factorial designs are used mostly to deal with main effects. If we really want to analyze interactions in three-level factorial designs, full factorials have to be used.

12.6   Summary

1. There are two main bodies of knowledge in DOE: experimental design and experimental data analysis.

2. Two types of experimental design strategy are discussed in this chapter: full factorial and fractional factorial. A full factorial design can obtain more information from the experiment, but the experiment size will increase exponentially with the number of experiment factors and levels. A fractional factorial design will obtain less information from the experiment, but the increase in its experiment size will be much slower than that of the full factorial. In addition, we can adjust the resolution of fractional factorial design to obtain needed information while keeping the experiment in manageable size. Therefore, fractional factorial design becomes the "workhorse" of DOE in the industrial application.

3. The main DOE data analysis tools include analysis of variance (ANOVA), empirical model building, and main-effects and interaction

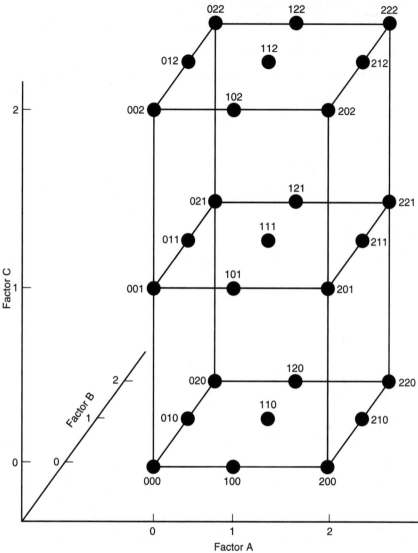

Figure 12.16 A 3^3 design schematic.

charts. ANOVA is able to identify the set of significant factors and interactions, and rank the relative importance of each effect and interaction in terms of their effect on process output. The empirical model, main-effects chart, and interaction chart show the empirical relationship between process output and process factors and can also be used to identify optimal factor level settings and corresponding optimal process performance level.

TABLE 12.18   Experiment Layout and Data for Example 12.8

	Factors		Response = measurement − standard		
Run number	Wind	Temperature	1	2	3
1	Low	Low	0.4	−0.8	0.6
2	Low	Mid	−0.7	0.5	0.3
3	Low	High	2.6	3.2	2.8
4	Mid	Low	−1.0	0.8	−0.7
5	Mid	Mid	−0.5	1.3	0.6
6	Mid	High	3.6	2.5	3.5
7	High	Low	2.1	−1.6	−0.8
8	High	Mid	−1.3	0.5	1.6
9	High	High	1.5	4.3	2.6

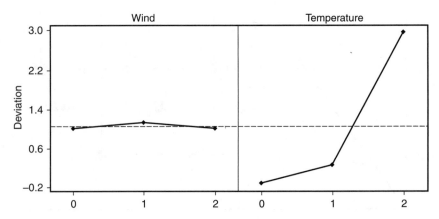

Figure 12.17   Main-effects chart of Example 12.8—LS means for deviation.

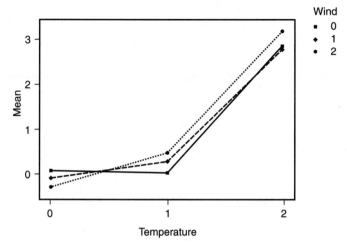

Figure 12.18 Interaction chart of Example 12.9—LS means for deviation.

13

Taguchi's Orthogonal Array Experiment

13.1 Taguchi's Orthogonal Arrays

The Taguchi method is a comprehensive quality strategy that builds robustness into a product/process during its design stage. The Taguchi method is a combination of sound engineering design principles and Taguchi's version of design of experiment, called an *orthogonal array experiment,* discussed in this chapter. Other aspects of the Taguchi method are discussed in subsequent chapters.

In Taguchi's experimental design system, all experimental layouts are derived from about 18 standard orthogonal arrays. An *orthogonal array* is a fractional factorial experimental matrix that is orthogonal and balanced. Let's look at the simplest orthogonal array, the L_4 array in Table 13.1.

TABLE 13.1 $L_4(2^3)$ Orthogonal Array

Experiment no.	Column		
	1	2	3
1	1	1	1
2	1	2	2
3	2	1	2
4	2	2	1

Linear Graph for L4

The values inside the array (i.e., 1 and 2) represent two different levels of a factor. By simply using -1 to substitute for 1, and $+1$ to substitute for 2, we can find that this L_4 array becomes

	Column		
Experiment no.	1	2	3
1	-1	-1	-1
2	-1	1	1
3	1	-1	1
4	1	1	-1

Clearly, this is a 2^{3-1} fractional factorial design, with defining relation $I = -ABC$. Where column 2 of L_4 is equivalent to the A column of the 2^{3-1} design, column 1 is equivalent to the B column of the 2^{3-1} design, and column 3 is equivalent to C column of 2^{3-1} design, with $C = -AB$.

In each of Taguchi's orthogonal arrays, there are one or more accompanying linear graphs. A *linear graph* is used to illustrate the interaction relationships in the orthogonal array. For example, in Table 13.1, the numbers 1 and 2 represent columns 1 and 2 of the L_4 array; the number 3 is above the line segment connecting 1 and 2, which means that the interaction between column 1 and column 2 is confounded with column 3, which is perfectly consistent with $C = -AB$ in the 2^{3-1} fractional factorial design.

For larger orthogonal arrays, there are not only linear graphs but also interaction tables to explain intercolumn relationships. Examples are Tables 13.2 and 13.3 for the L_8 array.

Again, if we change 1 to -1, and 2 to $+1$ in the L_8 array, it is clear that this is a 2^{7-4} fractional factorial design, where column 4 of L_8 corresponds to the A column of a 2^{7-4}, column 2 of L_8 corresponds to the B column of a 2^{7-4}, and column 1 of L_8 corresponds to the C column of a 2^{7-4}. Also,

Linear Graphs for L8

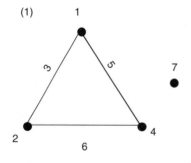
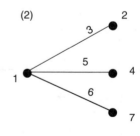

TABLE 13.2 $L_8(2^7)$ **Orthogonal Array**

Experiment no.	Column						
	1	2	3	4	5	6	7
1	1	1	1	1	1	1	1
2	1	1	1	2	2	2	2
3	1	2	2	1	1	2	2
4	1	2	2	2	2	1	1
5	2	1	2	1	2	1	2
6	2	1	2	2	1	2	1
7	2	2	1	1	2	2	1
8	2	2	1	2	1	1	2

TABLE 13.3   **Interaction Table for L_8**

Column	Column						
	1	2	3	4	5	6	7
1	(1)	3	2	5	4	7	6
2		(2)	1	6	7	4	5
3			(3)	7	6	5	4
4				(4)	1	2	3
5					(5)	3	2
6						(6)	1
7							(7)

we can easily see that column 3 is equivalent to $-BC$, column 5 is equivalent to $-AC$, column 6 is equivalent to $-BC$, and so on. Those are consistent with linear graph (1). Linear graph (1) indicates that the interaction between columns 1 and 2 is confounded with column 3, the interaction between columns 1 and 4 is confounded with column 5, and the interaction between columns 2 and 4 is confounded with column 6.

However, we know that a 2^{7-4} has four generators, so each main effect will be confounded with many two-factor interactions. So, each linear graph shows only a subset of interaction relationships.

The interaction table provides more information about interaction relationships. For example, if we look at the number in the first row and the second column of the interaction table, then it is 3, which means that the interaction between columns 1 and 2 is confounded with column 3. But we also see there is a 3 in row 5 and column 6, and in row 4 and column 7. Therefore, column 3 is also confounded with the interaction between columns 5 and 6 and between columns 4 and 7.

In the notation of orthogonal array, for example, $L_8(2^7)$, a 2 means two levels, an 8 means that the orthogonal array has 8 runs, and a 7 means that up to seven factors can be accommodated in this array.

TABLE 13.4 $L_9(3^4)$ (Array

	Column			
Experiment no.	1	2	3	4
1	1	1	1	1
2	1	2	2	2
3	1	3	3	3
4	2	1	2	3
5	2	2	3	1
6	2	3	1	2
7	3	1	3	2
8	3	2	1	3
9	3	3	2	1

Linear Graph for L9

Taguchi's orthogonal arrays also include three-level arrays and mixed-level arrays. The simplest one is an L_9 array as in Table 13.4.

The linear graph of L_9 indicates that columns 3 and 4 are both confounded with the interaction effects of columns 1 and 2.

More orthogonal arrays are listed in the chapter appendix.

13.2   Taguchi Experimental Design

There are many similarities between "regular" experimental design and Taguchi's experimental design. However, in a Taguchi experiment, only the main effects and two-factor interactions are considered. Higher-order interactions are assumed to be nonexistent. In addition, experimenters are asked to identify which interactions might be significant before conducting the experiment, through their knowledge of the subject matter.

After these two steps, the total degrees of freedom of the experimental factors should be determined in the Taguchi experimental design. The *degrees of freedom* are the relative amount of data needed in order to estimate all the effects to be studied. The determination of the degree of freedom is based on the following rules:

13.2.1   Degree-of-freedom (DOF) rules

1. The overall mean always uses 1 degree of freedom.

2. For each factor, $A,B,\ldots$; if the number of levels are $n_A, n_B, \ldots$, for each factor, the degree of freedom = number of levels -1; for example, the degree of freedom for factor $A = n_A - 1$.

3. For any two-factor interaction, for example, AB interaction, the degree of freedom $= (n_A - 1)(n_B - 1)$.

Example 13.1   In an experiment, there is 1 two-level factor, A, and 6 three-level factors, B,C,D,E,F,G, and 1 two-factor interaction, AB. Then, the total degree of freedom is as follows:

Factors	Degree of freedom
Overall mean	1
A	$2 - 1 = 1$
B,C,D,E,F,G	$6 \times (3 - 1) = 12$
AB	$(2 - 1)(3 - 1) = 2$
Total DOF	16

13.2.2   Experimental design

Taguchi experimental design follows a three-step procedure:

- Step 1: Find the total degree of freedom (DOF).
- Step 2: Select a standard orthogonal array using the following two rules:

  Rule 1: The number of runs in the orthogonal array $\geq$ total DOF.

  Rule 2: The selected orthogonal array should be able to accommodate the factor level combinations in the experiment.

- Step 3: Assign factors to appropriate columns using the following rules:

  Rule 1: Assign interactions according to the linear graph and interaction table.

  Rule 2: Use special techniques, such as dummy level and column merging, when the original orthogonal array is not able to accommodate the factor levels in the experiment.

  Rule 3: Keep some column(s) empty if not all columns can be assigned

In selecting orthogonal arrays, Table 13.5 can be used as a reference.

Example 13.2   In an experiment, there are seven factors. We will consider main effects only. First, we compute DOF $= 1 + 7(2 - 1) = 8$. Therefore, the selected orthogonal array should have at least eight runs. By examining Table 13.5, we find that the L_8 array can accommodate 7 two-level factors. Therefore, we can use L_8 and assign those seven factors to seven columns of L_8.

Example 13.3   In an experiment, there is one two-level factor A, and 6 three-level factors, B,C,D,E,F,G. First, DOF $= 1 + (2 - 1) + 6(3 - 1) = 14$.

TABLE 13.5   Basic Information on Taguchi Orthogonal Arrays

Orthogonal array	Number of runs	Maximum number of factors	Maximum number of column at these levels			
			2	3	4	5
L_4	4	3	3			
L_8	8	7	7			
L_9	9	4		4		
L_{12}	12	11	11			
L_{16}	16	15	15			
L'_{16}	16	5			5	
L_{18}	18	8	1	7		
L_{25}	25	6				6
L_{27}	27	13		13		
L_{32}	32	31	31			
L'_{32}	32	10	1		9	
L_{36}	36	23	11	12		
L'_{36}	36	16	3	13		
L_{50}	50	12	1			11
L_{54}	54	26	1	25		
L_{64}	64	63	63			
L'_{64}	64	21			21	
L_{81}	81	40		40		

Therefore, we have to use an array that has more than 14 runs. L_{16} has 16 runs, but it has only two-level columns, so it cannot accommodate 6 three-level columns. L_{18} has 1 two-level column and 7 three-level columns, so it can be used to accommodate all the factors in this example. The experimental layout is as follows:

Experiment no.	Factors							
	A	B	C	D	E	F	G	e
1	1	1	1	1	1	1	1	1
2	1	1	2	2	2	2	2	2
3	1	1	3	3	3	3	3	3
4	1	2	1	1	2	2	3	3
5	1	2	2	2	3	3	1	1
6	1	2	3	3	1	1	2	2
7	1	3	1	2	1	3	2	3
8	1	3	2	3	2	1	3	1
9	1	3	3	1	3	2	1	2
10	2	1	1	3	3	2	2	1
11	2	1	2	1	1	3	3	2
12	2	1	3	2	2	1	1	3
13	2	2	1	2	3	1	3	2
14	2	2	2	3	1	2	1	3
15	2	2	3	1	2	3	2	1
16	2	3	1	3	2	3	1	2
17	2	3	2	1	3	1	2	3
18	2	3	3	2	1	2	3	1

The e in column 8 means empty, so no factor will be assigned to column 8.

Example 13.4 In an experiment there are 9 two-level factors, A,B,C,D,E,F,G,H,I and the interactions AB, AC, AD, and AF are believed to be significant. First, DOF $= 1 + 9(2 - 1) + 4(2 - 1)(2 - 1) = 14$.

Array L_{16} has 16 runs, and it can accommodate up to 15 two-level factors. So, we will consider L_{16}. But first we need to work out how to deal with the four interactions. By examining the linear graph (3) of L_{16}, we could assign columns as follows:

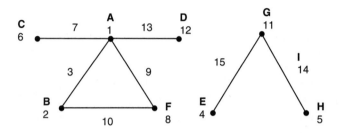

Column 3, 7, 9, and 13 are left empty to avoid confounding any other main effects with interactions AB, AC, AD, and AF. Columns 10 and 15 are also empty. The column assignments to an L_{16} array are as follows:

							Column assignments								
	A	B	AB	E	H	C	AC	F	AF	e	G	D	AD	I	E
Experiment no.	1	2	3	4	5	6	7	8	9	10	11	12	13	14	15
1	1	1	1	1	1	1	1	1	1	1	1	1	1	1	1
2	1	1	1	1	1	1	1	2	2	2	2	2	2	2	2
⋮	⋮	⋮	⋮	⋮	⋮	⋮	⋮	⋮	⋮	⋮	⋮	⋮	⋮	⋮	⋮

Example 13.5 In an experiment, there are 6 three-level factors, A,B,C,D,E,F as well as interactions AB, AC, and BC.

First, DOF $= 1 + 6(3 - 1) + 3(3 - 1)(3 - 1) = 25$. L_{27} has 27 runs, and ᵢit can accommodate 13 three-level factors. It is a plausible choice. By examining its linear graph, we have the following column assignments:

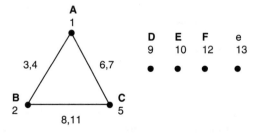

Columns 3, 4, 6, 7, 8, and 11 are left empty to avoid confounding of main effects with interactions AB, AC, and BC.

However, by using the standard orthogonal arrays in their original forms, we are not able to design a good experiment in many cases. Many standard arrays only have two- or three-level columns. The number of levels should be the same for all factors in the array. Sometimes, we would like to have factors with different levels in the same experiment, and we could have some factors with four or more levels.

In Taguchi's experiment, several special techniques are used to deal with these issues.

13.3   Special Techniques

13.3.1   Column merging method

The *column merging method* merges several low-level columns into a high-level column. Four cases of column merging methods are used in Taguchi's experiment.

1. *Creation of a four-level column by using two-level columns.* We need 3 two-level columns to create an eight-level column, because each two-level column has 1 degree of freedom, but each four-level column has 3 degrees of freedom, so 3 two-level columns are needed. These 3 two-level columns should be the two main-effects columns and another column corresponding to their interaction.

We will illustrate this by using the examples in Table 13.6.

Assuming that we want to create a four-level column in the L_8 array in Table 13.6a, we select the main-effects columns, columns 1 and 2, and column 3, which is the interaction column of columns 1 and 2. These three columns are to be combined to create a new four-level column. There are four levels in the new column: levels 1 to 4. The following table illustrates how new column levels are created by combining two-level columns. Column 3 is the interaction between columns 1 and 2; it has to be left unused for other factors. Otherwise, the use of column 3 by other factors could lead to confounding of that factor with our new four-level factor.

Column 1 levels	Column 2 levels	New column levels
1	1	1
1	2	2
2	1	3
2	2	4

Example 13.6   There are two factors in an experiment, A,B, where A is a four-level factor, and B is a two-level factor. AB interaction may also be significant. Again, first we calculate DOF $= 1 + (4 - 1) + (2 - 1) + (4 - 1)(2 - 1) = 8$. L_8 has eight runs.

TABLE 13.6   Examples of Column Merging

(a)

Experiment no.	Column						
	1	2	3	4	5	6	7
1	1	1	1	1	1	1	1
2	1	1	1	2	2	2	2
3	1	2	2	1	1	2	2
4	1	2	2	2	2	1	1
5	2	1	2	1	2	1	2
6	2	1	2	2	1	2	1
7	2	2	1	1	2	2	1
8	2	2	1	2	1	1	2

(b)

Experiment no.	New column	Column			
		4	5	6	7
1	1	1	1	1	1
2	1	2	2	2	2
3	2	1	1	2	2
4	2	2	2	1	1
5	3	1	2	1	2
6	3	2	1	2	1
7	4	1	2	2	1
8	4	2	1	1	2

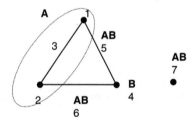

By examining this linear graph, it is clear that we can create column A by combining columns 1 to 3 into a four-level column. B can be assigned to column 4. By computing the degree of freedom of AB, it is $(4 - 1)(2 - 1) = 3$, so AB should use three columns; columns 5 and 6 are obviously related to the AB interaction. Another column relating to the AB interaction is the interaction between columns 3 and 4, because column 3 is part of column A. By checking the interaction table of L_8, we can see that the interaction between columns 3 and 4 is column 7. The detailed column assignment is

	Column				
	A	*B*	*AB*	*AB*	*AB*
Experiment no.		4	5	6	7
1	1	1	1	1	1
2	1	2	2	2	2
3	2	1	1	2	2
4	2	2	2	1	1
5	3	1	2	1	2
6	3	2	1	2	1
7	4	1	2	2	1
8	4	2	1	1	2

2. *Creation of an eight-level column using two-level columns.* We need 7 two-level columns to create a four-level column, because each two-level column has 1 degree of freedom, and each eight-level column has 3 degrees of freedom, so 7 two-level columns are needed. These 7 two-level columns should be three main-effects columns and four columns corresponding to their interactions.

Assume that we want to create an eight-level column by using the columns in an L_{16} array, by observing the following linear graph of L_{16},

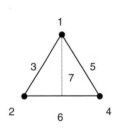 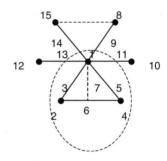

We can merge three main-effects columns, columns 1, 2, and 4, and their interactions, column 3, 5 and 6, as well as column 7, where column 7 is actually the interaction between columns 1 and 6, which can be identified from interaction tables (see chapter appendix). Table 13.7 illustrates the column assignment (numbering scheme) of this eight-level column merging example.

Table 13.8 illustrates how new column levels are created by combining two-level columns.

3. *Creation of a nine-level column by using three-level columns.* We need 4 three-level columns to create a nine-level column; because each three-level column has 2 degrees of freedom, and each nine-level column has 8 degrees of freedom. These four columns should have two main-effects columns, and two interaction columns of those two main effects.

TABLE 13.7   Column Assignment for the Eight-Level Column

Experiment no.	New column by merging (1,2,4,3,5,6,7)	Other columns							
		8	9	10	11	12	13	14	15
1	1	1	1	1	1	1	1	1	1
2	1	2	2	2	2	2	2	2	2
3	2	1	1	1	1	2	2	2	2
4	2	2	2	2	2	1	1	1	1
5	3	1	1	2	2	1	1	2	2
6	3	2	2	1	1	2	2	1	1
7	4	1	1	2	2	2	2	1	1
8	4	2	2	1	1	1	1	2	2
9	5	1	2	1	2	1	2	1	2
10	5	2	1	2	1	2	1	2	1
11	6	1	2	1	2	2	1	2	1
12	6	2	1	2	1	1	2	1	2
13	7	1	2	2	1	1	2	2	1
14	7	2	1	1	2	2	1	1	2
15	8	1	2	2	1	2	1	1	2
16	8	2	1	1	2	1	2	2	1

TABLE 13.8   Column Levels

Column 1 levels	Column 2 levels	Column 4 levels	New column levels
1	1	1	1
1	1	2	2
1	2	1	3
1	2	2	4
2	1	1	5
2	1	2	6
2	2	1	7
2	2	2	8

Example 13.7   In an experiment there are six factors, A,B,C,D,E,F, where A is a nine-level factor and the others are three-level factors. First, DOF = $1 + (9 - 1) + 5(3 - 1) = 19$. Since L_{27} can accommodate up to 13 three-level factors and it has 27 runs, we consider the following linear graph of L_{27}:

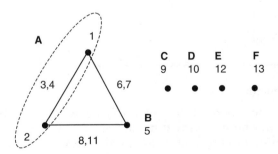

We can combine columns 1 to 4 into a nine-level column for the A factor, and B,C,D,E,F can be assigned to other columns, such as column 5,9,12,13 as shown above. The column assignments in L_{27} are shown below:

	Column assignment									
	A	B	e	e	e	C	D	e	E	F
Experiment no.	(1,2,3,4)	5	6	7	8	9	10	11	12	13
1	1	1	1	1	1	1	1	1	1	1
2	1	2	2	2	2	2	2	2	2	2
3	1	3	3	3	3	3	3	3	3	3
4	2	1	1	1	2	2	2	3	3	3
5	2	2	2	2	3	3	3	1	1	1
6	2	3	3	3	1	1	1	2	2	2
7	3	1	1	1	3	3	3	2	2	2
8	3	2	2	2	1	1	1	3	3	3
9	3	3	3	3	2	2	2	1	1	1
⋮	⋮	⋮	⋮	⋮	⋮	⋮	⋮	⋮	⋮	⋮

The following table illustrates how new column levels are created by combining three-level columns:

Column 1 levels	Column 2 levels	New column levels
1	1	1
1	2	2
1	3	3
2	1	4
2	2	5
2	3	6
3	1	7
3	2	8
3	3	9

4. *Creation of a six-level column by using two-level and three-level columns.* In both L_{18} and L_{36} arrays, there are both two- and three-level columns. We can merge 1 two-level column with a three-level column to create a six-level column. The following table illustrates how new column levels are created by combining a two-level column and a three-level column:

Two-level column	Three-level column	Six-level column
1	1	1
1	2	2
1	3	3
2	1	4
2	2	5
2	3	6

13.3.2 Dummy-level technique

The "dummy-level technique" is used to assign a factor with m levels to a column with n levels, where n is greater than m.

We can apply the dummy-level technique to assign a three-level factor to a two-level factor orthogonal array.

Example 13.8 In an experiment there are 1 two-level factor, A, and 3 three-level factors, B,C,D. Its DOF $= 1 + (2 - 1) + 3(3 - 1) = 8$. L_8 won't be able to accommodate those factors because it has only two-level columns. L_9 has nine runs and can accommodate up to 4 three-level factors; here we can use the dummy-level technique to assign a two-level factor, A, to a three-level column, and assign other three-level factors, B,C,D, to other 3 columns, as follows:

Experiment no	A	B	C	D
	1	2	3	4
1	1	1	1	1
2	1	2	2	2
3	1	3	3	3
4	2	1	2	3
5	2	2	3	1
6	2	3	1	2
7	1′	1	3	2
8	1′	2	1	3
9	1′	3	2	1

In this array, 1′ means that we assign level 1 to the place of level 3 in column 1. We could also assign level 2. The level we selected to duplicate should be the level we would like to use to get more information.

If a three-level column is to be assigned into an orthogonal array with all two-level factors, then we can first merge 3 two-level columns to form a four-level column, and then assign this three-level column to the four-level column.

Example 13.9 In an experiment we have one three-level factor, A, and 7 two-level factors, B,C,D,E,F,G,H, as well as interactions BC, DE, and FG.

First, DOF $= 1 + (3 - 1) + 7(2 - 1) + 3(2 - 1)(2 - 1) = 13$. L_{16} has 16 runs and can accommodate up to 15 two-level factors. The following linear graph seems very appropriate to accommodate a four-level factor by merging columns 1, 2, and 3, which will be used for A, and all other two-level factors and interactions.

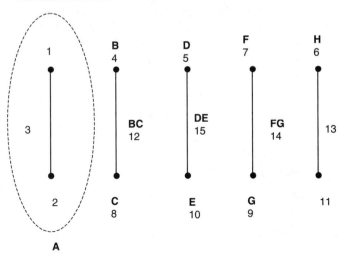

Then we can use the dummy-level technique on the four-level factor column to accommodate A.

							Column assignments						
	A	B	D	H	F	C	G	E	e	BC	e	FG	DE
Experiment no.	(1–2–3)	4	5	6	7	8	9	10	11	12	13	14	15
1	1	1	1	1	1	1	1	1	1	1	1	1	1
2	1	1	1	1	1	2	2	2	2	2	2	2	2
3	1	2	2	2	2	1	1	1	1	2	2	2	2
4	1	2	2	2	2	2	2	2	2	1	1	1	1
5	2	1	1	2	2	1	1	2	2	1	1	2	2
6	2	1	1	2	2	2	2	1	1	2	2	1	1
7	2	2	2	1	1	1	1	2	2	2	2	1	1
8	2	2	2	1	1	2	2	1	1	1	1	2	2
9	3	1	2	1	2	1	2	1	2	1	2	1	2
10	3	1	2	1	2	2	1	2	1	2	1	2	1
11	3	2	1	2	1	1	2	1	2	2	1	2	1
12	3	2	1	2	1	2	1	2	1	1	2	1	2
13	1′	1	2	2	1	1	2	2	1	1	2	2	1
14	1′	1	2	2	1	2	1	1	2	2	1	1	2
15	1′	2	1	1	2	1	2	2	1	2	1	1	2
16	1′	2	1	1	2	2	1	1	2	1	2	2	1

13.3.3 Compound factor method

The compound factor method is used when the number of factors exceeds the number of columns in the orthogonal array.

Example 13.10 There are 2 two-level factors, A and B, and 3 three-level factors, C, D, and E, and the management allows no more than nine experimental runs. Assuming that we selected an L_9 array, only four factors can be assigned to L_9, so we are trying to assign these 2 two-level factors, A and B, into 1 three-level column.

There are four combinations of A and B: A_1B_1, A_1B_2, A_2B_1, and A_2B_2. Because each three-level column has only three levels. We could select only three combinations, such as $(AB)_1 = A_1B_1$, $(AB)_2 = A_1B_2$, and $(AB)_3 = A_2B_1$. The compound factor (AB) can be assigned to a three-level column:

		Column		
	AB	C	D	E
Experiment no.	1	2	3	4
1	$(AB)_1$	1	1	1
2	$(AB)_1$	2	2	2
3	$(AB)_1$	3	3	3
4	$(AB)_2$	1	2	3
5	$(AB)_2$	2	3	1
6	$(AB)_2$	3	1	2
7	$(AB)_3$	1	3	2
8	$(AB)_3$	2	1	3
9	$(AB)_3$	3	2	1

However, in the compound factor method, there is a partial loss of orthogonality. The two compound factors are not orthogonal to each other, but each of them is orthogonal to other factors in the experiment.

13.4   Taguchi Experiment Data Analysis

There are many similarities between the data analysis of the Taguchi experiment and "classical" design of experiment.

In Taguchi experimental data analysis, the following three items are very important:

1. Analysis of variance (ANOVA)

2. Main-effects chart and interaction chart

3. Optimization and prediction of expected response

13.4.1   Analysis of variance

There is actually no difference between analysis of variance of classical DOE and Taguchi DOE. First, we compute the sum of squares (SS), then the *mean squares* (MS), where an MS is computed by dividing the SS by the degree of freedom. In Taguchi DOE, the F test is not as important as that of classical DOE. Sometimes, the relative importance of each factor is computed by its percentage contribution to the total sum of squares.

For each column of an orthogonal array, assume that there are k levels, and for each level t, the total sum of response at tth level is represented by T_t, the total sum of responses is represented by T, the total number of runs is N, and the number of replicates is n; then for each column, the sum of squares is

$$\text{SS} = \frac{k}{N \times n} \sum_{t=1}^{k} T_t^2 - \frac{T^2}{N \times n} \qquad (13.1)$$

Example 13.11   A truck front fender's injection-molded polyurethane bumpers suffer from too much porosity. So a team of engineers conducted a Taguchi experiment design project to study the effects of several factors to the porosity:

	Factors	Low	High
A	Mold temperature	A_1	A_2
B	Chemical temperature	B_1	B_2
D	Throughput	D_1	D_2
E	Index	E_1	E_2
G	Cure time	G_1	G_2

Interactions AB and BD are also considered

The following L_8 orthogonal array is used for each run, and two measurements of porosity are taken; for porosity values, the smaller, the better:

	Experimental factors							Porosity measurements	
	A	B	AB	D	E	BD	G		
Experiment no.	1	2	3	4	5	6	7	1	2
1	1	1	1	1	1	1	1	26	38
2	1	1	1	2	2	2	2	16	6
3	1	2	2	1	1	2	2	3	17
4	1	2	2	2	2	1	1	18	16
5	2	1	2	1	2	1	2	0	5
6	2	1	2	2	1	2	1	0	1
7	2	2	1	1	2	2	1	4	5
8	2	2	1	2	1	1	2	5	3

Then we can compute

$$SS_A = \frac{2}{8 \times 2}(T_{A_1}^2 + T_{A_2}^2) - \frac{T^2}{8 \times 2}$$

where $T_{A1} = 26 + 38 + 16 + 6 + 3 + 17 + 18 + 16 = 140$, $T_{A2} = 0 + 5 + 0 + 1 + 4 + 5 + 5 + 3 = 23$, and $T = 26 + 38 + 16 + \cdots + 5 + 3) = 163$ and $SS_A = 2516.125 - 1660.5625 = 855.5625$. Similarly

$$SS_B = \frac{2}{16}(T_{B_1}^2 + T_{B_2}^2) - \frac{T^2}{16} = \frac{2}{16}[(26 + 38 + 16 + 6 + 0 + 5 + 0 + 1)^2$$

$$+ (3 + 17 + 18 + 16 + 4 + 5 + 5 + 3)^2]$$

$$- \frac{163^2}{16} = 27.56$$

and

$$SS_{AB} = 115.56$$

$$SS_D = 68.06$$

$$SS_E = 33.06$$

$$SS_{BD} = 217.56$$

$$SS_G = 175.56$$

and

$$SS_T = \sum_{i=1}^{N} \sum_{j=1}^{n} y_{ij}^2 - \frac{T^2}{N \times n} \qquad (13.2)$$

where y_{ij} are individual observations, in this example:

$$SS_T = (26^2 + 38^2 + \cdots + 5^2 + 3^2) - \frac{163^2}{16} = 1730.44$$

Actually, MINITAB will give exactly the same result. The ANOVA table is given as follows:

```
Analysis of Variance for Porosity, using Adjusted SS for Tests
Source      DF     Seq SS     Adj SS     Adj MS        F        P
A            1     855.56     855.56     855.56     28.82    0.001
B            1      27.56      27.56      27.56      0.93    0.363
AB           1     115.56     115.56     115.56      3.89    0.084
D            1      68.06      68.06      68.06      2.29    0.168
E            1      33.06      33.06      33.06      1.11    0.322
BD           1     217.56     217.56     217.56      7.33    0.027
G            1     175.56     175.56     175.56      5.91    0.041
Error        8     237.50     237.50      29.69
Total       15    1730.44
```

In Taguchi experimental analysis, *percentage contributions* are often used to evaluate the relative importance of each effect. In this example, it is clear that

$$SS_T = SS_A + SS_B + SS_{AB} + SS_D + SS_E + SS_{BD} + SS_G + SS_{error}$$

Therefore

$$\frac{SS_A}{SS_T} + \frac{SS_B}{SS_T} + \cdots + \frac{SS_G}{SS_T} + \frac{SS_{error}}{SS_T} = 100\%$$

$SS_A/SS_T \times 100\%$ is the percentage contribution for main effect A and so on.

The effects which have higher percentage contribution are considered to be more influential for the response. For this example, we can obtain Fig. 13.1 for the percentage contribution.

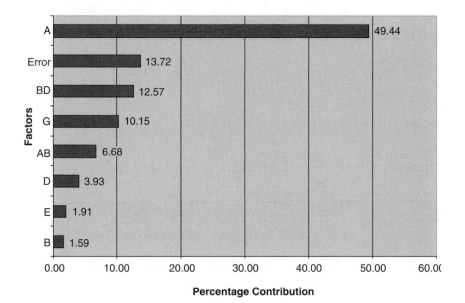

Figure 13.1 Percentage contribution for SS_T.

Clearly, A has the largest percentage contribution, followed by BD and G, AB, and so on.

Example 13.12 Three kinds of fertilizers are applied to soybeans (Park 1996): nitrogen (N), phosphoric acid (P_2O_5), and potash (K_2O), and the response of interest is the average yield in kilograms (kg) per plot of soybeans. The factors are assigned as follows:

	Levels, kg		
Factors	1	2	3
A Nitrogen	0.5	1.0	1.5
B Phosphoric acid	0.3	0.6	0.9
C Potash	0.4	0.7	1.0

An L_9 orthogonal array is used in this experiment. The experiment layout and data are listed as below:

	Factors				
	A	B	e	C	
Experiment no.	1	2	3	4	Response (soybean yield)
1	1	1	1	1	8
2	1	2	2	2	12
3	1	3	3	3	9
4	2	1	2	3	11
5	2	2	3	1	12
6	2	3	1	2	15
7	3	1	3	2	21
8	3	2	1	3	18
9	3	3	2	1	20

Again, using

$$\text{SS} = \frac{k}{N \times n} \sum_{t=1}^{k} T_t^2 - \frac{T^2}{N \times n}$$

and

$$\text{SS}_T = \sum_{i=1}^{N} \sum_{j=1}^{n} y_{ij}^2 - \frac{T^2}{N \times n}$$

$$\text{SS}_A = \frac{3}{9 \times 1} (T_{A_1}^2 + T_{A_2}^2 + T_{A_3}^2) - \frac{T^2}{9 \times 1}$$

$$= \frac{3}{9} [(8 + 12 + 9)^2 + (11 + 12 + 15)^2 + (21 + 18 + 20)^2]$$

$$- \frac{1}{9} (8 + 12 + 9 + 11 + 12 + 15 + 21 + 18 + 20)^2$$

$$= 158$$

$$SS_B = \frac{3}{9 \times 1}(T_{B_1}^2 + T_{B_2}^2 + T_{B_3}^2) - \frac{T^2}{9 \times 1}$$

$$= \frac{3}{9}\,[8 + 11 + 21)^2 + (12 + 12 + 18)^2 + (9 + 15 + 20)^2]$$

$$- \frac{1}{9}\,(8 + 12 + 9 + 11 + 12 + 15 + 21 + 18 + 20)^2$$

$$= 2.667$$

Similarly

$$SS_C = 18.667$$

$$SS_T = [8^2 + 12^2 + 9^2 + 11^2 + 12^2 + 15^2 + 21^2 + 18^2 + 20^2]$$

$$- \frac{1}{9}\,(8 + 12 + 9 + 11 + 12 + 15 + 21 + 18 + 20)^2$$

$$= 180$$

Using MINITAB, we get the following ANOVA table:

Source	DF	Seq SS	Adj SS	Adj MS	F	P
A	2	158.000	158.000	79.000	**	
B	2	2.667	2.667	1.333	**	
C	2	18.667	18.667	9.333	**	
Error	2	0.667	0.667	0.333		
Total	8	180.000				

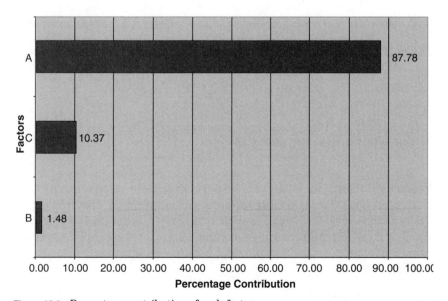

Figure 13.2 Percentage contribution of each factor.

Clearly, from Fig. 13.2, A is by far the most important variable and C is a distant second.

13.4.2   Main-effects chart and interaction chart

In Taguchi experimental data analysis, plots of main-effect charts and interaction charts are the same as those of classical experimental data analysis.

Main-effects chart.   The *main-effects chart* is a plot of average responses at different levels of a factor versus the factor levels.

Example 13.13   Continuing Example 13.11, for factor A, there are two levels: 1 and 2. For level 1

$$\overline{Y}_{A1} = \frac{26 + 38 + 16 + 6 + 3 + 17 + 18 + 16}{8} = 17.5$$

and for level 2

$$\overline{Y}_{A2} = \frac{0 + 5 + 0 + 1 + 4 + 5 + 5 + 3}{8} = 2.875$$

Plotting these two mean responses versus level, we get the main-effects plots shown in Fig. 13.3.

Example 13.14   Continuing Example 13.13, factor A has three levels and

$$\overline{Y}_{A1} = \frac{8 + 12 + 9}{3} = 9.67; \qquad \overline{Y}_{A2} = \frac{11 + 12 + 15}{3} = 12.67;$$

$$\overline{Y}_{A3} = \frac{21 + 18 + 20}{3} = 19.67$$

By plotting the three mean responses versus the factor level, we get the main-effects plot shown in Fig. 13.4.

For all factors in Example 13.12, we get the plots shown in Fig. 13.5.

Interaction chart.   In a Taguchi experiment, only two-factor interactions are considered, as higher-order interaction is assumed to be insignificant. For two-factor interaction, the interaction chart is plotted by displaying all factor level combinations for the relevant two factors. For example, in Example 13.13, BD is a significant interaction. We compute

$$\overline{Y}_{B1D1} = \frac{26 + 38 + 0 + 5}{4} = 17.25; \qquad \overline{Y}_{B1D2} = \frac{16 + 6 + 0 + 1}{4} = 5.75$$

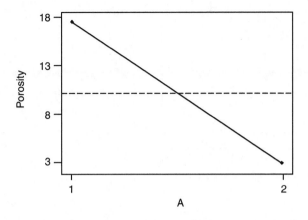

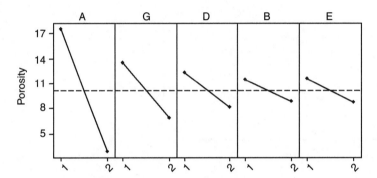

Figure 13.3   Main-effects plots—data means for porosity.

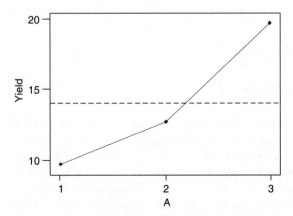

Figure 13.4   Main-effects plot—data means for yield.

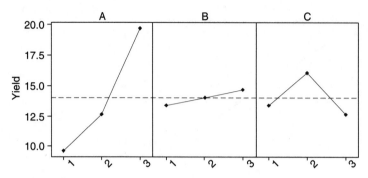

Figure 13.5   Main-effects plot—LS means for yield.

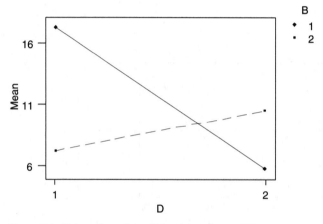

Figure 13.6   Interaction plot—data means for porosity.

$$\overline{Y}_{B_2D_1} = \frac{3 + 17 + 4 + 5}{4} = 7.25; \qquad Y_{B_2D_2} = \frac{18 + 16 + 5 + 3}{4} = 10.5$$

and obtain the interaction plot in Fig. 13.6.

13.4.3   Optimization and prediction of expected response

Optimization in the Taguchi experiment involves finding the factor level combination that gives the optimal response. Optimal response depends on what is the criterion for "best." In Examples 13.11 and 13.13, the porosity value is "the smaller, the better," so we want to find a set of factor level combinations that gives us minimum porosity.

First, we need to determine which factors are important. We can use the F test such as that of classical DOE, in the Taguchi experiment; sometimes we use the "top" factors and interactions that give us cumulative 90 percent of percentage contribution. From the F test, we can find that A, G, and interaction BD are statistically significant. (If we go with percentage contribution, then A,G,BD, error, and AB are top contributors, with a cumulative percentage contribution of 92.56 percent.)

The main-effects chart and BD interaction chart indicate that A and G should be set at level 2 (from main-effects chart). According to the BD interaction chart, B should be at least 1 and D at level 2.

The optimal response prediction in this problem is given by

$$\hat{y} = \bar{y}_{A2} + \bar{y}_{G2} + \bar{y}_{B_1D_2} - 2\bar{T} = 2.875 + 6.875 + 5.75 - 2$$

$$\times 10.188 = -4.873$$

where $\bar{T} = T/(N \times n) = 163/(8 \times 2) = 10.188$ is the average response for the experiment.

For Examples 13.12 and 13.14, we cannot use an F test because there is insufficient degree of freedom for error to conduct a meaningful test. If we use percentage contribution, factors A and C have a cumulative contribution of 98.15 percent, from the main-effects chart in Fig. 13.5, we can determine that A should be at level 3 and C should be at level 2. The estimated optimal yield is $\hat{y} = \bar{y}_{A3} + \bar{y}_{C2} - \bar{T} = 19.67 + 16 - 14 = 21.67$.

13.5 Summary

1. Taguchi experimental design uses standard orthogonal arrays, with the help of a linear graph, an interaction table, and special techniques. A Taguchi experiment considers only main effects and some predetermined two-factor interactions, as higher-order interactions are assumed to be nonexistent.

2. Taguchi experimental data analysis includes
   - ANOVA
   - Main-effects chart and interaction chart
   - Best factor level selection and optimal performance level prediction

Appendix: Selected Orthogonal Arrays

1. $L_4(2^3)$ orthogonal array:

Experiment no.	Column		
	1	2	3
1	1	1	1
2	1	2	2
3	2	1	2
4	2	2	1

Linear Graph for L4

1 ●——————————————● 2
        3

2. $L_8(2^7)$ orthogonal array:

Experiment no.	Column						
	1	2	3	4	5	6	7
1	1	1	1	1	1	1	1
2	1	1	1	2	2	2	2
3	1	2	2	1	1	2	2
4	1	2	2	2	2	1	1
5	2	1	2	1	2	1	2
6	2	1	2	2	1	2	1
7	2	2	1	1	2	2	1
8	2	2	1	2	1	1	2

Linear Graphs for L8

(1)

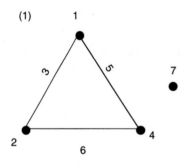

(2)

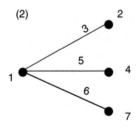

Interaction table for L_8:

Column	Column						
	1	2	3	4	5	6	7
1	(1)	3	2	5	4	7	6
2		(2)	1	6	7	4	5
3			(3)	7	6	5	4
4				(4)	1	2	3
5					(5)	3	2
6						(6)	1
7							(7)

3. $L_9(3^4)$ array:

Experiment no.	Column			
	1	2	3	4
1	1	1	1	1
2	1	2	2	2
3	1	3	3	3
4	2	1	2	3
5	2	2	3	1
6	2	3	1	2
7	3	1	3	2
8	3	2	1	3
9	3	3	2	1

Linear Graph for L9

4. $L_{12}(2^{11})$ *array*. Interactions are partially confounded with all 11 columns.

Experiment no.	Columns										
	1	2	3	4	5	6	7	8	9	10	11
1	1	1	1	1	1	1	1	1	1	1	1
2	1	1	1	1	1	2	2	2	2	2	2
3	1	1	2	2	2	1	1	1	2	2	2
4	1	2	1	2	2	1	2	2	1	1	2
5	1	2	2	1	2	2	1	2	1	2	1
6	1	2	2	2	1	2	2	1	2	1	1
7	2	1	2	2	1	1	2	2	1	2	1
8	2	1	2	1	2	2	2	1	1	1	2
9	2	1	1	2	2	2	1	2	2	1	1
10	2	2	2	1	1	1	1	2	2	1	2
11	2	2	1	2	1	2	1	1	1	2	2
12	2	2	1	1	2	1	2	1	2	2	1

5. $L_{16}(2^{15})$ array:

Experiment no.	Columns														
	1	2	3	4	5	6	7	8	9	10	11	12	13	14	15
1	1	1	1	1	1	1	1	1	1	1	1	1	1	1	1
2	1	1	1	1	1	1	1	2	2	2	2	2	2	2	2
3	1	1	1	2	2	2	2	1	1	1	1	2	2	2	2
4	1	1	1	2	2	2	2	2	2	2	2	1	1	1	1

| | Columns | | | | | | | | | | | | | | |
Experiment no.	1	2	3	4	5	6	7	8	9	10	11	12	13	14	15
5	1	2	2	1	1	2	2	1	1	2	2	1	1	2	2
6	1	2	2	1	1	2	2	2	2	1	1	2	2	1	1
7	1	2	2	2	2	1	1	1	1	2	2	2	2	1	1
8	1	2	2	2	2	1	1	2	2	1	1	1	1	2	2
9	2	1	2	1	2	1	2	1	2	1	2	1	2	1	2
10	2	1	2	1	2	1	2	2	1	2	1	2	1	2	1
11	2	1	2	2	1	2	1	1	2	1	2	2	1	2	1
12	2	1	2	2	1	2	1	2	1	2	1	1	2	1	2
13	2	2	1	1	2	2	1	1	2	2	1	1	2	2	1
14	2	2	1	1	2	2	1	2	1	1	2	2	1	1	2
15	2	2	1	2	1	1	2	1	2	2	1	2	1	1	2
16	2	2	1	2	1	1	2	2	1	1	2	1	2	2	1

Linear Graphs for L16

(1)

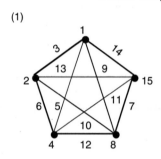

(2)

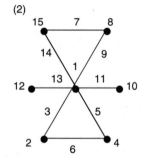

(3)

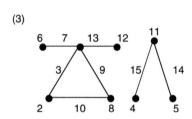

(4)

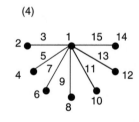

(5)

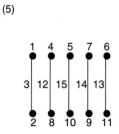

(6)

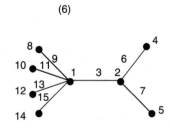

Interaction Table for L_{16}

Column	1	2	3	4	5	6	7	8	9	10	11	12	13	14	15
								Column							
1	(1)	3	2	5	4	7	6	9	8	11	10	13	12	13	14
2		(2)	1	6	7	4	5	10	11	8	9	14	15	12	13
3			(3)	7	6	5	4	11	10	9	8	15	14	13	12
4				(4)	1	2	3	12	13	14	15	8	9	10	11
5					(5)	3	2	13	12	15	14	9	8	11	10
6						(6)	1	14	15	12	13	10	11	8	9
7							(7)	15	14	13	12	11	10	9	8
8								(8)	1	2	3	4	5	6	7
9									(9)	3	2	5	4	7	6
10										(10)	1	6	7	4	5
11											(11)	7	6	5	4
12												(12)	1	2	3
13													(13)	3	2
14														(14)	1
15															(15)

6. $L_{18}(2^1 3^7)$ array:

Experiment no.	Factors							
	1	2	3	4	5	6	7	8
1	1	1	1	1	1	1	1	1
2	1	1	2	2	2	2	2	2
3	1	1	3	3	3	3	3	3
4	1	2	1	1	2	2	3	3
5	1	2	2	2	3	3	1	1
6	1	2	3	3	1	1	2	2
7	1	3	1	2	1	3	2	3
8	1	3	2	3	2	1	3	1
9	1	3	3	1	3	2	1	2
10	2	1	1	3	3	2	2	1
11	2	1	2	1	1	3	3	2
12	2	1	3	2	2	1	1	3
13	2	2	1	2	3	1	3	2
14	2	2	2	3	1	2	1	3
15	2	2	3	1	2	3	2	1
16	2	3	1	3	2	3	1	2
17	2	3	2	1	3	1	2	3
18	2	3	3	2	1	2	3	1

Linear Graph for $L_{18}(2^1 \times 3^7)$

Interaction between columns 1 and 2 is not confounded with that between other columns, but interactions are partially confounded with all columns.

7. $L_{27}(3^{13})$ array:

Experiment no.	Column												
	1	2	3	4	5	6	7	8	9	10	11	12	13
1	1	1	1	1	1	1	1	1	1	1	1	1	1
2	1	1	1	1	2	2	2	2	2	2	2	2	2
3	1	1	1	1	3	3	3	3	3	3	3	3	3
4	1	2	2	2	1	1	1	2	2	2	3	3	3
5	1	2	2	2	2	2	2	3	3	3	1	1	1
6	1	2	2	2	3	3	3	1	1	1	2	2	2
7	1	3	3	3	1	1	1	3	3	3	2	2	2
8	1	3	3	3	2	2	2	1	1	1	3	3	3
9	1	3	3	3	3	3	3	2	2	2	1	1	1
10	2	1	2	3	1	2	3	1	2	3	1	2	3
11	2	1	2	3	2	3	1	2	3	1	2	3	1
12	2	1	2	3	3	1	2	3	1	2	3	1	2
13	2	2	3	1	1	2	3	2	3	1	3	1	2
14	2	2	3	1	2	3	1	3	1	2	1	2	3
15	2	2	3	1	3	1	2	1	2	3	2	3	1
16	2	3	1	2	1	2	3	3	1	2	2	3	1
17	2	3	1	2	2	3	1	1	2	3	3	1	2
18	2	3	1	2	3	1	2	2	3	1	1	2	3
19	3	1	3	2	1	3	2	1	3	2	1	3	2
20	3	1	3	2	2	1	3	2	1	3	2	1	3
21	3	1	3	2	3	2	1	3	2	1	3	2	1
22	3	2	1	3	1	3	2	2	1	3	3	2	1
23	3	2	1	3	2	1	3	3	2	1	1	3	2
24	3	2	1	3	3	2	1	1	3	2	2	1	3
25	3	3	2	1	1	3	2	3	2	1	2	1	3
26	3	3	2	1	2	1	3	1	3	2	3	2	1
27	3	3	2	1	3	2	1	2	1	3	1	3	2

Linear Graphs for L27

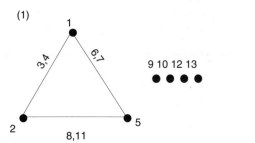

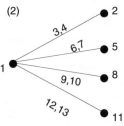

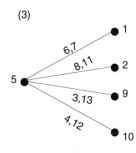

Design Optimization: Taguchi's Robust Parameter Design

14.1 Introduction

Taguchi's robust parameter design is a systematic methodology that applies statistical experimental design for detailing transfer functions and design optimization. Dr. Genichi Taguchi's development of robust design is a great engineering achievement (Clausing 1994).

The P-diagram paradigm in Fig. 14.1 is used as the basic model for the Taguchi method.

Figure 14.1 illustrates a general design model where $\mathbf{y} = (y_1,...,y_u)$ is the output vector representing system performance, or the left side of the transfer function vector; $\mathbf{x} = (x_1,...,x_n)$ is the design parameter (or the process variable) vector of the design; and $\mathbf{z} = (z_1,...,z_p)$ is the vector representing the uncontrollable factors, or noise factors of the design.

Obviously, both design parameters and noise factors will influence the transfer function output $\mathbf{y}$. In this chapter, we are working on just one particular requirement characteristic, say, y. In a practical situation, we may have to deal with many requirements at the same time. We can deal with multiple requirements by the response surface method (Chap. 17) or using Taguchi's ideal function (Chap. 15). The following transfer function can be used to represent the cause-and-effect relationship between output y and $\mathbf{x}$ and $\mathbf{z}$:

$$y = g(x_1,...,x_n, z_1,...,z_p) \qquad (14.1)$$

In some circumstances, we can assume that output y is determined by design parameters $\mathbf{x}$ and the design parameters are influenced by noise factors, $\mathbf{z}$. In this case:

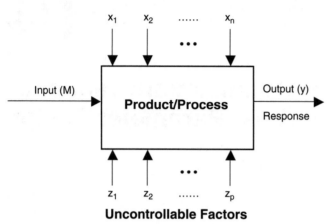

Figure 14.1   A design P-diagram.

$$y = f(x_1,\ldots,x_n) \tag{14.2}$$

Mathematically, $\mathbf{z} = (z_1,\ldots,z_p)$ is a random vector with unknown distribution. The design parameter $\mathbf{x} = (x_1,\ldots,x_n)$ is also a random vector, because in mass production, where we cannot guarantee production of a massive number of identical products, there will be piece-to-piece (piecewise) variation. We can assume $x_i \sim N(\mu_{x_i}, \sigma_{x_i}^2)$ for $i = 1\cdots n$. We further assume that $\mu_{x_i} = \mu_i$, which is the nominal value of the design parameter x_i. The nominal value of the design parameter can be chosen by the designer, and we can define $\sigma_i = \sigma_{x_i}$, which is the standard deviation of the design parameter, which can also be controlled (at a cost) at tolerance design stage.

Because both $\mathbf{x}$ and $\mathbf{z}$ are random variables, the requirement Y will of course be a random variable. In Taguchi's parameter design, it is assumed that there is a target value for Y specifically, T. For example, the main basic function of a power supply circuit is to provide electrical power to small appliances, and a key output performance is its output voltage. If the needed voltage for a smaller appliance is 6 V, then the target value $T = 6$.

14.2   Loss Function and Parameter Design

14.2.1   Quality loss function

In the Taguchi method, quality loss function is a basic starting point:

$$L = kE(Y - T)^2 \tag{14.3}$$

where L = quality loss
$\quad k$ = quality loss coefficient
$\quad E$ = expected value
$\quad T$ = target value of Y

The objective of the robust design is to minimize the loss function L. Taguchi concluded that any deviation of performance y from its target value will incur a quality loss.

Figure 14.2 can be used to show the meaning of the loss function. In the late 1970s, American consumers showed a preference for television sets made by Sony Japan over those made by Sony USA. The reason cited in the study was quality. However, both factories produced televisions using identical designs and tolerances. In its investigation report, the Asahi newspaper showed the distribution of color density for the sets made by two factories (see Fig. 14.2). In Figure 14.2, T is the target value, and $T \pm 5$ are the tolerance limits. The distribution of Sony USA was nearly uniform within tolerance limits with no units out of the tolerance limits. The distribution of Sony Japan was approximately normal, with approximately 0.3 percent out of the tolerance limits. Why did customers prefer Sony Japan? By looking into the distribution more closely, we can see that more units made by Sony Japan are A grade, that is, the color density is within $T \pm 1$, and less

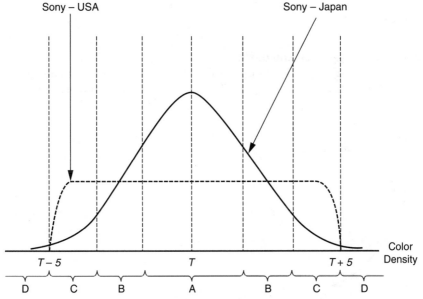

Figure 14.2 Distribution of color density in television sets. (*From The Asahi, April 17, 1979.*)

B or C grade than that of Sony USA. So, for an average customer, the average grade of television set made by Sony Japan is better than that of Sony USA.

Thus, in many cases, any deviation of key quality characteristics from their target values will create customer dissatisfaction. The more deviation, the more dissatisfaction. This is exactly the meaning of quality loss function.

Figure 14.3 gives a graphical view of quality loss function; when the performance level $y = T$, the quality loss is zero. Assume that $T + \Delta_0$ and $T - \Delta_0$ are the functional limits, that is, when either $y < T - \Delta_0$ or $y > T + \Delta_0$, the product will not function at all and the customer will demand replacement. We further assume that the replacement cost is A_0; then, according to Eq. (14.3), quality loss at $y = T + \Delta_0$ or at $y = T - \Delta_0$ is equal to A_0:

$$A_0 = k(\Delta_0)^2 \qquad \therefore \qquad k = \frac{A_0}{\Delta_0^2} \qquad (14.4)$$

Example 14.1 Television Set Color Density (Phadke 1989)   Suppose that the functional limits for the color density are $T \pm 7$; that is, that customers will demand that their TV sets be replaced or repaired when the color density is at or beyond $T \pm 7$. Assume that the replacement cost is $A_0 = \$98$; then using Eq. (14.4) we obtain $k = 98/7^2 = 2$. Also, by Eq. (14.3), the quality loss at $y = T+4$ is $L = 2(T + 4 - T)^2$ and at $y = T + 2$ it is $L = 2(2)^2 = \$8$.

Quality characteristics and quality loss functions.   The quality loss function given by Eq. (14.3) is applicable whenever the quality characteristic y has a finite target value T, which is usually nonzero, and the

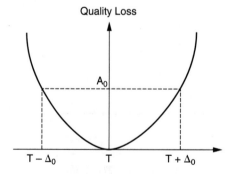

Figure 14.3   Quality loss function.

quality loss is symmetric on either side of the target value. This type of quality characteristic is called nominal-the-best (meaning "nominal is best"). An example is TV set color density (Example 14.1).

The following quality characteristics require different types of quality loss functions:

- *Smaller-the-better quality characteristics.* For some quality characteristics, such as percentage defect rate or radiation leakage from a microwave oven, the ideal target values are zero. Other examples of such quality characteristics, termed *smaller-the-better* (i.e., "the smaller, the better" or "smaller is better"), are response time of a computer, leakage current in electronic circuits, and pollution from automobile exhaust. The quality loss function in such cases can be obtained by letting $T = 0$ in Eq. (14.3).

$$L = kEY^2 \qquad (14.5)$$

- *Larger-the-better quality characteristics.* For some quality characteristics, such as welding bond strength, the ideal target value is infinity. These are *larger-the-better* (i.e., "the larger, the better" or "larger is better") quality characteristics. Performance level will progressively worsen if y decreases; the worst possible value is zero. It is clear that the behavior of this characteristic is the reciprocal of or inversely proportional to that of the smaller-the-better characteristic. Thus, we can substitute $1/Y$ in Eq. (14.5) to obtain the quality loss function in this case:

$$L = kE\left(\frac{1}{Y^2}\right) \qquad (14.6)$$

- In this case, if the functional limit is Δ_0, below which the product will fail and the replacement or repair cost is A_0, then, by Eq. (14.6), k can be determined as

$$k = A_0\Delta_0^2 \qquad (14.7)$$

These three kinds of quality characteristics and their loss functions are plotted in Fig. 14.4.

Components of quality loss.   Without losing the generality, we use the nominal-the best quality loss function in Eq. (14.3) to get

$$L = kE(Y - T)^2 = k(\mu_y - T)^2 + k\,\text{Var}(Y) = k(\mu_y - T)^2 + k\sigma_y^2 \qquad (14.8)$$

where $\mu_y = E(Y)$, which is the mean value of performance level Y, and $\text{Var}(Y)$ is the variance of performance level Y.

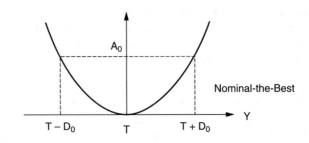

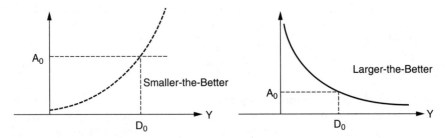

Figure 14.4 Quality loss functions.

So we can see that there are two components in quality loss:

1. *Off-target.* This component corresponds to the deviation of mean performance level from the target value and can be represented by $k(\mu_y - T)^2$.

2. *Variation.* This corresponds to the variation of performance level, represented by $k\,\mathrm{Var}(Y)$.

If we want to reduce the quality loss, we need to reduce both off-target and variation components.

14.2.2 Sources of variation

From Eq. (14.8)

$$L = kE(Y - T)^2 = k(\mu_y - T)^2 + k\,\mathrm{Var}(Y) = k(\mu_y - T)^2 + k\sigma_y^2$$

It is clear that in order to reduce the quality loss, we should reduce both off-target $k(\mu_y - T)^2$ and variation components, as illustrated in Figs. 14.5 and 14.6.

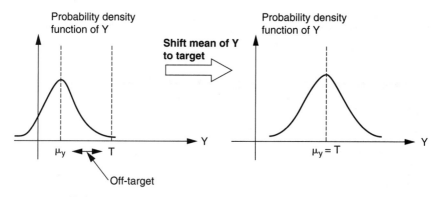

Figure 14.5 Reduce quality loss by shifting mean performance to target.

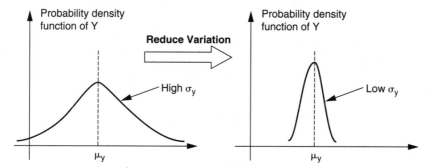

Figure 14.6 Reduce quality loss by variation reduction.

Usually, adjusting the mean performance μ_y to the target value T is relatively easy, because there are many design parameters $\mathbf{x} = (x_1, x_2,...,x_n)$ which can be adjusted to change μ_y.

However, the things are more complicated for reducing variation Var(Y), the variation of Y (Fig. 14.7), is caused by both the variation of design parameters $\mathbf{x} = (x_1, x_2,..., x_n)$ and the variation of noise factors $\mathbf{z} = (z_1, z_2, ...,z_n)$.

Overall, the sources of variation can be summarized by the following three categories:

- *External sources (usage and environment)*. Examples include temperature; use, misuse, and abuse by the consumer; and loading-related variation, which is mostly from the variation of noise factors $\mathbf{z} = (z_1, z_2,..., z_n)$.

- *Unit-to-unit sources (manufacturing and supplier variation)*. This is dimensional, assembly-related, or material property variation from

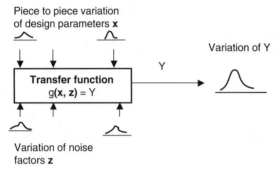

Figure 14.7   Sources of variation for *Y*.

target values. This unit-to-unit variation of design parameters is expressed as $\mathbf{x} = (x_1, x_2,..., x_n)$.

- *Deterioration sources (wearout)*. These are functional characteristics which degrade or deteriorate from the time the product is new, such as material fatigue or aging and wear, abrasion, and the general effects of usage over time. This source could be either the time-related variation of design parameters $\mathbf{x} = (x_1, x_2,..., x_n)$, where certain design parameters may degrade over time, or time-related variation of noise factors $\mathbf{z} = (z_1, z_2,..., z_n)$.

In Taguchi's approach, we can achieve quality improvement by reducing quality loss in the following three stages:

1. *System design*. This broadly corresponds to conceptual design in the generalized model of the design process. In the system design stage, the technology used in the design is finalized; the system configuration, the subsystem structures, and their relationships are determined. Robust design using the Taguchi method does not focus on the system design stage. However, different system designs certainly will have different sets of design parameters, noise factors, and transfer functions; therefore, system design will certainly affect quality loss.

2. *Parameter design*. This stage follows the system design stage. In the parameter design stage, the transfer function will already have been finalized (in the system design stage); but the nominal values of design parameters can still be adjusted, such as when we design power supply circuits for small appliances. After system design, the circuit diagram has already been finalized. But we still can change the design values of the circuit parameters, such as the designed value of resistors and transistors. Given the transfer function

$$Y = g(x_1, x_2,..., x_n; z_1, z_2,..., z_n).$$

we can get the following relationships by Taylor series approximation:

$$\Delta y = \frac{\Delta g}{\Delta x_1} \Delta x_1 + \frac{\Delta g}{\Delta x_2} \Delta x_2 + \cdots + \frac{\Delta g}{\Delta x_n} \Delta x_n$$

$$+ \frac{\Delta g}{\Delta z_1} \Delta z_1 + \frac{\Delta g}{\Delta z_2} \Delta z_2 + \cdots + \frac{\Delta g}{\Delta z_m} \Delta z_m \qquad (14.9)$$

and

$$\text{Var}(Y) = \sigma_y^2 \approx \left(\frac{\partial g}{\partial x_1} \right)^2 \sigma_{x1}^2 + \left(\frac{\partial g}{\partial x_2} \right)^2 \sigma_{x2}^2 + \cdots + \left(\frac{\partial g}{\partial x_n} \right)^2 \sigma_{xn}^2$$

$$+ \left(\frac{\partial g}{\partial z_1} \right)^2 \sigma_{z1}^2 \left(\frac{\partial g}{\partial z_2} \right)^2 \sigma_{z2}^2 + \left(\frac{\partial g}{\partial z_m} \right)^2 \sigma_{zm}^2 \qquad (14.10)$$

From this equation, it is clear that we can reduce Var(Y) by reducing either the sensitivities $\partial g/\partial x_i$, for $i = 1...n$, which are the sensitivities to the variation in design parameters, or $\partial g/\partial z_j$, for $j = 1...m$, which are sensitivities to noise factors. Fortunately, many of these sensitivities are influenced by the nominal values of design parameters. Figure 14.8 shows how parameter setting may influence the variation of Y. Both transfer function $y = g(\mathbf{x},\mathbf{z})$ and sensitivity, for example, $\partial g/\partial x_i$, can be nonlinear functions of parameter

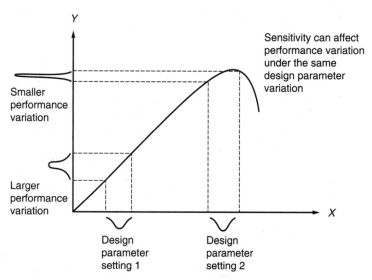

Figure 14.8 Adjusting sensitivity to reduce Var(Y).

settings; therefore, by choosing the parameter setting in a low-sensitivity area, we can reduce Var(Y) significantly, although we have the same variation of either design parameters or noise factors.

In summary, the tasks of parameter design are as follows:

a. Adjusting the nominal settings of design parameters $E(x_i) = \mu_i$, for $i = 1 \cdots n$, to reduce the sensitivities of functional performance y to the variations of design parameters and noise factors, and as a result, reducing the system functional performance variation Var(Y).

b. Adjusting the nominal settings of design parameters to shift mean performance level μ_y to its target value T.

The major objective of task a is to make the system insensitive or "robust" to all sources of variation. Therefore, Taguchi's parameter design is also called *robust parameter design*.

3. *Tolerance design.* Tolerance design is usually performed after parameter design. The major task of tolerance design is to set the acceptable variation level for design parameters. Specifically, tolerance design is to set a pair of limits for each design parameter x_i such that $\mu_i - \Delta_i \leq x_i \leq \mu_i + \Delta_i$, where usually $\Delta_i = 3C_p\sigma_i$, where C_p is the process capability index (when $C_p = 2$, that is a Six Sigma tolerance limit). By choosing high-grade parts, or more precise (expensive) manufacturing methods, we can reduce σ_i; by Eq. (14.10), we can reduce Var(Y) as well. However, this approach has the following inadequacies:

a. It is an expensive method.

b. We can usually control and reduce the variation of design parameters σ_i for $i = 1 \cdots n$, only. In other words, we can control piece to piecewise variation only by tightening tolerances for design parameters. We can seldom effectively control the external variation and deterioration.

Therefore, tightening tolerance is inadequate in variation reduction for system performance; robust parameter design is an important step in improving quality level by reducing variation.

Although minimizing the quality loss is the goal for Taguchi's robust parameter design method, that method does not work on loss function directly. Taguchi's parameter design method is an integration of the Taguchi orthogonal array experiment and signal-to-noise ratio analysis. The signal-to-noise ratio is related to the loss function, and we discuss it in detail in the next section.

14.3   Loss Function and Signal-to-Noise Ratio

In the field of communication engineering, *signal-to-noise ratio* (S/N), which is also related to loss function, has been used as a performance characteristic for many circuits or products. Taguchi, whose background is communication and electronic engineering, introduced the same concept in many earlier designs of experiment projects in Japan.

14.3.1 Nominal-the-best quality characteristics

In communication engineering, Fig. 14.9 is a typical representation of how the system works.

In a typical communication system, if an input signal is given, then an output signal y should be produced by the system. In an ideal situation, if there is no noise input, the output y will be consistent and have no variation. However, with noise input, the output y will not be consistent and will vary. For a good communication system, the noise effect should have minimal influence in comparison with the output y; therefore, the following signal-to-noise ratio (S/N) is often used as a quality characteristic of a communication system:

$$\frac{\text{Signal power}}{\text{Noise power}} = \frac{\mu^2}{\sigma^2} \qquad (14.11)$$

where $\mu = E(Y)$, $\sigma^2 = \text{Var}(Y)$. The larger the ratio given by Eq. (14.11), the more desirable the communication is. In communication engineering, the signal-to-noise ratio is actually 10 times the logarithm of the ratio in Eq. (14.11):

$$\text{S/N} = 10 \log\left(\frac{\mu^2}{\sigma^2}\right) \qquad (14.12)$$

Use of a logarithm of a signal-to-noise ratio is based on the fact that the logarithm will transform a multiplicative relationship into an additive relationship; thus the logarithm transformation will "smooth out" lots of nonlinearities and interactions, which is desirable in data analysis.

In real-world conditions, both μ and σ^2 can only be estimated statistically. If a set of observations of quality characteristic Y are given, that is, $y_1, y_2, \ldots, y_n$, the statistical estimate for μ is

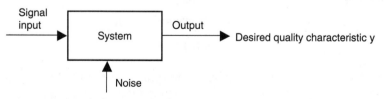

Figure 14.9 A typical communication system.

$$\hat{\mu} = \bar{y} = \frac{1}{n} \sum_{i=1}^{n} y_i \tag{14.13}$$

The statistical estimate for σ^2 is

$$\hat{\sigma}^2 = s^2 = \frac{1}{n-1} \sum_{i=1}^{n} (y_i - \bar{y})^2 \tag{14.14}$$

Therefore, the S/N ratio calculation for a set of real output performance data $y_1, y_2, ..., y_n$ is given by

$$S/N = 10 \log \left(\frac{\hat{\mu}^2}{\hat{\sigma}^2} \right) = 10 \log \left(\frac{\bar{y}^2}{s^2} \right) = 20 \log \left(\frac{\bar{y}}{s} \right) \tag{14.15}$$

Comparing this with the nominal-the-best quality loss function, we obtain

$$L(Y) = k(\mu - T)^2 + k\sigma^2 \tag{14.16}$$

Taguchi proposed a two-step optimization procedure:

1. Adjust design parameters to maximize the S/N ratio.
2. Find some other design parameters that have no effect on S/N ratio but affect the mean level of Y, $E(Y) = \mu$, which is called the *mean adjusting parameter,* and then use it to tune $E(Y)$ to the target value.

We can see that by applying these two steps, $L(Y)$ will be minimized. This two-step procedure was challenged and criticized by many people, including Box (1988), and alternative approaches have been proposed. However, in communication engineering, maximizing S/N first and then adjusting the mean to the target is a common practice.

For many other noncommutation engineering systems, the idea of minimizing the influence of "noise factors" and being robust to all sources of variation has also proved to be very valuable. Millions of Taguchi projects have been conducted worldwide since the Taguchi method became well known.

14.3.2 Smaller-the-better quality characteristics

The smaller-the-better quality loss function is given by

$$L(Y) = kE(Y^2)$$

In the real world, $E(Y^2)$ can only be estimated statistically. If a set of observations of quality characteristic Y are given, that is, $y_1, y_2, ..., y_n$, the statistical estimate of $E(Y^2)$ is

$$\text{MSD} = \frac{1}{n} \sum_{i=1}^{n} y_i^{2} \tag{14.17}$$

where MSD stands for the *mean-squared deviation* from the target value 0. The signal-to-noise ratio in this case is defined as

$$\text{S/N} = -10 \log \left(\frac{1}{n} \sum_{i=1}^{n} y_i^{2} \right) \tag{14.18}$$

Clearly, S/N is -10 times the logarithm of MSD; the smaller the MSD, the larger the S/N ratio. Therefore, for the smaller-the-better quality loss function, maximizing S/N is equivalent to minimizing the loss function.

14.3.3 Larger-the-better quality characteristics

The larger-the-better quality loss function is given by

$$L(Y) = kE \left(\frac{1}{Y^{2}} \right)$$

If a set of observations of quality characteristic Y are given, that is, $y_1, y_2, \ldots, y_n$, the statistical estimate of $E(1/Y^2)$ is

$$\text{MSD} = \frac{1}{n} \sum_{i=1}^{n} \frac{1}{y_i^{2}} \tag{14.19}$$

The corresponding S/N is

$$\text{S/N} = -+10 \log \left(\frac{1}{n} \sum_{i=1}^{n} \frac{1}{y_i^{2}} \right) \tag{14.20}$$

Again, maximizing S/N is equivalent to minimizing the quality loss function.

14.3.4 Robust parameter design using signal-to-noise ratio and orthogonal array experiment

In Taguchi's robust parameter design approach, a number of design parameters will be selected and an orthogonal array for the experiment will be selected. In each experimental run, several replicates of output performance observations will be collected as shown in Table 14.1.

TABLE 14.1   Taguchi Robust Parameter Design Setup

Experiment no.	Design parameters			Replicates			S/N
	A	B	...	1	...	n	η
1	1	1	...	y_{11}	...	y_{1n}	η_1
2	2	1	...	y_{21}	...	y_{2n}	η_2
⋮	1	2	...	⋮	⋮	⋮	⋮
	2	2	...				
	⋮	⋮					
N	2	2	...	y_{N1}	...	y_{Nn}	η_N

The signal-to-noise ratio is computed for each experimental run. If the quality characteristics y are either of smaller-the-better or larger-the-better type, then we will try to find the design parameter level combination to maximize S/N, thus minimizing the quality loss.

If the quality characteristic y is the nominal-the-best type characteristic, we will perform the following two steps:

1. Find and adjust significant design parameters to maximize S/N.

2. Find the mean adjustment design parameter to tune the mean response to the target value.

Example 14.2 (Harrell and Cutrell 1987)   The weather strip in an automobile is made of rubber. In the rubber industry, an extruder is used to mold the raw rubber compound into the desired shapes. Variation in output from the extruders directly affects the dimensions of the weather strip as the flow of the rubber increases or decreases. A Taguchi experiment is conducted in order to find appropriate control factor levels for smooth rubber extruder output. Table 14.2 gives the control factors and levels for this Taguchi experiment:

In this project, interaction is considered to be insignificant. An L_8 array is used. Output was measured by setting up the condition for each experimental run and the product produced during a 30-seconds time period was weighed.

TABLE 14.2   Control Factors for Example 14.2

Control Factors	Level 1	Level 2
A	Same	Different
B	Same	Different
C	Cool	Hot
D	Current level	Additional material
E	Low	High
F	Low	High
G	Normal range	Higher

Ten samples, representing 30 seconds of output each, were gathered for each experimental run. Table 14.3 represents data generated in this study.

In this project, the desired production output is neither smaller-the-better nor larger-the-better. The producer should determine an appropriate production output level if variation in output is well under control. Therefore, in Table 14.3, the signal-to-noise ratio is computed by using Eq. (14.15), that is, S/N for the nominal-the-best case:

$$S/N = \eta = 20 \log \left(\frac{\overline{y}}{s} \right)$$

By conducting analysis of variance on the S/N, we can obtain the following ANOVA table:

```
Analysis of Variance for S/N, using Adjusted SS for Tests
Source   DF    Seq SS    Adj SS    Adj MS     F    P
A         1     9.129     9.129     9.129    **
B         1     5.181     5.181     5.181    **
C         1     5.324     5.324     5.324    **
D         1    39.454    39.454    39.454    **
E         1     5.357     5.357     5.357    **
F         1     6.779     6.779     6.779    **
G         1     2.328     2.328     2.328    **
Error     0     0.000     0.000     0.000
Total     7    73.552
```

By using the percentage contribution calculation discussed in Chap. 13, we can obtain the chart in Fig. 14.10, where it is clear that D is by far the most important factor for S/N (see also Fig. 14.11). Factor G has only a 3.17 percent contribution and thus is considered to be insignificant.

On the basis of the main-effects chart for S/N in Fig. 14.4, since a higher S/N is desired, we should combine factors A, B, D, and E at level 1 with factors C and F at level 2; this combination should give us the highest S/N. We can denote this final factor level combination as $A_1B_1C_2D_1E_1F_2$.

The predicted S/N at optimal factor levels combination can be calculated by using the method described in Chap. 13:

$$S/N = \overline{A}_1 + \overline{B}_1 + \overline{C}_2 + \overline{D}_1 + \overline{E}_1 + \overline{F}_2 - 5\overline{T}$$
$$= 34.84 + 34.57 + 34.59 + 35.99 + 34.55 + 34.69 - 5 \times 33.77$$
$$= 40.38$$

where $\overline{T}$ is the average S/N ratio, $\overline{A}_1$ is the average S/N corresponding to A at level 1, and so on.

$$\overline{T} = \frac{37.48 + 32.17 + 38.27 + 31.43 + 34.26 + 34.38 + 33.94 + 28.22}{8}$$

$$= 33.77$$

$$\overline{A}_1 = \frac{37.48 + 32.17 + 38.27 + 31.43}{4} = 34.84$$

TABLE 14.3 Experimental Setup and Data for Example 14.2

Experiment no.	Control factors							Output per 30 second										S/N ratio η
	A	B	C	D	E	F	G	1	2	3	4	5	6	7	8	9	10	
1	1	1	1	1	1	1	1	268.4	262.9	268.0	262.2	265.1	259.1	261.5	267.4	264.7	270.2	37.48
2	1	1	1	2	2	2	2	302.9	295.3	298.6	302.7	314.4	305.5	295.2	286.3	302.0	299.2	32.17
3	1	2	2	1	1	2	2	332.7	336.5	332.8	342.3	332.2	334.6	334.8	335.5	338.2	326.8	38.27
4	1	2	2	2	2	1	1	221.7	215.9	219.7	221.2	221.5	230.1	228.3	228.3	214.6	213.2	31.43
5	2	1	2	1	2	1	2	316.6	326.5	320.4	327.0	311.4	310.8	314.4	319.3	310.0	314.5	34.26
6	2	1	2	2	1	2	1	211.3	222.0	218.2	218.6	218.6	216.5	214.8	217.4	210.8	223.9	34.38
7	2	2	1	1	2	2	1	210.7	210.0	211.6	211.7	210.1	206.5	203.4	207.2	208.0	219.3	33.94
8	2	2	1	2	1	1	2	287.5	299.2	310.6	289.9	290.0	294.5	294.2	297.4	293.7	325.6	28.22

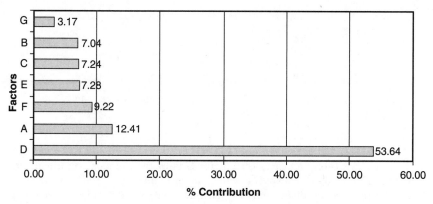

Figure 14.10 Percentage contribution of control factors for S/N.

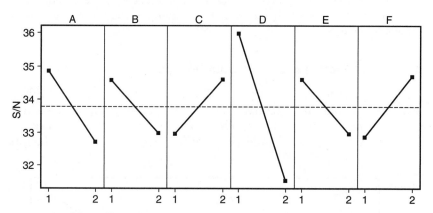

Figure 14.11 Main-effects chart—data means for S/N.

$$\overline{B}_1 = \frac{37.48 + 32.17 + 34.26 + 34.38}{4} = 34.57$$

Similarly, $\overline{C}_2 = 34.59$, $\overline{D}_1 = 35.99$, $\overline{E}_1 = 34.55$, $\overline{F}_2 = 34.69$.

We also need to determine an appropriate production output level; that is, we have to determine the appropriate $E(Y)$. In this project, an analysis variance is conducted for the original output data (y data with 10 replicates), and the following ANOVA table is obtained:

```
Analysis of Variance for Output/3, using Adjusted SS for Tests
Source    DF    Seq SS     Adj SS    Adj MS        F         P
A          1      7770       7770      7770    190.71    0.000
B          1      1554       1554      1554     38.15    0.000
C          1       366        366       366      8.99    0.004
D          1      9990       9990      9990    245.23    0.000
E          1      5521       5521      5521    135.52    0.000
F          1      1984       1984      1984     48.70    0.000
G          1    141742     141742    141742   3479.20    0.000
Error     72      2933       2933        41
Total     79    171861
```

This ANOVA table indicates that factor G is by far the most significant factor in influencing the output. Because factor G has an insignificant effect on S/N but a very significant effect on output itself, it is a perfect choice for the mean adjusting factor. The main-effects chart for G (Fig. 14.12) indicates that shifting G to a higher level will increase the production output. Since increasing G will increase the output but not affect S/N, and a higher production level will certainly increase the throughput, the optimal setting for G should be level 2.

14.4   Noise Factors and Inner-Outer Arrays

In the last section and Example 14.2, the "variation" portion of S/N is estimated by computing sample variance s^2 from a group of replicated output response observations. This variation in replicated observations will not adequately reflect the full effects of actual noise factors

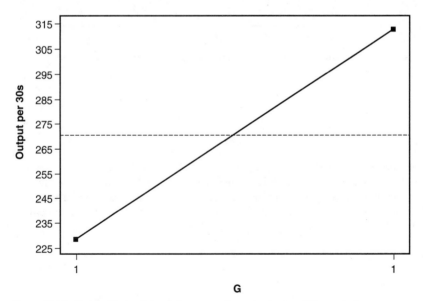

Figure 14.12 Main-effects plot—data means for output per a 30-s period.

on the product. As discussed earlier, when a product is in use, many noise factors will interact with the product and influence its performance. To be successful in the marketplace, it is very important that a product be able to withstand the effects of noise factors and perform well under their influence. There are four common approaches to noise factors:

1. *Ignore noise factors.* Clearly this "do nothing" approach is not helpful for a product to succeed.

2. *Control or eliminate noise factors.* Examples of this approach include tolerance design and standardization (control piecewise variation) and factories or workshops with artificially controlled environments (to control external noises). It is an expensive approach and is not applicable to many consumer products. (For example, how can you shield a car from the environment?)

3. *Compensate the effects of noise factors.* Examples of this approach include feedback control and adaptive control (both compensate disturbances for a process) and selective assembly (compensate piecewise variation).

4. *Minimize the effect of noise (Taguchi robust parameter design).* This approach is to artificially introduce noise factors into the early design stage, and the objective is to select the set of optimal design parameters that make the product perform well not only on small-scale prototype production under a controlled environment but also in mass production and actual adverse user environments.

Clearly, approach 4 is a proactive and inexpensive approach. It is highly desirable to conduct a good robust parameter design in the product design stage. If we can achieve a robust product design, we could either eliminate the need for approaches 2 and 3, or at least make them easier to implement and less costly.

In the early design stage, many design and development activities are conducted in either labs with controlled environments or in the form of computer models; there are very few actual noise factors. In parameter design, we have to artificially introduce, create, or simulate noise factors. It is highly desirable to select a sufficient number of noise factors to simulate what the product will encounter in real-world use. "Artificial" noise factors can be identified by following the checklist of three sources of variation:

- External variation (usage and environment)
- Unit-to-unit variation (manufacturing and supplier variation)
- Deterioration variation

After noise factors are identified, they will be assigned in the robust parameter design experiment, the paradigm of which is illustrated in Fig. 14.13.

The actual assignments of control factors and noise factors, the experimental layout and data collection follows the inner-array/outer-array template illustrated in Table 14.4.

Control factors are assigned to an orthogonal array called *inner array,* where each experimental run corresponds to a unique control factor level combination. At each inner array experimental run, we can vary noise factors according to the layout of another orthogonal array assigned to noise factors, called the *outer array.* Each run in the outer array corresponds to a unique combination of noise factor levels. For instance, if the outer array has N_2 runs, then for each inner-array experimental run, we will obtain N_2. Experimental observations in which each observation corresponds to a different noise factor level combination. For example, in the first run of inner array in Table 14.4, we get $y_{11}, y_{12}, ..., y_{1N_2}$, where each observation corresponds to the same control factor level combination but different noise factor level combinations, so the variation among $y_{11}, y_{12}, ..., y_{1N_2}$ is clearly caused by noise factors. S/N will be computed for each run of inner array; clearly, a higher S/N indicates less sensitivity to the effects of noise factors. Therefore, the optimal design parameter selection based on this kind of experimental study will have a far better chance to be robust in performance in real-world usage.

In many robust parameter design projects, because of the high cost of building the actual hardware prototypes, computer simulation models are used instead of real tests. In these projects, design parameters are varied in computer models and output responses are simulated by the

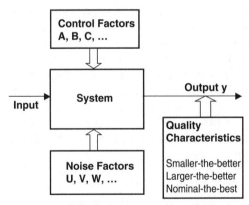

Figure 14.13 Taguchi robust parameter design process.

TABLE 14.4   Taguchi's Inner Array–Outer Array Layout

						Outer array				
						Experiment no.			Noise factors	
						1	2	... N_2		
	Inner array					1	1	2	U	
Experiment	Control Factors					1	2	2	:	S/N
no.	A	B	C ...	F		1	2	1	W	η
1	1	1	1	1		y_{11}	y_{12}	y_{1N_2}		η_1
2	2	1	2	1		y_{21}	y_{22}	Y_{2N_2}		η_2
:	:	:	:	:		:	:	:		:
N_1	2	2	2	2		$y_{N_1 1}$	$y_{N_1 2}$	$y_{N_1 N_2}$		η_{N_1}

computer. In this situation, S/N calculated as described in Sec. 14.3 will not work because the variation s^2 from a group of replicated output response observations will always be zero. Therefore, noise factors must be introduced for computer-simulated robust parameter design.

It is desirable to introduce a sufficient number of noise factors in robust parameter design to ensure future performance robustness toward many kinds of noise factors. However, if we use the inner/outer-array approach, the more noise factors we introduce, the larger the outer array will be, and the larger N_2 will be. In Table 14.4, it is clear that the total number of output response observations y_{ij} is equal to $N_1 \times N_2$; when N_2 increases, $N_1 \times N_2$ will be too high.

In order to reduce the number of experimental runs, Dr. Taguchi proposed a "compounded noise factor" strategy for introducing noise factors. Taguchi stated that to estimate the robustness indicated by signal-to-noise ratio, we do not need to run *all* possible noise combinations designated by the outer array. If a product design is robust toward a small number of extreme or "worst" noise factors level combinations, it will probably be robust to all "minor" or benign noise factor level combinations. The compounded noise factors means to find two extreme combinations of noise factors:

$N1$ negative-side extreme condition

$N2$ positive-side extreme condition

Or three combinations:

$N1$ Negative-side extreme condition

$N2$ Standard-side condition

$N3$ Positive-side condition

For example, in a power supply circuit design, if the output response is the output voltage, then we need to find one extreme noise factor combination which will make output voltage lower, that is, $N1$, and another set of noise factor combination which will make output voltage higher, that is, $N2$.

Example 14.3. Compounded Noise Factors (Phadke 1989)   Figure 14.14 illustrates the circuit diagram of a temperature control circuit, in which R_T is the thermistor resistance, which is inversely proportional to temperature. R_1, R_2, R_3, and R_T actually form a Wheatstone bridge. When temperature decreases and R_T increases, after R_T increases to a value $R_{T,on}$, the Wheatstone bridge is out of balance and will trigger the amplifier to activate the heater. After the heater is on for a while, the temperature increases and R_T decreases; after R_T decreases to a value $R_{T,off}$, the amplifier will shut off the heater and the process will continue.

In a robust parameter design experiment, $R_{T,on}$ is an output response y. The variation of circuit design parameters R_1, R_2, R_4, E_0, and E_z are noise factors. The noise factor levels are given in the following table:

	Levels		
Factor	1	2	3
R_1	2% below nominal value	Nominal value	2% above nominal value
R_2	2% below nominal value	Nominal value	2% above nominal value
R_4	2% below nominal value	Nominal value	2% above nominal value
E_0	2% below nominal value	Nominal value	2% above nominal value
E_Z	2% below nominal value	Nominal value	2% above nominal value

Through standard circuit analysis, $R_{T,on}$ can be expressed by the following simple mathematical equation:

$$R_{T,on} = \frac{R_3 R_2 (E_Z R_4 + E_0 R_1)}{R_1 (E_Z R_2 + E_Z R_4 - E_0 R_2)}$$

If we try to design an outer array for noise factors, we may have to use an L_{18} array to accommodate 5 three-level factors, which would be very expensive. So we would like to use compounded noise factors. From the equation,

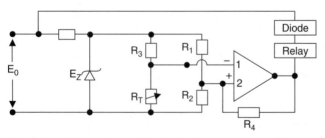

Figure 14.14 Temperature control circuit.

it is very easy (by evaluating the partial derivatives of $R_{T,\text{on}}$ versus R_1, R_2, etc.) to determine whether $R_{T,\text{on}}$ increases whenever R_1 decreases, R_2 increases, R_4 decreases, E_0 increases, or R_Z decreases and vice versa. Therefore, R_1 at low level, R_2 at high level, R_4 at low level, E_0 at high level, and R_Z at low level is one extreme combination of noise levels that will place $R_{T,\text{on}}$ on the high side. On the other hand, R_1 at high level, R_2 at low level, R_4 at high level, E_0 at low level, and R_Z at high level is another extreme combination of noise levels that will put $R_{T,\text{on}}$ on the low side. In summary, we can use the following three compounded noise factors:

$N1$ $(R_1)_1$, $(R_2)_3$, $(R_4)_1$, $(E_0)_3$, $(E_Z)_1$
$N2$ $(R_1)_2$, $(R_2)_2$, $(R_4)_2$, $(E_0)_2$, $(E_Z)_2$
$N3$ $(R_1)_3$, $(R_2)_1$, $(R_4)_3$, $(E_0)_1$, $(E_Z)_3$

where $(R_1)_1$ denotes that R_1 is at level 1; $(R_2)_3$, that R_2 is at level 3; and so on. By using this set of compounded noise factors, for each run in the inner array, we will obtain three output response observations under compounded noise conditions specified by $N1$, $N2$, and $N3$.

14.5   Parameter Design for Smaller-the-Better Characteristics

For smaller-the-better characteristics, the steps for a Taguchi parameter design will be as follows:

1. Select an appropriate output quality characteristic to be optimized.

2. Select control factors and their levels, identifying their possible interactions.

3. Select noise factors and their levels; if there are many noise factors, use compound noise factors to form two or three compounded noise combinations.

4. Select adequate inner and outer arrays; assign control factors to the inner array, and noise factors to the outer array.

5. Perform the experiment.

6. Perform statistical analysis based on S/N to identify the optimal control factor levels. Sometimes it is helpful to conduct a control factor–noise factor interaction study.

7. Predict optimal output performance level based on an optimal control factor level combination, and conduct a confirmation experiment to verify the result.

Example 14.4 illustrates these steps.

Example 14.4. Optimization of a Purification Method for Metal-Contaminated Wastewater (Barrado et al., 1996)   Wastewater that contains metal ions is very harmful because of its toxicity and nonbiodegradable nature. It is proposed to use hydrated iron oxides [Fe(II)] with an appropriate pH level to remove these harmful metal ions. The objective of this project is to remove as many metal ions as possible from the wastewater. The output quality characteristic is the *total remaining (metal) concentration* [TRC; in milligrams per liter (mg/L)], which is a smaller-the-better quality characteristic. The following four control factors are identified:

	Control factors	Level 1	Level 2	Level 3
F	Fe(II)/(total metal in solution)	2	7	15
T	Temperature, °C	25	50	75
H	Aging time, h	1	2	3
P	pH	8	10	12

The variability in the composition of wastewater is considered to be the noise factor. In this Taguchi parameter design project, artificially introduced potassium permanganate is used to simulate the noise factor as follows:

	Noise factor	Level 1	Level 2	Level 3
N	$KMnO_4$ concentration, mol/L	3.75×10^{-3}	3.75×10^{-2}	7.5×10^{-2}

It is assumed that there is no interaction between control factors; an L_9 orthogonal array is selected as the inner array. The experimental data are tabulated as follows:

Experiment no.	Control factors and levels				Output (TRC, mg/L) at different noise factor levels						S/N ratio η
					N1		N2		N3		
	T	P	F	H	Rep. 1	Rep. 2	Rep. 1	Rep. 2	Rep. 1	Rep. 2	
1	1	1	1	1	2.24	0.59	5.29	1.75	155.04	166.27	−39.35
2	1	2	2	2	1.75	5.07	1.05	0.41	0.38	0.48	−7.05
3	1	3	3	3	5.32	0.65	0.40	1.07	0.51	0.36	−7.05
4	2	1	2	3	0.37	0.32	0.34	0.68	4.31	0.65	−5.19
5	2	2	3	1	7.20	0.49	0.48	0.44	0.80	0.88	−9.54
6	2	3	1	2	39.17	27.05	46.54	25.77	138.08	165.61	−39.34
7	3	1	3	2	0.57	1.26	0.61	0.70	0.91	1.42	0.28
8	3	2	1	3	3.88	7.85	22.74	36.33	92.80	120.33	−36.20
9	3	3	2	1	15.42	25.52	35.27	48.61	67.56	72.73	−33.79

The signal-to-noise ratio is computed by the following formula:

$$\text{S/N} = -10 \log\left(\frac{1}{n} \sum_{i=1}^{n} y_i^2\right)$$

By using MINITAB, we can obtain the following ANOVA table for S/N ratio and Fig. 14.15 for percentage contribution of each control factor. Clearly, control factor F [Fe(II)/(total metal in solution)] is by far the most important factor in influencing S/N, contributing 74.91 percent to total S/N; control factor H (aging time) and P (pH) are also significant (more than 10 percent contribution). Control factor T has a negligible effect.

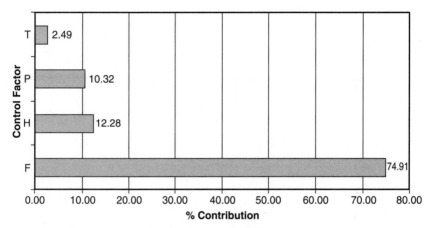

Figure 14.15 Percentage contribution for each control factor.

```
Analysis of Variance for S/N, using Adjusted SS for Tests
Source   DF   Seq SS    Adj SS    Adj MS     F     P
T         2    56.60     56.60     28.30    **
P         2   234.80    234.80    117.40    **
F         2  1704.78   1704.78    852.39    **
H         2   279.46    279.46    139.73    **
Error     0     0.00      0.00      0.00
Total     8  2275.64
```

Main-effects charts of S/N ratio are also plotted (Fig. 14.16), indicating that P at level 1, F at level 3, and H at level 2 will maximize the S/N ratio. Since T has a negligible effect on S/N, it can be set at any level. In summary, the optimal factor level combination is $P_1F_3H_2$.

We can also use original output (TRC, mg/L) data to plot interaction charts between control factors and the noise factor. Using MINITAB, we obtain the interaction charts in Fig. 14.17.

It is clear that control factors F, H, and P undergo significant interactions with noise factor N. Actually, these interactions can be used to improve the robustness toward noise factor; for example, for factor F at level 1, the TRC [total remaining (metal) concentration] will increase greatly as noise factor N increases, but for F at level 3, TRC will hardly change with change in noise factor N, indicating that F at level 3 will deliver robust metal ion removal despite the noise factors. We can also see that P at level 1 and H at level 2 give reasonable robustness.

The predicted S/N at optimal control factor level is

$$\bar{T} = \frac{-39.35 - 7.05 - 7.05 - 5.19 - 9.54 - 39.34 + 0.28 - 36.20 - 33.79}{9}$$

$$= -19.69$$

$$\bar{F}_3 = \frac{-7.05 - 9.54 + 0.28}{3} = -5.44$$

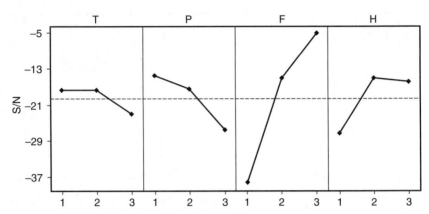

Figure 14.16 Main effects plot—LS means for S/N.

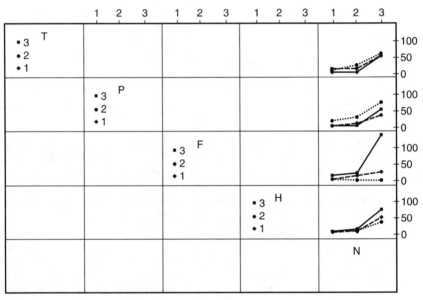

Figure 14.17 Interaction plot—LS means for TRC.

$$\overline{H}_2 = \frac{-7.05 - 39.34 + 0.28}{3} = -15.37$$

$$\overline{P}_1 = \frac{-39.35 - 5.19 + 0.28}{3} = -14.75$$

$$\text{S/N} = \overline{F}_3 + \overline{P}_1 + \overline{H}_2 - 2\overline{T} = -5.44 - 14.75 - 15.37 - 2(-19.69) = 3.82$$

14.6   Parameter Design for Nominal-the-Best Characteristics

For smaller-the-better characteristics, the steps for a Taguchi parameter design will be as follows:

1. Select an appropriate output quality characteristic to be optimized.
2. Select control factors and their levels, identifying their possible interactions.
3. Select noise factors and their levels; if there are many noise factors, use compound noise factors to form two or three compounded noise combinations.
4. Select adequate inner and outer arrays, assigning control factors to the inner array and noise factors to the outer array.
5. Perform the experiment.
6. Perform statistical analysis and the two-step optimization procedure.

   a. Select control factor levels to maximize S/N.

   b. Select mean adjusting factor(s) to adjust the mean to target value.
7. Predict optimal output performance level based on the optimal control factor level combination, and conduct a confirmation experiment to verify the result.

Example 14.5 illustrates these steps.

Example 14.5. Parameter Design for Consistent Polymer Melt Production in Injection Molding (Khoshooee and Coates 1998)   The injection-molding process consists of three overlapping subprocesses: polymer melt production, injection, and solidification. In the polymer melt production, solid polymer granules are melted and pumped by a reciprocating Archimedean screw in a temperature-controlled barrel. The injection shot weight (amount of polymer injected) is an important quality characteristic for the polymer melt production stage. For ideal quality, the shot weight should have the smallest variation possible from shot to shot. The following control factors and their levels are identified:

	Control factors	Level 1	Level 2	Level 3
BP	Backpressure, bar	40	100	160
SS	Screw speed, r/min	50	150	250
SMT	Set melt temperature, °C	210	240	270
DD	Decompression distance, mm	2	3	4
DV	Decompression velocity, mm/s	1	2	3

Injection stroke is selected as the noise factor used in this experiment. Two levels of injection stroke, 20 and 25 mm, are selected as the two levels of the noise factor, that is, N_1 and N_2; four replicates of shot weight are recorded under each noise level.

An L_{18} array is used as the inner array for control factors. Only columns 2 to 6 of L_{18} are used to accommodate five control factors; the other columns are left empty. The experimental layout and experimental data are listed in the following table:

Experi-ment no.	Control factors					Shot weight under noise factors								$\bar{y}$	S/N ratio η
						$N1$				$N2$					
	SMT	BP	SS	DD	DV	R_1	R_2	R_3	R_4	R_1	R_2	R_3	R_4		
1	1	1	1	1	1	5.89	5.88	5.91	5.90	7.38	7.37	7.59	7.41	6.67	18.12
2	1	2	2	2	2	6.26	6.10	6.31	6.24	7.86	7.64	7.79	7.80	7.00	18.52
3	1	3	3	3	3	6.50	6.40	6.46	6.50	8.02	8.08	8.09	8.07	7.27	18.57
4	2	1	1	2	2	6.04	6.04	6.05	6.08	7.74	7.73	7.68	7.80	6.89	17.67
5	2	2	2	3	3	6.35	6.41	6.37	6.42	8.09	8.09	8.03	8.08	7.23	18.09
6	2	3	3	1	1	6.73	6.55	6.56	6.75	8.15	8.15	8.16	8.18	7.40	19.20
7	3	1	2	1	3	6.46	6.44	6.36	6.37	8.00	8.00	8.00	8.01	7.20	18.53
8	3	2	3	2	1	6.79	6.83	6.81	6.80	8.09	8.12	8.10	8.10	7.45	20.64
9	3	3	1	3	2	6.63	6.63	6.61	6.61	8.06	8.10	8.01	8.10	7.34	19.54
10	1	1	3	3	2	6.21	6.01	6.21	6.15	7.93	7.88	7.86	7.93	7.02	17.46
11	1	2	1	1	3	6.06	6.06	6.03	6.00	7.54	7.50	7.57	7.49	6.78	18.61
12	1	3	2	2	1	6.29	6.36	6.34	6.34	7.95	7.90	7.93	7.91	7.13	18.47
13	2	1	2	3	1	6.23	6.24	6.22	6.23	8.01	8.01	8.06	7.99	7.12	17.45
14	2	2	3	1	2	6.51	6.49	6.46	6.50	8.13	8.14	8.14	8.11	7.31	18.42
15	2	3	1	2	3	6.38	6.35	6.37	6.34	8.07	8.05	8.06	8.06	7.21	17.99
16	3	1	3	2	3	6.64	6.62	6.65	6.66	8.05	8.04	8.04	8.05	7.34	19.82
17	3	2	1	3	1	6.31	6.39	6.37	6.37	8.01	8.04	8.03	8.03	7.19	18.14
18	3	3	2	1	2	6.57	6.62	6.55	6.62	8.07	8.07	8.07	8.06	7.33	19.35

In the table, S/N is calculated by the following formula:

$$\text{S/N} = 10 \log \left(\frac{\bar{y}^2}{s^2} \right)$$

By using MINITAB, the ANOVA table for the signal-to-noise ratio is as follows:

```
Analysis of Variance for S/N, using Adjusted SS for Tests
Source   DF    Seq SS    Adj SS    Adj MS     F       P
SMT       2    5.1007    5.1007    2.5503   7.89    0.016
BP        2    1.5696    1.5696    0.7848   2.43    0.158
SS        2    1.6876    1.6876    0.8438   2.61    0.142
DD        2    1.3670    1.3670    0.6835   2.12    0.191
DV        2    0.0949    0.0949    0.0475   0.15    0.866
Error     7    2.2614    2.2614    0.3231
Total    17   12.0812
```

The percentage contribution of control factors to S/N is given by Fig. 14.18. It is clear that SMT is the most important control factor influencing S/N; SS, BP, and DD also have significant contributions. Figure 14.19 is the main-effects chart for S/N.

Clearly, SMT at level 3, BP at level 3, SS at level 3, and DD at level 2 are the best selection for maximizing S/N ratio: $(\text{SMT})_1(\text{BP})_2(\text{SS})_3(\text{DD})_2$.

We can calculate the mean shot weights $\bar{y}$ as response in the DOE analysis using MINITAB to give the following ANOVA table:

```
Analysis of Variance for y-bar, using Adjusted SS for Tests
Source    DF      Seq SS      Adj SS      Adj MS       F       P
SMT        2    0.346259    0.346259    0.173130  124.61   0.000
BP         2    0.168626    0.168626    0.084313   60.69   0.000
SS         2    0.244219    0.244219    0.122110   87.89   0.000
DD         2    0.020491    0.020491    0.010245    7.37   0.019
DV         2    0.001519    0.001519    0.000760    0.55   0.602
Error      7    0.009725    0.009725    0.001389
Total     17    0.790840
```

Clearly, SMT, BP, and SS are important variables in influencing mean shot weight. We can also obtain the mean-effects chart for mean shot weight (Fig. 14.20). Clearly, setting SMT, BP, and SS at level 3 increases not only S/N ratio but also mean shot weight, which actually helps increase production throughput. Therefore, we select SMT, BP, and SS at level 3, and DD at level 2 as our optimal factor level selection.

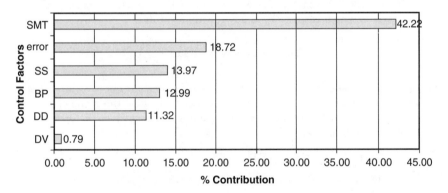

Figure 14.18 Percentage contribution on S/N.

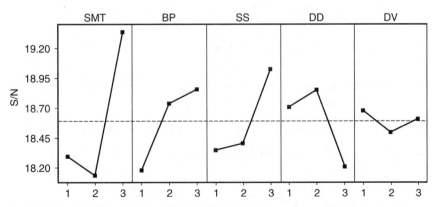

Figure 14.19 Main-effects plot—LS means for S/N.

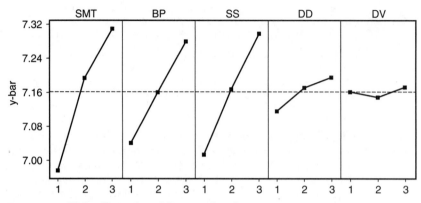

Figure 14.20 Main-effects plot—LS means for y-bar.

Instead of analyzing $\bar{y}$, many Taguchi method books propose analyzing sensitivity, which is given by the following formula:

$$\text{Sensitivity} = S = 10\ \log\left(\frac{T_i^2}{n}\right)$$

where T_i is the summation of output observations, that is, y_i values at each run of the inner array. Analyzing sensitivity will give the same conclusion as that obtained by analyzing $\bar{y}$. The predicted S/N ratio at the optimal factor level combination is

$$\text{S/N} = \overline{\text{SMT}}_3 + \overline{\text{BP}}_3 + \overline{\text{SS}}_3 + \overline{\text{DD}}_2 - 3\overline{T}$$
$$= 19.34 + 18.85 + 19.02 + 18.85 - 3 \times 18.59 = 20.29$$

The predicted mean shot weight $\bar{y}$ at the optimal factor level combination is

$$\bar{y} = \bar{y}_{\text{SMT3}} + \bar{y}_{\text{BP3}} + \bar{y}_{\text{SS3}} + \bar{y}_{\text{DD2}} - 3\bar{\bar{y}}$$
$$= 7.31 + 7.28 + 7.30 + 7.30 + 7.17 - 3 \times 7.16 = 7.58$$

14.7   Parameter Design for Larger-the-Better Characteristics

For larger-the-better characteristics, the steps for a Taguchi parameter design will be as follows:

1. Select an appropriate output quality characteristic to be optimized.
2. Select control factors and their levels, identifying their possible interactions.
3. Select noise factors and their levels; if there are many noise factors, use compound noise factors to form two or three compounded noise combinations.

4. Select adequate inner and outer arrays; assign control factors to the inner array and noise factors to the outer array.

5. Perform the experiment.

6. Perform statistical analysis based on S/N ratio; sometimes it is helpful to analyze the interactions between control factors and noise factors.

7. Predict optimal output performance level based on optimal control factor level combination, and conduct a confirmation experiment to verify the result.

Example 14.6 illustrates these steps.

Example 14.6. Maximize Pulloff Force for an Elastometric Connector to a Nylon Tube   This Taguchi parameter design project involves finding the optimal process parameters in assembling an elastometric connector to a nylon tube in an automobile engine application. It is required that the strength of the connection be strong. The quality characteristic measured in this project is the pulloff force, which is a larger-the-better characteristic. There are four control factors in this experiment:

	Control factors	Level 1	Level 2	Level 3
A	Interference	Low	Medium	High
B	Connector wall thickness	Thin	Medium	Thick
C	Insertion depth	Shallow	Medium	Deep
D	Percent adhesive in connector predip	Low	Medium	High

There are three noise factors as follows:

	Noise factors	Level 1	Level 2
U	Conditioning time, h	24	120
V	Conditioning temperature, °F	72	150
W	Conditioning relative humidity, %	25	75

In this experiment two compound noise factors are selected as follows:

$N1$ Set conditioning time and conditioning temperature at low level, and set humidity at high level, that is, U and V at level 1, and W at level 2; this noise combination will weaken the connection.

$N2$ Set conditioning time and conditioning temperature at high level and humidity at low level, that is, U and V at level 2 and W at level 1; this noise combination will strengthen the connection.

An L_9 array is used as the inner array. Pulloff forces are measured for each inner array run with replicates at two compounded noise combinations. The experimental layout and experimental data are listed in the following table:

Experiment	Control factors				Pulloff force measured at compounded noise levels			
no.	A	B	C	D	$N1$	$N2$	$\bar{y}$	S/N η
1	1	1	1	1	9.5	20.0	14.75	21.68
2	1	2	2	2	16.2	24.2	20.20	25.59
3	1	3	3	3	6.7	23.3	20.00	25.66
4	2	1	2	3	17.4	23.2	20.30	25.88
5	2	2	3	1	18.6	27.5	23.05	26.76
6	2	3	1	2	16.3	22.5	19.40	25.42
7	3	1	3	2	19.1	24.3	21.70	26.54
8	3	2	1	3	15.6	23.2	19.40	25.25
9	3	3	2	1	19.9	22.6	21.25	26.49

where the signal-to-noise ratio is computed by using the formula for the larger-the-better quality characteristic:

$$\text{S/N} = -10 \log\left(\frac{1}{n} \sum_{i=1}^{n} \frac{1}{y_i^2}\right)$$

By using MINITAB, we can obtain the following ANOVA table for S/N and can also compute the percentage contribution of control factors influencing S/N (Fig. 14.21):

```
Analysis of Variance for S/N, using Adjusted SS for Tests
Source    DF     Seq SS     Adj SS     Adj MS      F     P
A          2     6.1128     6.1128     3.0564     **
B          2     2.7057     2.7057     1.3528     **
C          2     8.4751     8.4751     4.2376     **
D          2     1.2080     1.2080     0.6040     **
Error      0     0.0000     0.0000     0.0000
Total      8    18.5016
```

Clearly, C is the most important factor; A and B also have significant effects on S/N.

From the main-effects chart on S/N (Fig. 14.22), we can see that C and A should be set at level 3 and B should be set at level 2.

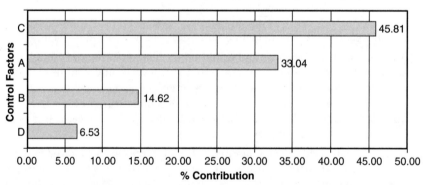

Figure 14.21   Percentage contribution for S/N.

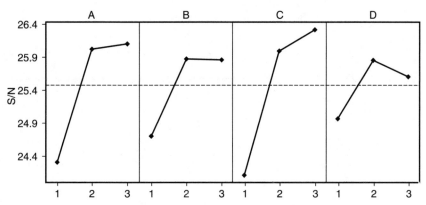

Figure 14.22 Main-effects plot—LS means for S/N.

The predicted signal-to-noise ratio at the factor level combination given above is

$$S/N = A_3 + C_3 + B_2 - 2T = 26.10 + 26.32 + 25.87 - 2 \times 25.48 = 27.33$$

The predicted mean pulloff force is $A_3 + C_3 + B_2 - 2T = 20.78 + 21.58 + 20.88 - 2 \times 20.0 = 23.24$.

Design Optimization: Advanced Taguchi Robust Parameter Design

15.1 Introduction

The parameter design discussed in Chap. 14 is also called the *static parameter design*; here *static* means "fixed target." In all static parameter design studies discussed in Chap. 14, the objective was to reduce the variability of a product around a fixed "performance target" caused by the noise factors such as environmental conditions, deterioration, and piece-to-piece (piecewise) variation. Those performance targets include zero, infinity, or a fixed nominal value. In Taguchi's terminology, this is called *nondynamic* or *static parameter design*.

After the 1980s, Taguchi (1993, 1994) added some new focuses on robust parameter design:

1. *What to measure.* Dr. Taguchi thinks that it is very important to select the right characteristic as a system performance measure, in other words, *y*, in transfer function detailing and optimization study. He defined four classes of quality measures:

- *Downstream quality. Downstream quality* (also known as *customer quality*) is the type of quality requirement that is noticed by customers, such as noise, vibration, percentage defective, and mean time to failure. Customer quality is obtained in phase 1 and phase 2 QFD within the DFSS algorithm.

- *Midstream quality. Midstream quality* (also known as *specified quality*) requirements are usually some key design characteristics for the product such as dimension and strength.

- *Upstream quality. Upstream quality* (also known as *robust quality*) is the static signal-to-noise ratio we discussed in Chap. 14; it is a measure of stability of a given requirement. Both midstream and upstream qualities are usually conducted in the process mapping within the DFSS algorithm.

- *Original quality. Original quality* (also known as *functional quality*) is related to the *generic function* of an individual product, a class of product, or a branch of technology. This kind of quality is represented in the physical structure of the DFSS algorithm.

Dr. Taguchi stated that in robust parameter design study, downstream quality is the worst choice as the performance measure. The best choice is the *dynamic signal-to-noise ratio,* which is related to the input-output relationship; this is called the "ideal function," introduced in Chap. 6. Ideal function is specified by the *generic function* of the product or product family and is often the *energy transfer* related to a design. These kinds of functions are derived using the zigzagging process within the DFSS algorithm. The second best choice for system performance characteristic is the upstream quality, or robust quality. Midstream quality is an acceptable but not preferred choice.

2. *Interaction.* If a dynamic S/N ratio based on ideal function is chosen as the performance characteristic in a robust parameter design study within the DFSS project, then the interaction between control factors (design parameters) is an indication of inconsistency and non-reproducibility; therefore, it has a harmful effect and indicates a deficiency of the design concept. However, the control factor–noise factor interaction can be utilized to enhance the robustness toward the influence of noise factors.

3. *Orthogonal array.* Dr. Taguchi thinks that L_{12}, L_{18}, and L_{36} are preferred orthogonal arrays because the interactions of main effects are evenly confounded among all columns, and the interaction effects can then be considered as a part of noise in subsequent S/N study. This is done in pursuit of an additive transfer function (Chap. 6).

4. *Robust technology development.* Dr. Taguchi and many Japanese industries believe that the robust parameter design should be brought into stage 0 (see Chaps. 1 and 5) of the product development cycle, the new technology development phase, by introducing noise factors and building robustness of the generic function of a new technology startup phase, thus reducing downstream hiccups in the product development cycle and significantly shortening product development cycle time.

For many people, these four points are not well understood. In this chapter, we explain and illustrate the following concepts and approaches:

1. Generic functions, functional quality, and ideal function to enforce the concepts introduced in Chaps. 5 and 6

2. Energy transfer and other transfer in design

3. Dynamic characteristic and dynamic signal-to-noise ratio

4. Functional quality, robust quality, specified quality, and customer quality

5. The interactions and robustness

6. Signal factor, noise factor, control factor, and dynamic robust parameter design layout and data analysis (Chap. 6)

7. Robust technology development

Statistics alone will not be able to explain these concepts and approaches, they are closely related to the knowledge base of engineering design. The works of Phal and Beitz (1988), Nam Suh's axiomatic design principle (Suh 1990), and TRIZ (Altshuller 1988) are very helpful in understanding these concepts and approaches. In the next section we review some works in the engineering design area and explain many concepts in Dr. Taguchi's new focuses.

15.2 Design Synthesis and Technical Systems

15.2.1 Static versus dynamic view of a design

The word *static* means exhibiting "little or no movement or change"; the word *dynamic* means "alive, functioning, and operative." The dynamic view of a product means that we see a product as a *process*. By the definition in Chap. 2, a process is a "continuous and regular action or succession of actions, taking place or carried on in a definite manner, and leading to the accomplishment of some result; a continuous operation or series of operations," or "a combination of inputs, actions and outputs." It is easy to understand that some products can be a process, such as a TV set, in which user intent, control, and electrical energy are input, whereas the functions provided by TV, such as image and sound, are output. But what about some simple products, such as a cup or a pencil—are they processes as well? Actually, they are, because the use of all products designed for human consumption inevitably involves human interaction. Therefore, we can also regard the whole product usage spectrum as a process. Regarding products in a process perspective is very helpful for building quality into products, because quality level is a user perception, which is built on the total experience of the product usage process.

There are three basic aspects of a process: the input, the output, and transformation. When we treat a product as a process, it will also have all these basic aspects. The technical system model discussed by Phal and Beitz (1988) is a good example of a product treated as a process.

15.2.2 Technical system models, functions, and transformations

According to Phal and Beitz (1988) and Hubka (1980), designs should be treated as technical systems connected to their environments by the means of inputs and outputs. A design can be divided into subsystems. *Technical systems* are designed to perform functions for their users. As discussed in Chap. 6, the process of delivering functions involves technical processes in which energy, material, and signals are channeled and/or converted. For example, a TV set delivers its functions, such as displaying images and playing sounds, utilizing many forms of energy transformation and signal transformation (electrical energy to optical images, electrical to acoustic energy, a lot of signal processing, conversion, etc.).

Energy can be transformed in many ways. An electric motor converts electrical into mechanical and thermal energy, an internal-combustion engine converts chemical into mechanical and thermal energy, and so on.

Material also can be transformed in many ways. It can be mixed, separated, dyed, coated packed, transported or reshaped. Raw materials are turned into part-finished products. Mechanical parts are given particular shapes, surface finishes, and so on. Information or data signals can also be transformed. Signals are received, prepared, compared, or combined with others; transmitted; changed; displayed; recorded; and so on.

Overall, technical systems can be represented by the block diagram in Fig. 15.1 and the synthesis process highlighted in Chap. 6.

In a particular technical design, one type of transformation (of energy, material, or signal) may prevail over others, depending on the design. In this case, the transformation involved is treated as the main trans-

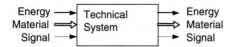

Figure 15.1 Technical system: conversion of energy, material, and signals.

formation. For example, in using an automobile, energy conversion is the main transformation.

In technical designs, the main transformation is usually accompanied by a second type, and quite frequently, all three transformations come into play. There can be no transformation of material or signals without an accompanying transformation of energy, however small. The conversion of energy is often associated with conversion of material. Transformations of signals are also very common in the form of control and regulate energy and/or the material transformation process. Every signal transformation is associated with transformation of energy, although not necessarily with material transformation.

Technical systems are designed to deliver functions. A technical system delivers a main function in order to accomplish the main task of the system, namely, the task that the system is supposed to accomplish for the user. The system also delivers auxiliary functions, or supporting functions, which contribute indirectly to the main function.

In a DFSS context, technical systems are designed to deliver functional requirements as derived by the zigzagging process. A technical system delivers high-level functions in order to accomplish the attributes of the voice of the customer. The system also delivers low-level functions in the physical structure.

Phal and Beitz (1988) use the notations and symbols shown in Fig. 15.2 to model the physical and process structures in a block diagram to effect some visual translation above the mathematical mappings obtained via the zigzagging process (Chaps. 5, 6, and 8).

Example 15.1. Tensile Testing Machine (Phal and Beitz 1988) *A tensile testing machine* is a measurement device that measures tensile strength of

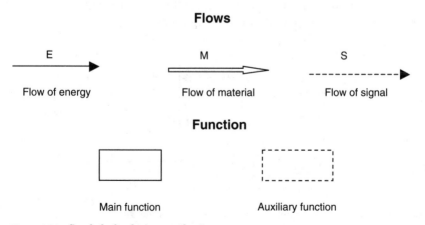

Figure 15.2 Symbols for design synthesis.

materials. The tensile strength of a material such as wire, rope, or stone is its ability to support a load without breaking. In a tensile strength test, a piece of specimen is loaded and fixed at the fixture of the testing machine; then a mechanical force is applied on the specimen and the magnitude of the force is gradually increased until the specimen is "broken" or has "failed." Both the "force at breaking point" (tensile strength) and the deformation of the specimen will be measured and recorded. Figure 15.3 is a rough function diagram for the tensile testing machine using the symbols defined in Fig. 15.2. It gives a good idea of design hierarchy and the roles of these three kinds of transformation on each level.

Figure 15.4 gives a more detailed block diagram for the tensile testing machine.

Clearly, in each block, several transformations can occur simultaneously, but there is usually one main type of transformation; this could be any one of the three transformation types. When the main mode of transformation is energy, or material transformation, signal transformation is often accompanied with it in the form of control (design parameter). The functions are accomplished by the transformation process.

The following checklist is helpful in identifying, recognizing, and designing different modes of transformation in the technical system:

Energy transformation:
- Changing energy (e.g., electrical to mechanical energy)
- Varying energy components (e.g., amplifying torque)
- Connecting energy with signals (e.g., switching on electrical energy)
- Channeling energy (e.g., transferring power)
- Storing energy (e.g., storing kinetic energy)

Material transformation:
- Changing matter (e.g., liquefying a gas)
- Varying material dimensions (e.g., rolling sheetmetal)
- Connecting matter with energy (e.g., moving part)

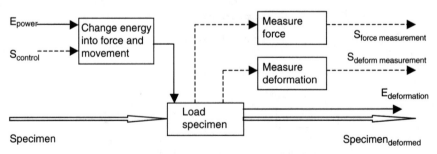

Figure 15.3 Functional diagram of a tensile testing machine.

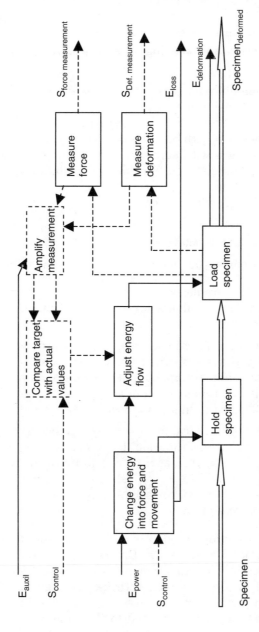

Figure 15.4 Detailed block diagram for tensile testing machine.

- Connecting matter with signal (e.g., cutting off stream)
- Connecting materials of different type (e.g., mixing or separating materials)
- Channeling material (e.g., mining coal)
- Storing material (e.g., keeping grain in a silo)

Signal transformation:
- Changing signals (e.g., changing analog signal to digital signal, paper signal to electronic signal)
- Varying signal magnitude (e.g., increasing a signal's amplitude)
- Connecting signals with energy (e.g., amplifying a measurement)
- Connecting signals with matter (e.g., marking materials)
- Transforming signals (e.g., filtering or translating a signal)
- Channeling signals (e.g., transferring data)
- Storing signals (e.g., in database)

15.2.3   Design and ideal function

From the discussions of the last section, we have the following facts:

1. Technical systems are designed to deliver functional requirements; there are high-level and low-level requirements in the physical structure hierarchy.

2. In technical design, functional requirements are accomplished by transformation processes; there are energy, material, and signal transformations. For example, an electric fan's main function is to "blow air," this function is accomplished mostly by energy transformation, from electrical energy to mechanical energy.

3. To accomplish any technical functional requirement, there is usually one major type of transformation, but other transformations often accompany it. For example, energy transformation is the major type of transformation used to accomplish the main function of an electric fan, but signal transformation also plays a role, in the form of "control the fan operation."

Taguchi's ideal function is closely related to the transformation process in technical systems. In Dr. Taguchi's more recent work (Taguchi 1993, 1994), ideal function is described as an ideal input-output relationship of a design (i.e., both fixed design parameters and absence of noise factors are assumed) related to energy transformation, and this relationship is related to the function of the product. Taguchi used the following example to illustrate the concept of ideal function. In his description, the function of the brake is to stop the car. The process of stopping the car is essentially an energy transformation process. "In the case of car brakes, the ideal function is that the input energy by foot is proportional to the output energy as brake torque" (Taguchi 1993). This is shown in Fig. 15.5.

Dr. Taguchi further stated that if a brake can work according to this linear ideal relationship "perfectly," that is, be repeatable and robust in the presence of all kinds of noise factors, such as temperature and moisture, then that brake is a "perfect" brake.

However, in a real-world situation, with the influence of noise factors, for many brakes, the actual relationship will resemble that illustrated in Fig. 15.6a, where there are many variations in different brake applications. If the brake system is robust toward all kinds of noise factors, the actual relationship will resemble that shown in Fig. 15.6b, in which the variation is smaller and the actual relationship is very close to the ideal relationship. Dr. Taguchi believes that accomplishing this kind of ideal relationship with a high degree of robustness should be the goal for any robust parameter design study.

If we examine this car brake problem closely, we can find that "stopping or slowing the car" is an energy transformation process in the block diagram of the design, in which the brake caliper moves to press the brake pad toward the rotor. This is the friction force between the pad and the rotor that transformed the mechanical energy (of the car moving) into heat and other energy, with some material loss (from the pad and the rotor). In today's car, the power that moves the pad is not from the foot force, and the foot force is actually a signal. The foot force actually controls the magnitude of the energy transformation that stops or slows down the car. So the real meaning of optimizing the ideal function with maximum robustness in the car brake example is as follows:

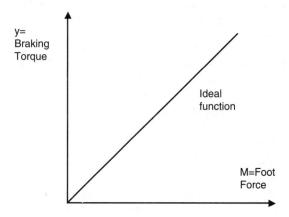

Figure 15.5 Ideal function of a brake.

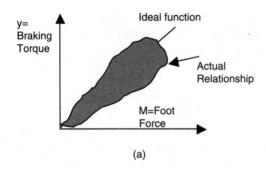

(a)

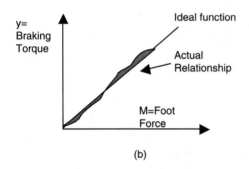

(b)

Figure 15.6 Actual relationships between foot force and braking torque. (*a*) Not robust, (*b*) robust.

1. To make sure that the main transformation (in the car brake case, it is an energy transformation process) process of the high-level function (stopping or slowing down the car) can perform robustly at different magnitudes of transformation.

2. To make sure that the brake system has excellent controllability for this transformation process; this is illustrated by pursuing the perfect linear relationship between the signal (M = foot force) and the magnitudes of transformation (measured by brake torque) with a high degree of robustness.

In summary, the "robustness at different magnitudes of transformation" and the "robustness at controllability" are the real focuses of Example 15.1.

Let's examine a few more examples of ideal functions.

Example 15.2 Example of Ideal Functions

1. Semiconductor oxidation process
   a. *Main function.* To oxidize a thin layer of material
   b. *Main transformation.* Material transformation (changing material with a chemical reaction)
   c. *Ideal function*

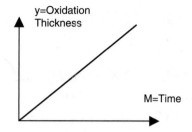

d. Analysis. Time is a signal; the more time the process runs, the more material transformation will occur. The robustness of ideal function is an indication of the robustness of

(1) *Different levels of transformation.* With different time settings, different amounts of oxidation will occur.

(2) *Controllability.* If a straight-line relationship is repeatable, that shows excellent controllability.

2. Design of an operational amplifier

   a. Main function. To produce a gain in a voltage.

   b. Main transformation. Energy transformation.

   c. Ideal function

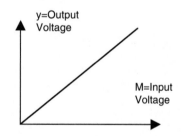

d. Analysis. An operational amplifier works in such a way that it draws electrical power from the power source and then creates the output signal that follows all "signatures" of the input signal. Therefore, input signal is again a "control." The robustness of an ideal function is the indication of the robustness for

(1) *Different levels of transformation.* With varying input, there is a varying output.

(2) *Controllability.* The output follows input in a straight line.

3. Measurement process

   a. Main function. To measure.

   b. Main transformation. This could be any of the three types—energy, material, or signal—depending on what to measure and the mechanism of measurement.

   c. Ideal function

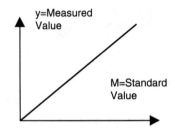

d. *Analysis.* The ideal measurement system should be able to give exactly the same value of the characteristic as the true value; for example, an ideal scale should show one gram (1 g) if the item weighs truly 1 g. When we use the ideal function shown above to benchmark a measurement system, for example, using several different standard weights to benchmark a scale, the standard weight is again a "control." The robustness of ideal function is the indication of the robustness for
(1) Different levels of transformation
(2) Controllability

4. Injection-molding process
   a. *Main function.* Making parts to the mold shape
   b. *Main transformation.* Material transformation
   c. *Ideal function*

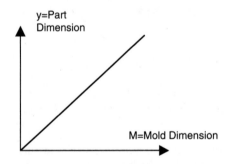

d. *Analysis.* Injection molding transforms polymer granules into shaped parts, so the main transformation is obviously a material transformation. The mold is a "control" for the shape. The linear relationship in the ideal function shown above means that when the mold changes, say, into a larger mold, then the magnitude of material transformation also changes. This linear relationship also indicates controllability. Therefore, the robustness of the ideal function is an indication of the robustness for
(1) Different levels of transformation
(2) Controllability

All of these examples have shown the following common features of ideal functions:

1. An ideal function is directly related to the main transformation process of the main function of the technical system under study. Sometimes it is impossible to measure the transformation directly; the things measured in ideal functions are directly related to this transformation. For example, case 2 of Example 15.2, the operational amplifier, clearly is an energy transformation process; the output that we measured is the output voltage, which is not energy, but clearly that input voltage–output voltage relationship is a direct surrogate of this energy transformation.

2. If a system is robust in delivering its ideal function, that is, if the actual input-output relationship is very close to the ideal function and has very low variation under the influence of the noise factors, for example, such as in the case indicated in Fig. 15.6b, then it indicates the following:

a. The transformation process is repeatable and robust at different levels of transformation, with the presence of noise factors. For example, for an energy transformation, the process is robust at high, low, and mid energy levels.

b. The controllability of the transformation process is robust. If we change the signal, the transformation process will respond with high precision and high repeatability, even in the presence of noise factors.

How to identify and define the ideal function.   Given a technical system, we can follow the following step-by-step guidelines to identify and define the ideal function:

1. Identify the main function of the system. If it is hard to identify, drawing a functional diagram similar to Figs. 15.3 and 15.4 may help.

2. Identify the main type of transformation behind the main function; again, if it is hard to identify, drawing a functional diagram may help.

3. Identify the "control signal" for the main transformation; usually it is very rare that the transformation is uncontrolled in a man-made technical system.

4. Identify the measurable input and output, where input should be the control signal itself or a direct surrogate, and the output should be directly related to the magnitude of transformation. Sometimes we have to install additional measurement devices to measure input and output.

5. Identify the ideal linear relationship between output and input.

Clearly, for any product or technical system, achieving robustness in delivering its ideal function is highly desirable. Dr. Taguchi developed a systematic approach to achieve this goal. This approach is the robust parameter design for dynamic characteristics, and it will be discussed in the next section.

15.3   Parameter Design for Dynamic Characteristics

15.3.1   Signal-response system and ideal function

Dynamic characteristic is also called *signal-response design*. Figure 15.7 gives a block diagram of a signal-response design.

Dr. Taguchi's use of this signal-response system for quality improvements started with communication and measurement systems, both of which can be represented by the signal-response system well. For a measurement system, the signal is the measurement sample; the output response is the measurement value of a sample characteristic. For a good measurement system, the following conditions are required:

1. From the same sample, the measurements must be repeatable no matter who is doing the measuring and how many times the sample is measured.

2. From two different samples, a small difference in true characteristics should be detected by the measurement system; that is, it is desirable to have high sensitivity.

3. A measurement system must be easy to calibrate; a linear relationship between the true value of the characteristic and measurement value is ideal.

The ideal relationship between signal and output response (e.g., a functional requirement) is illustrated in Fig. 15.8.

In a later application of the Taguchi method, the signal-response system serves as the paradigm for many technical systems. In most cases, the ideal signal-response relationship is the ideal function that we discussed in the last section. The robust parameter design for dynamic characteristics actually becomes the robust parameter design for ideal functional quality.

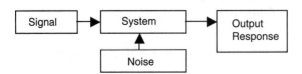

Figure 15.7   A P-diagram with input signal.

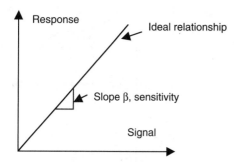

Figure 15.8 Ideal signal-response relationship.

15.3.2 Parameter design layout and dynamic signal-to-noise ratio

Robust parameter design for dynamic characteristics is carried out by using the following inner-outer array layout, which is illustrated by Table 15.1. In the layout, we can see that the control factors are assigned to the inner array and the signal factor and noise factors are assigned to the outer array.

The signal factor is the "input signal" treated as an experimental factor. In the array in Table 15.1, we use M to represent the signal factor. In the experiment, a number of levels for the signal factor, say, k levels, will be selected. We denote them as $M_1, M_2, \ldots, M_k$. At each level of the signal factor, several combinations of noise factors, say, $N1, N2, \ldots$, are assigned, as illustrated in Table 15.1. Therefore, for each run of the inner array, the signal factor will be varied k times, at each signal factor level several noise factor combinations will be attempted, and under each signal-noise combination a functional requirement (FR), say, y_{ij}, will be measured. Because we expect that as the signal factor increases, the response will also increase; a typical complete inner-array run of output responses (e.g., an FR vector) data will resemble the scatterplot in Fig. 15.9.

Dr. Taguchi proposed using the following dynamic signal-to-noise ratio:

$$\text{S/N} = 10 \log \left(\frac{\beta_1^2}{\hat{\sigma}^2} \right) = 10 \log \left(\frac{\beta_1^2}{\text{MSE}} \right) \qquad (15.1)$$

where β_1 is the linear regression coefficient for slope and MSE is the mean-squared error for the linear regression.

As a measure of robustness for a signal-response system, the greater the S/N ratio, the better the system robustness will be. Specifically, for each run of the inner array, we will get the following FR observations under the corresponding signal-noise combination as given in Table 15.2, assuming that there are k levels for the signal factor and m levels for the noise factor.

TABLE 15.1 Inner-Outer Array Layout for Parameter Design of Dynamic Requirement

	Inner array					Outer array				
	Control factors					Experiment no.				
						1	2	...	N_2	Signal factor
						M_1		...	M_k	
						Noise factors				
Experiment no.	A	B	C	...	F	$N1$	$N2$	$N2$	$N3$ ...	Output response y / S/N η
1	1	1	1	...	1	y_{11}	y_{12}	y_{13}	... y_{1N_2}	η_1
2	2	1	2	...	1	y_{21}	y_{22}	y_{22}	... y_{2N_2}	η_2
...	...	...	...	...	...	...	...	...	...	...
N_1	2	2	1	...	2	y_{N_11}	y_{N_12}	y_{N_13}	... $y_{N_1N_2}$	η_{N_1}

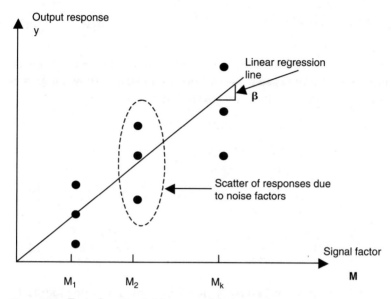

Figure 15.9 Typical scatterplot for a complete inner-array run.

TABLE 15.2 Complete Data Set for Each Inner-Array Run

Noise factor levels	Signal factor levels			
	M_1	M_2	...	M_k
$N1$	y_{11}	y_{12}	...	y_{1k}
$N2$	y_{21}	y_{22}	...	y_{2k}
⋮	⋮	⋮		⋮
Nm	y_{m1}	y_{m2}	⋮	y_{mk}

By fitting a linear regression equation using the standard procedure, we obtain

$$y = \beta_0 + \beta_1 M + \varepsilon \quad \text{where} \quad \text{Var}(\varepsilon) = \sigma^2 \quad (15.2)$$

where

$$\hat{\beta}_1 = \frac{\sum\limits_{j=1}^{k} \sum\limits_{i=1}^{m} M_j(y_{ij} - \bar{y}_{...})}{m \sum\limits_{j=1}^{k} (M_j - \overline{M})^2} \quad (15.3)$$

$$\hat{\beta}_0 = \bar{y}_{...} - \hat{\beta}_1 \overline{M} \quad (15.4)$$

$$\hat{\sigma}^2 = \text{MSE} = \frac{1}{mk - 2} \sum_{j = 1}^{k} \sum_{i = 1}^{m} (y_{ij} - \hat{\beta}_0 - \hat{\beta}_1 M_j)^2 \qquad (15.5)$$

Sometimes we want to fit a regression line passing the origin; that is, $\beta_0 = 0$ is assumed. This case arises where we need $y = 0$, when $M = 0$ (no signal, no response). Then the regression equation is

$$y = \hat{\beta}_1 M + \varepsilon \qquad (15.6)$$

$$\hat{\beta}_1 = \frac{\displaystyle\sum_{j = 1}^{k} \sum_{i = 1}^{m} M_j y_{ij}}{m \displaystyle\sum_{j = 1}^{k} M_j^2} \qquad (15.7)$$

$$\hat{\sigma}^2 = \text{MSE} = \frac{1}{mk - 1} \sum_{j = 1}^{k} \sum_{i = 1}^{m} (y_{ij} - \hat{\beta}_1 M_j)^2 \qquad (15.8)$$

All these calculations can be easily performed by a standard statistical package such as MINITAB.

Clearly, the signal-to-noise ratio (S/N) is inversely proportional to MSE, and S/N is proportional to β_1. MSE is the mean squared error for linear regression. For a perfect linear regression fit, there is no scattering around the linear regression line, the MSE will be equal to zero, and S/N will be infinity, which will be an ideal S/N.

Figure 15.10 shows what will affect S/N in a signal-response system. For Fig. 15.10a and b it is obvious that S/N is inversely proportional to variation, because high variation will increase MSE, thus decreasing S/N. For Fig. 15.10c, since S/N is evaluated according to linear model assumption, a nonlinear relationship between signal and response, even if it is a perfect one, will create residuals for a linear fit. Therefore, MSE will increase while S/N will decrease (i.e., S/N and MSE are inversely proportional to nonlinearity, as in Fig. 15.10c). Actually, Taguchi's dynamic S/N ratio will penalize the nonlinearity. The justification of this criterion is discussed in the next section.

For Fig. 15.10e and f, from Eq. (15.1), clearly the higher the sensitivity b, the higher the S/N. For some application, higher sensitivity means that the "signal" can adjust the level of main transformation more effectively and therefore is desirable. For example, for a measurement system, higher sensitivity and lower variation indicate that the measurement system has high resolution and high repeatability, high reproducibility, and low measurement error; so it is very desirable. For some other applications, there is a target value for sensitivity.

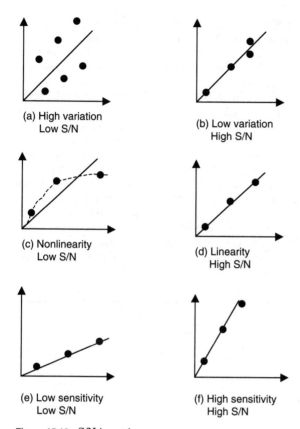

(a) High variation
Low S/N

(b) Low variation
High S/N

(c) Nonlinearity
Low S/N

(d) Linearity
High S/N

(e) Low sensitivity
Low S/N

(f) High sensitivity
High S/N

Figure 15.10 S/N in various cases.

15.3.3   Two-step optimization procedure and examples

Taguchi dynamic robust parameter design proposes the following preparation and two-step optimization procedure:

Step 0 (preparation). Using the layout described in Table 15.1, run the experiment and collect data. For each inner-array run, compute the sensitivity β and S/N.

Step 1. Run a complete Taguchi DOE data analysis; using dynamic S/N as response, find the control factor combination to maximize S/N.

Step 2. Run another Taguchi DOE data analysis; using sensitivity β as response, find sensitivity adjustment factor(s), which do(es) not affect S/N, and use it (them) to tune β to the target value.

We will use the following examples to illustrate this two-step optimization procedure.

Example 15.3. Improve a Strain Gauge *Strain* is the amount of deformation of a body due to an applied force. More specifically, strain *e* is defined as the fractional change in length. In most applications, the magnitude of measured strain is very small. Therefore, strain is often expressed as *microstrain* μ*e*, which is *e* × 10^{-6}. Strain can be measured by a *strain gauge*, a device whose electrical resistance varies in proportion to the amount of strain in the device; this electrical resistance change is usually extremely small in the magnitude of microhms (μΩ). Therefore, the key requirement for strain gauge is to measure the small resistance with high degrees of accuracy, repeatability, and reproducibility. The output signal is a voltage in the unit of millivolts (MV), which should be proportional to the resistance change.

In this experiment, the signal factor *M* is the electrical resistance change and the functional requirement *y* is the output voltage. Three levels are signal factors, selected as follows: $M_1 = 10$ μΩ, $M_2 = 100$ μΩ, and $M_3 = 1000$ μΩ.

Certain noise factors, such as environmental temperature, may affect the measurement process. Two levels of compounded noise factor, N_1 and N_2, are used in the experiment. Nine control factors, all of them design parameters of the strain gauge, are selected in the experiment, and their levels are listed in the following table:

	Control factors	Level 1	Level 2	Level 3
A	Foil alloy type	Type 1	Type 2	
B	Foil thickness	5	10	15
C	Specimen material type	6 nylon	6/6 nylon	6/6 nylon+glass
D	Bonding method	Method 1	Method 2	Method 3
E	Outer-grid line width	45/60	40/55	35/40
F	Inner-grid line width	35	25	
G	Coating thickness	9	3	
H	End-loop dimension	50	70	80
I	Specimen thickness	10	25	40

An L_{18} array is used to accommodate control factors. All control factors, the noise factor, and the signal factor are assigned to the inner-outer array layout described in Table 15.3.

In assigning *F* and *G* factors, the compound factor method is used. A compound factor termed *FG* is defined as follows:

$$(FG)_1 = F_1 G_1, \qquad (FG)_2 = F_2 G_1, \qquad (FG)_3 = F_2 G_2$$

Because a strain gauge is a measurement system, zero signal–zero output is required, and we will use the linear Eq. (15.6): $y = \hat{\beta}_1 M + \varepsilon$. Equations (15.7) and (15.8) are used to compute the sensitivity and S/N ratio. For example, for the first run

$$\hat{\beta}_1 = \frac{10 \times 10.0 + 10 \times 10.0 + 100 \times 249.0 + 100 \times 258.0 + 1000 \times 2602.0 + 1000 \times 2608.0}{2(10^2 + 100^2 + 1000^2)}$$

$$= 2.60$$

From MINITAB, by fitting the first row of data versus *M* values, it is very easy to get:

TABLE 15.3 Experimental Layout and Experimental Data of Example 15.3

| Experiment no. | Inner-array control factors | | | | | | | | Outer-array control factors | | | | | | | |
	A	B	C	D	E	FG	H	I	M_1 N1	M_1 N2	M_2 N1	M_2 N2	M_3 N1	M_3 N2	β	S/N η
1	1	1	1	1	1	1	1	1	10.0	10.0	249.0	258.0	2602.0	2608.0	2.60	−12.97
2	1	1	2	2	2	2	2	2	9.0	9.0	249.0	256.0	2490.0	2501.0	2.50	−12.93
3	1	1	3	3	3	3	3	3	8.0	9.0	241.0	243.0	2420.0	2450.0	2.43	−15.10
4	1	2	1	1	2	2	3	3	10.0	11.0	260.0	259.0	2710.0	2751.0	2.73	−16.77
5	1	2	2	2	3	3	1	1	11.0	12.0	280.0	291.0	2996.0	3011.0	3.00	−14.59
6	1	2	3	3	1	1	2	2	9.0	10.0	260.0	265.0	2793.0	2800.0	2.79	−15.23
7	1	3	1	2	1	3	2	3	9.0	9.0	229.0	231.0	2400.0	2456.0	2.43	−19.04
8	1	3	2	3	2	1	3	1	8.0	9.0	272.0	276.0	2702.0	2738.0	2.72	−15.67
9	1	3	3	1	3	2	1	2	9.0	9.0	270.0	275.0	2761.0	2799.0	2.78	−15.90
10	2	1	1	3	3	2	2	1	10.0	11.0	267.0	279.0	2809.0	2903.0	2.85	−21.27
11	2	1	2	1	1	3	3	2	16.0	16.0	240.0	251.0	2616.0	2699.0	2.66	−21.10
12	2	1	3	2	2	1	1	3	8.0	9.0	241.0	248.0	2406.0	2499.0	2.45	−22.07
13	2	2	1	2	3	1	3	2	16.0	17.0	291.0	301.0	3002.0	3100.0	3.05	−20.66
14	2	2	2	3	1	2	1	3	12.0	13.0	259.0	272.0	2622.0	2699.0	2.66	−19.89
15	2	2	3	1	2	3	2	1	10.0	11.0	250.0	261.0	2699.0	2702.0	2.70	−14.51
16	2	3	1	3	2	3	1	2	9.0	10.0	298.0	299.0	3010.0	3052.0	3.03	−15.91
17	2	3	2	1	3	1	2	3	11.0	12.0	190.0	198.0	2094.0	2100.0	2.10	−15.14
18	2	3	3	2	1	2	3	1	13.0	14.0	241.0	258.0	2582.0	2632.0	2.61	−17.61

```
y = 2.60 M
Analysis of Variance
Source            DF         SS         MS          F          P
Regression        1     13700163   13700163   102203.58    0.000
Residual Error    5          670        134
Total             6     13700833
```

So

$$\text{S/N} = \eta_1 = 10 \log\left(\frac{\beta_1^2}{\text{MSE}}\right) = 10 \log\left(\frac{2.60^2}{134}\right) = -12.97$$

Similarly, we can compute sensitivities and S/N for the other rows in Table 15.3.

By using the Taguchi two-step optimization procedure, we first conduct an ANOVA analysis for the S/N ratio using MINITAB:

```
Analysis of Variance for S/N, using Adjusted SS for Tests
Source      DF      Seq SS     Adj SS     Adj MS       F         P
A            1      49.87      49.87      49.87      1.84     0.308
B            2       3.23       3.23       1.61      0.06     0.944
C            2       5.16       5.16       2.58      0.10     0.913
D            2       9.43       9.43       4.72      0.17     0.852
E            2       5.38       5.38       2.69      0.10     0.910
FG           2       1.45       1.45       0.73      0.03     0.974
H            2       6.59       6.59       3.30      0.12     0.892
I            2      10.85      10.85       5.42      0.20     0.833
Error        2      54.30      54.30      27.15
Total       17     146.26
```

By calculating percentage contribution of each control factor, we get Fig. 15.11.

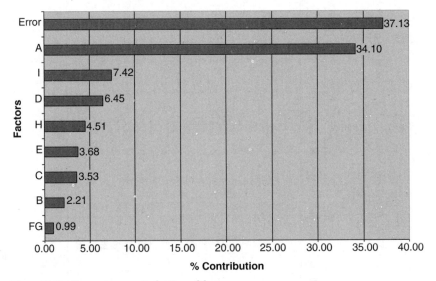

Figure 15.11 Percentage contribution of factors.

It is clear that B and FG have very little contribution on the S/N ratio. Using the main-effect's chart for other factors, we get Fig. 15.12.

Clearly, $A_1C_2D_1E_2H_2I_1$ is the best control factor setting for maximizing S/N. MINITAB can also compute the following average S/N for different control factor levels:

```
Least Squares Means for S/N
A       Mean    SE Mean
1      -15.36    1.737
2      -18.68    1.737
C
1      -17.77    2.127
2      -16.55    2.127
3      -16.74    2.127
D
1      -16.07    2.127
2      -17.82    2.127
3      -17.18    2.127
E
1      -17.64    2.127
2      -16.31    2.127
3      -17.11    2.127
H
1      -16.89    2.127
2      -16.35    2.127
3      -17.82    2.127
I
1      -16.10    2.127
2      -16.96    2.127
3      -18.00    2.127
```

The mean of S/N is

```
Mean of S/N = -17.020
```

The predicted S/N for optimal setting, $A_1C_2D_1E_2H_2I_1$ is

$$S/N = \overline{A}_1 + \overline{C}_2 + \overline{D} + \overline{E}_2 + \overline{H}_2 + \overline{I}_1 - 5\overline{T} = -15.36 - 16.55 - 16.07$$
$$- 16.31 - 16.35 - 16.1 - 5(-17.02) = -11.64$$

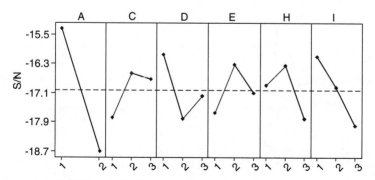

Figure 15.12 Main-effects plot—data means for S/N.

We then analyze the effect of control factors on sensitivity, β; we get the main-effects chart in Fig. 15.13.

The control factor level means for sensitivity is

```
Least Squares Means for beta
A       Mean    SE Mean
1       2.664   0.04589
2       2.679   0.04589
B
1       2.582   0.05621
2       2.822   0.05621
3       2.612   0.05621
C
1       2.782   0.05621
2       2.607   0.05621
3       2.627   0.05621
D
1       2.595   0.05621
2       2.673   0.05621
3       2.747   0.05621
E
1       2.625   0.05621
2       2.688   0.05621
3       2.702   0.05621
FG
1       2.618   0.05621
2       2.688   0.05621
3       2.708   0.05621
H
1       2.753   0.05621
2       2.562   0.05621
3       2.700   0.05621
I
1       2.747   0.05621
2       2.802   0.05621
3       2.467   0.05621
```

The mean of beta is

```
Mean of beta = 2.6717
```

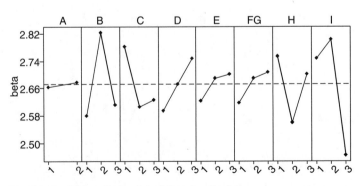

Figure 15.13 Main-effects plot—LS means for beta.

We can see that factor B has a significant effect on sensitivity but very little effect on S/N. Since higher sensitivity means high resolution, B2 is selected to improve sensitivity. The final control factor level selection is $A_1B_2C_2D_1E_1H_2I_1$. The predicted ß is

$$\hat{\beta} = \overline{A}_1 + \overline{B}_2 + \overline{C}_2 + \overline{D}_1 + \overline{E}_2 + \overline{H}_2 + \overline{I}_1 - 6\overline{T}$$
$$= 2.66 + 2.82 + 2.61 + 2.60 + 2.69 + 2.56 + 2.75 - 6(2.67) = 2.76$$

Example 15.4. Automobile Weather-Strip Parameter Design (Pack et al. 1995)   A *weather strip* is the rubber strip around the door ring of an automobile. There are two commonly used technical requirements for the weather-strip system: the closing effort and the wind noise. *Closing effort* is the magnitude of force needed to shut the door. If the weather strip is too soft, it is very easy to close the door, but the door seal will not be good, so the *wind noise* will be large. If the weather strip is too hard, a lot of force will be needed to shut the door, but the door may seal pretty well. Many efforts have been made to enhance closing effort and wind noise. However, a study to improve closing effort may lead to a design change that deteriorates the door seal. Then another study launched to improve wind noise will lead to another design change that leads to high closing effort.

In this Taguchi parameter design study, the team spent lots of time discussing the right ideal function before jumping into the experiment. Finally, the team came up with the ideal function illustrated in Fig. 15.14 and the testing fixture illustrated in Fig. 15.15. It was determined that the basic function of weather stripping is to "seal door ring without leak." When we press the door shut, we expect the weather strip to seal. "How well the weather strip can seal" is measured by the air pressure gauge illustrated in Fig. 15.15.

We expect that the more downward the displacement M is, the tighter the seal should be, so the higher the air pressure gauge reads, the higher the load on the load cell in Fig. 15.15 will read; this load is an index of the closing effort. In this experiment, the functional requirement y is the ratio of *air pressure / load ratio*; clearly, the higher this ratio, the higher the efficiency with which the weather-strip system utilizes the elastic displacement effect to seal the door ring.

The signal factor of this experiment M is the downward weather-strip "bulb" displacement; $M_1 = 0.05$, $M_2 = 0.25$, and $M_3 = 0.45$.

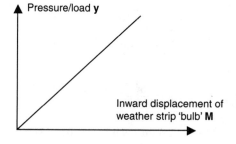

Pressure/load **y**

Inward displacement of weather strip 'bulb' **M**

Figure 15.14 Ideal function of weather-strip example.

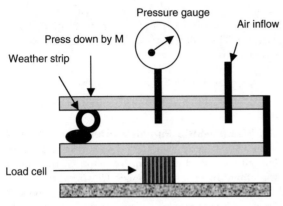

Figure 15.15 Test fixture for weather-strip example.

The noise factors are as follows: N_1 = wet, N_2 = dry, Q_1 = low airflow rate, Q_2 = medium airflow rate, Q_3 = high airflow rate. Six control factors are given in the following table:

Control Factors	Level 1	Level 2	Level 3	Level 4	Level 5	Level 6
A Material	Sponge low	Sponge medium	Sponge high	Dense low	Dense medium	Dense high
B Coating	None	Type 1	Type 2			
C Corner shape	Small radii	Large radii	Flat			
D Venting	None	Large venting	Small venting			
E Bulb shape	Round	Square	Triangle			
F Attachments	Tape	Pin	Carrier			

The L_{18} array in Table 15.4 is used to assign control factors.

The outer-array layout and pressure/load ratio experiment data are given in Table 15.5. In this experiment, we use the following linear regression model (zero signal, zero output):

$$y = \hat{\beta}_1 M + \varepsilon$$

For each inner-array run, we use the same formula to calculate the sensitivity β_1 and S/N as that of Example (15.3):

$$\hat{\beta}_1 = \frac{\sum_{j=1}^{k} \sum_{i=1}^{m} M_j y_{ij}}{m \sum_{j=1}^{k} M_j^2}, \qquad \hat{\sigma}^2 = \text{MSE} = \frac{1}{mk-1} \sum_{j=1}^{k} \sum_{i=1}^{m} (y_{ij} - \hat{\beta}_1 M_j)^2$$

$$\text{S/N} = 10 \log \left(\frac{\beta_1^2}{\hat{\sigma}^2} \right) = 10 \log \left(\frac{\beta_1^2}{\text{MSE}} \right)$$

Table 15.6 gives the sensitivity and S/N ratio for each inner-array run.

TABLE 15.4   Inner Array

| Experiment no. | Control factors | | | | | |
	A	B	C	D	E	F
1	1	1	1	1	1	1
2	1	2	2	2	2	2
3	1	3	3	3	3	3
4	2	1	1	2	2	3
5	2	2	2	3	3	1
6	2	3	3	1	1	2
7	3	1	2	1	3	3
8	3	2	3	2	1	1
9	3	3	1	3	2	2
10	4	1	3	3	2	1
11	4	2	1	1	3	2
12	4	3	2	2	1	3
13	5	1	2	3	1	2
14	5	2	3	1	2	3
15	5	3	1	2	3	1
16	6	1	3	2	3	2
17	6	2	1	3	1	3
18	6	3	2	1	2	1

We again follow the two-step optimization procedure. First, we conduct ANOVA analysis on S/N by using MINITAB. We get the ANOVA table as follows:

```
Analysis of Variance for S/N, using Adjusted SS for Tests
Source   DF   Seq SS   Adj SS   Adj MS     F     P
A         5    89.06    89.06    17.81   0.74  0.662
B         2     3.80     3.80     1.90   0.08  0.927
C         2    95.83    95.83    47.92   1.98  0.336
D         2     9.04     9.04     4.52   0.19  0.843
E         2    32.12    32.12    16.06   0.66  0.601
F         2    54.56    54.56    27.28   1.13  0.470
Error     2    48.41    48.41    24.21
Total    17   332.82
```

The chart in Fig. 15.16 gives percentage contribution of each factor toward S/N.

Clearly, factors C,A,F,E are important factors contributing to S/N. The following MINITAB data and the main-effects chart in Fig. 15.17 show that $A_5 C_2 E_1 F_1$ gives the highest signal-to-noise ratio:

```
Least Squares Means for S/N
A     Mean    SE Mean
1    14.01     2.841
2    17.38     2.841
3    12.44     2.841
4    16.49     2.841
5    18.60     2.841
6    13.48     2.841
```

TABLE 15.5 Pressure/Load Ratio Data from a Weather-Strip Experiment

L_{18}	M_1						M_2						M_3					
	Q_1		Q_2		Q_3		Q_1		Q_2		Q_3		Q_1		Q_2		Q_3	
	N1	N2	N1	N2	N1	N2	N1	N2	N1	N2	N1	N2	N1	N2	N1	N2	N1	N2
1	1.78	9.15	1.78	9.86	2.37	12.68	2.52	3.63	4.72	6.65	6.29	8.16	4.40	5.84	6.48	8.63	6.25	9.64
2	0.0	0.0	0.00	0.00	0.00	0.00	3.62	4.59	3.62	4.59	3.82	4.59	3.38	4.72	3.38	4.72	3.38	4.72
3	4.11	7.52	3.38	10.9	3.62	10.34	7.07	7.43	6.34	10.31	6.88	11.75	4.14	4.26	4.53	5.91	4.92	6.73
4	4.67	5.04	4.67	9.24	5.61	9.24	10.68	26.53	11.17	28.57	11.65	29.59	6.65	9.74	7.25	10.49	7.25	10.86
5	0.00	0.44	1.36	0.44	2.17	0.88	2.65	0.96	4.13	1.35	3.54	4.43	3.58	2.85	3.93	4.07	4.29	4.61
6	1.69	0.00	1.41	0.00	1.69	0.74	2.08	1.15	2.34	1.91	2.73	2.67	4.74	3.42	4.74	3.85	4.90	3.85
7	2.33	3.60	2.62	4.40	3.21	5.20	2.62	4.25	3.42	4.69	3.30	4.83	3.52	4.81	3.52	4.68	3.44	4.50
8	2.47	10.75	3.57	9.81	3.02	9.35	6.60	11.72	6.60	12.68	8.00	12.92	8.35	25.27	8.13	28.69	8.46	30.62
9	0.00	2.27	0.00	3.79	0.00	3.79	0.00	10.36	0.00	11.71	0.00	13.51	0.00	5.79	0.00	6.84	1.34	9.47
10	1.02	0.00	1.28	0.00	2.30	0.24	4.35	0.00	6.72	0.00	7.51	0.00	6.22	8.77	8.11	9.47	8.38	11.23
11	1.68	3.20	1.68	2.49	1.68	3.55	6.06	20.15	8.13	21.17	8.29	21.17	5.96	13.06	6.54	12.84	6.69	14.22
12	0.00	3.70	0.00	3.70	0.00	3.70	1.37	2.56	3.66	6.25	5.26	5.97	4.61	9.78	6.66	10.37	8.70	11.74
13	1.28	4.19	1.26	4.63	1.26	4.33	9.38	12.16	9.81	13.41	9.29	13.83	12.51	13.95	13.77	15.00	14.53	14.91
14	0.00	0.00	0.00	0.00	0.00	0.00	0.00	0.00	0.00	0.00	0.12	0.00	0.69	3.25	1.43	3.77	1.80	4.29
15	0.00	0.00	0.74	0.00	1.23	0.00	2.79	5.00	4.18	4.29	3.76	7.86	4.58	8.11	7.20	10.07	6.46	10.07
16	0.00	0.00	0.00	0.00	0.00	0.00	0.40	1.17	1.11	3.16	2.11	4.09	1.10	9.42	2.90	9.80	3.31	10.79
17	4.98	3.55	6.16	4.48	5.64	4.78	6.82	3.07	7.64	6.39	8.47	6.65	3.28	1.81	5.52	3.39	6.38	4.07
18	0.00	1.94	0.00	2.90	0.00	3.23	1.21	4.84	2.52	5.46	4.03	6.40	2.99	3.38	3.12	4.96	3.03	5.75

TABLE 15.6   Sensitivity and S/N

Experiment no.	Control factors						Sensitivity ß	S/N η
	A	B	C	D	E	F		
1	1	1	1	1	1	1	17.25	11.72
2	1	2	2	2	2	2	10.61	19.47
3	1	3	3	3	3	3	17.09	10.85
4	2	1	1	2	2	3	33.37	10.65
5	2	2	2	3	3	1	9.32	19.00
6	2	3	3	1	1	2	9.31	22.50
7	3	1	2	1	3	3	10.99	14.20
8	3	2	3	2	1	1	40.68	15.21
9	3	3	1	3	2	2	11.84	7.92
10	4	1	3	3	2	1	17.57	17.45
11	4	2	1	1	3	2	28.89	13.38
12	4	3	2	2	1	3	18.71	18.65
13	5	1	2	3	1	2	34.78	23.01
14	5	2	3	1	2	3	4.20	11.89
15	5	3	1	2	3	1	17.39	20.89
16	6	1	3	2	3	2	12.18	13.55
17	6	2	1	3	1	3	13.58	11.40
18	6	3	2	1	2	1	10.47	15.05

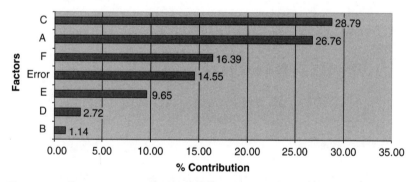

Figure 15.16 Percentage contribution to S/N.

```
B      Mean    SE Mean
1      15.10    2.009
2      15.06    2.009
3      16.05    2.009
C
1      12.66    2.009
2      18.31    2.009
3      15.24    2.009
D
1      14.87    2.009
2      16.40    2.009
3      14.94    2.009
```

```
E      Mean     SE Mean
1     17.08      2.009
2     13.81      2.009
3     15.31      2.009
F
1     16.63      2.009
2     16.64      2.009
3     12.94      2.009
Mean of S/N = 15.402
```

However, the main-effects chart for sensitivity or β gives the plot in Fig. 15.18 and MINITAB analysis gives the following:

```
Least Squares Means for Beta
A      Mean     SE Mean
1     15.02     12.098
2     17.33     12.098
3     21.17     12.098
4     21.72     12.098
5     18.79     12.098
6     12.08     12.098
B
1     21.04      8.554
2     17.88      8.554
3     14.14      8.554
C
1     20.40      8.554
2     15.81      8.554
3     16.84      8.554
D
1     13.54      8.554
2     22.16      8.554
3     17.36      8.554
E
1     22.40      8.554
2     14.68      8.554
3     15.98      8.554
```

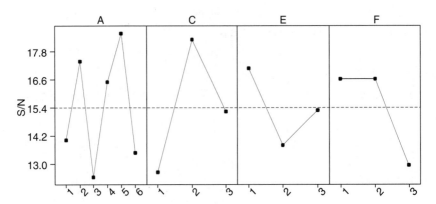

Figure 15.17 Main-effects plot—data means for S/N.

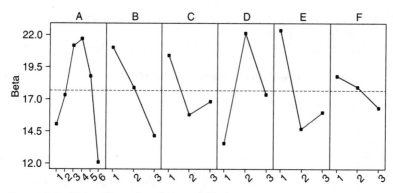

Figure 15.18   Main-effects plot—LS means for beta.

```
F        Mean      SE Mean
1        18.80       8.554
2        17.94       8.554
3        16.32       8.554
Mean of Beta  =  17.685
```

In our case, higher sensitivity means that when the weather-strip bulb displacement M increases, the pressure/load ratio will also increase, indicating that the ability to seal, measured by pressure, is increasing on unit-load force. So seal efficiency will drastically increase with a small increase in closing effort. Therefore, the higher the sensitivity, the better, and since B_1D_2 significantly increases b, we will also choose $A_5B_1C_2D_2E_1F_1$ as the final optimal control factor setting.

The predicted S/N is

$$\text{S/N} = \overline{A}_5 + \overline{B}_1 + \overline{C}_2 + \overline{D}_2 + \overline{E}_1 + \overline{F}_1 - 5\overline{T}$$
$$= 18.6 + 15.1 + 18.31 + 16.4 + 17.08 + 16.63 - 5 \times 15.4 = 25.12$$

The predicted β_1 is

$$\hat{\beta}_1 = \overline{A}_5 + \overline{B}_1 + \overline{C}_2 + \overline{D}_2 + \overline{E}_1 + \overline{F}_1 - 5\overline{T}$$
$$= 18.79 + 21.04 + 15.81 + 22.16 + 22.4 + 18.8 - 5 \times 17.69 = 30.55$$

A confirmation experiment is run to verify the optimal solution and it gives S/N = 20.90 and sensitivity = 32.90. In comparison with the old design, in which S/N = 15.38 and sensitivity = 17.74, the selected optimal solution is a huge improvement.

15.3.4   Double-signal factors

In many dynamic signal-response systems, there is only one signal factor. However, in certain cases there could be more than one signal factor. As we discussed earlier, a *signal factor* is often a "signal" that affects the magnitude of transformation. There are also cases where two signals work together to control the magnitude of transformation, as illustrated by Fig. 15.19.

The combined signal is a function of two separate signals:

$$\text{Combined signal} = h\ (M, M^*) \qquad (15.9)$$

The two most common types of $h(M, M^*)$ are

1. $h(M, M^*) = MM^*$
2. $h(M, M^*) = M/M^*$ or $h(M, M^*) = M^*/M$

We will use the following example to illustrate how to find the correct $h(M, M^*)$ and compute sensitivity and S/N.

Example 15.5 Chemical Formula for Body Warmers (Taguchi et al. 2000) Disposable *body warmers* are chemical pockets that can be fit into sports or outdoor outfits. The chemicals inside the body warmer will slowly react and generate heat. The heat generated will keep the wearer warm in outdoor conditions. The functional requirement (FR) of body warmer y is the heat generated. There are two signals:

1. M = heat-generating time. It has four levels: $M_1 = 6$, $M_2 = 12$, $M_3 = 18$, $M_4 = 24$ (hours).
2. M^* = amount of ingredients. It has three levels: $M_1^* = 70$ percent, $M_2^* = 100$ percent, $M_3^* = 130$ percent standard amount.

Clearly, more time will generate more heat, and more chemicals will also generate more heat; in this example $h(M, M^*) = MM^*$ is a good choice. Table 15.7 gives a complete output data set for an inner-array run.

Using $h(m, M^*) = MM^*$, we can transform Table 15.7 into a data set with combined signal factor MM^* as shown in Table 15.8.

Using Eqs. (15.6) to (15.8) and MINITAB, we find that the regression equation is

$$\text{Amount of heat} = 15.6\ MM^*$$

$$\text{MSE} = 3738$$

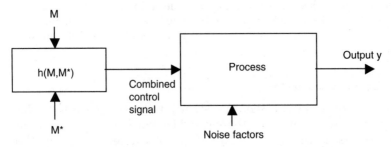

Figure 15.19 Two signals.

TABLE 15.7   Complete Inner-Array Run of Body Warmer Data

		M			
M^*	Noise factor	$M_1 = 6$	$M_2 = 12$	$M_3 = 18$	$M_4 = 24$
M_1^*	$N1$	93.82	160.44	199.06	221.76
70%	$N2$	104.16	199.92	263.34	305.76
M_2^*	$N1$	105.00	183.6	219.6	232.80
100%	$N2$	142.80	264.0	351.0	400.80
M_3^*	$N1$	152.10	277.68	341.64	358.80
130%	$N2$	165.36	322.92	435.24	483.60

So,

$$\text{S/N} = 10 \log \left(\frac{15.6^2}{3738} \right) = -11.86$$

15.4   Functional Quality and Dynamic S/N Ratio

After the thorough description of the ideal function, dynamic signal-to-noise ratio, and robust parameter design for dynamic characteristics (requirement) in the last few sections, we are ready to discuss many in-depth issues in advanced Taguchi methods.

15.4.1   Why is functional quality preferred in parameter design?

Functional quality is expressed by the ideal function. The ideal function is directly related to the main transformation process; whether it is energy, material, or signal transformation, the main transformation is the most important transformation for the system's main function. In a product/process design stage, the most important goal is to make sure that the product will deliver its main function correctly and consistently. Therefore, functional quality addresses the most important quality issue in the design stage.

If we focus on customer quality, such as percentage failure and noise, at parameter design stage, we could overreact to symptoms of the poor design and ignore the fundamental issues of the design. We could react to one type of symptom, such as vibration, and make some design changes, which may "cure" that symptom but worsen other aspects of the design. Next time, we will have to react to another symptom and we could run into a cycle of rework and contradiction.

If we have defined a good ideal function and use dynamic S/N as the robustness measure, together with a sound choice of noise factors and

TABLE 15.8 Combined Signal Factor MM^*

$MM^*=$	4.2	6	7.8	8.4	12	12.6	15.6	16.8	18	23.4	24	31.2
$N1$	93.82	105.00	152.10	160.44	183.6	199.06	277.68	221.76	219.6	341.64	232.80	358.80
$N2$	104.16	142.80	165.36	199.92	264.0	263.34	322.92	305.76	351.0	435.24	400.80	483.60

design parameters in a robust parameter design, we will actually kill several birds with one stone. This is because

1. Higher dynamic S/N means that the system will follow the ideal function with the least amount of variation. This means that the main transformation process behind the system high-level function can consistently perform at different transformation levels and has excellent controllability and thus will guarantee that the high-level function of the system will perform well and consistently.

2. The goal of the robust parameter design is to make the system perform its main function consistently under the influence of noise factors. If robustness is achieved, it will eliminate many potential symptoms in the downstream stages. Because focusing on symptom(s) is equivalent to working on one noise factor at a time, it is highly inefficient and time-consuming and we could run into cycles.

3. In summary, focusing functional quality and using dynamic S/N will ensure real quality and save product development time.

15.4.2   Why do we use linear ideal function and dynamic S/N?

Dynamic S/N is derived from *linear* ideal function. From Fig. 15.10, using dynamic S/N as the robustness measure clearly will penalize

- Variation
- Nonlinearity
- Low sensitivity

It is easy to understand why variation is undesirable. We have already discussed at great length the need to reduce the variation. But why is nonlinearity undesirable? Nonlinearity is a form of complexity, from an axiomatic design viewpoint. The second axiomatic design principle stated that if there are several design concepts and all of them can deliver required functions, the design with the least complexity will be preferred. Robust parameter design is equivalent to selecting a good design by using dynamic S/N as a benchmark, and penalizing nonlinearity will help select a good design. Linearity is a proportionality property; for a signal, it is a desirable property.

Also, if we conduct a robust parameter design at an early stage of product development, we may conduct it on a small laboratory scale or on a computer. If many nonlinearities already exist in the signal-response relationship, this relationship may become even more complicated in large-scale production; the whole signal-response relation may become uncontrollable or unpredictable.

In many applications, high sensitivity is a desired property. If a target sensitivity is required, that is, if sensitivity should be neither too high nor too low, the two-step optimization procedure can be used to tune the sensitivity to target.

15.4.3   Why are interactions among control factors not desirable?

Dr. Taguchi stated that the interactions among control factors are not desirable. In a dynamic robust parameter design study, Dr. Taguchi proposes the use of L_{12}, L_{18}, and L_{36}, because in those arrays, the interactions will be evenly confounded among all columns. Dr. Taguchi treats those interaction effects as noises. This statement has drawn a lot of criticism from statisticians.

However, if we study this issue from an axiomatic design perspective, we realize that interaction is again a form of complexity. The system that has interactions is definitely more complex than one without interactions. In other words, *interaction is not a design choice.* Because Dr. Taguchi uses only an "additive model" among control factors and S/N based only on linear ideal function, the S/N for the control factor level combination with severe nonlinearity and nonadditivity will definitely be very low and so it will not be selected in parameter design.

Again, robust parameter design is equivalent to selecting a good design by using dynamic S/N as a benchmark, and penalizing nonadditivity will help in selecting a good design with less complexity.

15.5   Robust Technology Development

Robust technology development means building robustness into newly developed generic technology, or new technology at its infancy. The examples of such new generic technology include new memory chips, new electronic bonding technology, and new materials. New technologies are usually developed at research laboratories under ideal conditions, in small batches and on a small scale. After a generic new technology is developed, product developers will try to integrate it into new products. But usually there are a lot of hiccups in this integration process; the new technology that works well in the lab may not work well after integration, and its performance may not be stable and up to people's expectations. It usually takes many trials and errors to "make it work right."

Robust technology development is a strategy that tries to streamline and expedite the integration of new technology with products and production. Robust technology development proposes conducting robust parameter design on the new technology when the new technology is still in the research lab. This robust design must be a dynamic design that focuses on improving its main, generic function. The ideal function has to be carefully selected in order to build robustness into the main transformation process that underlies the high-level function.

In a regular lab environment, there are very few noise factors. Robust technology development study will artificially bring many noise factors into the environment. Design and redesign will take place for the new technology until robustness in functional performance against noise factors is achieved.

If a new technology is "tested out" in the lab for many potential noise factors, its success in future products and production will be greatly improved.

16

Tolerance Design

16.1 Introduction

In the DFSS algorithm, the objective of tolerance design is to gain further optimization beyond what has been gained via parameter optimization and transfer function detailing to achieve a Six Sigma–capable design. In this step, the DFSS team determines the allowable deviations in design parameters and process variables, tightening tolerances and upgrading only where necessary to meet the functional requirements. Where possible, tolerances may also be loosened. This is the objective of step 11 of the DFSS algorithm in the optimization phase (O) of the ICOV process (Chap. 5). In tolerance design, the purpose is to assign tolerances to the part, assembly, or process, identified in the physical and process structures, based on overall tolerable variation in the functional requirements, the relative influence of different sources of variation on the whole, and the cost-benefit trade-offs.

Tolerance design can be conducted analytically on the basis of the validated transfer function obtained empirically or via testing. In either case, the inputs of this step are twofold: the DFSS team should have a good understanding of (1) the product and process requirements and (2) their translation into product and process specifications using the QFD.

The going-in position in the DFSS algorithm is to initially use tolerances that are as wide as possible for cost considerations, and then to optimize the function of the design and process through a combination of suitable design parameters (DPs). Following this, it is necessary to identify those customer-related functional requirements (FRs) that are not met through parameter design optimization methods. Tightening tolerances and upgrading materials and other parameters will usually be required to meet Six Sigma FR targets.

Systematic application of DFSS principles and tools such as QFD allows the identification of customer-sensitive requirements and the development of target values for these characteristics to meet customer expectations. It is vital that these characteristics be traced down to lowest-level mappings, and that appropriate targets and ranges be developed.

By definition, *tolerance* is the permissible deviation from a specified value or a standard. Tolerances have different meanings at different stages of the design process. Figure 16.1 illustrates the design tolerances at different design mappings (Chap. 8).

For a product or service, customers often have explicit or implicit requirements and allowable requirement variation ranges, called *customer tolerance*. In the next stage, customer requirements and tolerances will be mapped into design functional requirements and functional tolerances. For a design to deliver its functional requirements to satisfy functional tolerances, the design parameters must be set in correct nominal values and their variations must be within design parameter tolerances. For design development, the last stage is to develop manufacturing process variable set points and tolerances.

Example 16.1: Power Supply Circuit   The high-level functional requirement of a power supply circuit is to provide electrical power for small appliances; output voltage is one customer requirement. Although few customers will bother to measure the output voltage, the excessive deviation of output voltage will affect the functional requirements of small appliances. The larger the deviation, the more customers will notice the requirement degradations of the appliances. Customer tolerance is usually defined as the tolerance limit such that 50 percent of customers will be unsatisfied. For example, the nominal value of a power supply circuit could be 6 V, or $T = 6$, but if we assume that when the actual output voltage y is either <5.5 or >6.5 V, 50 percent of customers will be unsatisfied, then the customer tolerance will be 6 ±0.5 V. A power supply circuit consists of many components, such as resistors, transistors, and capacitors. Setting the nominal values for

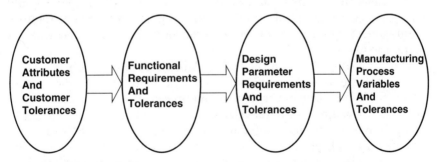

Figure 16.1   Tolerance development process.

these component parameters are the tasks for parameter design, and setting the tolerance limits for those parameters are the tasks for the tolerance design at the development stage. For example, if a resistor's nominal value is 1000 Ω, the tolerance of that resistor could be ±10 Ω, or ±1 percent of its nominal value. These are called *component specifications*. The company that makes the resistor will then have to set nominal values and tolerances for materials, process variables. This example described the tolerance development process illustrated in Fig. 16.1.

Example 16.2. Automobile Body Panel Dimensions and Tolerances

Automobile body panels, such as doors and hoods, have many functional requirements, such as protecting the customers, passenger compartments, engine compartment, and providing stylish looks. Many design requirements for body panels are dimensional requirements.

Customer attributes and tolerances. Customer requirements are usually expressed in very vague terms, such as "the car should look cool, body panels should be nicely put together." However, it is very easy to make the customer unhappy if the body panel dimensions have even moderate variations. For example, if a car door's dimension is off by just several millimeters, the door will either be difficult to open or close or show some large gaps, and will certainly make customers very unhappy.

Functional requirements (FR) tolerances. After analyzing customer requirements and "how the body panel works," we can gradually derive functional tolerances for body panels. For example, for door panels, the car door should be easy to open and close and "look good and fit," and the gaps between door and door frame should be small enough for the door to seal well and subsequently minimize wind noise when driving. By analyzing all these requirements, we may end up with the final tolerance that all actual body panel dimensional variations should be within ±2 mm from the designed dimensions.

Design parameter (DP) tolerances. Automobile body panels are made by welding many small sheetmetal panels together; the small sheetmetal panels are fabricated by a stamping process. In this example, developing design parameter tolerances is to set tolerances on small sheetmetal panels, called *subassemblies*.

Process variable (PV) tolerances. Process variables in the automobile body panel fabrication process include die requirements, fixture requirements, and stamping requirements. Developing process variable tolerances includes setting tolerances on these requirements.

Example 16.3: Automobile Paint Finishes

Automobile paint has many functions, such as protecting the metal from corrosion and rust and providing color.

Customer attribute and tolerances. Customers' requirements on paint finishes are usually ambiguous, such as "uniform and fresh look," "doesn't fade," "no defects," and "does not rust after extended time." However, customers do compare paint finishes of cars from different brand names and manufacturers. For example, if the paint in a customer's car fades sooner than that on another competitor's car, the customer will be unsatisfied.

Functional requirements tolerances.   There are many layers of paint for automobile paint finishes. The layers have different functions; some layers' main functions may be rust prevention, while other layers' functions may be mainly to "provide" or "protect" color. Automobile paint finish functional requirements have many categories, such as thickness requirements, uniformity requirements, and appearance requirements.

Design parameter tolerances.   The paint "recipe," the paint material specifications for each layer of paint, are among design parameter tolerances.

Process variable tolerances.   In this example, there are many process variables, and their tolerances need to be determined. Paint is first sprayed on the car body, so there are many process variables such as paint flow rate through paint application equipment and the patterns of paint application. After each layer of paint, the painted car body will enter the "paint oven" because heat is needed to "cure" the paint. *Curing* is a chemical reaction that bounds the paint molecules. The process variables include oven zone temperature set points and tolerances.

In the tolerance development process, FR tolerance development is the first important step. It translates ambiguous, mostly verbal customer tolerances into clear and quantitative functional tolerances. This step involves the use of functional analysis as obtained in the physical structure via the zigzagging method (Chap. 8), customer attribute analysis, and competitors' benchmarking. Phase 2 QFD is certainly a valuable tool for translating customers' requirements into functional specifications.

However, most of the work in tolerance design involves determination of design parameter tolerances and process variable tolerances, given that the functional tolerances have already been determined. The term *tolerance design* in most industries actually means the determination of design parameter tolerances and process variable tolerances.

If a product is complicated, with extremely coupled physical and process structures, tolerance design is a multistage process. After functional tolerances are determined, system and subsystem tolerances are determined first, and then the component tolerances are determined on the basis of system and subsystem tolerances. Finally, process variable tolerances are determined with respect to component tolerances. Each actual stage of the tolerance design is illustrated in Fig. 16.2.

In Fig. 16.2, the "high-level requirement and tolerances" were derived at the previous stage. For example, FRs and tolerances are at "high level" in comparison with the DP requirements and tolerances, and system requirements and tolerances are at "higher level" in comparison with component requirements and tolerances. $Y = f(x_1, x_2, \ldots, x_n)$ is the

Low-level requirements and tolerances

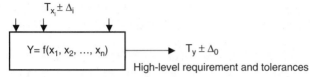

Figure 16.2 A typical stage of tolerance design.

transfer function relationship between higher- and lower-level parameters. In a typical stage of tolerance design, the main task is, given the target requirement of y and its tolerance (i.e., $T_y \pm \Delta_0$), how to assign the tolerances for x_i values. There are three major issues in tolerance design:

1. Manage variability.

2. Achieve functional requirements satisfactorily.

3. Keep life-cycle cost of design at low level.

From an economical point of view, functional requirements should be satisfied with minimum variation. From the earlier discussion of customer tolerances, *customer tolerance* is defined as "the tolerance limits for which 50 percent of customers will be unhappy if they [the limits] are exceeded." For each individual customer, tolerance varies from person to person; a very selective customer will not tolerate any deviation of requirement from its ideal state. If a design requirement is at the boundary of customer tolerance, there are already 50 percent of customers unhappy with the design. Minimizing functional variations will maximize customer satisfaction and will also certainly reduce rework, warranty cost, and aftersale service cost. On the other hand, it is also highly desirable that design parameter tolerances and process variable tolerances be set at wider intervals. Obviously, loose tolerances of design parameters and process variables will make manufacturing easier and cheaper. Taguchi's parameter design is trying to minimize requirement variation with the presence of noise factors; the noise factors also include piece-to-piece (piecewise) variation. Therefore, a very successful parameter design could "loosen up" some tolerances. However, if a parameter design is insufficient to limit the FR variation, tolerance design is very essential.

In tolerance design, cost is an important factor. If a design parameter or a process variable is relatively easy and cheap to control, a tighter tolerance is desirable; otherwise, a looser tolerance is desirable. Therefore, for each stage of tolerance design, the objective is to effectively ensure low functional variation by economically setting appropriate tolerances on design parameters and process variables.

The situation described in Fig. 16.2 is very common in most tolerance design circumstances. However, subject-matter knowledge also plays a role in different circumstances. For example, if the design parameters are all dimension-related, such as automobile body dimensions, or mechanical parts dimensions, there is a full body of knowledge regarding this, called *geometric dimensioning and tolerance* (GDT), where the fundamental idea about tolerance design is really the same as those for any other tolerance designs. However, many special methods and terminologies are purely dimensional in nature.

In this chapter, we discuss all major tolerance design methods under the tolerance design paradigm illustrated in Figs. 16.1 and 16.2. Subject-germane tolerance design aspects, such as GDT, are not included.

There are two classes of tolerance design methods: the traditional tolerance design method and Taguchi's tolerance design method. *Traditional tolerance design methods* include worst-case tolerance analysis, statistical tolerance analysis, and cost-based tolerance analysis. *Taguchi tolerance design methods* include the relationship between customer tolerance and producer's tolerance, and tolerance design experiments. All these methods are discussed in subsequent sections.

16.2   Worst-Case Tolerance

Worst-case tolerance is a tolerance design approach that is against the *worst-case scenario*. Specifically, let us assume that the transfer function between a higher-level requirement y with lower-level characteristics is $x_1, x_2, \ldots x_i, \ldots, x_n$ is

$$y = f(x_1, x_2, \ldots, x_i, \ldots, x_n) \tag{16.1}$$

and it is further assumed that the target value for y is T, and the tolerance limit for y is Δ_0 such that y is within specification if $T - \Delta_0 \leq y \leq T + \Delta_0$. Sometimes it is possible for the tolerance limits to be asymmetric, with y in the specification limit if $T - \Delta_0' \leq y \leq T + \Delta_0$, where Δ_0' is the left tolerance limit and Δ_0 is the right tolerance limit.

For each x_i, $i = 1 \cdots n$, the target value for x_i is T_i and the tolerance limit for x_i is Δ_i, such that x_i is within specification if $T_i - \Delta_i \leq x_i \leq T_i + \Delta_i$ is satisfied. Sometimes it is also possible for the tolerance limit for x_i to be asymmetric, with x_i within specification if $T_i - \Delta_i' \leq x_i \leq T_i + \Delta_i$ is satisfied.

The worst-case tolerance design rule can be expressed by the following formula:

$$T - \Delta_0' = \min_{x_i \in (T_i - \Delta_i, T_i + \Delta_i) \forall i} f(x_1, x_2, \ldots, x_i, \ldots, x_n) \tag{16.2}$$

and

$$T + \Delta_0 = \underset{x_i \in (T_i - \Delta_i, T_i + \Delta_i)\forall_i}{\text{Max}} f(x_1, x_2, \ldots, x_i, \ldots, x_n) \qquad (16.3)$$

where $\forall_i$ means for all i.

Example 16.4: Assembly Tolerance Stackup A pile of 10 metal plates are assembled together as shown in Fig. 16.3.

The total thickness of the pile $y = x_1 + x_2 + \cdots + x_i + \cdots + x_{10}$ is of concern. If the target value for x_i is T_i and the tolerance limit for x_i is Δ_i, $i = 1, \ldots, 10$, the target value for y is T, and the tolerance limit for y is Δ_0, and assuming that $T = T_1 + T_2 + \cdots + T_{10}$, then, according to Eqs. (16.2) and (16.3), the relationship between high- and low-level tolerances is

$$T + \Delta_0 = \text{Max}(x_1 + x_2 + \cdots + x_i + \cdots + x_{10})$$
$$= T_1 + T_2 + \cdots + T_{10} + \Delta_1 + \Delta_2 + \cdots + \Delta_i + \cdots + \Delta_{10}$$

$$T - \Delta_0 = \text{Min}(x_1 + x_2 + \cdots + x_i + \cdots + x_{10})$$
$$= T_1 + T_2 + \cdots + T_{10} - \Delta_1 - \Delta_2 - \cdots - \Delta_i - \cdots - \Delta_{10}$$

Obviously:

$$\Delta_0 = \Delta_1 + \Delta_2 + \cdots + \Delta_i + \cdots + \Delta_{10}$$

Specifically, if for each metal plate i, the nominal thickness $T_i = 0.1$ in, tolerance limit $\Delta_i = 0.002$ in, for $i = 1 \cdots 10$, then the tolerance limit for the pile $\Delta_0 = 0.02$ in.

16.2.1 Tolerance analysis and tolerance allocation

In Example 16.4, the tolerances of low-level characteristics, that is, Δ_i values, are given by applying tolerance rules such as worst-case tolerance, specified by Eqs. (16.2) and (16.3), and the high-level tolerance Δ_0 is obtained. Deriving tolerance limits for high-level requirements from tolerances of low-level characteristics is called *tolerance analysis*. On the other hand, if the tolerance limit of a high-level requirement is given, assigning appropriate tolerance limits for low-level characteristics is called *tolerance allocation*.

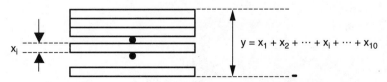

Figure 16.3 Assembly tolerance stackup.

In Example 16.4, suppose that the required tolerance for the pile thickness is $\Delta_0 = 0.01$. If we use worst-case tolerance rules, the tolerance of each plate, $\Delta_i = 0.002$ in, is too wide. Since $\Delta_0 = \Delta_1 + \Delta_2 + \cdots + \Delta_i + \cdots + \Delta_{10}$, we can multiply 0.5 on each Δ_i, and then $\Delta_i = 0.001$, for each $i = 1 \cdots 10$, which will make $\Delta_0 = 0.01$. This is an example of tolerance allocation. Multiplying a constant factor on each old lower-level tolerance limits is called *proportional scaling* (Chase and Greenwood 1988).

Example 16.5: Assembly Clearance   Figure 16.4 shows the assembly relationships for segments A, B, and C. We assume that the target value and tolerance limits for A are 2.000 ± 0.001 in and for B, 1.000 ± 0.0001 in.

Assuming that the clearance $y = C - A - B$ must be between 0.001 and 0.006 in, we are asked to design target dimensions and tolerance limits for C, that is, T_C, Δ'_C, and Δ_C. According to Eqs. (16.2) and (16.3), we have

$$T - \Delta'_0 = 0.001 = \text{Min}(C - A - B) = T_C - \Delta'_C - 2.001 - 1.001$$

$$T + \Delta_0 = 0.006 = \text{Max}(C - B - A) = T_C + \Delta_C - 1.999 - 0.999$$

So

$$T_C - \Delta'_C = 3.003$$

$$T_C + \Delta_C = 3.004$$

If a symmetric tolerance limit is selected for C, then $T_C = 3.0035$ and $\Delta'_C = \Delta_C = 0.0005$.

In both Examples 16.4 and 16.5, all transfer functions in the relationship $y = f(x_1, x_2, \ldots, x_i, \ldots, x_n)$, are linear. Example 16.6 describes a case in which the transfer function is nonlinear.

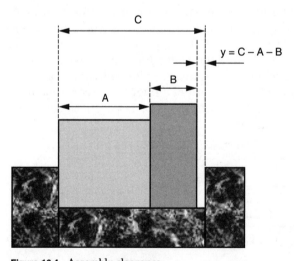

Figure 16.4   Assembly clearance.

Example 16.6. RL Circuit A 100-V, f-Hz power supply across a resistance R in series with an inductance L will result in a current of y amperes:

$$y = \frac{100}{\sqrt{R^2 + (2\pi f L)^2}}$$

Assuming that $f = 50$ Hz, the nominal value for the resistor is $T_R = 9.5\ \Omega$, the current tolerance $\Delta_R = 1.0\ \Omega$, the nominal value for the inductor, $T_L = 0.01$ H (henry), and the current tolerance $\Delta_L = 0.006$ H.

If the customer's tolerance for the circuit is $y = 10.0 \pm 1.0$ A, we would like to ascertain whether the design parameter tolerances are adequate. From Eqs. (16.2) and (16.3)

$$\text{Max}(y) = \text{Max}\left(\frac{100}{\sqrt{R^2 + (2\pi f L)^2}}\right)$$

$$= \frac{100}{\sqrt{(9.5 + 1.0)^2 + (2 \times 3.1416 \times 50 \times (0.01 + 0.006))^2}} = 11.64$$

$$\text{Min}(y) = \text{Min}\left(\frac{100}{\sqrt{R^2 + (2\pi f L)^2}}\right)$$

$$= \frac{100}{\sqrt{(9.5 - 1.0)^2 + (2 \times 3.1416 \times 50 \times (0.01 - 0.006))^2}} = 8.59$$

$$\text{E}(y) = E\left(\frac{100}{\sqrt{R^2 + (2\pi f L)^2}}\right)$$

$$= \frac{100}{\sqrt{(9.5)^2 + (2 \times 3.1416 \times 50 \times (0.01))^2}} = 9.99$$

Clearly, from a worst-case tolerance perspective, the design parameter tolerance is not adequate to ensure customer tolerance.

16.2.2 Nonlinear worst-case tolerance analysis

If the transfer function equation $y = f(x_1, x_2, \ldots, x_i, \ldots, x_n)$ is nonlinear, the tolerance analysis is difficult. From the Taylor expansion formula (Chap. 6) we obtain

$$\Delta y \cong \frac{\partial f}{\partial x_1}\,\Delta x_1 + \frac{\partial f}{\partial x_2}\,\Delta x_2 + \cdots + \frac{\partial f}{\partial x_i}\,\Delta x_i + \cdots + \frac{\partial f}{\partial x_n}\,\Delta x_n \quad (16.4)$$

According to Chase and Greenwood (1988), the worst-case tolerance limit in the nonlinear case is

$$\Delta_0 \cong \left|\frac{\partial f}{\partial x_1}\right|\Delta_1 + \left|\frac{\partial f}{\partial x_2}\right|\Delta_2 + \cdots + \left|\frac{\partial f}{\partial x_i}\right|\Delta_i + \cdots + \left|\frac{\partial f}{\partial x_n}\right|\Delta_n \quad (16.5)$$

In (our) Example 16.6

$$\left|\frac{\partial f}{\partial R}\right| = \frac{100R}{(R^2 + (2\ \pi fL)^2)^{3/2}}, \qquad \left|\frac{\partial f}{\partial L}\right| = \frac{100(2\ \pi f)^2 L}{(R^2 + (2\pi fL)^2)^{3/2}}$$

When $R = 9.5$, $L = 0.01$, and

$$\left|\frac{\partial f}{\partial R}\right| = 0.948, \qquad \left|\frac{\partial f}{\partial L}\right| = 98.53$$

So

$$\Delta_0 \cong \left|\frac{\partial f}{\partial R}\right|\Delta_R + \left|\frac{\partial f}{\partial L}\right|\Delta_L = 0.948 \times 1.0 + 98.53 \times 0.006 = 1.54$$

This is very close to the actual calculation in Example 16.6. If we want to reduce Δ_0 to 1.0, we can multiply a proportion $p = 1.0/1.54 = 0.65$ to both Δ_R and Δ_L; then

$$\Delta_R = 0.65 \times 1.0 = 0.65, \qquad \Delta_L = 0.006 \times 0.65 = 0.004$$

In complex, hard-to-derive transfer functions, the numerical estimation may be useful following the derivation and assumptions steps of Eqs. (6.15) and (6.16). Modification to these equations may be necessary to accommodate different Δ values per a given parameter.

16.3 Statistical Tolerance

The worst-case tolerance design can ensure that high-level tolerance limits are satisfied on all combinations of lower-level characteristics, even in extreme cases. However, this approach will create very tight tolerances for low-level characteristics, and tight tolerance usually means high cost in manufacturing. On the other hand, those low-level characteristics, such as part dimensions and component parameters, are usually random variables. The probability that all low-level characteristics are equal to extreme values (all very low or very high) simultaneously is extremely small. Therefore, the worst-case tolerance method tends to overdesign the tolerances; worst-case tolerance design is used only if the cost of nonconformance is very high for the high-level requirement and the cost to keep tight tolerances on low-level characteristics is low.

The statistical tolerance design method treats both the high-level requirement and low-level characteristics as random variables. The objective of statistical tolerance design is to ensure that the high-level requirement will meet its specification with very high probability.

Low-level characteristics are often assumed to be independent random variables. This assumption is quite valid because low-level characteristics, such as part dimensions and part parameter values often originate in different, unrelated manufacturing processes. Normal distribution is the most frequently used probability model for low-level characteristics. If a low-level characteristic such as a part dimension or component parameter is produced by the existing manufacturing process, historical statistical process control data can be used to estimate its mean and standard deviation.

In this chapter, we assume that each low-level characteristic x_i is a normally distributed random variable, that is, $x_i \sim N(\mu_i, \sigma_i^2)$ for $i = 1 \cdots n$. We also assume that the higher-level requirement, y, is also a normally distributed variable, $y \sim N(\mu, \sigma^2)$.

16.3.1 Tolerance, variance, and process capabilities

Recall the definition of process capability C_p, which we discussed in Chap. 2 (where USL, LSL = upper, lower specification limits):

$$C_p = \frac{\text{USL} - \text{LSL}}{6\sigma}$$

If the process is centered, or in other words, if the target value is equal to the mean of a characteristic [say, x_i, $T_i = E(x_i)$, and the specification limit is symmetric, $\Delta_i = \Delta_i'$], then it is clear that

$$C_p = \frac{\text{USL} - \text{LSL}}{6\sigma_i} = \frac{\text{USL} - T_i}{3\sigma_i} = \frac{T_i - \text{LSL}}{3\sigma_i} = \frac{\Delta_i}{3\sigma_i}$$

So

$$\Delta_i = 3C_p\sigma_i \qquad (16.6)$$

For each low-level characteristic, x_i, $i = 1 \cdots n$. Similarly, for high-level requirement y

$$\Delta_0 = 3C_p\sigma \qquad (16.7)$$

If a Six Sigma quality is required, then $C_p = 2$.

16.3.2 Linear statistical tolerance

If the transfer function equation between the high-level requirement and low-level parameters or variables $x_1, x_2, \ldots, x_i, \ldots, x_n$, is a linear function

$$y = f(x_1, x_2, \ldots, x_i, \ldots, x_n) = a_1 x_1 + a_2 x_2 + \cdots + a_i x_i + \cdots + a_n x_n \quad (16.8)$$

then, we have the following relationship:

$$\mathrm{Var}(y) = \sigma^2 = a_1^2\sigma_1^2 + a_2^2\sigma_2^2 + \cdots + a_i^2\sigma_i^2 + \cdots + a_n^2\sigma_n^2 \quad (16.9)$$

Equation (16.9) gives the relationship between the variance of the high-level requirement and the variances of low-level parameters. Equations (16.6) and (16.7) provide the relationship between tolerance, variances, and process capabilities of both high- and low-level characteristics. From Eqs. (16.6) to (16.9), we can derive the following step-by-step (stepwise) linear statistical tolerance design procedure:

Step 1. Identify the exact transfer function (Chap. 6) between high-level requirement y and low-level parameters or variables; that is, identify Eq. (16.8).

Step 2. For each low-level characteristic x_i, $i = 1 \cdots n$, identify its σ_i, C_p, and Δ_i. This can be done by analyzing sampling or historical process control data, if x_i is created by an existing process. Otherwise, one should make an initial allocation of its σ_i, C_p, and Δ_i, from the best knowledge available.

Step 3. Calculate σ^2, the variance of y, by using Eq. (16.9).

Step 4. From Eq. (16.7), it is clear that $C_p = \Delta_0/3\sigma$. Use this equation to calculate the current C_p for the high-level requirement; if this C_p meets the requirement, stop. If not, go to step 5.

Step 5. Select a desirable C_p level; for example, if Six Sigma level is required, then $C_p = 2$.

Compute the required high-level variance by

$$\sigma_{\mathrm{req}}^2 = \left(\frac{\Delta_0}{3C_p}\right)^2 \quad (16.10)$$

In order to achieve this high-level variance requirement, we need to "scale down" low-level variances. If proportional scaling is used, we can use the following formula to find the scaling factor p:

$$\sigma_{\mathrm{req}}^2 = p^2 \sum_{i=1}^{n} a_i^2\sigma_i^2 \quad (16.11)$$

So

$$p = \frac{\sigma_{\mathrm{req}}}{\sqrt{\sum_{i=1}^{n} a_i^2\sigma_i^2}} \quad (16.12)$$

Then the lower-level variance and tolerance can be determined by

$$\sigma_{i_{\text{new}}} = p\sigma_i \tag{16.13}$$

$$\Delta_i = 3C_p\sigma_{i_{\text{new}}} \tag{16.14}$$

Example 16.7. Assembly tolerance stackup revisited Recall Example 16.4 where 10 metal plates are stacked in a pile (Fig. 16.5).

The pile height $y = x_1 + x_2 + \cdots + x_i + \cdots + x_{10}$. Clearly, this is a linear function, so the linear statistical tolerance design method can be used.

In this example we assume that the target value of y, $T = 1.0$ in, $\Delta_0 = 0.02$ is required, and $C_p = 2$ is also required for pile height. For each metal plate, the height x_i is assumed to be normally distributed, and $T_i = 0.1$ in, $\Delta_i = 0.002$ in, and $C_p = 1.33$ for the metal plate fabrication process, $i \cdots 10$.

First

$$\sigma_i = \frac{\Delta_i}{3C_p} = \frac{0.002}{3 \times 1.33} = 0.0005$$

Then

$$\text{Var}(y) = \sigma^2 = a_1^2\sigma_1^2 + a_2^2\sigma_2^2 + \cdots + a_i^2\sigma_i^2 + \cdots + a_n^2\sigma_n^2$$
$$= \sum_{i=1}^{10} \sigma_i^2 = 10 \times 0.0005^2 = 0.0000025$$
$$\sigma = 0.00158$$

For y

$$C_p = \frac{\Delta_0}{3\sigma} = \frac{0.02}{3 \times 0.00158} = 4.21$$

This is a very high C_p. Even if we reduce Δ_0 to 0.01, C_p will still be 2.105. A quality exceeding Six Sigma is achieved. This calculation has shown that worst-case tolerance would overdesign the tolerances.

Example 16.8. Assembly Clearance Example Revisited We revisit Example 16.5, and assume that the target value and tolerance limits for A and B in Fig. 16.6 are 2.000 ± 0.001 in for A and 1.000 ± 0.001 in for B and that $C_p = 1.33$.

Assuming that the clearance $y = C - A - B$ should be between 0.001 and 0.005 in, we are asked to design a target dimension and tolerance limits for

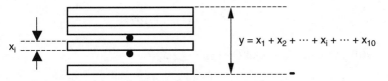

Figure 16.5 Assembly tolerance stackup.

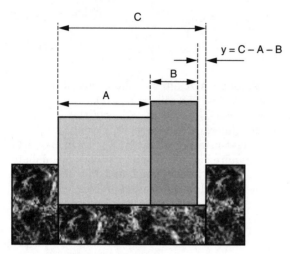

Figure 16.6 Assembly clearance.

C, that is, T_C, Δ'_C and Δ_C, such that the clearance requirement is satisfied with $C_p = 2.00$.

The centerpoint of clearance $E(y)$ is $(0.001 + 0.005)/2 = 0.003$. Because $E(y) = E(C) - E(A) - E(B)$, it follows that $0.0025 = T_c - 2.0 - 1.0$, $T_c = 3.003$. Then

$$\sigma_A = \frac{\Delta_A}{3C_p} = \frac{0.001}{3 \times 1.33} = 0.00025$$

Similarly, $\sigma_B = 0.00025$. The variance of y is $\sigma^2 = \sigma_A^2 + \sigma_B^2 + \sigma_C^2 = 2 \times 0.00025^2 + \sigma_C^2$. For y, if $C_p = 2$, the required standard deviation for clearance is

$$\sigma = \frac{\Delta_0}{3C_p} = \frac{0.002}{3 \times 2} = 0.00033333$$

Because $\sigma^2 = 0.0003333^2$ is less than the current $\sigma_A^2 + \sigma_B^2 = 2 \times 0.00025^2$, there will be no feasible σ_C and Δ_C, unless the tolerances for A and B are changed. If we can change these tolerances, we can find feasible tolerance for C. If we assume $\sigma_A = \sigma_B = \sigma_C$, then

$$\sigma^2 = \sigma_A^2 + \sigma_B^2 + \sigma_C^2 = 3\sigma_A^2 = 0.000333333^2$$

$$\sigma_A = \sqrt{\frac{0.000333333^2}{3}} = 0.0001924$$

If C_p is still 1.333 for A, B, and C, then

$$\Delta_A = \Delta_B = \Delta_C = 3C_p\sigma_A = 3 \times 1.333 \times 0.0001924 = 0.00077$$

16.3.3   Nonlinear statistical tolerance

If the transfer function equation between high-level requirement y and low-level characteristics $x_1, x_2, \ldots, x_i, \ldots, x_n$, is not a linear function, then

$$y = f(x_1, x_2, \ldots, x_i, \ldots, x_n) \tag{16.15}$$

is not a linear function, Then, we have the following approximate relationship:

$$\text{Var}(y) = \sigma^2 \cong \left(\frac{\partial f}{\partial x_1} \right)^2 \sigma_1^2 + \left(\frac{\partial f}{\partial x_2} \right)^2 \sigma_2^2 + \cdots + \left(\frac{\partial f}{\partial x_i} \right)^2 \sigma_i^2$$

$$+ \cdots + \left(\frac{\partial f}{\partial x_n} \right)^2 \sigma_n^2 \tag{16.16}$$

Equation (16.16) gives the approximate relationship between the variance of the high-level requirement and the variances of low-level characteristics. Equations (16.6) and (16.7) can still provide the relationship between tolerance, variances, and process capabilities of both high- and low-level characteristics.

The transfer function $y = f(x_1, x_2, \ldots, x_i \ldots, x_n)$ is seldom a closed-form equation. In the design stage, computer simulation models are often available for many products/processes, such as the FEA model for mechanical design and electronic designs for electric circuit simulators. Many of these computer simulation models can provide sensitivities, which is essentially $\Delta y / \Delta x_i$. These sensitivities can be used to play the roles of partial derivatives, $\partial f / \partial x_i$.

Here we can develop the following step-by-step procedure for nonlinear statistical tolerance design:

Step 1.   Identify the exact transfer function between high-level requirement y and low-level characteristics; that is, identify Eq. (16.15). If the equation is not given in closed form, we can use a computer simulation model, or an empirical model derived from a DOE study.

Step 2.   For each low-level characteristic (parameter), x_i, $i = 1 \cdots n$, identify its σ_i, C_p, and Δ_i. This can be done by looking into historical process control data, if x_i is created by an existing process. Otherwise, make an initial allocation of its σ_i, C_p, and Δ_i, from the best knowledge available.

Step 3.   Calculate σ^2, the variance of y; with Eq. (16.16), sensitivities can be used to substitute partial derivatives.

Step 4.   From Eq. (16.7), it is clear that

$$C_p = \frac{\Delta_0}{3\sigma}$$

Use this equation to calculate current C_p for the high-level requirement; if this C_p meets the requirement, stop. If not, go to step 5.

Step 5. Select a desirable C_p level. For example, if Six Sigma level
is required, then $C_p = 2$.

Compute the required high-level variance by

$$\sigma_{\text{req}}^2 = \left(\frac{\Delta_0}{3C_p}\right)^2$$

In order to achieve this high-level variance requirement, we need to
"scale down" low-level variances. If proportional scaling is used, we
can use the following formula to find the scaling factor p:

$$\sigma_{\text{req}}^2 = p^2 \sum_{i=1}^{n} \left(\frac{\partial f}{\partial x_i}\right)^2 \sigma_i^2 \qquad (16.17)$$

So

$$p = \frac{\sigma_{\text{req}}}{\sqrt{\displaystyle\sum_{i=1}^{n} \left(\frac{\partial f}{\partial x_i}\right)^2 \sigma_i^2}} \qquad (16.18)$$

Then the lower-level variance and tolerance can be determined by

$$\sigma_{i\text{new}} = p\sigma_i \qquad (16.19)$$

$$\Delta_i = 3C_p\sigma_{i\text{new}} \qquad (16.20)$$

Example 16.9. *RL* Circuits Revisited   Recall that in Example 16.6, a 100-
V, f-Hz power supply across a resistance R in series with an inductance L
will result in a current of y amperes:

$$y = \frac{100}{\sqrt{R^2 + (2\pi f L)^2}}$$

Assuming that $f = 50$ Hz, the nominal value for the resistor $T_R = 9.5\ \Omega$, the
current tolerance $\Delta_R = 1.0\ \Omega$, the nominal value for the inductor $T_L = 0.01$
H, and the current tolerance $\Delta_L = 0.006$ H. Assume that for both R and L,
$C_p = 1.33$.

The customer's tolerance for the circuit current is $y = 10.0 \pm 1.0$ A, and
we would like to satisfy this requirement with $C_p = 2.0$. So the required
standard deviation for the current y is

$$\sigma_{\text{req}} = \frac{\Delta_0}{3C_p} = \frac{1.0}{3 \times 2.0} = 0.1667$$

The actual variance of circuit current is approximately

$$\sigma^2 \cong \left(\frac{\partial f}{\partial R}\right)^2 \sigma_R^2 + \left(\frac{\partial f}{\partial L}\right)^2 \sigma_L^2$$

where

$$\sigma_R = \frac{\Delta_R}{3C_p} = \frac{1.0}{3 \times 1.333} = 0.25$$

$$\sigma_L = \frac{\Delta_L}{3C_p} = \frac{0.006}{3 \times 1.333} = 0.0015$$

$$\left(\frac{\partial f}{\partial R}\right)^2 = \frac{(100R)^2}{(R^2 + (2\pi f L)^2)^3}, \qquad \left(\frac{\partial f}{\partial L}\right)^2 = \frac{(100(2\pi f)^2 L)^2}{(R^2 + (2\pi f L)^2)^3}$$

when $R = 9.5$, $L = 0.01$, and

$$\left(\frac{\partial f}{\partial R}\right)^2 = 0.899, \qquad \left(\frac{\partial f}{\partial L}\right)^2 = 9708.2$$

So

$$\sigma^2 \cong \left(\frac{\partial f}{\partial R}\right)^2 \sigma_R^2 + \left(\frac{\partial f}{\partial L}\right)^2 \sigma_L^2 = 0.899 \times 0.25^2 + 9708.2 \times 0.0015^2 = 0.078$$

$$\sigma = \sqrt{0.078} = 0.279$$

Because $\sigma > \sigma_{req}$, the current circuit will not be able to satisfy the customer requirement with $C_p = 2$. We can use the proportional scaling factor p:

$$p = \frac{\sigma_{req}}{\sigma} = \frac{0.1667}{0.279} = 0.597$$

Therefore, the new tolerance limits for R and L are

$$\Delta_R = 3C_p p \sigma_R = 3 \times 1.333 \times 0.597 \times 0.25 = 0.597 \ \Omega$$

$$\Delta_L = 3C_p p \sigma_L = 3 \times 1.333 \times 0.597 \times 0.0015 = 0.00358 \ H$$

16.4   Cost-Based Optimal Tolerance

The purpose of tolerance design is to set tolerance limits for the design parameters. The tolerance limits are usually in the form of $T_i - \Delta_i \leq x_i \leq T_i + \Delta_i$ for each design variable x_i, $i...n$, where $\Delta_i = 3C_p\sigma_i$. Essentially, the tolerance limits will limit the magnitude of σ_i^2. The variance of the high-level requirement y, $\mathrm{Var}(y) = \sigma^2$, follows the following relationships:

$$\text{Var}(y) = \sigma^2 \cong \left(\frac{\partial f}{\partial x_i}\right)^2 \sigma_1^2 + \left(\frac{\partial f}{\partial x_2}\right)^2 \sigma_2^2 + \cdots + \left(\frac{\partial f}{\partial x_i}\right)^2 \sigma_i^2 + \cdots +$$

$$+ \cdots + \left(\frac{\partial f}{\partial x_n}\right)^2 \sigma_n^2$$

For a nonlinear relationship $y = f(x_1, x_2, \ldots, x_i, \ldots, x_n)$ and

$$\text{Var}(y) = \sigma^2 = a_1^2 \sigma_1^2 + a_2^2 \sigma_2^2 + \cdots + a_i^2 \sigma_i^2 + \cdots + a_n^2 \sigma_n^2$$

For a linear transfer function $y = a_1 x_1 + a_2 x_2 + \cdots + a_i x_i + \cdots + a_n x_n$. Clearly, the reduction of $\text{Var}(y) = \sigma^2$ can be achieved by reducing σ_i^2 for $i = 1 \cdots n$. However, the reduction of σ_i^2 values will incur cost. This variance reduction cost might be different for different low-level characteristics, that is, x_i. However, the impact of reduction for each variance σ_i^2, made on the reduction of $\text{Var}(y)$, σ^2, depends on the magnitude of sensitivities $|\partial f/\partial x_i|$. The greater the sensitivity, the greater the impact of the reduction of σ_i^2 on the reduction of σ^2. Therefore, it is more desirable to tighten the tolerances for those parameters that have high sensitivity and low tolerance tightening costs. The objective of a cost-based optimal tolerance design is to find an optimum strategy that results in a minimum total cost (variability reduction cost + quality loss).

The tolerance reduction cost is usually a nonlinear curve illustrated by Fig. 16.7.

Much work has been done (Chase 1988) to the cost-based optimal tolerance design. In this chapter, we discuss the cost-based optimal tolerance design approach proposed by Yang et al. (1994). In this approach, the tolerance design problem is formulated as the following optimization problem (Kapur 1993):

$$\text{Minimize:} \quad \text{TC} = \sum_{i=1}^{n} C_i(\sigma_i) + k\sigma^2$$

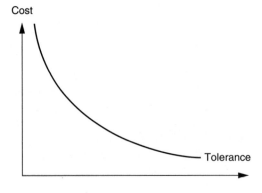

Figure 16.7 Cost versus tolerance.

$$\text{Subject to:} \qquad \sigma^2 \leq \sigma^2_{\text{req}} \qquad (16.21)$$

where σ^2_{req} is the required variance for y. $C_i(\sigma_i)$ is the tolerance control cost for x_i, which should be a decreasing function of σ_i. $k\sigma^2$ is the Taguchi quality loss due to variation, and TC stands for total cost.

Using Eq. (16.16), the optimization problem (16.21) becomes

$$\text{Minimize:} \qquad \sum_{i=1}^{n} C_i(\sigma_i) + k \sum_{i=1}^{n} \left(\frac{\partial f}{\partial x_i} \right)^2 \sigma_i^2$$

$$\text{Subject to:} \qquad \sum_{i=1}^{n} \left(\frac{\partial f}{\partial x_i} \right)^2 \sigma_i^2 \leq \sigma^2_{\text{req}} \qquad (16.22)$$

The optimal tolerances can be derived by using the Karush-Kuhn-Tucker condition (KKT condition) for (16.22) as follows:

$$\frac{dC_i(\sigma_i)}{d\sigma_i} + 2(k + \lambda) \left(\frac{\partial f}{\partial x_i} \right)^2 \sigma_i = 0$$

$$\lambda \geq 0; \qquad \lambda(\sigma^2_{\text{req}} - \sigma^2) = 0; \qquad \sigma^2 \leq \sigma^2_{\text{req}} \qquad (16.23)$$

where λ is the Lagrange multiplier. By solving the KKT condition, we can obtain the optimal tolerances for all x_i, $i = 1,...,n$:

$$\Delta_i = \frac{3C_p}{2(k + \lambda)} \frac{\left| \dfrac{dC_i(\sigma_i)}{d\sigma_i} \right|}{\left(\dfrac{\partial f}{\partial x_i} \right)^2} = \frac{3C_p}{2(k + \lambda)} \frac{\delta C_i}{(\delta f)_i^2} \sigma_i \qquad (16.24)$$

where δC_i is the unit tolerance reduction cost (per unit change in the tolerance of x_i) and $(\delta f)_i$ is the incremental change in requirement y for each unit change in x_i. It is difficult to use Eq. (16.24) directly to solve for optimal tolerances, because the Lagrange multiplier λ is difficult to get. However, in Eq. (16.24), $3C_p/2(k+\lambda)$ is the same for all x_i values, and we can use $p_i = \delta C_i/(\delta f)_i^2$ as a scale factor for optimal tolerance reduction and the optimal tolerance tightening priority index. The low-level characteristic, x_i, with a smaller p_i index, indicates that it is more appropriate to control the tolerance of x_i. Since x_i has a relatively small unit tolerance control cost and relatively high sensitivity for the requirement variation, the reduction in x_i variability will result in a larger reduction in variability of Y with a relatively low tolerance reduction cost. From this discussion, we can develop the following cost-based optimal tolerance design procedure:

Step 1. Identify the exact transfer function between high-level requirement y and low-level characteristics; that is, identify the equation $y = f(x_1, x_2, \ldots, x_i, \ldots, x_n)$. If the equation is not given in closed form, we can use a computer simulation model, or an empirical model derived from a DOE study.

Step 2.   For each low-level characteristic x_i, $i = 1\cdots n$, identify its σ_i, C_p, and Δ_i. This can be done by analyzing historical process control data, if x_i is made by an existing process. Otherwise, make an initial allocation of its σ_i, C_p, and Δ_i from the best knowledge available.

Step 3.   Calculate σ^2, the variance of y, by using Eq. (16.16); sensitivities can be used to substitute partial derivatives.

Step 4.   From Eq. (16.7), it is clear that $C_p = \Delta_0/3\sigma$. Use this equation to calculate current C_p for the high-level characteristic. If this C_p meets the requirement, stop. If not, go to step 5.

Step 5.   Select a desirable C_p level. For example, if Six Sigma level is required, then $C_p = 2$. Compute the required high level variance by:

$$\sigma^2_{req} = \left(\frac{\Delta_0}{3C_p} \right)^2$$

For each x_i, compute

$$p_i = \frac{\delta C_i}{(\delta f)^2_i} \tag{16.25}$$

In order to achieve this high-level variance requirement, we need to scale down low-level variances:

$$\sigma^2_{req} = p^2 \sum_{i=1}^{n} p_i^2 \left(\frac{\partial f}{\partial x_i} \right)^2 \sigma_i^2 \tag{16.26}$$

So

$$p = \frac{\sigma_{req}}{\sqrt{\sum_{i=1}^{n} p_i^2 \left(\frac{\partial f}{\partial x_i} \right)^2 \sigma_i^2}} \tag{16.27}$$

Then the lower-level variance and tolerance can be determined by

$$\Delta_i = 3C_p p p_i \sigma_i \tag{16.28}$$

Example 16.10. *RL* Circuits Again   Recall that in Example 16.9, a 100-V f-Hz power supply across a resistance R in series with an inductance L will result in a current of y amperes:

$$y = \frac{100}{\sqrt{R^2 + (2\pi f L)^2}}$$

Assume that $f = 50$ Hz, the nominal value for the resistor $T_R = 9.5\ \Omega$, the tolerance $\Delta_R = 1.0\ \Omega$, the nominal value for the inductor $T_L = 0.01$ H, and the tolerance $\Delta_L = 0.006$ H. Assume that for R and L, $C_p = 1.33$. Finally, assume that the tolerance Δ_R can be reduced by 0.5 Ω with an additional cost of $0.15 and that Δ_L can be reduced by 0.003 H with an additional cost of 20 cents.

The customer's tolerance for the circuit current is $y = 10.0 \pm 1.0$ A, and we would like to satisfy this requirement with $C_p = 2.0$. So the required standard deviation for the current y is

$$\sigma_{\text{req}} = \frac{\Delta_0}{3C_p} = \frac{1.0}{3 \times 2.0} = 0.1667$$

The actual variance of circuit current is approximately

$$\sigma^2 \cong \left(\frac{\partial f}{\partial R} \right)^2 \sigma_R^2 + \left(\frac{\partial f}{\partial L} \right)^2 \sigma_L^2$$

where

$$\sigma_R = \frac{\Delta_R}{3C_p} = \frac{1.0}{3 \times 1.333} = 0.25$$

$$\sigma_L = \frac{\Delta_L}{3C_p} = \frac{0.006}{3 \times 1.333} = 0.0015$$

$$\left(\frac{\partial f}{\partial R} \right)^2 = \frac{(100R)^2}{(R^2 + (2\pi f L)^2)^3}, \qquad \left(\frac{\partial f}{\partial L} \right)^2 = \frac{(100(2\pi f)^2 L)^2}{(R^2 + (2\pi f L)^2)^3}$$

When $R = 9.5$, $L = 0.01$ and

$$\left(\frac{\partial f}{\partial R} \right)^2 = 0.899, \qquad \left(\frac{\partial f}{\partial L} \right)^2 = 9708.2$$

So

$$\sigma^2 \cong \left(\frac{\partial f}{\partial R} \right)^2 \sigma_R^2 + \left(\frac{\partial f}{\partial L} \right)^2 \sigma_L^2 = 0.899 \times 0.25^2 + 9708.2 \times 0.0015^2 = 0.078$$

$$\sigma = \sqrt{0.078} = 0.279$$

Because $\sigma > \sigma_{\text{req}}$, the current circuit will not be able to satisfy the customer requirement with $C_p = 2$.

We can compute

$$p_R = \frac{\delta C_R}{(\delta f)_R^2} = \frac{0.15}{0.474} = 0.316$$

$$p_L = \frac{\delta C_L}{(\delta f)_L^2} = \frac{0.20}{0.295} = 0.678$$

Because

$$(\delta f)_R \cong \left| \frac{\partial f}{\partial R} \right| \delta R = 0.948 \times 0.5 = 0.474 \, \text{A}$$

$$(\delta f)_L \cong \left| \frac{\partial f}{\partial L} \right| \delta L = 0.98.53 \times 0.003 = 0.295 \, \text{A}$$

Therefore, reducing R is more cost-effective than reducing L. Then

$$p = \frac{\sigma_{\text{req}}}{\sqrt{\sum\limits_{i=1}^{n} p_i^2 \left(\frac{\partial f}{\partial x_i} \right)^2 \sigma_i^2}}$$

$$= \frac{0.1667}{\sqrt{0.316^2 \times 0.899 \times 0.25^2 + 0.678^2 \times 9708.2 \times 0.0015^2}} = 1.332$$

Therefore, the new tolerance limits for R and L are

$$\Delta_R = 3C_p \, pp_R \sigma_R = 3 \times 1.333 \times 1.332 \times 0.316 \times 0.25 = 0.42 \, \Omega$$

$$\Delta_L = 3C_p \, pp_L \sigma_L = 3 \times 1.333 \times 1.332 \times 0.678 \times 0.0015 = 0.0054 \, \text{H}$$

16.5   Taguchi's Loss Function and Safety Tolerance Design

Dr. Taguchi developed a unique approach to tolerance design and tolerance analysis. His tolerance design approach includes a cost-based tolerance design and allocation. The most important consideration in cost is the quality loss due to requirement deviation from the ideal requirement level.

16.5.1   Customer tolerance and producer tolerance

Nominal-the-best requirement.   The quality loss can be expressed by the Taguchi quality loss function that we discussed in detail in Chap. 14. Figure 16.8 shows the quadratic curve for the quality loss function for the "nominal-the-best" case.

The quality loss function can be expressed as follows:

$$L(y) = \frac{A_0}{\Delta_0^2} (y - T)^2 \tag{16.29}$$

Quality loss

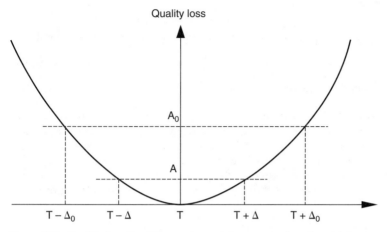

Figure 16.8 Quality loss function, customer tolerance, and producer tolerance.

where Δ_0 is the customer tolerance limit and A_0 is the cost incurred to the customer when the requirement level y is at the boundary of the customer tolerance limit. Taguchi reasons that if the design can be repaired or fixed before shipping to customers at a lower cost, say, A and $A < A_0$, then the *producer tolerance* Δ, which is the internal tolerance limit for shipping inspection, should be set at a narrower level than the customer tolerance limits. Taguchi proposes that the specific producer tolerance limit Δ should be set according to the quality loss function. Specifically, he thinks that at producer tolerance, the quality loss should break even with the internal repair cost. Thus, when $y = T + \Delta$ or $y = T - \Delta$, by Eq. (16.29), the quality loss

$$L(y) = \frac{A_0}{\Delta_0^2} \Delta^2$$

Then, setting the quality loss $L(y)$ equal to the repair cost A, we obtain

$$A = \frac{A_0}{\Delta_0^2} \Delta^2$$

Therefore, the producer tolerance limit Δ can be computed by

$$\Delta = \sqrt{\frac{A}{A_0}} \, \Delta_0 = \frac{\Delta_0}{\sqrt{\frac{A_0}{A}}} = \frac{\Delta_0}{\phi} \qquad (16.30)$$

where

$$\phi = \sqrt{\frac{A_0}{A}} = \sqrt{\frac{\text{loss of exceeding functional limit}}{\text{loss at factory for exceeding factory standard}}}$$

(16.31)

where ϕ is called the safety factor.

Example 16.11. TV Color Density   Recall Example 13.1. Assuming that the customer tolerance limit Δ_0 for color density y is equal to 7, if the color density y is either greater than $T + 7$ or less than $T - 7$, 50 percent of the customers will be unhappy and will demand replacement of their TV sets, and the replacement cost A_0 will be \$98. However, if the TV is repaired within the factory, the repair cost will be \$10, and then the producer tolerance Δ should be

$$\Delta = \sqrt{\frac{A}{A_0}} \Delta_0 = \sqrt{\frac{10}{98}} \Delta_0 = \frac{\Delta_0}{\sqrt{98/10}} = \frac{\Delta_0}{3.13} = \frac{7}{3.13} = 2.24$$

that is, the producer tolerance Δ should be 2.24, and the safety factor ϕ in this example is 3.13.

Justification for safety factor and tighter producer tolerance.   From Fig. 16.9, we can see that if a company does not have any aftersale customer service, the cost for customers due to poor quality will follow the quadratic curve. If the company cannot control the variation in design requirement, the cost for customers will be high. Clearly this is very bad for customers as well as for the company. Actually, the company may lose even more than the quality loss cost, because customer dissatisfaction, bad reputation, and bad publicity will hurt the company even more than they hurt the customers. For example, if a famous restaurant chain had one incidence of food poisoning, the loss to the customer will be medical cost, a few days or weeks without pay, and so on, but the cost for the restaurant chain will be bad publicity, litigation, loss of consumer confidence, and big loss in sales, which could very easily reach a magnitude higher than the customer loss.

Having a customer tolerance limit of Δ_0 and aftersale service, the company can reduce the maximum customer loss to below the external repair cost A_0. If the company can set a tighter producer tolerance limit Δ, and practice "preventive repair" at the cost A, then the maximum loss to the customers will be reduced to the internal repair cost A.

Smaller-the-better requirement.   For the "smaller-the-better" requirement, the quality loss function is

$$L(y) = \frac{A_0}{\Delta_0^2} y^2$$

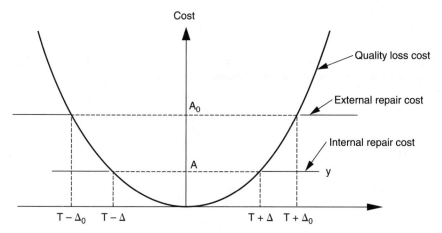

Figure 16.9 Quality loss cost versus repair costs.

If the internal repair cost is A at producer tolerance Δ, by letting the repair cost equal to quality loss, then

$$A = \frac{A_0}{\Delta_0^2} \Delta^2$$

Therefore

$$\Delta = \frac{\Delta_0}{\sqrt{A_0/A}} = \frac{\Delta_0}{\phi} \qquad (16.32)$$

Example 16.12. Bacteria Count in a Meat Item   Clearly the bacteria count y in a portion of meat is a smaller-the-better requirement. Assume that if the bacteria count in a meat pack is more than 8000, then the person who eats the meat will get sick. Suppose that the average cost of medical treatment and lost pay is equal to \$500. If the meat is determined to exceed the limit within the meat-packing plant, the meat will be discarded, and the cost will be \$3.00. We can suggest the internal inspection limit for bacteria count Δ to be

$$\Delta = \frac{\Delta_0}{\sqrt{\dfrac{A_0}{A}}} = \frac{8000}{\sqrt{500/3}} = \frac{8000}{12.9} = 620$$

In other words, the meat-packing factory will inspect the meat packs, and if the pack contains more than 620 bacteria count, the meat will be discarded. The safety factor is 12.9.

Larger-the-better requirement.   For the "larger-the-better" requirement, the quality loss function is

$$L(y) = A_0\Delta_0^2 \frac{1}{y^2}$$

If the internal repair cost is A at producer tolerance Δ, by letting repair cost equal to quality loss, we obtain

$$A = A_0\Delta_0^2 \frac{1}{\Delta^2}$$

Therefore

$$\Delta = \sqrt{\frac{A_0}{A}} \Delta_0 = \phi\Delta_0 \qquad (16.33)$$

Example 16.13. Strength of a Cable Suppose that a cable is used to hang a piece of equipment, and that the equipment could produce a pull of 5000 kgf (kilograms-force) to the cable. If the cable breaks and the equipment falls off, the cost would be \$300,000. The cable's strength is proportional to the cross-sectional area at 150 kgf/mm². Assume that the cost of cable is proportional to its cross-sectional area at \$60/mm².

Letting the area be x, then $A = 60x$, and $\Delta = 150x$, and using Eq. (16.33), we obtain

$$150x = \sqrt{\frac{300,000}{60x}} \, 5000$$

Then $(150x)^2 \, 60x = 5000^2 \times 300,000$. We can solve for x: $x = 177$. So the cable cost $A = 60x = \$10,620$ and cable strength $D = 150\Delta = 150x = 150 \times 177 = 26,550$. The safety factor

$$\phi = \sqrt{\frac{300,000}{10,620}} = 5.31$$

16.5.2 Determination of tolerance of low-level characteristic from high-level tolerance

Given the transfer function of a high-level requirement y and low-level characteristics $x_1, x_2, \ldots, x_i, \ldots, x_n$, we have

$$y = f(x_1, x_2, \ldots, x_i, \ldots, x_n)$$

Taguchi tolerance design also has its own approach for determining low-level tolerances. If the transfer function expressed above can be linearized, or sensitivities can be found, then for each low-level characteristic x_i, for $i = 1 \cdots n$, we have

$$y - T \approx \frac{\partial f}{\partial x_i} (x_i - T_i) \cong a_i (x_i - T_i) \tag{16.34}$$

Then the Taguchi loss function is approximately

$$L(y) = \frac{A_0}{\Delta_0^2} (y - T)^2 \approx \frac{A_0}{\Delta_0^2} [a_i (x_i - T_i)]^2 \tag{16.35}$$

Assume that when a low-level characteristic x_i exceeds its tolerance limit Δ_i, the cost for replacing it is A_i; then, equating A_i with quality loss in Eq. (16.35), we get

$$A_i = \frac{A_0}{\Delta_0^2} [a_i \Delta_i]^2 \tag{16.36}$$

$$\Delta_i = \sqrt{\frac{A_1}{A_0}} \frac{\Delta_0}{|a_i|} \tag{16.37}$$

Example 16.14. Power Supply Circuit The specification of the output voltage of a power supply circuit is 9 ± 1.5 V. If a power supply circuit is out of specification, the replacement cost is \$2.00. The resistance of a resistor affects its output voltage; every 1 percent change in resistance will cause output voltage to vary by 0.2 V. If the replacement cost for a resistor is 15 cents, what should be the tolerance limit for the resistor (in percentage)? By using Eq. (16.37), we get

$$\Delta_i = \sqrt{\frac{A}{A_0}} \frac{\Delta_0}{|a_i|} = \sqrt{\frac{0.15}{2.0}} \frac{1.5}{0.2} = 2.05$$

So the tolerance limit for the resistor should be set at about ±2 percent.

16.5.3 Tolerance allocation for multiple parameters

Given the transfer function $y = f(x_1, x_2, \ldots, x_i, \ldots x_n)$, if we want to design tolerance limits for all low-level characteristics $x_1, x_2, \ldots, x_i, \ldots x_n$ in Taguchi's approach, we can simply apply Eq. (16.37) to all parameters:

$$\Delta_1 = \sqrt{\frac{A_1}{A_0}} \frac{\Delta_0}{|a_1|}; \quad \Delta_2 = \sqrt{\frac{A_2}{A_0}} \frac{\Delta_0}{|a_2|}, \ldots, \Delta_n = \sqrt{\frac{A_n}{A_0}} \frac{\Delta_0}{|a_n|}$$

Therefore, the square of the range of the output y caused by the variation of $x_1, x_2, \ldots, x_i, \ldots, x_n$ is

$$\Delta^2 = (a_1 \Delta_1)^2 + (a_2 \Delta_2)^2 + \cdots + (a_n \Delta_n)^2$$

$$= \left(\sqrt{\frac{A_1}{A_0}} \Delta_0 \right)^2 + \left(\sqrt{\frac{A_2}{A_0}} \Delta_0 \right)^2 + \cdots + \left(\sqrt{\frac{A_n}{A_0}} \Delta_0 \right)^2 \quad (16.38)$$

$$= \frac{A_1 + A_2 + \cdots + A_k}{A_0} \Delta_0$$

or

$$\Delta = \sqrt{\frac{\sum\limits_{i=1}^{n} A_n}{A_0}} \Delta_0 \quad (16.39)$$

If $x_1, x_2, \ldots, x_i, \ldots x_n$, are n components that form a system, then in Eqs. (16.38) and (16.39), A_0 stands for the cost to bring the system back to specification, and $A_1 + A_2 + \cdots + A_n$ stands for the total cost of purchasing these n components, then we have the following three cases:

Case 1: $A_1 + A_2 + \cdots + A_n \ll A_0$ This means that the total cost of components is much smaller than the cost of rectifying the system. This could certainly happen if the whole system has to be scrapped when it is out of specification. From Eq. (16.39), it is clear that Δ, the actual "producer tolerance" for the assembly, will be much smaller than the customer tolerance Δ_0. For example, if

$$A_1 + A_2 + \cdots + A_n = \frac{1}{4} A_0$$

That is, the total component cost is one-fourth of the system replacement cost, then by Eq. (16.39)

$$\Delta = \sqrt{\frac{1}{4}} \Delta_0 = \frac{1}{2} \Delta_0$$

That is, the producer tolerance should be half of the customer tolerance.

Case 2: $A_1 + A_2 + \cdots + A_n \gg A_0$ This means that the total cost of components is much greater than the cost of rectifying the system, and this could certainly happen if the system can be easily adjusted back to specification. From Eq. (16.39), it is clear that Δ, the actual producer tolerance for the assembly, will be much larger than the customer tolerance Δ_0.

Case 3: $A_1 + A_2 + \cdots + A_n \approx A_0$ This means that the total cost of components is about the same as the cost of rectifying the system. This case does not occur very often. From Eq. (16.39), it is clear that Δ, the actual producer tolerance for the assembly, will be about the same as the customer tolerance Δ_0.

We can see clearly that there is a fundamental philosophical difference between traditional tolerance allocation and Taguchi tolerance allocation. Traditional tolerance allocation will assign tolerances $\Delta_1, \Delta_2, \ldots, \Delta_n$ for low-level characteristics $x_1, x_2, \ldots, x_i, \ldots, x_n$, such that the actual tolerance for y, say, Δ, is about the same as that of customer tolerance Δ_0. On the other hand, Taguchi tolerance allocation will assign tolerances $\Delta_1, \Delta_2, \ldots, \Delta_n$ for low-level characteristics $x_1, x_2, \ldots, x_i, \ldots, x_n$, such that the actual tolerance for y, say, Δ, could be either much smaller than the customer tolerance Δ_0 (if the cost of correcting error at a high level is much greater than that at low level) or Δ could be much larger than the customer tolerance Δ_0 (if the cost of correcting error at a high level is very small).

16.6 Taguchi's Tolerance Design Experiment

Given a system where a high-level requirement y is related to a group of low-level characteristics, $x_1, x_2, \ldots, x_i, \ldots, x_n$, with an *unknown* transfer function:

$$y = f(x_1, x_2, \ldots, x_i, \ldots, x_n)$$

The tolerance design methods discussed in Sec. 16.5 cannot be applied directly. In this case, Dr. Taguchi (1986) proposed using a tolerance design experiment to determine the impact of the variability of low-level characteristics, or factors, $x_1, x_2, \ldots, x_i, \ldots, x_n$, on the high-level requirement y. The experimental data analysis is able to prioritize the low-level factors' tolerance adjustments on the basis of their impact and cost.

A tolerance design experiment is conducted after the parameter design experiment, when the nominal level of control factors, that is, the target values of $x_1, x_2, \ldots, x_i, \ldots, x_n$, have already been determined. In the tolerance design experiment, Taguchi (1986) recommended that the experimental factor levels be set by the following rules:

- Two-level factors:
  First level = target value $T_i - \sigma_i$
  Second level = target value $T_i + \sigma_i$

- Three-level factors:
  First level = $T_i - \sqrt{3/2}\,\sigma_i$
  Second level = T_i
  Third level = $T_i + \sqrt{3/2}\,\sigma_i$

By using these settings, for two-level factors, the two levels are located at the 15th and 85th percentile points in the variation range. For three-level factors, the three levels are located at the 10th, 50th, and 90th percentile points in the variation range. Clearly, these levels are set at "reasonable" ends of the variation range.

Tolerance design is a regular orthogonal array design experiment in which the regular functional requirement y is observed at each run of the experiment. According to the equation

$$\text{Var}(Y) = \sigma^2 \approx \sum_{i=1}^{n} \left(\frac{\delta y}{\delta x_i} \right)^2 \sigma_i^2 \qquad (16.40)$$

if we set $\delta x_i \approx \sigma_i$, then

$$\text{Var}(Y) = \sigma^2 \approx \sum_{i=1}^{n} (\delta y)_i^2 \qquad (16.41)$$

Equation (16.41) actually indicates that the mean-square total MS_T in the DOE is a good statistical estimator of $\text{Var}(Y)$. The percentage contribution of each factor for sum-of-squares total SS_T can be used to prioritize the tolerance reduction effort. If a factor has a large percentage contribution for SS_T and it is also cheap to reduce the tolerance, then that factor is a good candidate for variation reduction. We will illustrate the Taguchi tolerance design experiment by the following example.

Example 16.15. Tolerance Design Experiment for Process Variables   A composite material is processed by a heat-curing process, and the tensile strength of the material is the key characteristic. There are four process variables: A—baking temperature, B—baking time, C—fiber additive quantity, and D—stirring rate.

After parameter design, the nominal values of these process variables have already been determined: A, 300°F; B, 30 min; C, 15 percent; and D, 300 r/min. These process variables cannot be controlled precisely; historical data indicate that the long-term standard deviations for each of those variables are σ_A, 10°F; σ_B, 1.8 min; σ_C, 1.6 percent; and σ_D, 12 rpm. Using

First level $= T_i - \sqrt{3/2}\,\sigma_i$

Second level $= T_i$

Third level $= T_i + \sqrt{3/2}\,\sigma_i$

we get

Factors	Level 1	Level 2	Level 3
A, °F	288	300	312
B, min	27.8	30	32.2
C, %	13	15	17
D, r/min	285.4	300	314.6

An L_9 array is used, and the experimental layout and tensile strength data turn out as follows:

| Experiment no. | Factors | | | | Tensile strength |
	A	B	C	D	
1	1	1	1	1	243
2	1	2	2	2	295
3	1	3	3	3	285
4	2	1	2	3	161
5	2	2	3	1	301
6	2	3	1	2	260
7	3	1	3	2	309
8	3	2	1	3	274
9	3	3	2	1	198

By using MINITAB, we obtain the following ANOVA table:

```
Analysis of Variance for Strength, using Adjusted SS for Tests
Source  DF   Seq SS   Adj SS   Adj MS    F   P
A        2   1716.2   1716.2    858.1   **
B        2   4630.9   4630.9   2315.4   **
C        2   9681.6   9681.6   4840.8   **
D        2   4011.6   4011.6   2005.8   **
Error    0      0.0      0.0      0.0
Total    8  20040.2
```

$$\hat{\sigma}^2 = \text{MST} = \frac{\text{SST}}{8} = 2505; \qquad \hat{\sigma} = \sqrt{2505} = 50.0$$

The percentage contribution of each factor is computed as shown in Fig. 16.10.

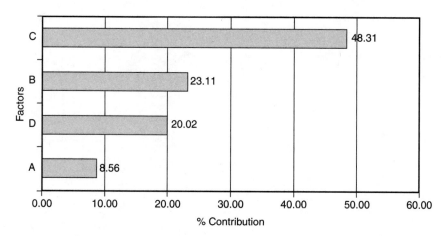

Figure 16.10 Percentage contribution to variance.

Clearly, factors C and B are the top two contributors for the variance. If we are able to reduce the standard deviation of C and B by 50 percent, then the new variance for y should be

$$\hat{\sigma}^2_{new} = 2505\left(0.0856 + 0.2311 \times \left(\frac{1}{2}\right)^2 + 0.4831 \times \left(\frac{1}{2}\right)^2 + 0.2002\right)$$

$$= 1163.2$$

$$\hat{\sigma}_{new} = \sqrt{1163.2} = 34.1$$

17

Response Surface Methodology

17.1 Introduction

Response surface methodology is a specialized DOE technique that may be used to detail and optimize transfer functions of a DFSS project (Chap. 6). The method can be used in the optimization (O) phase of the DFSS algorithm. *Response surface methodology* (RSM) is a combination of statistical and optimization methods that can be used to model and optimize designs. It has many applications in design and improvement of products and processes. Again, similar to treatment in previous chapters, the P-diagram paradigm is the paradigm for RSM, as illustrated in Fig. 17.1.

17.1.1 The role of RSM in design improvement and optimization

For many new design developments, our understanding of the design is insufficient in the beginning stage. We are able to list the output responses, or key requirements, $y_1, y_2, \ldots, y_n$, and we are also able to list

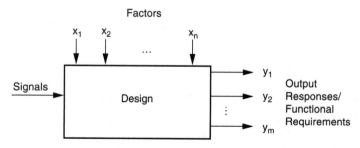

Figure 17.1 A design P-diagram.

many factors, or variables, $x_1, x_2, \dots, x_n$. However, the exact transfer functions between $y_1, y_2, \dots, y_n$ and $x_1, x_2, \dots, x_n$ are not well known. So the P-diagram illustrated in Fig. 17.1 is almost like a black box.

It is critically important to find the exact transfer functions between output requirements and design parameters, that is, to cause the P-diagram (Fig. 17.1) to become a "transparent box" in order to find optimal settings for the variables to ensure that the design achieves optimal performance. When the number of variables $x_1, x_2, \dots, X_n$ is high, finding the precise transfer functions between variables and output responses is difficult, and there will be too much work. It is desirable to start with a *screening experiment,* which is usually a low-resolution, fractional two-level factorial experiment (see Chap. 14). The purpose of the screening experiment is to find the subset of important variables.

Example 17.1. Screening Experiment   In a technology development project for a powder paint application process, there are several important output responses, or key requirements:

y_1 *Paint film thickness*—it is desirable to paint the object with a film of paint with a given thickness.

y_2 *Paint film uniformity*—it is desirable to ensure a very uniform film of paint over the object to be painted.

y_3 *Transfer efficiency*—the percentage of paint that is actually applied to the object being painted. There is always a percentage of paint that does not reach the object or falls off the object; clearly, 100 percent transfer efficiency is the ideal goal.

y_4 *Appearance*—this is a score that measures the aesthetic appearance of the painted surface.

Many factors could affect these output responses:

x_1 *Spray applicator voltage*—this variable is needed because electrostatic force is used to "hold" paint to the object.

x_2 *Distance from applicator to object.*

x_3 *Applicator spacing.*

x_4 *Applicator shaping air pressure.*

x_5 *Applicator airflow.*

x_5 *Paint powder flow rate.*

x_6 *Nozzle type.*

x_7 *Paint pattern overlap.*

x_8 *Conveyer speed.*

x_9 *Number of passes of paint application*—this variable indicates how many times the object is painted.

x_{10} *Oscillation speed.*

x_{11} *Applicator type.*

Out of these factors, x_6, nozzle type; x_7, paint pattern overlap; x_9, number of passes; x_{10}, oscillation speed (only two speeds); and x_{11}, applicator type, are discrete variables; the others are continuous variables. Clearly there are many variables and four output responses, and conducting a detailed study on all the transfer functions is not feasible. A screening experiment is conducted in which a 2^{15-11}_{111} (L_{16} array) fractional factorial experiment is used, with y_1, y_2, y_3, y_4 as responses. This screening experiment is feasible because only 16 experimental runs are needed. Figure 17.2 shows a portion of the main-effects chart in which y_3, transfer efficiency, is the response. The main-effects chart shows that oscillation, distance, and overlap are insignificant variables. Pass number and nozzle type are somewhat significant, since they are discrete variables and there are only two choices for each variable. Since the higher the transfer efficiency, the better, the second level of both these variables gives higher transfer efficiency, so we can "fix" these two variables at the second level in future investigation. From Fig. 17.2, it is clear that voltage, flow rate, and conveyor speed are very significant continuous variables, so we will study them further in future investigation.

Therefore, in this screening experiment, we narrowed down the many variables to three important ones.

In summary, the screening experiment can accomplish the following tasks:

1. Eliminate insignificant variables from further investigation.
2. Determine optimal settings of many discrete variables.
3. Identify a small number of important variables for further investigation.

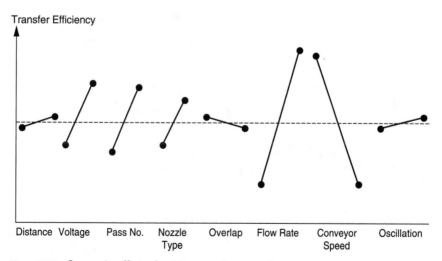

Figure 17.2 One main-effects chart in screening experiment.

After a small number of important variables are identified, we have a "reduced" problem. Usually we will conduct subsequent experiments to study those important variables in detail. However, in these subsequent experiments, two-level fractional factorial designs are rarely good choices. Especially when those variables are continuous variables, two-level factorial experiments are simply not able to find optimal variable settings, or an optimal solution. In cases where the optimal solution may be out of the experimental region (see Fig. 17.3*a*), two-level factorial experiments can only identify "a better solution" inside the region. Even if the optimal solution is inside the experimental region only (see Fig. 17.3*b*), if the true transfer functions between y and $x_1, x_2, \ldots, x_n$ are nonlinear, the two-level factorial model is still not able to find an optimal solution.

The response surface method (RSM) is designed to find an optimal solution for the "reduced problem" after a screening experiment, in

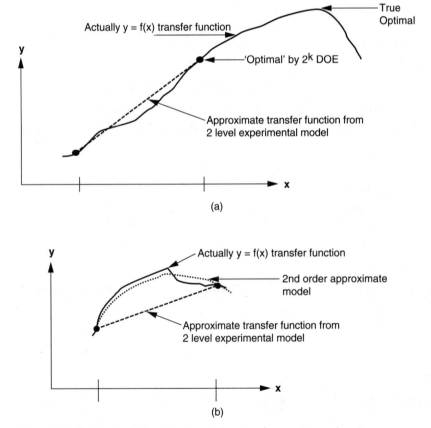

Figure 17.3   Optimal solution (*a*) out of and (*b*) inside experimental region.

which the output response(s) is (are) related to a small number of important continuous variables.

17.1.2 RSM strategy

We now describe the step-by-step procedure for the response surface method:

Step 0: Screening experiment.   After the screening experiment, a small number of important continuous variables are identified for further study.

Step 1: Determine if optimal solution is located inside the current experimental region.   The response surface method will first determine whether the current experimental region contains the optimal solution. If it does, go to step 3. Otherwise, go to step 2.

Step 2: Search the region that contains the optimal solution.   The response surface method will use a search method to move toward the region that contains the optimal solution; if it is determined that the region is identified, go to step 3.

Step 3: Design and conduct a response surface experiment, and collect more data to establish a nonlinear empirical model, usually a second order model.   RSM will augment the current experiment design to include more experimental points, with the addition of more experimental points and more data, a nonlinear empirical model will be fit.

Step 4: Identify the optimal solution.   Optimization methods will be applied to identify the optimal solution.

17.2   Search and Identify the Region That Contains the Optimal Solution

Screening experiments are usually fractional two-level factorial experiments. At the end of the screening experiment, we will have identified a small number of important continuous variables. At the beginning stage of the response surface method (RSM), we usually conduct a two-level fractional experiment with centerpoints, using this small number of important continuous variables. The purpose of adding several centerpoints is to conduct a statistical test on "curvature," or nonlinearity. Because the RSM assumes that if the current experimental region is remote from the optimal region, the output response tends to be linearly related to variables; if the experimental region contains the optimal solution, the output response (requirement) tends to be nonlinearly related to variables. Therefore, statistical testing on curvature for a two-level factorial experiment with centerpoints is the test for an optimal region in RSM.

17.2.1 2^k or 2^{k-p} factorial design with centerpoints

Usually in two-level factorial or fractional factorial design, $+1$ denotes the high level and -1 denotes the low level. For a continuous variable, if we choose a value exactly at the midpoint between these two levels, we can call it the *middle level* and denote it as 0. For example, if a variable A represents temperature, the high level is $300°C$, for which $A = +1$, and the low level is $200°C$, for which $A = -1$, then the middle level will be $250°C$, for which $A = 0$. A centerpoint in a 2^k or 2^{k-p} factorial design is the point for which *all* variables are at the middle level.

Figure 17.4 explains the concept of centerpoints in a 2^2 factorial design.

In Fig. 17.4, four "corner points" are the standard design points for a 2^2 factorial design. We can also call them *factorial design points*. A *centerpoint* in this case is the point corresponding to $A = 0$ and $B = 0$. In response surface design, there could be more than one centerpoint.

17.2.2 Curvature test

When we conduct a 2^k or 2^{k-p} experiment with centerpoints, a hypothesis testing procedure is available (Montgomery 1997) to test if there is significant curvature in the fitted regression model. Specifically, for

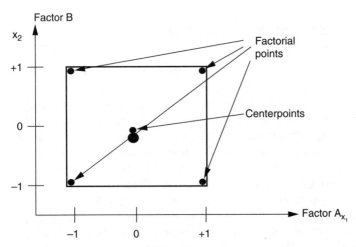

Figure 17.4 Centerpoints and factorial points.

a 2^k or 2^{k-p} experiment with centerpoints, the fitted regression model is as follows:

$$y = \beta_0 + \sum_{j=1}^{k} \beta_j x_j + \sum_{i<j} \beta_{ij} x_i x_j + \sum_{j=1}^{k} \beta_{jj} x_j^2 + \varepsilon \qquad (17.1)$$

where $\sum_{j=1}^{k} \beta_{jj} x_j^2$ corresponds to "curvatures." The statistical hypothesis test for curvature is as follows:

$$H_0 \quad \sum_{j=1}^{k} \beta_{jj} = 0 \quad \text{(no curvature effects)}$$

$$H_1 \quad \sum_{j=1}^{k} \beta_{jj} \neq 0 \quad \text{(curvature effect is significant)}$$

An F-test procedure is established for this hypothesis test. In this F-test procedure, a single-degree-of-freedom sum of squares for curvature is computed:

$$\text{SS}_{\text{curvature}} = \frac{n_F n_C (\bar{y}_F - \bar{y}_C)^2}{n_F + n_C} \qquad (17.2)$$

where $\bar{y}_F$ = average output response at factorial design points
$\bar{y}_C$ = average output response at centerpoints
n_F = number of factorial points
n_C = number of centerpoints

If there is no curvature, then $\bar{y}_F$ should be very close to $\bar{y}_C$. If there is significant curvature, there would be significant deviation between $\bar{y}_F$ and $\bar{y}_C$; the larger the deviation, the larger the $\text{SS}_{\text{curvature}}$ will be, and the more significant the curvature effects will be. The F-test statistic is the ratio of $\text{SS}_{\text{curvature}}$ over MSE:

$$F_0 = \frac{\text{SS}_{\text{curvature}}}{\text{MSE}} \qquad (17.3)$$

If $F_0 > F_{\alpha, 1 n_e}$, then H_0 will be rejected, and there is significant curvature effect, where n_e is the degree of freedom for MSE.

Example 17.2. Testing for Curvature A chemical engineer is studying the yield of a process. There are two variables, A = temperature, B = pressure, and the experimental layout and yield data are as shown by Fig. 17.5. There are five centerpoints and four factorial design points, so $n_C = 5$, $n_F = 4$:

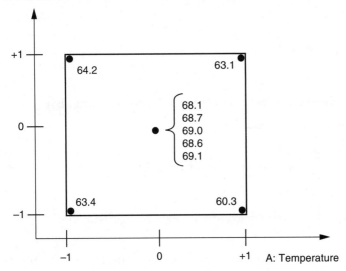

B: Pressure

Figure 17.5 Yield data for Example 17.2.

$$\bar{y}_F = \frac{64.1 + 63.1 + 63.4 + 60.3}{4} = 62.7,$$

$$\bar{y}_C = \frac{68.1 + 68.7 + 69.0 + 68.6 + 69.1}{5} = 68.7$$

Therefore

$$\text{SS}_{\text{curvature}} = \frac{n_F n_C (\bar{y}_F - \bar{y}_C)^2}{n_F + n_C} = \frac{4 \times 5 \times (62.7 - 68.7)^2}{4 + 5} = 78.7$$

An F test on curvature is conducted using MINITAB, and clearly there is very significant curvature in this case:

```
Analysis of Variance for Yield (coded units)

Source               DF   Seq SS   Adj SS   Adj MS        F      P
Main Effects          2   7.6500   7.6500   3.8250    24.68  0.006
2-Way Interactions    1   1.0000   1.0000   1.0000     6.45  0.064
Curvature             1  78.6722  78.6722  78.6722   507.56  0.000
Residual Error        4   0.6200   0.6200   0.1550
 Pure Error           4   0.6200   0.6200   0.1550
Total                 8  87.9422
```

In response surface methodology, if the curvature test indicates that there is a significant curvature effect, we can conclude that at least a local optimal solution is very likely inside the current experimental region.

17.2.3 Steepest-ascent method for searching the optimal region

When the curvature test indicates that there is no significant curvature effect in the experimental regression model, the optimal solution is unlikely to be found in the current experimental region. Then we need to find the location of the optimal region. There are many optimization search methods that search for the optimal region. For RSM, the steepest-ascent method, or gradient search method, is the method of choice for searching for the optimal region.

The steepest-ascent method assumes that the optimal value for output requirement y is "the larger, the better," if y is "the smaller, the better," then we can simply use the method in the opposite direction, that is, the steepest-descent method.

The steepest ascent method starts with establishing a linear model between y and $x_1, x_2, \ldots, x_k$ as follows:

$$y = \beta_0 + \beta_1 x_1 + \beta_2 x_2 + \cdots + \beta_k x_k \qquad (17.4)$$

This linear model can be established easily by conducting a 2^k or 2^{k-p} factorial experiment and subsequently fitting experimental data.

The steepest-ascent method is a *line search method,* that is, a method which searches along a line toward an improved direction for y. A line search method has two essential elements: (1) search direction and (2) search step length.

In the steepest-ascent method, the term *steepest ascent* means that the search direction in which the direction in which the rate of increase in y is greatest. Mathematically, this direction is specified by the gradient. On the other hand, the steepest-descent direction is the direction specified by negative gradient where y is "smallest the best." For a linear model [Eq. (17.4)] the gradient is simply

$$\frac{\nabla y}{\nabla x} = (\beta_1, \beta_2, \ldots, \beta_k)^T \qquad (17.5)$$

where T means transpose.

Step length is "how far each step will go," which is usually decided by the experimenter. Usually we use the following formula to denote step length:

$$x_i^{new} = x_i^{old} + \rho \, \frac{\beta_i}{\sqrt{\sum_{i=1}^{k} \beta_k^2}} \qquad (17.6)$$

for $i = 1 \cdots k$, where x_i^{old} is the current value for x_i, x_i^{new} is the next value for x_i after one search step, and ρ is the step length. If $\rho = 1$, it is called *unit step length,* and it is a popular choice for step length.

After each search step, we get a new set of values for $x_1, x_2, \ldots, x_k$, and a new test will be run to obtain a new output response value of y. If the search direction is good, the new value of y should be significantly better (greater) than that of the previous step. After a number of steps, the improvement of y for each step may become very low; then a new search direction should be found by running another factorial experiment with centerpoints. Figure 17.6 illustrates the process of line searches. For each new factorial experiment with centerpoints, curvature will be tested. If there is no significant curvature, a new search direction will be found by using gradients. If there is a significant curvature effect, an optimal solution is likely inside the current experimental region, and the problem should go to the optimization phase.

Example 17.3. Steepest-Ascent Method   After a screening experiment, it is concluded that the yield (y, in percent) of a chemical process is affected mainly by the temperature (X_1, in degrees Celsius) and by the reaction time (X_2, in minutes). A factorial experiment with centerpoints is conducted with the layout and data shown in Table 17.1.

In Table 17.1, the coded variable is a normalized variable in which the high level is +1, the low level is −1, and the center is 0. The natural variable is the variable in its original unit. We can use the following formula to translate between coded variables x_i, $i = 1 \cdots k$, to natural variables X_i, $i = 1 \cdots k$:

$$x_i = \frac{X_i - (X_{\text{low}} + X_{\text{high}})/2}{(X_{\text{high}} - X_{\text{low}})/2} \tag{17.7}$$

and

$$X_i = \frac{X_{\text{low}} + X_{\text{high}}}{2} + x_i \frac{X_{\text{high}} - X_{\text{low}}}{2} \tag{17.8}$$

where X_{low} and X_{high} correspond to the low-level and high-level settings of X_i, respectively.

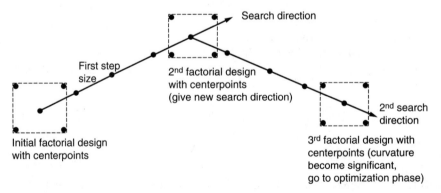

Figure 17.6   A sequence of steepest line searches in RSM.

TABLE   17.1 Experimental Layout and Data for Example 17.3

Coded variables		Natural variables		
x_1	x_2	X_1 (temperature, °C)	X_2 (reaction time, minutes)	Yield y
−1	−1	170	150	32.79
+1	−1	230	150	24.07
−1	+1	170	250	48.94
+1	+1	230	250	52.49
0	0	200	200	38.89
0	0	200	200	48.29
0	0	200	200	29.68
0	0	200	200	46.50
0	0	200	200	44.15

Using MINITAB, we can get the following regression equation for Example 17.3:

$$\hat{y} = 39.57 - 1.293x_1 + 11.14x_2$$

Therefore, the gradient direction is

$$\frac{\nabla y}{\nabla x} = \begin{pmatrix} \beta_1 \\ \beta_2 \end{pmatrix} = \begin{pmatrix} -1.293 \\ 11.14 \end{pmatrix}$$

By Eq. (17.6) and set $\rho = 1$, we get

$$x_1^{new} = x_1^{old} + \rho \dfrac{\beta_1}{\sqrt{\sum_{i=1}^{k} \beta_k^2}} = 0 + 1 \dfrac{-1.293}{\sqrt{(-1.293)^2 + 11.14^2}} = -0.115$$

$$x_2^{new} = x_2^{old} + \rho \dfrac{\beta_2}{\sqrt{\sum_{i=1}^{k} \beta_k^2}} = 0 + 1 \dfrac{11.14}{\sqrt{(-1.293)^2 + 11.14^2}} = -0.993$$

For natural variables:

$$X_1^{new} = \frac{X_{low} + X_{high}}{2} + x_1^{new} \frac{X_{high} - X_{low}}{2}$$

$$= \frac{170 + 230}{2} - -0.115 \frac{230 - 170}{2} = 196.5°C$$

$$X_2^{new} = \frac{X_{low} + X_{high}}{2} + x_2^{new} \frac{X_{high} - X_{low}}{2}$$

$$= \frac{150 + 250}{2} + 0.993 \frac{250 - 150}{2} = 249.6 \text{ min}$$

Therefore, a new test under the new experimental points, $(X_1, X_2) =$ (196.5°C, 249.6 min), will be conducted, and we expect the yield y to be improved. After this we perform another search step, and the search process will continue by following the strategy outlined in Fig. 17.6.

17.3   Response Surface Experimental Designs

In response surface methodology, when results from the curvature test demonstrate a significant curvature effect, this indicates that the actual transfer function between y and $x_1, x_2, ..., x_k$ is nonlinear and there is strong possibility that the optimal solution, at least a local optimal, is within the region.

In most industrial applications, a second-order empirical model suffices as an approximate model for the actual transfer functions. Numerically, it is very easy to find the optimal solution for a second-order model. However, the 2^k or 2^{k-p} factorial design with centerpoints is not an adequate design from which the experimental data can be fitted into a second-order model.

When k is small, say $k = 2$ or $k = 3$, 3^k or 3^{k-p} designs can be used to generate experimental data to fit a second-order model. However, when k is larger, the number of runs in 3^k or 3^{k-p} will be excessive.

Response surface experimental design layouts can be used to generate sufficient experimental data to fit second-order models. Compared with 3^k and 3^{k-p} designs, response surface designs use significantly fewer runs when the number of variables k is large. The most commonly used response surface designs are the central composite design, Box-Behnken design, and D-optimal design.

17.3.1   Central composite design

Central composite design (CCD) contains an embedded factorial or fractional factorial design with centerpoints that is augmented with a group of *axial points* that allow estimation of second-order effects. If the distance from the center of the design space to a factorial point is ±1 unit for each factor, the distance from the center of the design space to a star point is ±α with $|\alpha| > 1$. The precise value of α depends on certain properties desired for the design and the number of factors involved. Figure 17.7 shows a central composite design for two factors.

Table 17.2 gives the experimental layout of a CCD with two factors, and we can see that there are three types of design points in central composite design: factorial design points, centerpoints, and axial points. The number of factorial points n_F depends on what kind of 2^k factorial or 2^{k-p} fractional factorial design is selected. The number of centerpoints n_C depends on our requirement for the prediction error of

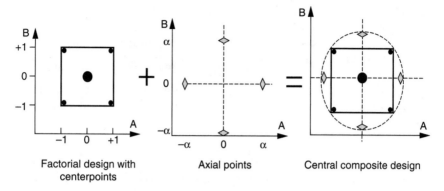

Figure 17.7 Central composite design for two factors.

TABLE 17.2 Experimental Layout for CCD with Two Factors

Point type	Experiment points	Factors A: x_1	B: x_2	Response y
Factorial design points	1	-1	-1	
	2	1	-1	
	.	-1	1	
	n_F	1	1	
Centerpoints	1	0	0	
	2	0	0	
	.	.	.	
	n_C	0	0	
Axial points	1	$-\alpha$	0	
	2	α	0	
	.	0	$-\alpha$	
	n_A	0	α	

the response surface model; usually the more centerpoints we have, the lower the prediction error will be. The number of axial points n_A is always equal to $2k$.

The value of α will depend on whether we prefer that the RSM design have rotatability property. The design that has the rotatability property is also called *rotatable design*. In a rotatable design, the variance of the predicted values of y, or prediction variance of y, is a function of the distance of a point from the center of the design and is not a function of the axis or the direction from the point to the center. The prediction variance is a measure of the prediction error. Before a study begins, there may be little or no knowledge about the region that contains the optimal response. Therefore, the experimental design matrix should not bias an investigation in any direction. In a rotatable design, the contours associated with the variance of the predicted values are concentric circles as shown in Fig. 17.8.

Figure 17.8 implies that in a rotatable design, the centerpoint has the smallest prediction variance. The prediction variance will increase uniformly as the distance to the centerpoint increases.

To achieve rotatability, the value of α depends on the number of experimental runs in the factorial portion of the central composite design, specifically

$$\alpha = (n_F)^{1/4} \tag{17.9}$$

Table 17.3 gives the appropriate selection criteria for factorial design points and α values for rotatable central composite design for two to six factors.

However, one possible problem with rotatable central composite design is that there will be five levels for each factor, that is, $(-\alpha, -1, 0, +1, +\alpha)$. For some industrial applications, it is difficult to set each factor at too many levels. If we set $\alpha = 1$, then the corresponding design is called *face-centered central composite design,* in which there are only three levels for each factor. A sketch of a two-factor face-centered central composite design is illustrated in Fig. 17.9. Obviously, this design is not rotatable.

17.3.2   Box-Behnken design

The *Box-Behnken design* is an independent quadratic design in that it does not contain an embedded factorial or fractional factorial design. In this experimental design the treatment combinations are at the midpoints of edges of the design space and at the center. These designs are rotatable (or nearly rotatable) and require three levels of each fac-

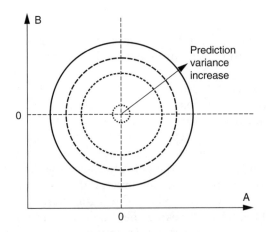

Figure 17.8 Prediction variance contour of a rotatable design.

TABLE   17.3 Composition of CCD Design Points and α Values

Number of factors	Factorial portion	Value for α
2	2^2	$2^{2/4} = 1.414$
3	2^3	$2^{3/4} = 1.682$
4	2^4	$2^{4/4} = 2.000$
5	2^{5-1}	$2^{4/4} = 2.000$
5	2^5	$2^{5/4} = 2.378$
6	2^{6-1}	$2^{5/4} = 2.378$
6	2^6	$2^{6/4} = 2.828$

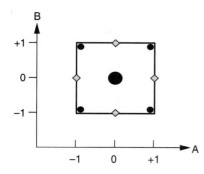

Figure 17.9 A two-factor face-centered central composite design.

tor. Figure 17.10 provides a graphical sketch of the experimental layout of a Box-Behnken design with three factors.

The design matrices of the Box-Behnken design for $k = 3$ factors and $k = 4$ factors are listed in Tables 17.4 and 17.5.

The advantages of Box-Behnken design include

1. It uses only three levels for each factor.

2. It is a near-rotatable design.

3. It needs fewer number of experimental runs than does a central composite design when $k = 3$ or 4; however when $k \geq 5$, this advantage disappears.

17.3.3   D-optimal design

The D-optimal design is the most popular computer-generated experimental design. *Computer-generated designs* are experimental designs that are based on a particular optimality criterion. One popular criterion is *D-optimality,* which seeks to maximize $|X'X|$, the determinant of the *information matrix $X'X$* of the design. This criterion results in minimizing the generalized variance of the parameter estimates according to a prespecified model. Therefore, D-optimal design involves

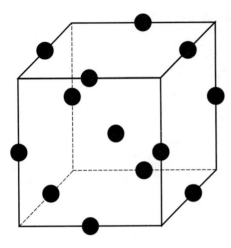

Figure 17.10 A Box-Behnken design for three factors.

TABLE 17.4 Box-Behnken Design Matrix for Three Factors

x_1	x_2	x_3
-1	-1	0
-1	1	0
1	-1	0
1	1	0
-1	0	-1
-1	0	1
1	0	-1
1	0	1
0	-1	-1
0	-1	1
0	1	-1
0	1	1
0	0	0
$\vdots$	$\vdots$	$\vdots$
0	0	0

a search for an experimental design such that its overall prediction error on model parameters is smallest.

Computer-generated experimental designs, such as the D-optimal design, have some advantages over traditional response surface designs such as the central composite design and Box-Behnken design. One major advantage is much greater flexibility in selecting response surface model types and the number of experimental runs. For example, for a three-factor response surface experiment, the following second-order model is the standard model for both CCD and Box-Behnken design:

$$y = \beta_0 + \beta_1 x_1 + \beta_2 x_2 + \beta_3 x_3 + \beta_{11} x_1^2 + \beta_{22} x_2^2$$
$$+ \beta_{33} x_3^2 + \beta_{12} x_1 x_2 + \beta_{13} x_1 x_3 + \beta_{23} x_{23} + \varepsilon$$

TABLE 17.5 Box-Behnken Design Matrix
for Four Factors

x_1	x_2	x_3	x_4
−1	−1	0	0
−1	1	0	0
1	−1	0	0
1	1	0	0
−1	0	−1	0
−1	0	1	0
1	0	−1	0
1	0	1	0
−1	0	0	−1
−1	0	0	1
1	0	0	−1
1	0	0	1
0	−1	−1	0
0	−1	1	0
0	1	−1	0
0	1	1	0
0	−1	0	−1
0	−1	0	1
0	1	0	−1
0	1	0	1
0	0	−1	−1
0	0	−1	1
0	0	1	−1
0	0	1	1
0	0	0	0
.	.	.	.
0	0	0	0

If we use a central composite design, then, according to Table 17.3, a 2^3 factorial design (8 runs), plus 6 axial points and a number of centerpoints (at very minimum, 1 centerpoint; several centerpoints are usually preferred), an experimental run of 15 or more is definitely needed. If the practical situation can allow only 14 experimental runs, central composite design will not be able to do the work.

For a D-optimal design, the user can select his/her own model; for example, if the terms $\beta_{33}x_3^2$, $\beta_{13}x_1x_{31}$ and $\beta_{23}x_2x_3$ do not seem to be relevant, the user can specify the model:

$$y = \beta_0 + \beta_1x_1 + \beta_2x_2 + \beta_3x_3 + \beta_{11}x_1^2 + \beta_{22}x_2^2 + \beta_{12}x_1x_2$$

as the model in the experiment. D-optimal design is *model-specified*; that is, it is designed to fit a given model, and when the model gets simpler, the number of runs needed in the experiment can also be reduced. In the situation described above, a D-optimal model can design an experiment with less than 14 experimental runs.

Given the total number of runs for an experiment and a specified model, the computer algorithm chooses the optimal set of design runs from a *candidate set* of possible design treatment runs. This candidate set of treatment runs usually consists of all possible combinations of various factor levels that one wishes to use in the experiment.

In other words, the candidate set is a collection of treatment combinations from which the D-optimal algorithm chooses the treatment combinations to include in the design. The computer algorithm generally uses a stepping and exchanging process to select the set of treatment runs.

Example 17.4. D-Optimal Design Using Design-Expert Software *Design-Expert* is a software program produced by Stat-Ease that is capable of designing D-optimal response surface design. In this example, we are going to design a three-factor D-optimal design with the following model form:

$$y = \beta_0 + \beta_1 x_1 + \beta_2 x_2 + \beta_3 x_3 + \beta_{11} x_1^2 + \beta_{22} x_2^2 + \beta_{12} x_1 x_2$$

A 14-run experiment is to be designed. Design-Expert will generate the following D-optimal design:

x_1	x_2	x_3
1	1	0
1	0	1
1	−1	−1
−1	1	1
0	−1	1
−1	−1	−1
0	0	−1
−1	1	−1
−1	0	0
0.5	−0.5	0
1	1	0
1	−1	−1
−1	−1	−1
0	−1	1

17.4 Response Surface Experimental Data Analysis for Single Response

In response surface methodology, after we have located the optimal region by curvature test, we will design a response surface experiment in order to obtain a second-order experimental model.

If there are k variables, the standard second-order model has the following form:

$$y = \beta_0 + \sum_{i=1}^{k} \beta_i x_i + \sum_{i=1}^{k} \beta_{ii} x_i^2 + \sum_{i<j} \beta_{ij} x_i x_j + \varepsilon \qquad (17.10)$$

If there are n sets of experimental observations, the matrix of (17.10) is:

$$\mathbf{y} = \mathbf{X}\boldsymbol{\beta} + \varepsilon \qquad (17.11)$$

where $\mathbf{y}' = (y_1, y_2, ..., y_n)$, $\mathbf{X}$ is the "information matrix" whose elements consist of powers and cross-products of the design settings for the independent variables, and ε is a random vector representing model-fitting errors.

Using the least-squares (LS) fit for multiple linear regression, the LS estimator of β is given by:

$$\hat{\beta} = (\mathbf{X}'\mathbf{X})^{-1}\mathbf{X}'\mathbf{y} \qquad (17.12)$$

where $E(\hat{\beta} - \beta) = 0$ and $\mathrm{Var}(\hat{\beta}) = (\mathbf{X}'\mathbf{X})^{-1}\sigma^2$.

Example 17.5. Chemical Process Response Surface Experiment In a chemical process, there are two important variables: X_1, temperature (°C) and X_2, the reaction time (minutes). A response surface experiment is conducted in order to find the optimal temperature and reaction time setting to maximize the yield (y). A central composite design is used, and we get the following experimental data, where $\alpha = 1.414$:

Coded variables		Natural variables		
x_1	x_2	X_1 (temperature, °C)	X_2 (reaction time, min)	Yield y
−1	−1	170	300	64.33
+1	−1	230	300	51.78
−1	+1	170	400	77.30
+1	+1	230	400	45.37
0	0	200	350	62.08
0	0	200	350	79.36
0	0	200	350	75.29
0	0	200	350	73.81
0	0	200	350	69.45
−1.414	0	157.58	350	72.58
+1.414	0	242.42	350	37.42
0	−1.414	200	279.3	54.63
0	+1.414	200	420.7	54.18

In this data set

$$
\mathbf{y} = \begin{bmatrix} 64.33 \\ 51.78 \\ 77.30 \\ 45.37 \\ 62.08 \\ 79.36 \\ 75.29 \\ 73.81 \\ 69.45 \\ 72.58 \\ 37.42 \\ 54.63 \\ 54.18 \end{bmatrix}, \quad
\mathbf{X} = \begin{bmatrix}
 & x_1 & x_2 & x_1^2 & x_2^2 & x_1 x_2 \\
1 & -1 & -1 & 1 & 1 & 1 \\
1 & 1 & -1 & 1 & 1 & -1 \\
1 & -1 & 1 & 1 & 1 & -1 \\
1 & 1 & 1 & 1 & 1 & 1 \\
1 & 0 & 0 & 0 & 0 & 0 \\
1 & 0 & 0 & 0 & 0 & 0 \\
1 & 0 & 0 & 0 & 0 & 0 \\
1 & 0 & 0 & 0 & 0 & 0 \\
1 & 0 & 0 & 0 & 0 & 0 \\
1 & -1.414 & 0 & 2 & 0 & 0 \\
1 & 1.414 & 0 & 2 & 0 & 0 \\
1 & 0 & -1.414 & 0 & 2 & 0 \\
1 & 0 & 1.414 & 0 & 2 & 0
\end{bmatrix}
$$

So

$$\hat{\beta} = (\mathbf{X'X})^{-1}\mathbf{X'y} = (72.0, -11.78.0.74, -7.27, -7.55, -4.85)$$

Therefore, the fitted second-order model is

$$\hat{y} = 72.0 - 11.78x_1 + 0.74x_2 - 7.25x_1^2 - 7.55x_2^2 - 4.85x_1x_2$$

17.4.1   Finding optimal solution in second-order model

The fitted second-order model can be expressed in the following matrix form:

$$\hat{y} = \hat{\beta}_0 + \sum_{i=1}^{k} \hat{\beta}_i x_i + \sum_{i=1}^{k} \hat{\beta}_{ii} x_i^2 + \sum_{i<j} \hat{\beta}_{ij} x_i x_j = \hat{\beta}_0 + \mathbf{x'b} + \mathbf{x'Bx} \quad (17.13)$$

where

$$\mathbf{x} = \begin{pmatrix} x_1 \\ x_2 \\ \vdots \\ x_k \end{pmatrix}, \quad \mathbf{b} = \begin{pmatrix} \hat{\beta}_1 \\ \hat{\beta}_2 \\ \vdots \\ \hat{\beta}_k \end{pmatrix}, \quad \mathbf{B} = \begin{bmatrix} \hat{\beta}_{11} & \hat{\beta}_{12}/2 & \cdots & \hat{\beta}_{1k}/2 \\ \hat{\beta}_{12}/2 & \hat{\beta}_{22} & & \hat{\beta}_{2k}/2 \\ \vdots & & \ddots & \vdots \\ \hat{\beta}_{1k} & & & \hat{\beta}_{kk} \end{bmatrix}$$

The optimal solution that optimizes the predicted response in Eq. (17.13), whether it is maximum or minimum, should be a stationary point of Eq. (17.13); that is, it should be a point $\mathbf{x'}_0 = (x_{10}, x_{20}, \ldots, x_{k0})$ at which all partial derivatives are equal to zero:

$$\frac{\partial \hat{y}}{\partial \mathbf{x}} = \begin{bmatrix} \dfrac{\partial \hat{y}}{\partial x_1} \\ \dfrac{\partial y}{\partial x_2} \\ \vdots \\ \dfrac{\partial y}{\partial x_k} \end{bmatrix} = \frac{\partial}{\partial \mathbf{x}}(\hat{\beta}_0 + \mathbf{x'b} + \mathbf{x'Bx}) = \mathbf{b} + 2\mathbf{Bx} = 0 \quad (17.14)$$

By solving Eq. (17.14), the stationary point is

$$\mathbf{x}_0 = -\frac{1}{2}\mathbf{B}^{-1}\mathbf{b} \quad (17.15)$$

The stationary point could be a maximum, minimum, or saddle point for a predicted response depending on the eigenvalues of the $\mathbf{B}$ matrix

(Myers and Montgomery 1995). Assume that $\lambda_1, \lambda_2, ..., \lambda_k$ are the eigenvalues of $\mathbf{B}$; then

1. If $\lambda_1, \lambda_2, ..., \lambda_k$ are all positive, then $\mathbf{x}_0$ is a minimum point.
2. If $\lambda_1, \lambda_2, ..., \lambda_k$ are all negative, then $\mathbf{x}_0$ is a maximum point.
3. If $\lambda_1, \lambda_2, ..., \lambda_k$ have different signs, then $\mathbf{x}_0$ is a saddle point.

In the case of a saddle point, a search method can be developed from canonical analysis to locate the optimal region (Myers and Montgomery 1995).

Example 17.6. Finding the Optimal Solution for Chemical Process (Example 17.4)   For Example 17.4

$$\hat{y} = 72.0 - 11.78x_1 + 0.74x_2 - 7.25x_1^2 - 7.55x_2^2 - 4.85x_1x_2$$

$$= 72.0 + (x_1, x_2)\begin{pmatrix} -11.78 \\ 0.74 \end{pmatrix} = (x_1, x_2)\begin{bmatrix} -7.25 & -2.425 \\ -2.425 & -7.55 \end{bmatrix}\begin{pmatrix} x_1 \\ x_2 \end{pmatrix}$$

Therefore, the stationary point

$$\mathbf{x}_0 = -\frac{1}{2}\mathbf{B}^{-1}\mathbf{b} = -\frac{1}{2}\mathbf{B}^{-1}\mathbf{b}$$

$$= -\frac{1}{2}\begin{bmatrix} -7.25 & -2.425 \\ -2.425 & -7.55 \end{bmatrix}^{-1}\begin{pmatrix} -11.78 \\ 0.74 \end{pmatrix} = \begin{pmatrix} -0.669 \\ -0.030 \end{pmatrix}$$

Because the eigenvalues for $\mathbf{B}$ are $\lambda_1 = -9.83$ and $\lambda_2 = -4.97$, $x_1 = -0.669$ and $x_2 = -0.03$ are the maximum solution that maximize the yield. The optimal yield

$$y_0 = 72.0 - 11.78(-0.669) + 0.74(-0.03) - 7.25(-0.669)^2$$
$$- 7.55(-0.03)^2 - 4.85(-0.669) \times (-0.03) = 76.50$$

By changing coded temperature and reaction time to natural variables, the optimal temperature and reaction time are

$$X_1 = \frac{X_{low} + X_{high}}{2} + x_1 \frac{X_{high} - X_{low}}{2} = \frac{170 + 230}{2} +$$

$$+ (-0.669)\frac{230 - 170}{2} = 180.0°C$$

$$X_2 = \frac{X_{low} + X_{high}}{2} + x_2 \frac{X_{high} - X_{low}}{2} = \frac{300 + 400}{2} +$$

$$+ (-0.03)\frac{400 - 300}{2} = 348.5 \text{ min}$$

Many statistical software programs, such as MINITAB, can work efficiently on response surface experimental data. A selected computer printout from MINITAB in Example 17.6 is given as follows:

Response surface regression: yield versus temperature, reaction time. The analysis was done using coded units.

```
Estimated Regression Coefficients for Yield
Term                      Coef    SE Coef        T        P
Constant                 72.00     2.580    27.902    0.000
Temperat                -11.78     2.040    -5.772    0.001
Reaction                  0.74     2.040     0.363    0.727
Temperat*Temperat        -7.25     2.188    -3.314    0.013
Reaction*Reaction        -7.55     2.188    -3.450    0.011
Temperat*Reaction        -4.85     2.885    -1.679    0.137
S = 5.77    R-Sq = 89.0%   R-Sq(adj) = 81.1%
```

```
Analysis of Variance for Yield

Source           DF    Seq SS    Adj SS    Adj MS       F       P
Regression        5   1881.73   1881.73    376.35   11.30   0.003
  Linear          2   1113.67   1113.67    556.84   16.73   0.002
  Square          2    674.17    674.17    337.08   10.13   0.009
  Interaction     1     93.90     93.90     93.90    2.82   0.137
Residual Error    7    233.04    233.04     33.29
  Lack-of-Fit     3     59.86     59.86     19.95    0.46   0.725
  Pure Error      4    173.18    173.18     43.29
Total            12   2114.77
```

Response optimization

```
Global Solution

Temperature = -0.66865
Reaction tim = -0.02964

Predicted Responses

Yield    =  76.5049, desirability = 0.60842
```

MINITAB can also provide the graphical plot of response surface and contour plot as shown in Figs. 17.11 and 17.12.

The optimal yield y is 76.5. In many industrial applications, people are more interested in obtaining an *operation window,* a region for process variables where a bottom-line requirement can be achieved. In the MINITAB output shown in Fig. 17.13, the white region represents an operation window for temperature and the reaction time region within which a yield of 75 percent or above will be achieved.

17.5 Response Surface Experimental Data Analysis for Multiple Responses

In many practical design improvement and optimization problems, more than one output response is involved. In Example 17.1, there are

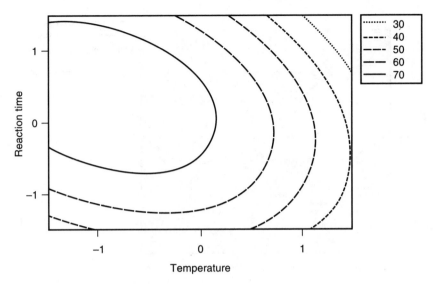

Figure 17.11 Contour plot of yield.

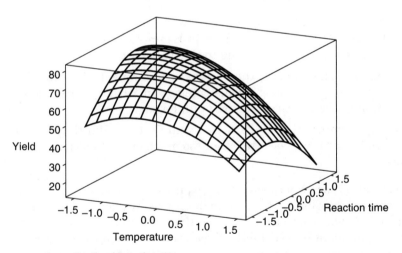

Figure 17.12 Surface plot of yield.

four output responses, or key requirements for the powder paint application process; y_1—paint film thickness; y_2—paint film uniformity; y_3—transfer efficiency; and y_4—appearance. We would like to find a set of process variables to ensure that satisfactory performances can be achieved for all four output requirements. Mathematically, this kind of problem is called *multiobjective optimization*. Each output requirement may have different optimization criteria. In the response surface method, there are three kinds of optimization criteria:

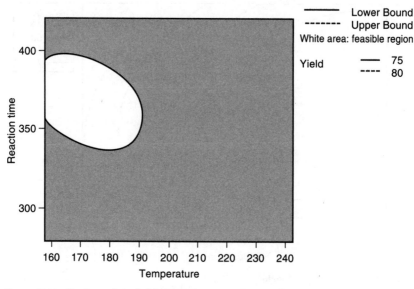

Figure 17.13 Contour plot of yield showing operation window.

1. *Maximum.* The larger, the better. In Example 17.1, appearance y_4 and transfer efficiency y_3 belong to this category.

2. *Minimum.* The smaller, the better. In Example 17.1, paint film uniformity y_2 may belong to this category, if variance or standard deviation is used as the measure of uniformity.

3. *Target.* Nominal the (is) best. In Example 17.1, film thickness y_1 belongs to this type, because either too much or too little paint is no good.

There are also cases where an output requirement must satisfy a given upper and/or lower limit. Thus, for an output requirement y, there is an upper limit U and a lower limit L such that as long as $L \leq y \leq U$, the output requirement is acceptable. We call this situation a *constraint*.

In the response surface method, there are three approaches to deal with multiple responses (requirements):

1. Mathematical programming methods

2. Desirability function method

3. Graphical optimization method

No matter which method we use, we will first fit a model to demonstrate the transfer functions between each output requirement y_i and variables $x_1, x_2, \ldots, x_k$:

$$y_1 = f_1(x_1x_2,...,x_k)$$

$$y_2 = f_2(x_1x_2,...,x_k)$$

$$\vdots$$

$$y_p = f_p(x_1x_2,...,x_k)$$

Example 17.7. Multiple Response Surface Analysis   Derringer and Suich (1980) present the following multiple-requirement experiment arising in the development of a tire-tread compound. The controllable factors are x_1, hydrated silica level; x_2, silane coupling agent level; and x_3, sulfur level. The four requirements to be optimized and their desired ranges are

y_1 Abrasion index, where y_1 is required to be larger than 120, and is "the larger, the better."

y_2 200 percent modulus; y_2 should be larger than 1000, and it is "the larger, the better."

y_3 Elongation at break—it is required that $400 \leq y_3 \leq 600$, and $y_3 = 500$ is desired.

y_4 Hardness—it is required that $60 \leq y_4 \leq 75$, and $y_4 = 67.5$ is desired.

The following experiments are conducted by using a central composite design, and the experimental layout and experimental data are listed in Table 17.6.

TABLE 17.6 Experimental layout and data for Example 17.7

Experiment no	x_1	x_2	x_3	y_1	y_2	y_3	y_4
1	−1	−1	−1	102	900	470	67.5
2	1	−1	−1	120	860	410	65.0
3	−1	1	−1	117	800	570	77.5
4	1	1	−1	198	2294	240	74.5
5	−1	−1	1	103	490	640	62.5
6	1	−1	1	132	1289	270	67.0
7	−1	1	1	132	1270	410	78.0
8	1	1	1	139	1090	380	70.0
9	−1.63	0	0	102	770	590	76.0
10	1.63	0	0	154	1690	260	70.0
11	0	−1.63	0	96	700	520	63.0
12	0	1.63	0	163	1540	380	75.0
13	0	0	−1.63	116	2184	520	65.0
14	0	0	1.63	153	1784	290	71.0
15	0	0	0	133	1300	380	70.0
16	0	0	0	133	1300	380	68.5
17	0	0	0	140	1145	430	68.0
18	0	0	0	142	1090	430	68.0
19	0	0	0	145	1260	390	69.0
20	0	0	0	142	1344	390	70.0

Using the response surface modeling technique discussed in Sec. 17.4 [Eq. (17.12)], we get the following fitted transfer functions:

$$\hat{y}_1 = 139.12 + 16.49x_1 + 17.88x_2 + 2.21x_3 - 4.01x_1^2 - 3.45x_2^2 - 1.57x_3^2$$
$$+ 5.12x_1x_2 - 7.88x_1x_3 - 7.13x_2x_3$$

$$\hat{y}_2 = 1261.13 + 268.15x_1 + 246.5x_2 - 102.63x_3 - 83.5x_1^2 - 124.82x_2^2$$
$$+ 199.2x_3{}^2 + 69.3x_1x_2 - 104.38x_1x_3 - 94.13x_2x_3$$

$$\hat{y}_3 = 417.5 - 99.67x_1 - 31.4x_2 - 27.42x_3$$

$$\hat{y}_4 = 68.91 - 1.41x_1 + 4.32x_2 + 0.21x_3 + 1.56.5x_1^2 + 0.058x_2^2 - 0.32x_3^2$$
$$-1.62x_1x_2 + 0.25x_1x_3 - 0.12x_2x_3$$

Similar to the single requirement case, the next step of data analysis is to search for the optimal solution. Since there are multiple requirements and the optimality criterion for each requirement might be different, the optimization step in the multiple-requirement case will be more complicated. As discussed earlier, there are three approaches to deal with multiple-response optimization; we will discuss them one by one.

17.5.1 Desirability function approach

The desirability function approach is one of the most popular methods used in the optimization of multiple-response surfaces. Assume that there are p output requirements. In the desirability function approach for each transfer function $y_i (\mathbf{x}) = f_i(x_1, x_2, \ldots, x_k)$, $i = 1, \ldots, p$, a desirability function $d_i = d_i(y_i) = d_i (y_i(\mathbf{x}))$ will assign values between 0 and 1 to the possible values of y_i, with $d_i(y_i) = 0$ for the "most undesirable values" of y_i and $d_i(y_i) = 1$ for the "most desirable values" of y_i, where $d_i(y_i)$ is the individual desirability for requirement y_i. The following geometric mean of all individual desirabilities D is used to represent the overall desirability for the whole multiple-response problem:

$$D = (d_1(y_1)d_2(y_2) \cdots d_p(y_p))^{1/p} \tag{17.16}$$

Clearly, higher overall desirability D should indicate a higher overall satisfaction for all responses (requirements).

For each individual desirability function $d_i(y_i)$, *the definition depends on the optimality criterion for a particular y_i.* There are four types of individual desirability functions:

The larger, the better. In this case a maximization of y_i is the most desirable result, and $d_i(y_i)$ is defined as follows:

$$d_i(y_i) = 0 \qquad\qquad y_i \leq L_i$$

$$d_i(y_i) = \left(\frac{y_i - L_i}{U_i - L_i} \right)^{w_i} \qquad L_i \leq y_i \leq U_i \tag{17.17}$$

$$d_i(y_i) = 1 \qquad\qquad y_i \geq U_i$$

where L_i is the lower bound for y_i, U_i is the upper bound for y_i, and w_i is the weight, or importance factor, for y_i. The higher the w_i relative to other weight factors, the more important the role that $d_i(y_i)$ plays in overall desirability D.

The following figure shows the shape of the $d_i(y_i)$ function for this case; clearly, for different weight values w_i, the shape of the function curve will vary:

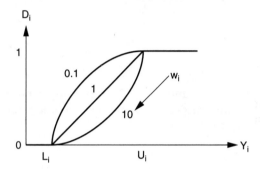

The smaller, the better. In this case a minimization of y_i is the most desirable result. The function $d_i(y_i)$ is defined as follows:

$$d_i(y_i) = 1 \qquad\qquad y_i \leq L_i$$

$$d_i(y_i) = \left(\frac{U_i - y_i}{U_i - L_i} \right)^{w_i} \qquad L_i \leq y_i \leq U_i \qquad (17.18)$$

$$d_i(y_i) = 0 \qquad\qquad y_i \geq U_i$$

where L_i is the lower bound for y_i, U_i is the upper bound for y_i, and w_i is the weight, or importance factor, for y_i. The following diagram shows the shape of $d_i(y_i)$ for this case:

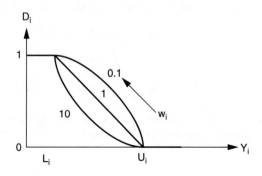

Nominal-the-best.   In this case, there is a target value T_i for y_i; the most desirable result is $y_i = T_i$. The function $d_i(y_i)$ is defined as follows:

$$d_i(y_i) = 0 \qquad\qquad\qquad y_i \leq L_i$$

$$d_i(y_i) = \left(\frac{y_i - L_i}{T_i - L_i}\right)^{w1_i} \qquad L_i \leq y_i \leq T_i$$

$$(17.19)$$

$$d_i(y_i) = \left(\frac{y_i - U_i}{T_i - U_i}\right)^{w2_i} \qquad T_i \leq y_i \leq U_i$$

$$d_i(y_i) = 0 \qquad\qquad\qquad y_i \geq U_i$$

where L_i is the lower bound for y_i, U_i is the upper bound for y_i, and w_i is the weight, or importance factor, for y_i. The following diagram shows the shape of $d_i(y_i)$ for this case:

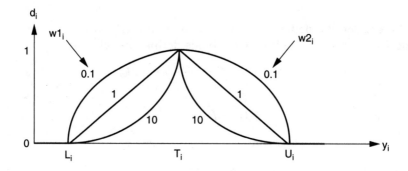

Constraint.   In this case, it is desirable as long as y_i is within the constraint $L_i \leq y_i \leq U_i$. The function $d_i(y_i)$ is defined as follows:

$$d_i(y_i) = 0 \qquad y_i \leq L_i$$

$$d_i(y_i) = 1 \qquad L_i \leq y_i \leq U_i \qquad (17.20)$$

$$d_i(y_i) = 0 \qquad y_i \geq U_i$$

where L_i is the lower bound for y_i and U_i is the upper bound for y_i. The shape of desirability function is as follows:

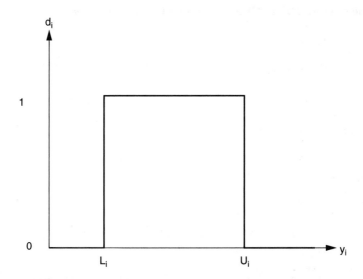

Many statistical software programs, such as MINITAB and Design-Expert, can effectively analyze multiple response surface experimental data by using the desirability function. When using these packages, we need to do the following steps:

1. Define each individual desirability function for each response (requirement).
2. Maximize the overall desirability D with respect to variables $x_1, x_2, \ldots, x_k$.

Example 17.8. Desirability Function Approach for Example 17.7   For the experimental data obtained from Example 17.7, we define the desirability functions for the four responses as follows:

$d_1(y_1)$ We will use the larger-the-better desirability function, with $L_1 = 120$, $U_1 = 170$ (because 170 is larger than all the y_1 data in the experiment), and weight = 1.

$d_2(y_2)$ We will use the larger-the-better desirability function, with $L_2 = 1000$, $U_2 = 2300$ (because 2300 is larger than all the y_2 data in the experiment), and weight = 1.

$d_3(y_3)$ We will use the nominal-the-best desirability function, with $L_3 = 400$, $U_3 = 600$, $T_3 = 500$, and weights = 1.

$d_4(y_4)$ We will use the nominal-the-best desirability function, with $L_4 = 60$, $U_4 = 75$, $T_4 = 67.5$, and weights = 1.

The selected MINITAB computer output is given below:

```
Global Solution

x1        = -0.17043
x2        = 0.33550
x3        = -1.35850

Predicted Responses

y1        = 137.04, desirability  = 0.34088
y2        = 1804.84, desirability = 0.61911
y3        = 449.09, desirability  = 0.49087
y4        = 69.99, desirability   = 0.66842

Composite Desirability = 0.51298
```

The two-dimensional contour plot and three-dimensional response surface plot for the overall desirability function D is given as follows:

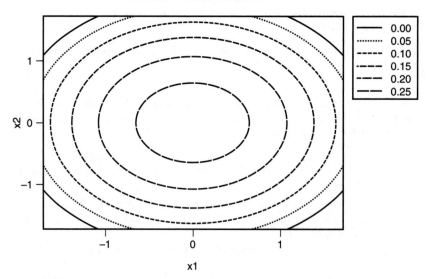

Contour Plot of DESIR1

Hold values: x3: −1.36

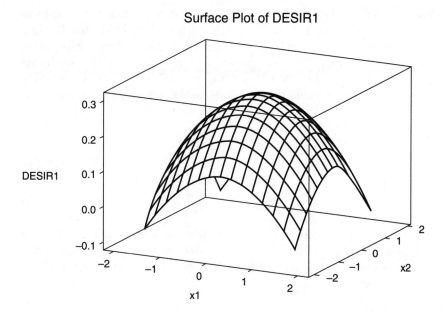

Surface Plot of DESIR1

Hold values: x3: −1.36

17.5.2 Mathematical programming approach

In a mathematical programming approach, only one output response (requirement) is selected as the main objective, and it will be either maximized or minimized. The other responses will be treated as constraints. Specifically, the mathematical programming approach can be expressed as follows:

Maximize: $\quad y_0 = f_0(x_1, x_2, \ldots, x_k)$

Subject to: $\quad L_1 \leq y_1 = f_1(x_1, x_2, \ldots, x_k) \leq U_1$

$$L_2 \leq y_2 = f_2(x_1, x_2, \ldots, x_k) \leq U_2 \qquad (17.21)$$

$$\vdots$$

$$L_k \leq y_1 = f_k(x_1, x_2, \ldots, x_k) \leq U_k$$

Actually, since type 4 desirability is about constraints, a mathematical programming case can be treated as a special case of the desirability function.

Graphical optimization method.   This method is to use the graphical capabilities of the statistical software, such as MINITAB and Design-Expert, to draw an operating window, if the lower and upper limits for each response variables are given. For example, in the multiple-response problem discussed in Examples 17.7 and 17.8, if we define the following "acceptable performance levels" for y_1, y_2, y_3, y_4 as

$$y_1 \geq 130$$

$$y_2 \geq 1700$$

$$450 \leq y_3 \leq 550 \tag{17.22}$$

$$65 \leq y_4 \leq 70$$

then MINITAB can plot the operating window shown in Fig. 17.14 for variables x_1 and x_2, such that as long as they are within the white area, the acceptable performance defined by Eq. (17.22) can be achieved.

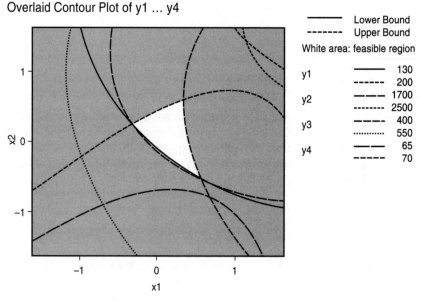

Figure 17.14   An example of graphical optimization.

18

Design Validation

18.1 Introduction

Design validation is a process that verifies whether the optimized product and process designs perform at a level that is consistent with customer-driven specifications. In a product development situation, it includes the three major tasks outlined in Secs. 18.1.1 to 18.1.3.

18.1.1 Product design validation

We need to verify the product design with respect to the following aspects:

1. *Functional performance validation.* This verifies whether the product can deliver all its functional requirements. For example, for a TV set, functional performance validation will verify if the TV set can receive TV signals, create aesthetically pleasing TV images, or produce good sound effects. For a pipeline, functional performance validation will verify if the pipeline can transmit liquid at designed volume within a given time period, and if the pipeline can withstand fluid pressure and so on.

2. *Operation environmental requirements validation.* This verifies whether the product can deliver its function in diverse environmental conditions, such as high and low temperatures, shocks and vibrations, humidity, wind, salt, and dust.

3. *Reliability requirements validation.* This verifies whether the product can perform its functions in an extended period of usage. Many products are designed for usage for an extended period of

time; for example, people expect a car to be in relatively good condition for at least 7 years. This validation should include useful life validation and functional degradation validation.

4. *Usage requirements validation.*   This verifies whether the product can deliver its functions under various usage conditions—sometimes, abusive usage conditions. For example, a copier manufacturer would test to see if the copier can still make good copies for smaller-size papers, or thick or thin papers.

5. *Safety requirements validation.*   This verifies whether the product can meet the safety requirements. For example, a toy manufacturer would have to verify that the toys they produce have no hazard for children. A bridge should be verified to ensure that it can withstand excessive wind, waves, stress, and fatigue so that people who cross the bridge would have zero risk of accident.

6. *Interface and compatibility validation.*   If a product or equipment has to work with other products or equipment, then we need to verify whether they can work well together (i.e., are compatible).

7. *Maintainability requirement validation.*   This verifies whether the necessary maintenance work can be performed conveniently, how well the maintenance can "refresh" the product, benchmark mean time between maintenance, mean corrective maintenance time, mean preventive maintenance time, and so on.

Not all products need to perform all these validations. For different products, the validation requirements and relative importance of each type of validation activity will be quite different; a validation requirements analysis should be conducted to compile the lists for all validation items.

18.1.2 Manufacturing process validation

The purpose of this task is to verify whether the manufacturing process can produce the product that meets design intent with sufficient process capability. Manufacturing process validation includes at least the following activities:

1. *Product specification validation.*   The objective is to verify whether the manufacturing process can produce the products that satisfy the design intent. For example, in a new engine plant, we need to verify if the process can process the engine that can deliver the specified power.

2. *Process capability validation.*   The purpose is to verify if the manufacturing process can produce the product with satisfactory process

capability. For example, at each model changeover for an automobile assembly plant, we need to verify whether the body assembly process can produce the "body in whites" with specified dimensional specifications with low variation.

18.1.3 Production validation

The objective of this task is to confirm that the final mass production can deliver good products with low cost, high throughput, and Six Sigma quality. Production validation includes at least the following activities:

1. *Process capability validation.* This verifies whether satisfactory process capability can still be achieved in mass production.

2. *Production throughput validation.* The purpose here is to verify that the mass production process can produce the products with sufficient quantity and productivity, with a satisfactorily low level of downtime and interruptions.

3. *Production cost validation.* The objective is to verify if the mass production can produce the product with sufficiently low cost.

Sections 18.1.4 to 18.1.6 outline the step-by-step procedures for product design validation, manufacturing process validation, and production validation.

18.1.4 A step-by-step procedure for product design validation

In a good product development strategy, the product design validation actually starts well before the final stage in design, because validating a design only at a final stage will cause many potential problems, such as costly redesign, and long delay in time to market. In the early stage of design, computer model simulation, early-stage prototypes (called *alpha prototypes*), and so on can be used for concept validation. Here we give a typical step-by-step design validation procedure:

Step 0: Early-stage validation. This step will start at concept generation stage and will spread out the whole design stages; computer simulation models and alpha prototypes will be used. A computer simulation model can be used on functional performance validation, and safety requirement validation. The Taguchi robust design method can be used on computer simulation models, as well as alpha prototypes, together will simulated noise effects, to participate in operation environment requirement validation, usage requirement validation, and reliability requirement validation, because operation variation, usage variation, environmental variation, and degradation effects can be modeled as noise factors in the Taguchi parameter design project. For a creative design, early-stage validation activities are especially important, because the new concept has not been tested before.

Step 1: Conduct design reviews and design requirement analysis.

Step 2: Build one or several productionlike prototype(s) (also called beta prototypes).

Step 3: Perform design verification testing on prototype. This is an important step in design validation. The tests need to be carefully designed and cover all aspects of product design validation, such as functional performance validation and reliability requirement validation. For a complicated product, these tests may be divided into component testing, subsystem testing, and system testing.

Step 4: Evaluate and verify prototype performances. The purpose of this step is to analyze the results from design verification testing and benchmark the prototype performances. If the prototype shows satisfactory performances, go to step 6. Otherwise, go to step 5.

Step 5: Resolve performance concerns and build more prototypes for testing. This step is the "fix" and then "build" in the "build-test-fix" cycle. In an ideal situation, we would not go to this step. However, many times we have to. Early-stage activities, such as robust parameter design, more upfront computer simulation validation, and alpha prototype validation, will help reduce the need to stick with build-test-fix cycles for too long. After this step, go to step 3 for more testing.

Step 6: Sign off design.

18.1.5 A step-by-step procedure for manufacturing process validation

This procedure provides a confirmation of the manufacturing capability. It covers the installation of the manufacturing equipment and the manufacture and evaluation of a significant number of preproduction prototypes that are produced under the conditions of mass production. Manufacturing process validation testing will be conducted to verify that the process is producing a product that meets the design intent. A functional evaluation of the end product by the combined team of customers and producers will provide additional verification that design intent has been achieved.

Similar to product design validation, early-stage validation activities in process validation will be very helpful to reduce the hiccups in process development.

We give a typical step-by-step manufacturing validation procedure:

Step 0: Early-stage validation.   This step should start in the manufacturing process design stage. Practicing design for manufacturing/design for assembly, and concurrent engineering will certainly be very helpful.

Step 1: Conduct process validation requirements analysis.   This step consists in compiling a checklist of necessary requirements for the manufacturing process to perform its functions under mass production conditions. This requirements list is the basis for designing process validation testing.

Step 2: Install machines in house and complete training of operators.

Step 3: Conduct process validation testing.

Step 4: Improve the process.

Step 5: Move to product launch.

18.1.6   A step-by-step procedure for production validation

Production validation includes product launch, rampup, and final confirmation of mass production. A good production validation should ensure function, cost, and Six Sigma quality objectives. Implementation of robust processes, upfront launch planning, and adequate employee training are among the important factors to enable a smooth launch and rampup to production speed.

The following is a typical step-by-step manufacturing validation procedure:

Step 0: Early-stage validation.   Early-stage production validation activities are also helpful. For example, using a computer simulation model to predict and analyze production flows and bottleneck points will be very helpful in improving production facility design. *Variation simulation modeling* is an approach used to simulate the effects of low-level parameter variations on the system variation. It can be used to simulate an assembly process to predict assembly process capability and analyze the major contributors for assembly variation.

Step 1: Develop a launch support plan.   This step includes development of

- Manufacturing engineering and maintenance launch support plan and support team
- Product engineering support plan and support team

- Launch/rampup strategy
- Launch concerns/engineering change plan

Step 2: Develop a marketing launch plan.

Step 3: Develop a product aftersale service support plan.

Step 4: Implement a production launch plan.

Step 5: Conduct mass production confirmation and process capability evaluation.

Step 6: Continuously improve product and process at a rate faster than that of competitors.

In subsequent sections, we discuss some important issues in design validation, such as validation testing design, prototype construction, early-stage validation activities, and process capability evaluation.

18.2   Design Analysis and Testing

The basic objective of design validation is to confirm that all the design requirements, such as functional performance requirements, operation environmental requirements, and reliability requirements (see Sec. 18.1.1), are confirmed. The methods for confirmation can be classified into two categories: design analysis and testing. *Design analysis* involves the use of analytical means, such as a mathematical model, a computer model, or a conceptual model (such as FMEA) to verify some aspects of the design requirements. For many product development situations, testing is the "hardware test" in which physical prototypes or actual products will be incorporated in well-designed tests. For software development, there is well-planned rigorous testing on the testing version or final version of the software. For the service process, testing may mean a trial market test. Design analysis is usually conducted in the earlier stage of the product/process development cycle. Testing is conducted at a later stage of the development cycle. Compared with testing, design analysis is usually easier, cheaper, and quicker to perform, but it may not be able to give a sufficient confirmation for the design requirements. Testing may give a sufficient confirmation with respect to some design requirements, but it is usually harder to perform and more expensive and time-consuming. Design analysis and testing are also related. If design analysis is very effective on a design requirement, then the final confirmation testing on that requirement might be reduced in test sample size, changed to a cheaper testing, or simply canceled. For example, in an automobile crash safety analysis project, if we do a good job on computer-aided crash simulation analysis, we may reduce the number of actual crash dummy tests. On the other hand, if the design analysis reveals some potential weaknesses in the design, we may also conduct more design-related testing to ensure that potential weakness are resolved in redesign. Planning of design validation activities heavily depends on the design requirements. Figure 18.1 illustrates the relationship between design requirements, design analysis, and testing.

From Fig. 18.1, it is very clear that compiling a good list of design requirements is the starting point for design validation. Design analysis is the first pass for validation; usually it can only partially confirm the design requirements. Testing is the next pass of validation. Sometimes, the validation process can go backward as well; a design analysis or testing may indicate that the current design cannot meet a design requirement, so we have to either redesign or adjust that design requirement. Previous results and other sources of knowledge are also very important in the validation process. There is no need to reinvent the wheel. Past test results on similar products, relevant knowledge in the public domain, government publications, and university research results can all be part of the knowledge base. More relevant information on the knowledge base for a design requirement will lead to less design analysis and less testing on that requirement. For example, if an automobile bumper is using the same steel material but a shape slightly different from that of last model year's design, then many past design analysis and testing results can be used to validate the current design.

In this section, we discuss all major aspects of the validation process, beginning with the *design requirements analysis,* which is an

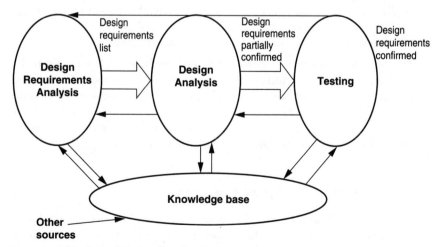

Figure 18.1   Design validation flow diagram.

information collection process that complies the list of design requirements (Grady 1998).

18.2.1 Design requirements analysis

The purpose of design requirement analysis is to compile a comprehensive list of design requirements. This list should be complete, accurate, and specific. For each item in the list, an importance index should also be provided. For example, in automobile or aircraft design, any safety-related design requirement should give a very high importance rating.

A checklist of design requirements is given in the following paragraphs.

Functional performance requirements

Definition. These are a collection of all functional performance target values and specifications.

Where to collect data. In the concept design stage and detailed design stage, these data should be available from design and development teams and the engineering department.

Level of detail needed. The more detail, the better. For major functional performance requirements, numerical target values and numerical specifications are highly desirable. In a complex product, there might be a hierarchy of functional performance requirements, such as system, subsystem, and component requirements and specifications.

Itemized priority index. This is needed. An itemized priority index is a rating of the relative importance of a functional performance requirement. For example, the "power output" requirement for an electric motor should have a high-priority index, because that is a very basic main functional performance requirement.

Previous validation results and knowledge base. This needs to be checked, and all relevant information should be retrieved. We need to determine how much information on previous validation results and relevant data in the knowledge base can be used directly and/or indirectly in our validation process and what additional information we need to obtain in new design analysis and new testing in order to validate the requirement.

Operation environmental requirements validation

Definition. This is a collection of operation and environmental profiles when the product is delivering its intended functions, such as high and low temperatures, shocks and vibrations, humidity, wind, salt, and dust. This also includes the functional performance requirement and other requirement target values and specifications under these operational and environmental profiles. For example, a battery may have to operate at high and low temperature, or high or low humidity; we need to know the range of those temperatures and humidity profiles in detailed specifications, such as -50 to $+50°C$ or 0 to 100% relative humidity and under those conditions, what is the required voltage level, battery life, and so on.

Where to collect data.   Sources include previous experience, competitors' current capability, knowledge base, and customer surveys.

Level of detail needed.   The more detail, the better.

Itemized priority index.   This is needed.

Previous validation results and knowledge base.   These need to be checked, and all relevant information should be retrieved. We need to determine how much information on the previous validation results and relevant data in the knowledge base can be used directly and/or indirectly in our validation process and what additional information we need to obtain in new design analysis and new testing to validate the requirement.

Reliability requirements validation

Definition.   This is a collection of requirements on useful life, functional performance degradation level, and operational faults. *Useful life* is the expected life for the product to function normally. *Functional performance degradation level* is the acceptable level of performance after a given period of usage. For example, a new car may have very high gas mileage, while a car several years old may have a somewhat lower gas mileage; the acceptable gas mileage for an old car is usually somewhat lower than that of a new car. Operational faults are "bugs" in the system; reliability in the software industry is usually related only to this measure.

Where to collect data.   Sources include previous experience, competitors' current capability, knowledge base, and customer surveys.

Level of detail needed.   The more detail, the better.

Itemized priority index.   This is needed.

Previous validation results and knowledge base.   These need to be checked, and all relevant information should be retrieved. We need to determine how much information on earlier validation results and relevant data in the knowledge base can be used directly and/or indirectly in our validation process and what additional information we need to obtain in new design analysis and new testing to validate the requirement.

Usage requirements validation

Definition.   This is a collection of usage profiles when the product is delivering its intended functions, such as "how abusively the consumers can use the products."

Where to collect data.   Sources include previous experience, competitors' current capability, knowledge base, and customers.

Level of detail needed.   The more detail, the better.

Itemized priority index.   This is needed.

Previous validation results and knowledge base.   These need to be checked, and all relevant information should be retrieved. We need to determine how much information on earlier validation results and relevant data in the knowledge base can be used directly and/or indirectly in our validation process and what additional information we need to obtain in new design analysis and new testing to validate the requirement.

Safety requirements validation

Definition.   A collection of requirements on safety related to the product, structure, and usage process.

Where to collect data.   Sources include government regulations and standards, company internal standards, previous experience, competitors' current capability, knowledge base, and customer surveys.

Level of detail needed.   The more detail, the better.

Itemized priority index.   This is needed.

Previous validation results and knowledge base.   These need to be checked, and all relevant information should be retrieved. We need to determine how much information on earlier validation results and relevant data in the knowledge base can be used directly and/or indirectly in our validation process and what additional information we need to obtain in new design analysis and new testing to validate the requirement.

Interface and compatibility

Definition.   These are a collection of requirements on the interface and compatibility when the product has to work with other products or equipment.

Where to collect data.   In the concept design stage and detailed design stage, those data should be available from design and development teams and the engineering department. If the other products or equipment are provided by another company or companies, we may have to get these requirements from them. Previous experience, competitors' current capability, knowledge base, customer surveys, and other sources may also help.

Level of detail needed. The more detail, the better.

Itemized priority index. This is needed.

Previous validation results and knowledge base. These need to be checked, and all relevant information should be retrieved. We need to determine how much information on earlier validation results and relevant data in the knowledge base can be used directly and/or indirectly in our validation process and what additional information we need to obtain in new design analysis and new testing to validate the requirement.

Maintainability requirement validation

Definition. This is a collection of requirements on maintenance.

Where to collect data. Sources include previous experience, competitors' current capability, knowledge base, and customer surveys.

Level of detail needed. The more detail, the better.

Itemized priority index. This is needed.

Previous validation results and knowledge base. These need to be checked, and all relevant information should be retrieved. We need to determine how much information on earlier validation results and relevant data in the knowledge base can be used directly and/or indirectly in our validation process and what additional information we need to obtain in new design analysis and new testing to validate the requirement.

In design requirements analysis, it is highly desirable to compile the list of design requirements as early as possible in as much detail as possible and utilize the knowledge base as much as possible. So the design validation process can be launched earlier in the product development cycle, and a prudent design analysis and testing plan can be formulated and executed with high efficiency and effectiveness, thus shortening the product development cycle time. However, in many practical situations, especially the creative design, several iterations of "requirements-analysis-testing" may be needed to establish all feasible design requirements. In this circumstance, the goal is to "get as much information as possible" in each iteration, so fewer cycles of requirements-analysis-testing are needed.

18.2.2 Design analysis

Design analysis (O'Connor 2001) is a collection of analytical methods that can be used to analyze design requirements, suggest design changes, and validate or partially validate design requirements.

There are many design analysis methods. We can roughly classify design analysis methods into several categories: design evaluation and review methods, mathematical models, and computer simulation models.

Design evaluation and review methods. The examples of such methods include QFD, FMEA, and formal design reviews. All these methods provide clearly defined procedures and templates. They could systematically guide team members to look into current design in detail to determine its strengths and weaknesses. These methods are more system-oriented, more comprehensive, and more subjective than other design analysis methods. They seldom can provide a "solid" validation.

Mathematical models. The most frequently used mathematical models used in design analysis include

1. Mechanism-based mathematical models such as mechanical stress, strength-strain math model; math models for electrical voltage, current, resistance, and so on; logical math models; financial math models; math model for three-dimensional geometric positions; and so on.

2. Mathematical and statistical software such as Mathematica, Microsoft Excel, optimization software, and MINITAB.

These methods can be used to model and analyze simple to moderately complex designs. For some applications, this can give pretty good validation capability. For example, a mathematical model can predict the relationship among current, voltage, resistance, and other parameters. However, it seldom provides good validation on operational environment requirements, reliability requirements (without testing or computer simulation data; only math won't do much for reliability validation), and other requirements.

Computer simulation models. Increasing choices of computer simulation models are available for many applications. There are two categories of computer simulation models mechanism-based simulation models and Monte Carlo simulation models.

Mechanism-based simulation models. *Mechanism-based simulation models* are usually commercial computer packages designed for specific fields of applications, such as mechanical and electrical engineering and electronics. The computer packages follow the established scientific principles or laws to model the relationships for specific classes of applications. We will briefly describe some of the most commonly used mechanism-based computer simulation models.

1. *Mechanical simulation models.* Mechanical components and system design can be created and analyzed using computer-aided design (CAD) software. Analysis capability includes geometric dimensions and tolerances (GDT), three-dimensional views, and animation. CAD software is usually the starting point for computer-aided engineering (CAE) analysis, such as stress and vibration analysis. *Finite element analysis* (FEA) is a computer-based technique that can evaluate the many key mechanical relationships, such as the relationships between force, deformation, material property, dimension, and structural strength on a design. In FEA, the original continuous 3D geometric shapes are subdivided into many small units, called *finite element meshes.* The precise differential equation forms of basic relationships in mechanics are approximated by a large-scale linear equation system. FEA methods can study mechanical stress patterns in static or dynamic loaded structures, strain, response to vibration, heat flow, and fluid flow. The use of FEA to study the fluid flow is called *computational fluid dynamics* (CFD). CFD can be used to analyze airflow, fluid flow, and other properties to analyze aerospace design, automobile engine design, ventilation systems, and other types of design.

2. *Electrical and electronics.* Electrical and electronic circuits can be analyzed by using electrical design automation (EDA) software that is based on the mechanism models in electrical engineering, such as Ohm's law and logical circuit models. In EDA software, component parameters, input electrical signals, such as power, waveform, and voltage, and circuit diagrams are input into the program. The program can perform many analyses, such as evaluation of circuit performance, sensitivity analysis, and tolerance analysis. Electromagnetic effects can be analyzed by using a special kind of FEA software that is based on Maxwell's law.

3. *Others.* There are also numerous mechanism-based computer simulation models for other applications, such as chemical engineering, financial operation, and economics.

Monte Carlo simulation model. *Monte Carlo simulation* is a technique that simulates large numbers of random events. For example, transactions in a bank branch office are random events. A random number of customers will enter the office at random arrival times, and each customer will make a random type of transaction, such as deposit, withdraw, or loan, with a random amount of money. The Monte Carlo simulation model can simulate such events. Monte Carlo simulation starts with generating a large number of random numbers that follow a prespecified probability distribution. For example, the customer arrival time to the bank branch can be assumed to be a Poisson arrival process, in which the key parameter, the mean interarrival time, can be set by the analyst or estimated by old data. After we set all the parameters for the probability models, such as mean and variance of transaction amount, average transaction time, and system interaction relationships, the Monte Carlo simulation can generate a virtual bank transaction process for a large number of virtual customer arrivals. The simulation model can then provide key system performance statistics in the form of a chart and histogram. In the bank case, these key performance measures could include customer waiting time, percentage of time that a clerk is idle, and daily transaction amount.

Monte Carlo simulation is very helpful in analyzing random events in service processes, factory flows, banks, hospitals, and other systems. Monte Carlo simulation can be integrated with process mapping, value stream mapping, and process management to analyze and benchmark a large variety of processes. It can be used in tolerance analysis, which is also called *variation simulation modeling* or *variation simulation analysis.* In this application, a large number of random numbers are generated that simulate a large number of low-level parameters, such as component dimensions. By giving the relationship equation between high- and low-level characteristics, the variation simulation model can provide a histogram distribution of high-level characteristic variation. For example, the variation simulation model can generate a large number of random component dimensions; if the mathematical relationship between component dimensions and system dimension is given, variation simulation modeling can generate a large number of "assembled system dimensions" and the histogram of the system dimensions will be provided. Therefore, system process capability can be estimated. The Monte Carlo simulation model can also be integrated with some mechanism-based simulation to analyze scientific and engineering systems.

Computer simulation models with other methods. Computer simulation models can be integrated with many other analysis methods, such as design of experiment (DOE), the Taguchi method, and response surface methods (RSMs). These kinds of integration can provide powerful tools for design analysis for validation of a great number of design requirements.

For example, by using the Taguchi method and computer simulation model, design parameters can be changed and functional performances evaluated. Noise factors can be approximated by computer model parameter variation. In many practical cases, these noise factors include operational environmental variation and piecewise variation; even some reliability-related variation could be approximated as computer model parameter variation. So these "analytical Taguchi projects" can partially serve the purpose of design requirements validation.

18.2.3 Validation testing

Validation testing is an important part of the design process (O'Connor 2001). Validation tests are performed to determine if the design requirements are met by the design. There are many aspects in validation testing, and we will discuss them in detail.

Validation testing objectives. All validation tests are designed to validate some design requirements, such as functional performance requirements, and reliability requirements. We can roughly divide the validation testing into the following categories.

Functional testing. The objective of functional testing is to validate the functional performance requirements. For many products, functional testing can be at least partially conducted by design analysis methods. For example, one important functional performance requirement of a power supply circuit is "to provide power with given voltage." Because computer simulation models can be very accurate in predicting circuit behaviors, computer simulation results can accomplish most of the functional testing. In such a case, after computer simulation, a small number of hardware measurements usually suffice to validate functional performance requirements.

Reliability testing. The objective of reliability testing is to validate the reliability requirements.

Reliability testing methods include life testing and degradation testing. A *life test* is designed to "test until failure," so that the time to failure can be estimated. *Degradation testing* monitors system performance degradation with time so that the rate of degradation can be estimated.

For reliable products, estimating reliability requirements is difficult. For example, if the designed life for a TV set is 8 years, nobody will want to test it for 8 years to validate this reliability requirement. Therefore, accelerated reliability tests are preferred in which "stress" or "failure cause" will be added to induce quick failures. Usually, some approximate math models are available to translate accelerated "time to failure" into actual (real-time) time to failure.

The "stress" used in reliability testing is related to the mechanisms that cause failure. For product development situation, for mechanical-type products, the causes of failure are often stress, fatigue, creep, vibration, temperature, wear, corrosion, and material failure. For an electrical or electronic product, there are electrical stresses, such as current, voltage, power stresses, and failure of wire bonding. For software products, the causes of failure are mostly design fault, or "bugs." Therefore, the type of "stresses" in reliability testing should be carefully selected on the basis of the failure mechanism.

Highly accelerated life testing (HALT) is a very popular type of reliability test. In HALT, a much more amplified stress level, that is, the stress level much higher than that of the operation environment, will be applied to greatly accelerate the failure process. On the other hand, a very detailed failure monitoring mechanism will be in place so that the cause and process of failure can be recorded and analyzed for design improvement purposes.

Design analysis can provide limited information on reliability requirement validation. However, many failure mechanisms are unknown and difficult to model. Actual hardware or prototype testing is very important in reliability requirement validation.

Safety and regulation–related testing. The purpose of this type of testing is to verify if the safety requirements and government regulations can be met by the design. Usually, very rigid requirements and regulations are provided by the government to guide these kinds of testing.

Testing for variation. This type of testing verifies whether the system can provide robust functional performance on diverse usage conditions, environmental conditions, and other types of variation. The Taguchi method described in very close detail in Chaps. 13 to 15 will fit perfectly for this kind of test.

Testing for interface and compatibility. The purpose of this type of test is to validate whether the product or equipment can work well with other systems. Specially designed prototypes or actual products will be used to verify those requirements.

Validation testing strategy. Validation testing strategy deals with how to subdivide the entire validating testing task into manageable pieces of subtasks. Ideally, if many of those subtasks can be executed in parallel, then the cycle time for design validation can be greatly reduced.

System, subsystem, and components testing. If a product can be subdivided into several subsystems, then each subsystem can be further divided into many components. We may want to conduct many small tests at different levels of the system. For example, we can conduct many small, inexpensive component tests concurrently. These tests can be used to sort out problems at the component level. When all components in a subsystem are validated, we can conduct the subsystem testing, and finally, we will conduct a system validation test.

In this approach, we have to notice that "not all the components are created equal." The ones with new technology or new design should be tested more carefully and vice versa. We may not have to test the components that are well known in our knowledge base.

Materials testing. Sometimes it is more appropriate to conduct very focused testing, such as testing new material, if the material is the only significant unknown factor in a component or system. This approach not only saves time and money but will also enrich our knowledge base.

New technology testing. If there are new technologies in our product, it is very important to conduct extensive design analysis and validation testing on them. Taguchi experiments and reliability testing are very helpful in new technology testing.

Validation activity planning. The basic objective of design validation is to make sure that our new product will comply with all design requirements. The main design validation tools are design analysis and validation testing. There are many possibilities in selecting tools in this type of testing. For a good design validation practice, it is very desirable to accomplish the following goals:

1. All design requirements are adequately validated.

2. The total cost of design validation is low.

3. The cycle time of design validation is short.

The cost and cycle time of each validation activity will definitely depend on the following factors:

- Design analysis versus testing; testing is usually more expensive.
- Prototype building; this could be expensive.
- Type of testing.
- Duration of testing.
- Testing sample size.
- Precedence of different tests, parallel tests, or one test depends on others.

Therefore, the whole design validation activity should be carefully designed in order to achieve optimal effectiveness, cost, and cycle time.

A QFD template can be a very useful tool in planning design validation activities. We will use the following example to illustrate how to use the QFD template to plan validation activities.

Example 18.1. QFD Planning on Design Validation A new product has many design requirements (see Table 18.1). It has a number of functional performance requirements, a number of reliability requirements, several safety requirements demanded by the government, several operation environment requirements, usage requirements, and other requirements. The product development team has decided to subdivide the validation activity into two phases: design analysis and validation testing. In the design analysis phase, the team decided to conduct three finite element analysis projects, three Taguchi parameter design projects on computer models, and one variation simulation modeling project on tolerance analysis. In the validation testing phase, the team decided to conduct two functional tests, two reliability tests, and one small Taguchi experiment on prototypes.

Table 18.1 is a QFD template that relates each validation activity to design requirements. In the table, the rows represent different items of design requirements; for example, the first row represents "functional requirement 1." Each column represents a design validation project; for example, the first column represents "FEA model project 1" and the last column represents "a Taguchi experiment type test." In this table, we use the "relevance index" commonly employed in QFD, where a 9 means very relevant. In Table 18.1, this means that the corresponding design validation activity can fully validate that design requirement. For example, FEA project 1 can fully validate safety requirement 1. Similarly, a 3 means "partially relevant," a 1 means "very little relevance," and a 0 means "no relevance." In this QFD-like template, if we add all the entries in each row and get a number that is greater than 9, and if there is at least one entry which is 9, then the corresponding design requirement will have been completely validated, because a 9 means "hard validation." On the other extreme, if the sum of entries in a row is small, then the corresponding design requirement will not have been validated. Therefore, this QFD planning tool can be used to check if our validation activities can fully validate all the design requirements.

Design validation activities, whether they are design analysis activities or validation tests, must be conducted on *prototypes*. There are many kinds of prototypes with vastly different properties. Prototype building, selection, and usage are also very important in design validation.

18.3   Prototypes

Prototypes are trail models for the product. Ulrich and Eppinger (2000) defined prototype as "an approximation of the product along one or more dimensions of interest." In this context, a prototype can be a drawing, a computer model, a plastic model, or a fully functional prototype fabricated at a pilot plant.

TABLE 18.1 Design Validation Planning Matrix

	Validation activities										
	Design analysis						Validation testing				
	FEA modeling project		Analytical Taguchi experiment			Tolerance simulation	Functional testing		Reliability testing		Taguchi experiment
Design requirements	1	2	1	2	3	1	1	2	1	2	1
Functional 1	3	0	3	0	0	1	9				3
2	3	3	0	3	0	1		9			3
...	0	1	0	0	3	0	9	9			3
Reliability requirement 1	1	0	3	1	3	1	0	0	9		3
2	0	1	3	1	3	0	0	0	9		3
3	3	0	1	3	3	0	0	0		9	3
...											
Safety requirement 1	9	0	0	0	0	0	0	0	3	3	3
2	3	3	3	0	0	0	0	0	9	9	3
3	3	0	1	0	0	0	0	0	9	9	3
...											
Operation environment 1	0	0	3	3	3	0	3	3	1	1	9
2	0	0	3	3	1	0	3	3	3	3	9
3	0	0	3	3	1	0	3	3	3	3	9
...											
Usage requirements 1	0	0	3	3	3	0	3	3	1	1	9
2	0	0	3	3	0	0	3	3	1	1	9
3	0	0	0	3	3	0	3	3	1	1	9
...											
Other 1	0	0	1	3	0	3	1	1	0	0	3
2	0	0	1	3	0	3	1	1	0	0	3
...											

18.3.1   Types of prototypes

Ulrich and Eppinger also further defined a prototype as being analytical or physical. An *analytical* prototype represents the product in a mathematical or computational form. Many aspects of the product can be analyzed. For example, an FEM (finite element model) model can be used to analyze a mechanical part about force stress, deformation, and other parameters. A Monte Carlo simulation model can be used to simulate service flow, waiting time, and patient processing rate in a clinic. A physical prototype is a "real look-alike" prototype made of either substitute materials or actual materials designed for the product. For example, a prototype of an automobile manifold may be made of "easy to form" plastic instead of aluminum, the actual manifold material. In this case, we are interested only in studying the geometric aspects of the manifold, not its heat resistance and strength aspects.

A prototype can be also *focused* or *comprehensive* (Ulrich and Eppinger 2000). A focused prototype represents only a part, or subset of, product functions or attributes. A comprehensive prototype represents most of the product functions and attributes; some prototypes represent all product functions and attributes.

All analytical prototypes are focused prototypes, at least at the current technological level, because it is impossible to build a virtue model which can represent all product functions and attributes. For example, an FEM model represents only the mechanical aspects of the product characteristics; it cannot represent chemical properties such as corrosion. Physical prototypes can be very comprehensive; a preproduction prototype has all the required product functions and attributes. Physical prototypes can also be very focused; a plastic prototype may be able to represent only the geometric aspects of a part, not material properties or functions.

The four most commonly used physical prototypes are experimental, alpha, beta, and preproduction prototypes:

- *Experimental prototypes.*   These are very focused physical prototypes, designed and made to test or analyze a very well-defined subset of functions and attributes. For example, a plastic prototype of an automobile manifold is built to enable the engineering team to study the geometric aspects of manifold functions.

- *Alpha prototypes.*   These are used in product functional performance validation. An experimental prototype made in the lab or a concept car model is one such example. Usually an alpha prototype can deliver all the intended functions of a product. The materials and components used in alpha prototypes are similar to what will be used in actual production. However, they are made in a prototype process, not by the mass-production-based manufacturing process.

- *Beta prototypes.* These are used to analyze the reliability requirements validation, usage requirements validation, product specification validation, and so on. They may also be used to test and "debug" the manufacturing process. The parts in beta prototypes are usually made by actual production processes or supplied by the intended part suppliers. But they seldom are produced at an intended mass production facility. For example, an automobile door panel beta prototype might be made by the same machinery as that at an assembly plant, but it is made by a selected group of validation engineers and technicians, not by hourly workers.

- *Preproduction prototypes.* These are the first batch of products made by the mass production process, but at this point in time the mass production process is not operating at full capacity. These prototypes are generally used to verify production process capability, as well as test and debug the mass production process.

Figure 18.2 illustrates prototype types and classifications.

18.3.2   Usefulness of prototypes

In product development, prototypes can serve many purposes. They can be used for design analysis, validation testing and debugging, and interface and compatibility testing, as well as in communication and for demonstration purposes. Following are the main applications of prototypes:

Design analysis.   This purpose can be served mainly by analytical prototypes. For a prototype that is analytical and focused, such as a Monte Carlo simulation model for a clinic, or FEM model for a mechanical component, a subset of selected product functional performances or some very specific properties of the product will be analyzed. If the expected performance goals cannot be met, the design change will be made to improve the performance.

Validation testing and debugging.   Usually this purpose can be served much better by physical prototypes, because real performances are delivered by physical products. For reliability life testing and confirmation, a physical prototype is a must. The real value of a true physical prototype is that some unexpected failures, bugs, and other defects can be found only by physical prototype testing.

Interface and compatibility testing.   Prototypes are often used to ensure that the components and/or subsystems can be assembled well. Also, we may want to test to see if different subsystems can work well after being assembled together as a system.

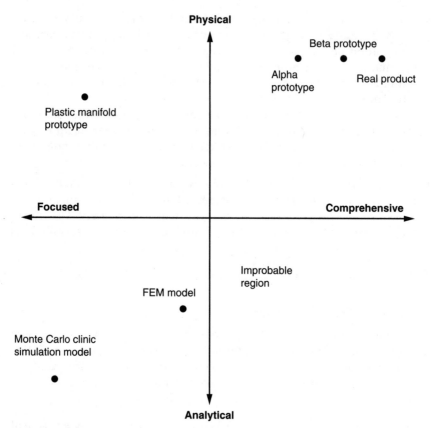

Figure 18.2 Types and classifications of prototypes.

Communication and demonstration. Prototypes, especially physical prototypes, can enhance the power of communication and demonstration tremendously. "A picture is worth a thousand words"; by the same token, "a real thing is worth a thousand pictures."

18.3.3 Principles of prototyping

According to Ulrich and Eppinger, several "principles of prototyping" are useful in guiding the uses of prototypes in the product development process.

1. *Analytical prototypes are generally more flexible than physical prototypes.* It is usually much easier to adjust the "design parameters" in a mathematical model or a computer model, and compute the new output values. It usually takes much longer to build a physical

prototype, and after it has been built, it is usually quite difficult to change the design parameters.

2. *Physical prototypes are required to detect unexpected phenomena.* An analytical prototype can display the "properties" only within the assumptions, mathematical models, and mechanisms on which the analytical prototypes are based; it can never reveal phenomena that are not part of its assumptions and mechanisms. For example, a Monte Carlo simulation prototype for a clinic is based on a stochastic process model and given probability distributions; it can never, for instance, represent the actual physician-patient interactions. No matter how well the computer simulation model works, the testing on true physical prototypes often detects some unexpected bugs.

3. *A prototype may reduce the risk of costly iterations.* Analytical prototypes can be analyzed conveniently, and design change can be simulated and improvements made. Physical prototypes can be tested, and some unexpected phenomena, good or bad, can be revealed. Eliminating design disadvantage in a timely manner will definitely shorten the product development cycle time.

However, building prototypes also takes time and costs money, planning of prototyping activities should be based on the balance of risk and development cost. Usually for an item that is inexpensive, with a low failure cost, and/or with known technology, less effort should be spent in prototyping. For an item that is expensive, with a high failure cost, and/or with new or unknown technology, more effort should be spent on prototyping activities. Figure 18.3 illustrates the economical factors affecting prototype-building decisions.

18.3.4 Prototyping activity management

Prototyping activity management involves making the following kinds of decisions:

1. How many prototypes should be built
2. How many analytical prototypes and how many physical prototypes should be constructed
3. When each prototype should be built
4. What we should do on each prototype, such as testing (if so, which tests) or demonstration

Clearly, prototyping activity management is closely related to the overall design validation activity planning that we discussed in Sec. 18.2. Therefore, prototyping activity and design validation activity should be planned together. In other words, the prototype decision

Technical or Market Risk

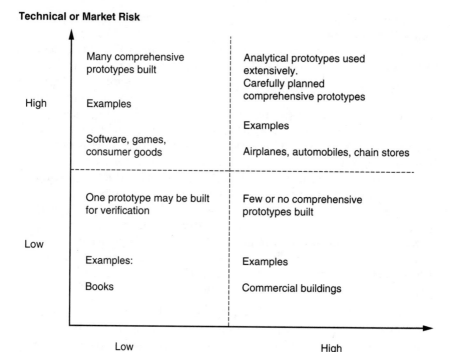

Figure 18.3 Economical factors affecting prototypes built. [*Adapted from Ulrich and Eppinger (2000).*]

should be design-validation-activity-driven. For example, if we use an FEA simulation as the first step in design analysis, then the relevant prototype must be an FEA model. If a validation test is to "verify assembly with other parts," then a "look-alike" prototype made of cheap plastic is probably enough. If another design validation activity is a reliability test, then several real physical prototypes might be needed, and the number of prototypes to be built will depend on reliability test requirement.

Design requirements validation sufficiency, cost, and cycle time are the most important factors in planning decisions. Design requirement validation sufficiency is a measure of whether all design requirements are sufficiently validated. The QFD template can be used to facilitate the planning activities in this aspect, as we have illustrated in Example 18.1.

Many *rapid prototyping* technologies (Ulrich and Eppinger 2000) are available to build prototypes in a very short time. These technologies could help reduce cycle time.

18.4   Process and Production Validation

There are many similarities between product design validation and process/production validation. In the design-development phase of a manufacturing process, the validation tasks are almost the same as those of the product design. However, in any mass production, defects and variation are generated from the manufacturing process. For a good production process, it is very important to keep defects and variation at a low level. Variation level is measured by *process capability.* The concept of process capability is discussed extensively in Chap. 2. One of the most important and distinct process/production validation activities is *process capability validation,* which is discussed in this section. The step-by-step procedure for the whole process/production validation cycle was discussed in Sec. 18.1.

18.4.1   Process capability validation flowchart

Figure 18.4 illustrates a flowchart for process capability validation. It can be used as a step-by-step roadmap. We will discuss some important steps.

Select key characteristics. These key characteristics are either key *product* characteristics that are very important for customers, such as the key dimensions of a high-precision part, or key *process* characteristics that are important in determining product quality. For example, in the automobile paint application–curing process, oven temperature is a key process variable that determines the paint's finish quality.

For each key characteristic, we need to determine its target value and functional specification limits. For two-sided specification limits, there are an upper spec limit (USL) and a lower spec limit (LSL).

Establish goal for capability. If we strive for Six Sigma quality, the goal for process capability will be $C_p = 2$, or $C_{pk} = 2$.

Verify measurement system. The key characteristics are measured by gauges, sensors, instruments, and so on. It is highly desirable that the measurement error be much smaller than the variation level of the key characteristics. In an ideal situation, the measurement error should be no more than 10 percent of that of the variation level of the key characteristics. In most practical cases, this requirement is 30 percent. The measurement system evaluation is discussed in many books [e.g., see Montgomery (2000)].

Select appropriate control chart. Many control charts are available for different types of characteristics. For continuous variables, x-bar and

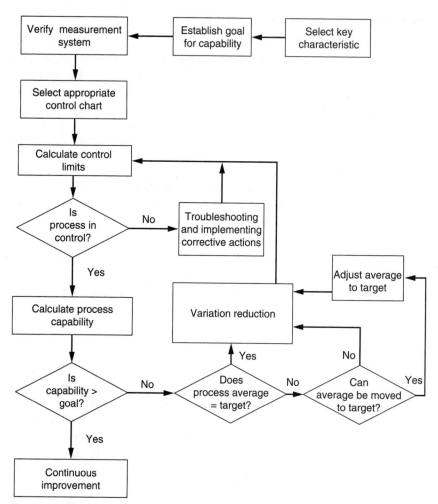

Figure 18.4 Process capability validation flowchart.

R charts are the most frequently used control charts. Control charts are discussed extensively by Montgomery (2000).

Out-of-control condition and instability. "Out of control" indicates that the process is unstable. In statistical process control, the first priority is to identify and eliminate process instability. Process capability cannot be benchmarked until process stability is achieved. We briefly discuss process instability in Sec. 18.4.2, because it is very common in new process validation.

Calculate process capability. After the process has achieved stability, we can calculate the process capability and compare that with our goal.

One important issue in calculating process capability is the distinction between short-term capability and long-term capability. For a new process without much data, this distinction is often unclear. We will discuss short- and long-term capabilities in Sec. 18.4.3.

Troubleshooting and variation reduction. If a process is either not stable or excessive in variation, we need to troubleshoot and reduce variation. Actually, in many cases, *the data patterns displayed in an instability situation can be used for troubleshooting.* Many statistical methods can provide help on these tasks, such as the multi-vari chart (Perez-Wilson 2002) and multivariate statistical analysis (Yang 1996, Jang et al. 2003).

18.4.2 Process instability

Process instability prohibits the benchmarking of process capability. *Process capability* is a predictive measure about how reliable the process is in producing good products consistently. For example, $C_p = 2$ means Six Sigma quality, or 3.4 defects per million units. Obviously, one precondition for such a quality claim should be that the process be *predictable.* When a process is out of control, or displays some non-random patterns, we call this process *unstable.* For example, Fig. 18.5a displays a nonrandom pattern, although all the points are within control limits. One question that many people will ask is "What will happen next?" Clearly, *unstable* means unpredictable. (If a few people claim that their company has a Six Sigma quality with this control chart as proof, do you believe them?)

Specifically, from the statistical process control viewpoint, a process can be judged as "out of control" or unstable if the control charts demonstrate any of the following patterns:

1. One or more points outside the control limits

2. Two out of three consecutive points outside the 2σ warning limits but still inside the control limit

3. Four out of five consecutive points beyond the 1σ limits

4. A run of eight consecutive points on one side of the centerline

Over the years, many rules have been developed to detect nonrandom patterns within the control limits. Grant and Leavenworth (1980) suggested that nonrandom variations are likely to be presented if any one of the following sequences of points occurs in the control charts:

1. Seven or more consecutive points on the same side of the centerline.

2. At least 10 out of 11 consecutive points on the same side of the centerline

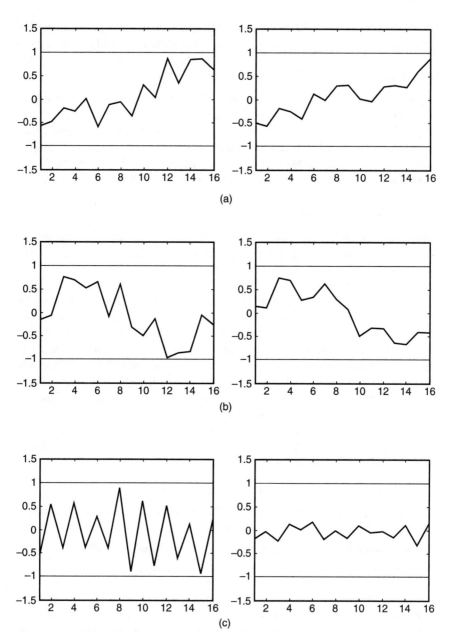

Figure 18.5 Selected patterns of process instability. (a) Upward trends, (b) downward trends, (c) systematic patterns.

3. At least 12 out of 14 consecutive points on the same side of the centerline

4. At least 14 out of 17 consecutive points on the same side of the centerline

A good and credible process capability estimate can be obtained only after we have achieved statistical process control.

However, process instability may provide valuable clues for new process troubleshooting. Unfortunately, process instability is usually the rule, not the exception, in a new process startup, because there are usually many "bugs" in the new processes. The process will not be stable until all these bugs have been eliminated.

However, the special patterns displayed in stability may serve as clues for finding these bugs. For example, in metal cutting, an increasing or decreasing pattern might be caused by gradual overheating or wearing out of a machine tool. For cyclic patterns, their frequency may provide clues as to where they are coming from. Under the pattern recognition approach, numerous research studies (Western Electric 1956) have defined several types of out-of-control patterns (e.g., trends, cyclic pattern, mixture) with a specific set of possible causes. When a process exhibits any of these nonrandom patterns, those patterns may provide valuable information for process improvement. Therefore, once any nonrandom patterns are identified, the scope of process diagnosis can be greatly narrowed down to a small set of possible root causes that must be investigated.

Many other techniques, such as the multi-vari chart (Perez-Wilson 2002) and multivariate statistical analysis (Yang 1996, Jang et al. in press), can work very effectively in troubleshooting problems in early process startups.

18.4.3   Short- and long-term process capabilities

In the production process capability validation process, another important issue is that we need to distinguish between long- and short-term process capabilities. In the remainder of this chapter, we discuss only the continuous performance characteristics. In statistical process control, variable control charts, such as x-bar and R charts and x-bar and S charts, will be used in this case. The short- and long-term capabilities of other types of performance characteristics and control charts are thoroughly discussed by Bothe (1997).

A typical set of variable SPC (statistical process control) sample data will look like that of Table 18.2, where n is the subgroup size and K is the number of samples.

TABLE 18.2   A Typical Set of Variable Data in SPC

Sample no.	X_1	X_2	$\cdots$	X_n
1	x_{11}	x_{12}		x_{1n}
2	x_{21}	x_{22}		x_{2n}
3	$\vdots$	$\vdots$		$\vdots$
K	x_{K1}	x_{K2}	$\cdots$	x_{Kn}

If we use x-bar and R charts, for each sample I, $i = 1,2,...,K$, we obtain

$$\overline{X}_i = \frac{1}{n} \sum_{j=1}^{n} x_{ij} \qquad \text{(x-bar in ith sample)} \qquad (18.1)$$

$$\overline{\overline{X}} = \frac{1}{K} \sum_{i=1}^{K} \overline{X}_i \qquad \text{(grand mean)} \qquad (18.2)$$

$$R_i = \text{Max}(x_{i1},...,x_{in}) - \text{Min}(x_{i1},...,x_{in}) \quad \text{(range of ith sample)} \qquad (18.3)$$

$$\overline{R} = \frac{1}{K} \sum_{i=1}^{K} R_i \qquad \text{(average range)} \qquad (18.4)$$

If we use x-bar and S charts, then

$$S_i = \sqrt{\frac{1}{n-1} \sum_{j=1}^{n} (x_{ij} - \overline{X}_i)^2} \quad \text{(standard deviation of ith sample)} \qquad (18.5)$$

$$\overline{S} = \frac{1}{K} \sum_{i=1}^{K} S_i \qquad \text{(average standard deviation)} \qquad (18.6)$$

In estimating C_p or C_{pk}, we generally used the following formulas; first we estimate process standard deviation (sigma):

$$\hat{\sigma} = \frac{\overline{R}}{d_2} \qquad (18.7)$$

or

$$\hat{\sigma} = \frac{\overline{S}}{c_4} \qquad (18.8)$$

where d_2 and c_4 are constant coefficients depending on sample size. Then we will calculate process capability index estimates:

$$\hat{C}_p = \frac{\text{USL}-\text{LSL}}{6\hat{\sigma}} \tag{18.9}$$

$$\hat{C}_{pk} = \text{Min} \left(\frac{\text{USL} - \overline{X}}{3\hat{\sigma}}, \frac{\overline{X}-\text{LSL}}{3\hat{\sigma}} \right) \tag{18.10}$$

However, the $\hat{\sigma}$ (sigma) calculated by Eq. (18.7) and (18.8) is only a *short-term* process standard deviation, because it is calculated on the basis of variation within subgroups. So actually

$$\hat{\sigma} = \hat{\sigma}_{\text{ST}} \tag{18.11}$$

in Eqs. (18.7) and (18.8), where $\hat{\sigma}_{\text{ST}}$ stands for short-term standard deviation.

In the production validation process, using $\hat{\sigma}_{\text{ST}}$ to estimate sigma level and using process capability could lead to severe overestimation of process capability.

Figure 18.6 illustrates that when a process is drifting or moving, the long-term variation spread is much larger than the short-term variation spread. Using short-term sigma $\hat{\sigma}_{\text{ST}}$ will greatly overestimate our process capability.

Use of long-term sigma $\hat{\sigma}_{\text{LT}}$ is recommended for estimation of the long-term process capability, where

$$\hat{\sigma}_{\text{LT}} = \sqrt{\frac{1}{Kn} \sum_{i=1}^{K} \sum_{j=1}^{n} (x_{ij} - \overline{X})^2} \quad \text{(standard deviation of pooled data)} \tag{18.12}$$

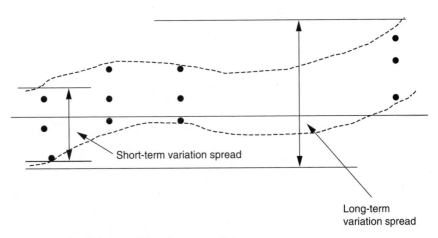

Figure 18.6 Short-term and long-term variation.

The long-term capability indexes can be computed by

$$\hat{C}_p = \frac{\text{USL} - \text{LSL}}{6\hat{\sigma}_{\text{LT}}} \tag{18.13}$$

$$\hat{C}_{pk} = \text{Min}\left(\frac{\text{USL} - \overline{X}}{3\hat{\sigma}_{\text{LT}}}, \quad \frac{\overline{X} - \text{LSL}}{3\hat{\sigma}_{\text{LT}}}\right) \tag{18.14}$$

Example 18.2. Short-Term and Long-Term Process Capabilities   The specification for a quality characteristic is $(-10, 10)$, and target value is 0. Fifteen samples of SPC data are collected as follows:

Sample no.	X_1	X_2	X_3	X_4	X_5
1	1	3	3	4	1
2	1	4	3	0	3
3	0	4	0	3	2
4	1	1	0	2	1
5	−3	0	−3	0	−4
6	−7	2	0	0	2
7	−3	−1	−1	0	−2
8	0	−2	−3	−3	−2
9	2	0	−1	−3	−1
10	0	2	−1	−1	2
11	−3	−2	−1	−1	2
12	−8	2	0	−4	−1
13	−6	−3	0	0	−8
14	−3	−5	5	0	5
15	−1	−1	−1	−2	−1

Using MINITAB, we obtain

$$\overline{X} = -0.3867; \qquad \hat{\sigma}_{\text{ST}} = \frac{4.933}{2.326} = 2.12$$

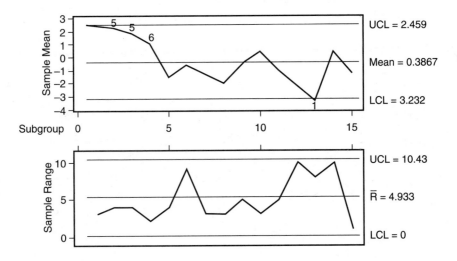

Xbar/R Chart for C1 - C5

Short-term capabilities are

$$\hat{C}_p = \frac{\text{USL} - \text{LSL}}{6\hat{\sigma}_{\text{ST}}} = \frac{20}{6 \times 2.12} = 1.57 \qquad (4.71 \text{ sigma level})$$

$$\hat{C}_{pk} = \text{Min}\left(\frac{\text{USL} - \overline{X}}{3\hat{\sigma}_{\text{ST}}}, \frac{\overline{X} - \text{LSL}}{3\hat{\sigma}_{\text{ST}}}\right) = \text{Min}\left(\frac{10 + 0.3867}{3 \times 2.12}, \frac{10 - 0.3867}{3 \times 2.12}\right)$$

$$= 1.151 \qquad (4.53 \text{ sigma level})$$

However, if we use long-term sigma $\hat{\sigma}_{\text{LT}}$

$$\hat{\sigma}_{\text{LT}} = \sqrt{\frac{1}{Kn}\sum_{i=1}^{K}\sum_{j=f}^{n}(x_{ij} - \overline{X})^2} = 2.72572$$

the long-term capabilities are

$$\hat{C}_p = \frac{\text{USL} - \text{LSL}}{6\hat{\sigma}_{\text{LT}}} = \frac{20}{6 \times 2.726} = 1.22 \ (3.67 \text{ sigma level})$$

$$\hat{C}_{pk} = \text{Min}\left(\frac{\text{USL} - \overline{X}}{3\hat{\sigma}_{\text{LT}}}, \frac{\overline{X} - \text{LSL}}{3\hat{\sigma}_{\text{LT}}}\right) = \text{Min}\left(\frac{10 + 0.3867}{3 \times 2.726}, \frac{10 - 0.3867}{3 \times 2.726}\right)$$

$$= 1.18 \qquad (3.54 \text{ sigma level})$$

Obviously, long-term process capabilities are noticeably lower than those of short-term ones.

Acronyms*

ABC	activity-based cost
AD	axiomatic design
AGV	automated guided vehicle
AI	artificial intelligence
AIAG	Automotive Industry Action Group
ANOVA	analysis of variance (MANCOVA = analysis of covariance; MANOVA = multivariate ANOVA)
APQP	advanced product quality planning
ASQ	American Society for Quality
BB	black belt (BBMO = black belt momentum; MBB = master black belt)
B/L	bill of lading
BM	business metric
BP	backpressure
CA	customer attribute
CAD/CAE	computer-aided design/engineering
CCD	central composite design
CDI	customer desirability index
CFD	computational fluid dynamics
CIRP	College Internationale de Recherches pour la Production
CLT	central-limit theorem
CNC	computer numerically controlled
CPM	critical path method
CSM	customer satisfaction metric
CTC	critical to cost (CTD = critical to delivery; CTQ = critical to quality; CTS = critical to satisfaction)
DD	decomposition distance
DFA	design for assembly
DFE	design for environment
DFM	design for maintainability

*Very common or well-known acronyms (IR, MIT, NSF, UV, etc.) omitted from list.

DFMA	design for manufacture and assembly
DFR	design for reliability
DFS	design for serviceability
DFSS	design for Six Sigma
DFX	design for X
DM	deployment mass
DMAIC	design-measure-analyze-improve-control
DOE	design of experiment
DOF	degree(s) of freedom
DPMO	defects per million opportunities
DPFMEA	design and process failure mode–effect analysis
DPO	defect per opportunity
DV	deployment velocity
EDA	electrical design automation
EE	electrical energy
FA	financial analyst
FEA	finite element analysis
FEAD	front-end accessory drive
FMEA	failure mode–effect analysis (DFMEA, PFMEA = design process FMEA)
FR	functional requirement
FTA	fault-tree analysis
GB	green belt
GDT	geometric dimensioning and tolerance
HALT	highly accelerated life testing
HOQ	house of quality
ICAD	International Conference on Axiomatic Design
ICOM	input(s), control(s), output(s), mechanism(s)
ICOV	identify (requirements), characterize, optimize, and verify (the design)
IDEF	international DEFinition
IFR	ideal final result
IPPD	integrated product and process design
IRR	internal rate of return
KKT	Karush-Kuhn-Tucker
LCC	life-cycle cost
LMP	Laboratory for Manufacturing and Productivity
LS	least squares (LSE = least-squares error)
LSL, USL	lower, upper specification limits

MARR	minimum attractive rate of return
MSD	mean-squared deviation
MSE	mean-squared error
MTBF	mean time between failure(s)
MTTF	mean time to failure
MTTR	mean time to repair
NPW	net present worth
NVH	noise, vibration, harshness
OSTN	object state transition network
PCB	printed-circuit board
PD	pitch diameter
pdf	probability density function
PE	potential energy
PERT	program evaluation and review technique
PMI	Project Management Institute
PMS	program management system
PV	potential variation; process variable
QFD	quality function deployment
R&R	repeatability and reproduceability
RPN	risk priority number
RSM	response surface method(ology)
RTY	rolled throughput yield
SIPOC	supplier-input-process-output-customer
SMT	set melt temperature
SOP	standard operating procedure
SPC	statistical process control
SS	screw speed; sum of squares
SS_A	sum of squares due to A
SS_E	sum of squares due to error
SS_T	total sum of squares
TC	total cost
TIPS	theory of inventive problem solving
TQM	total quality management
TRC	total remaining (metal) concentration
TRIZ	teoriya resheniya izobreatatelskikh zadatch (see TIPS)
VOC	voice of customer
WDK	Workshop Design-Construcktion
WIP	work in progress

References

Abbatiello, N. (1995), *Development of Design for Service Strategy,* M.S. thesis, University of Rhode Island.

Alexander, C. (1964), *Notes on the Synthesis of Form,* Harvard University Press, Cambridge, Mass.

Altshuller, G. S. (1988), *Creativity as Exact Science,* Gordon & Breach, New York.

Altshuller, G. S. (1990), "On the Theory of Solving Inventive Problems," *Design Methods and Theories,* vol. 24, no. 2, pp. 1216–1222.

Anjard, R. P. (1998), "Process Mapping: A Valuable Tool for Construction Management and Other Professionals," *MCB University Press,* vol. 16, no. 3/4, pp. 79–81.

Arciszewsky, T. (1988), "ARIZ 77: An Innovative Design Method," *Design Methods and Theories,* vol. 22, no. 2, pp. 796–820.

Arimoto, S., T. Ohashi, M. Ikeda, and S. Miyakawa (1993), "Development of Machining Productivity Evaluation Method (MEM)," *Annals of CIRP,* vol. 42, no. 1, pp. 119–122.

Ashby, W. R. (1973), "Some Peculiarities of Complex Systems," *Cybernetic Medicine,* no. 9, pp. 1–7.

Automotive Industry Action Group (AIAG) (2001), *Potential Failure Mode and Effects Analysis (FMEA) Reference Manual,* 3d ed., AIAG, Southfield, Mich.

Barrado, E., M. Vega, R. Pardo, P. Grande, and J. Del Valle (1996), "Optimization of a Purification Method for Metal-Conditioning Wastewater by Use of Taguchi Experimental Design," *Water Research,* vol. 30, no. 10, pp. 2309–2314.

Beitz, W. (1990), "Design for Ease of Recycling (Guidelines VDI-2243)," *ICED Proceedings 90,* Dubrovnik, Heurista, Zurich.

Boothroyd, G., and P. Dewhurst (1983), *Product Design for Assembly Handbook,* Boothroyd-Dewhurst Inc., Wakefield, R.I.

Boothroyd, G., P. Dewhurst, and W. Knight (1994), *Product Design or Manufacture and Assembly,* Marcel Dekker, New York.

Bothe, D. R. (1997), *Measuring Process Capabilities,* McGraw-Hill, New York.

Box, G. E. P., S. Bisgaard, and C. A. Fung (1988), "An Explanation and Critique of Taguchi's Contributions to Quality Engineering," *Quality and Reliability Engineering International,* vol. 4, pp. 123–131.

Brejcha, M. F. (1982), *Automatic Transmission,* 2d ed., Prentice Hall, Upper Saddle River, N.J.

Bremer, M. (2002), The Cumberland Group, "Value Stream Mapping," *Smoke Signals,* vol. 1, no. 8, May 28 (www.imakenews.com/rainmakers/e_article000046701.cfm).

Bussey, L. E. (1998), *The Economic Analysis of Industrial Projects,* Prentice Hall, Upper Saddle River, N.J.

Carnap, R. (1977), *Two Essays on Entropy,* University of California Press, Berkeley.

Caulkin, S. (1995), "Chaos Inc." *Across the Board,* pp. 32–36, July.

Cha, J. Z., and R. W. Mayne (1987), "Optimization with Discrete Variables via Recursive Quadratic Programming," *Advances in Design Automation—1987,* American Society of Mechanical Engineers, Sept.

Chase, K. W., and W. H. Greenwood (1988), "Design Issues in Mechanical Tolerance Analysis," *Manufacturing Review,* vol. 1, no. 1, pp. 50–59.

Chen, E. (2002), "Potential Failure Mode and Effect Analysis: Enhancing Design and Manufacturing Reliability," George Washington Center for Professional Development, course.

Clausing, D. P., and K. M. Ragsdell (1984), "The Efficient Design and Development of Medium and Light Machinery Employing State-of-the-Art Technology," *International Symposium on Design and Synthesis,* Tokyo, July 11–13.

Clausing, D. P. (1994), *Total Quality Development: A Step by Step Guide to World-Class Concurrent Engineering,* ASME Press, New York.

Cohen, L. (1988), "Quality Function Deployment and Application Perspective from Digital Equipment Corporation," *National Productivity Review,* vol. 7., no. 3, pp. 197–208.

Cohen, L. (1995), *Quality Function Deployment: How to Make QFD Work for You,* Addison-Wesley, Reading, Mass.

Deming, E. (1982), *Out of Crisis,* Massachusetts Institute of Technology, Center for Advanced Engineering Study, Cambridge, Mass.

Derringer, G., and R. Suich (1980), "Simultaneous Optimization of Several Response Variables," *Journal of Quality Technology,* vol. 12, pp. 214–219.

Dewhurst, P. (1988), "Cutting Assembly Costs with Molded Parts," *Machine Design,* July.

Dewhurst, P., and C. Blum (1989), "Supporting Analyses for the Economic Assessment of Die Casting in Product Design," *Annuals of CIRP,* vol. 28., no. 1., p. 161.

Dixon, J. R. (1966), *Design Engineering: Inventiveness, Analysis, and Decision Making,* McGraw-Hill, New York.

Domb, E. (1997), "Finding the Zones of Conflict: Tutorial", *TRIZ Journal,* June (www.triz-journal.com).

Domb, E., J. Terninko, J. Miller, and E. MacGran, (1999), "The Seventy-Six Standard Solutions: How They Relate to the 40 Principles of Inventive Problem Solving", *TRIZ Journal,* May (www.triz-journal.com).

Dorf, R. C., and R. H. Bishop (2000), *Modern Control Systems,* 9th ed., Prentice Hall, Upper Saddle River, N.J.

Dovoino, I. (1993), "Forecasting Additional Functions in Technical Systems," *Proceedings of ICED-93,* The Hague, vol. 1, pp. 247–277.

El-Haik, B. (1996), *Vulnerability Reduction Techniques in Engineering Design,* Ph.D. dissertation, Wayne State University, Mich.

El-Haik, B., and K. Yang (1999), "The Components of Complexity in Engineering Design," *IIE Transactions,* vol. 31, no. 10, pp. 925–934.

El-Haik, B., and K. Yang (2000a), "An Integer Programming Formulations for the Concept Selection Problem with an Axiomatic Perspective (Part I): Crisp Formulation," *Proceedings of the First International Conference on Axiomatic Design,* MIT, Cambridge, Mass., pp. 56–61.

El-Haik, B., and K. Yang (2000b), "An Integer Programming Formulations for the Concept Selection Problem with an Axiomatic Perspective: Fuzzy Formulation," *Proceedings of the First International Conference on Axiomatic Design,* MIT, Cambridge, Mass., pp. 62–69.

Fey, V., and E. Rivin (1997), *The Science of Innovation: A Managerial Overview of the TRIZ Methodology,"* TRIZ Group, West Bloomfield, Mich.

Foo, G., J. P. Clancy, L. E. Kinney, and C. R. Lindemudler (1990), "Design for Material Logistics," *AT&T Technical Journal,* vol. 69, no. 3, pp. 61–67.

Fowlkes, W. Y., and C. M. Creveling (1995), *Engineering Methods for Robust Product Design,* Addison-Wesley, Reading, Mass.

Fredriksson, B. (1994), "Holistic Systems Engineering in Product Development," *The Saab-Scania Griffin, 1994,* Saab-Scania, AB, Linkoping, Sweden, Nov.

Fuglseth, A., and K. Gronhaug (1997), "IT-Enabled Redesign of Complex and Dynamic Business Processes; the Case of Bank Credit Evaluation," *OMEGA International Journal of Management Science,* vol. 25, pp. 93–106.

Galbraith, J. R. (1973), *Designing Complex Organizations,* Addison-Wesley, Reading, Mass.

Gardner, S., and D. F. Sheldon (1995), "Maintainability as an Issue for Design," *Journal of Engineering Design,* vol. 6, no. 2, pp. 75–89.

Garvin, D. A. (1988), *Managing Quality: The Strategic and Competitive Edge,* Free Press, New York.

Gershenson, J., and K. Ishii (1991), "Life Cycle Serviceability Design," *Proceedings of ASME Conference on Design and Theory and Methodology,* Miami, Fla.

Grady, J. O. (1998), *System Validation and Verification,* CRC Press, Boca Raton, Fla.

Grant, E. L., and R. S. Leavenworth (1980), *Statistical Quality Control,* 5th ed., McGraw-Hill, New York.

Grossley, E. (1980). "Make Science a Partner," *Machine Design,* April 24.

Handerson, R., and K. B. Clark (1990), "Architectural Innovation: The Reconfiguration of Existing Product Technologies and the Failure of Established Firms," *Administrative Science Quarterly,* vol. 35, no. 1, pp. 9–30.

Harrell, W., and G. Cutrell (1987), "Extruder Output Control and Optimization for Rubber Weather-strips," *Fifth Symposium on Taguchi Method,* Oct. 1987, American Supplier Institute, Livonia, Mich.

Harry, M. J. (1994), *The Vision of 6-Sigma: A Roadmap for Breakthrough,* Sigma Publishing Company, Phoenix, Ariz.

Harry, M. J. (1998), "Six Sigma: A Breakthrough Strategy for Profitability," *Quality Progress,* pp. 60–64, May.

Harry, M., and R. Schroeder (2000), *Six Sigma: The Breakthrough Management Strategy Revolutionizing the World's Top Corporations,* Doubleday, New York.

Hartley, R. V. (1928), "Transmission of Information," *The Bell Systems Technical Journal,* no. 7, pp. 535–563.

Hauser, J. R., and D. Clausing (1988), "The House of Quality," *Harvard Business Review,* vol. 66, no. 3, pp. 63–73, May–June.

Hintersteiner, J. D., and A. S. Nain (1999), "Integrating Software into Systems: An Axiomatic Design Approach," *The Third International Conference on Engineering Design and Automation,* Vancouver, B.C., Canada, Aug. 1–4.

Hintersteiner, J. D. (1999), "A Fractal Representation for Systems," *International CIRP Design Seminar,* Enschede, the Netherlands, March 24–26.

Hornbeck, R. W. (1975), *Numerical Methods,* Quantum Publishers, New York, pp. 16–23.

Huang, G. Q., ed. (1996), *Design for X: Concurrent Engineering Imperatives,* Chapman & Hall, London.

Hubka, V. (1980), *Principles of Engineering Design,* Butterworth Scientific, London.

Hubka, V., and W. Eder (1984), *Theory of Technical Systems,* Springer-Verlag.

Hwan, M. (1996), "Robust Technology Development. An Explanation with Examples," *Second Annual Total Product Development Symposium,* Nov. 6–8, Pomona, Calif. American Supplier Institute.

Jang, K., K. Yang, and C. Kang (2003), "Application of Artificial Neural Network to Identify Nonrandom Variation Patterns on the Run Chart in Automotive Assembly Process," *International Journal of Production Research.*

Johnson, R. A., and D. W. Wichern (1982), *Applied Multivariate Statistical Analysis,* Prentice Hall, Upper Saddle River, N.J.

Kacker, R. N. (1985), "Off-line Quality Control, Parameter Design, and the Taguchi Method," *Journal of Quality Technology,* no. 17., pp. 176–188.

Kapur, K. C. (1988), "An Approach for the Development for Specifications for Quality Improvement," *Quality Engineering,* vol. 1, no. 1, pp. 63–77.

Kapur, K. (1993), "Quality Engineering and Tolerance Design," *Concurrent Engineering: Automation, Tools and Techniques,* pp. 287–306.

Kapur, K. C., and L. R. Lamberson (1977), *Reliability in Engineering Design,* Wiley, New York.

Keller, G. et al. (1999), *Sap R / 3 Business Blueprint: Understanding the Business Process Reference Model,* Prentice Hall, Upper Saddle River, N.J.

Khoshooee, N., and P. Coates (1998), "Application of the Taguchi Method for Consistent Polymer Melt Production in Injection Moulding," *Proceedings of the Institution of Mechanical Engineers,* vol. 212, Pt. B, pp. 611–620, July.

Kim, S. J., N. P. Suh, and S. G. Kim (1991), "Design of Software System Based on Axiomatic Design," *Annals of the CIRP,* vol. 40/1, pp. 165–170.

Knight, W. A. (1991), "Design for Manufacture Analysis: Early Estimates of Tool Costs for Sintered Parts," *Annals of CIRP.* vol. 40, no. 1, p. 131.

Kota, S. (1994), "Conceptual Design of Mechanisms Using Kinematics Building Blocks— A Computational Approach," *Final Report NSF Design Engineering Program,* Grant DDM 9103008, Oct.

Ku, H. H. (1996), "Notes on the Use of Propagation of Error Formulas," *Journal of Research of the National Bureau of Standards: C. Engineering and Instrumentation,* vol. 70, no. 4, pp. 263–273.

Kusiak, A., and E. Szczerbicki (1993), "Transformation from Conceptual to Embodiment Design," *IIE Transactions,* vol. 25, no. 4.

Lee, T. S. (1999), *The System Architecture Concept in Axiomatic Design Theory: Hypotheses Generation and Case-Study Validation,* S.M. thesis, Department of Mechanical Engineering, MIT, Cambridge, Mass.

Magrab, E. B. (1997), *Integrated Product and Process Design and Development,* CRC Press, LLC, Boca Raton, Fla.

Mann, D. L. (1999), "Axiomatic Design and TRIZ: Compatibility and Contradictions," *TRIZ Journal,* June/July.

Mann, D. (2002), *Hands on Systematic Innovation,* CREAX Press, Leper, Belgium.

Maskell, B. H. (1991), *Performance Measurement for World Class Manufacturing,* Productivity Press, Cambridge, Mass.

Matousek, R. (1957), *Engineering Design: A Systematic Approach,* Lackie & Son Ltd., London.

McCord, K. R., and S. D. Eppinger (1993), *Managing the Integration Problem in Concurrent Engineering,* MIT Sloan School of Management Working Paper 359-93-MSA.

Miles, B. L. (1989), "Design for Assembly: A Key Element within Design for Manufacture," *Proceedings of IMechE, Part D, Journal of Automobile Engineering,* no. 203, pp. 29–38.

Miles, L. D., (1961), Techniques of Value Analysis and Engineering, McGraw-Hill, New York.

Montgomery D. (1997), *Design and Analysis of Experiments,* Wiley, New York.

Montgomery, D. C. (2000), *Introduction to Statistical Quality Control,* Wiley, New York.

Mostow, J. (1985), "Toward Better Models of the Design Process," *The AI Magazine,* pp. 44–57.

Mughal, H., and R. Osborne (1995), "Design for Profit," *World Class Design to Manufacture,* vol. 2, no. 5, pp. 160–226.

Myers, J. D. (1984), *Solar Applications in Industry and Commerce,* Prentice Hall, Upper Saddle River, N.J.

Myers, R. H., and D. C. Montgomery (1995), *Response Surface Technology,* Wiley, New York.

Navichandra, D. (1991), "Design for Environmentality," *Proceedings of ASME Conference on Design Theory and Methodology,* New York.

Niebel, B. W., and E. N. Baldwin, 1957, *Designing for Production,* Irwin, Homewood, Ill.

Norlund, M., D. Tate, and N. P. Suh (1996), "Growth of Axiomatic Design through Industrial Practice," *Third CIRP Workshop on Design and Implementation of Intelligent Manufacturing Systems,* Tokyo, Japan, pp. 77–84, June 19–21.

O'Connor, P. T. (2001), *Test Engineering,* Wiley, New York.

O'Grady, P., and J. Oh (1991), "A Review of Approaches to Design for Assembly," *Concurrent Engineering,* vol. 1, pp. 5–11.

Oakland, J. S. (1994), *Total Quality Management: The Route to Improving Performance,* 2d ed., Butterworth-Heineman, Oxford.

Pack, J., I. Kovich, R. Machin, and J. Zhou (1995), "Systematic Design of Automotive Weaterstrips Using Taguchi Robust Design Methods," *ASI Symposium,* American Supplier Institute.

Papalambros, P. Y., and D. J. Wilde (1988), *Principles of Optimal Design,* Cambridge University Press, Cambridge, U.K.

Park, R. J. (1992), *Value Engineering,* R. J. Park & Associates, Inc., Birmingham, Mich.

Park, S. (1996), *Robust Design and Analysis for Quality Engineering,* Chapman & Hall, London.

Pech, H. (1973), *Designing for Manufacture, Topics in Engineering Design Series,* Pitman & Sons, London.

Penny, R. K. (1970), "Principles of Engineering Design," *Postgraduate,* no. 46, pp. 344–349.

Peppard, J., and P. Rowland (1995), *The Essence of Business Process Re-engineering,* Prentice-Hall Europe, Hemel Hempstead.

Perez-Wilson, M. (2002), *Multi-Vari Chart and Analysis,* Advanced Systems Consultants.

Peters, T. (1982), *In Search of Excellence,* HarperCollins, New York.

Phadke, M. (1989), *Quality Engineering Using Robust Design,* Prentice Hall, Upper Saddle River, N.J.

Phal, G., and W. Beitz (1988), *Engineering Design: A Systematic Approach,* Springer-Verlag.

Pimmler, T. U., and S. D. Eppinger (1994), "Integration Analysis of Product Decomposition," *Design Theory and Methodology,* vol. 68, pp. 343–351.

Pugh, S. (1991), *Total Design: Integrated Methods for Successful Product Engineering,* Addison-Wesley, Reading, Mass.

Pugh, S. (1996), in *Creating Innovative Products Using Total Design,* D. P. Clausing and Andrade, eds., Addison-Wesley, Reading, Mass.

Ramachandran, N., N. A. Langrana, L. I. Steinberg, and V. R. Jamalabad (1992), "Initial Design Strategies for Iterative Design," *Research in Engineering Design,* vol. 4, no. 3.

Rantanen, K. (1988), "Altshuller's Methodology in Solving Inventive Problems, ICED-88, Budapest. Aug. 23–25.

Reinderle, J. R. (1982), *Measures of Functional Coupling in Design,* Ph.D. dissertation, Massachusetts Institute of Technology, June.

Reklaitis, J. R., A. Ravindran, and K. M. Ragsdell (1983), *Engineering Design Optimization,* Wiley, New York.

Sackett, P., and A. Holbrook (1988), "DFA as a Primary Process Decreases Design Deficiencies," *Assembly Automation,* vol. 12, no. 2, pp. 15–16.

Sekimoto, S., and M. Ukai (1994), "A Study of Creative Design Based on the Axiomatic Design Theory," DE-Vol. 68, *Design Theory and Methodology-DTM'94 ASTM,* pp. 71–77.

Shannon, C. E. (1948), "The Mathematical Theory of Communication," *The Bell System Technical Journal,* no. 27, pp. 379–423, 623–656.

Sheldon, D. F., R. Perks, M. Jackson, B. L. Miles, and J. Holland (1990), "Designing for Whole-life Costs at the Concept Stage," *Proceedings of ICED,* Heurista, Zurich.

Simon, H. A. (1981), *The Science of the Artificial,* 2d ed. MIT Press, Cambridge, Mass.

Smith, G., and G. J. Browne (1993), "Conceptual Foundation of Design Problem Solving," *IEEE Transactions on Systems, Man, and Cybernetics,* vol. 23, no. 5, Sept./Oct.

Sohlenius, G., A. Kjellberg, and P. Holmstedt (1999), "Productivity System Design and Competence Management," *World XIth Productivity Congress,* Edinburgh.

Soliman, F. (1998), "Optimum Level of Process Mapping and Least Cost Business Process Reengineering," *International Journal of Operations & Production Management,* vol. 18, no. 9/10, pp. 810–816.

Spotts, M. F. (1973), "Allocation of Tolerance to Minimize Cost of Assembly," *Transactions of the ASME,* pp. 762–764.

Srinivasan, R. S., and K. L. Wood (1992), "A Computational Investigation into the Structure of Form and Size Errors Based on Machining Mechanics," *Advances in Design Automation,* Phoenix, Ariz., pp. 161–171.

Steinberg, L., N. Langrana, T. Mitchell, J. Mostow, and C. Tong (1986), *A Domain Independent Model of Knowledge-Based Design,* Technical Report AI/VLSI Project Working Paper No 33, Rutgers University.

Steward, D. V. (1981), *Systems Analysis and Management: Structure, Strategy, and Design,* Petrocelli Books, New York.

Suh, N. P. (1984), "Development of the Science Base for the Manufacturing Field through the Axiomatic Approach," *Robotics & Computer Integrated Manufacturing,* vol. 1, no. 3/4.

Suh, N. P. (1990), *The Principles of Design,* Oxford University Press, New York.

Suh, N. (1995), "Design and Operation of Large Systems," *Journal of Manufacturing Systems,* vol. 14, no. 3.

Suh, N. P. (1996), "Impact of Axiomatic Design," *Third CIRP Workshop on Design and the Implementation of Intelligent Manufacturing Systems,* Tokyo, Japan, June 19–22, pp. 8–17.

Suh, N. (2001), *Axiomatic Design: Advances and Applications,* Oxford University Press.

Suh, N. P. (1997), "Design of Systems," *Annals of CIRP,* vol. 46, no. 1, pp. 75–80.

Sushkov, V. V. (1994), "Reuse of Physical Knowledge in Creative Design," *ECAI-94 Workshop Notes,* Amsterdam, Aug.

Suzue, T., and A. Kohdate (1988), *Variety Reduction Programs: A Production Strategy for Product Diversification,* Productivity Press, Cambridge, Mass.

Swenson, A., and M. Nordlund (of Saab) (1996), *Axiomatic Design of Water Faucet,* unpublished report, Linkoping, Sweden.

Taguchi, G. (1986), *Introduction to Quality Engineering,* UNIPUB/Kraus International Publications, White Plains, N.Y.

Taguchi, G. (1993), *Taguchi on Robust Technology Development,* ASME Press, New York.

Taguchi, G. (1994), *Taguchi Methods, Vol. 1, Research and Development,* Japan Standard Association and ASI Press.

Taguchi, G., and Y. Wu (1979), *Introduction to Off-Line Quality Control,* Central Japan Quality Control Association, Magaya, Japan (available from American Supplier Institute, Livonia, Mich.).

Taguchi, G., and Y. Wu (1986), *Introduction to Off-Line Quality Control,* Central Japan Quality Control Association, Magaya, Japan.

Taguchi, G., E. Elsayed, and T. Hsiang (1989), *Quality Engineering in Production Systems,* McGraw-Hill, New York.

Taguchi, G., S. Chowdhury, and S. Taguchi (2000), *Robust Engineering,* McGraw-Hill, New York.

Tate, D., and M. Nordlund (1998), "A Design Process Roadmap as a General Tool for Structuring and Supporting Design Activities," *SDPS Journal of Integrated Design and Process Science,* vol. 2, no. 3, pp. 11–19.

Tayyari, F. (1993), "Design for Human Factors," in *Concurrent Engineering,* H. R. Parsaei and W. G. Sullivan, eds., Chapman & Hall, London, pp. 297–325.

Terninko, J., A. Zusman, and B. Zlotin (1998), *Systematic Innovation: An Introduction to TRIZ,* St. Lucie Press, Delray Beach, Fla.

The Asahi (1979), Japanese language newspaper, April 17.

Tsourikov, V. M. (1993), "Inventive Machine: Second Generation," *Artificial Intelligence & Society,* no. 7, pp. 62–77.

Ullman, D. G. (1992), *The Mechanical Design Process,* McGraw-Hill, New York.

Ulrich, K. T., and K. Tung (1994), "Fundamentals of Product Modularity," *ASME Winter Annual Meeting,* DE-Vol. 39, Atlanta, pp. 73–80.

Ulrich, K. T., and S. D. Eppinger (1995), *Product Design and Development,* McGraw-Hill, New York.

Ulrich, K. T., and S. D. Eppinger (2000), *Product Design and Development,* McGraw-Hill, New York.

Ulrich, K. T., and W. P. Seering (1987), "Conceptual Design: Synthesis of Systems Components," *Intelligent and Integrated Manufacturing Analysis and Synthesis,* American Society of Mechanical Engineers, New York, pp. 57–66.

Ulrich, K. T., and W. P. Seering (1988), "Function Sharing in Mechanical Design," *Seventh National Conference on Artificial Intelligence,* AAAI-88, Minneapolis, Aug. 21–26.

Ulrich, K. T., and W. P. Seering (1989), "Synthesis of Schematic Description in Mechanical Design," *Research in Engineering Design,* vol. 1, no. 1.

Vasseur, H., T. Kurfess, and J. Cagan (1993), "Optimal Tolerance Allocation for Improved Productivity," *Proceedings of the 1993 NSF Design and Manufacturing Systems Conference,* Charlotte, N.C., pp. 715–719.

Verno, A., and V. Salminen (1993), "Systematic Shortening of the Product Development Cycle," presented at the International Conference on Engineering Design, The Hague, the Netherlands.

Wagner, T. C., and P. Papalambros (1993), "A General Framework for Decomposition Analysis in Optimal Design," paper presented at the 23rd ASME Design Automation Conference, Albuquerque, N.M., Sept.

Wang, J., and T. Ruxton (1993), "Design for Safety of Make-to-Order Products," *National Design Engineering Conference of ASME,* vol. 93-DE-1. American Society of Mechanical Engineers, New York.

Warfield, J., and J. D. Hill (1972), *A United Systems Engineering Concept,* Battle Monograph no. 1, June.

Weaver, W. (1948), "Science and Complexity," *American Scientist,* no. 36, pp. 536–544.

Wei, C., D. Rosen, J. K. Allen, and F. Mistree (1994), "Modularity and the Independence of Functional Requirements in Designing Complex Systems," *Concurrent Product Design,* DE-Vol. 74, American Society of Mechanical Engineers, New York.

Western Electric Company (1956), *Statistical Quality Control Handbook,* Western Electric Co., Inc., New York.

Wirth, N. (1971), "Program Development by Stepwise Refinement," *Communications of the ACM,* vol. 14, pp. 221–227.

Wood, G. W., R. S. Srinivasan, I. Y. Tumer, and R. Cavin (1993), "Fractal-Based Tolerancing: Theory, Dynamic Process Modeling, Test Bed Development, and Experiment," *Proceedings of the 1993 NSF Design and Manufacturing Systems Conference,* Charlotte, N.C., pp. 731–740.

Wu, Y., and W. H. Moore (1985), *Quality Engineering Product and Process Design Optimization,* American Supplier Institute, Livonia, Mich.

Yang, K., W. Xie, and Y. He (1994), "Parameter and Tolerance Design in Engineering Modeling Stage," *International Journal of Production Resarch,* vol. 32, no. 12, pp. 2803–2816.

Yang, K., and J. Trewn (1999), "A Treatise on the Mathematical Relationship between System Reliability and Design Complexity," *Proceedings of IERC,* Phoenix, Ariz.

Yang, K. (1996), "Improving Automotive Dimensional Quality by Using Principal Component Analysis," *Quality and Reliability Engineering International,* vol. 12, pp. 401–409.

Zaccai, G. (1994), "The New DFM: Design for Marketability," *World-Class Manufacture to Design,* vol. 1, no. 6, pp. 5–11.

Zadeh, L. A. (1965), "Fuzzy Sets," *Information and Control,* no. 8, pp. 338–353.

Zenger, D., and P. Dewhurst (1988), *Early Assessment of Tooling Costs in the Design of Sheet Metal Parts,* Report 29, Department of Industrial and Manufacturing Engineering, University of Rhode Island.

Zhang, H. C., and M. E. Huq (1994), "Tolerancing Techniques: The State-of-the-Art," *International Journal of Production Research,* vol. 30, no. 9, pp. 2111–2135.

Zlotin, B., and A. Zusman (1999), "Mapping Innovation Knowledge," *TRIZ Journal,* April (www.triz-journal.com).

Zwicky, F. (1984), *Morphological Analysis and Construction,* Wiley Interscience, New York.

Index